# THE EARLY DAYS OF YEAST GENETICS

# THE EARLY DAYS OF YEAST GENETICS


*Edited by*

MICHAEL N. HALL
*Biocenter of the University of Basel*

PATRICK LINDER
*Biocenter of the University of Basel*

COLD SPRING HARBOR LABORATORY PRESS
1993

THE EARLY DAYS OF YEAST GENETICS


Book design by Emily Harste

*Photographs courtesy of*

*M. Hall:* (p. 1) R.K. Mortimer and D. Hawthorne. (p. 217) G. Schatz (Veronique Rochette Photographie).
*R.K. Mortimer:* (p. 1) Ø. Winge (Presshuset & Dansk Journalistforbunds); C.C. Lindegren and G. Magni; C.C. Lindegren.
*C. Roman:* (p. 1) H. Roman; B. Ephrussi and H. Roman.
*F. Sherman:* (p. 129) A. Nasim, R.K. Mortimer, G. Magni, and F. Sherman.
*R.H. Haynes:* (p. 129) C.A. Tobias; R.C. von Borstel. (p. 217) P.P. Slonimski and F. Kaudewitz; B.S. Cox. (p. 271) V.L. MacKay. (p. 305) D.H. Williamson; J.M. Mitchison and B. Stevens. (p. 415) R.E. Esposito and H. Roman.
*A. Goffeau:* (p. 217) P.P. Slonimski; P.P. Slonimski. (p. 359) R. Davis. (p. 415) H. Roman.
*Annual Review of Genetics* (*14:* 447 [1980]): (p. 1) B. Ephrussi.
*Cold Spring Harbor Laboratory Archives:* (p. 1) M. Fox, H. Roman, and U. Leupold; H. Ephrussi-Taylor, B. Ephrussi, and L. Szilard; B. Ephrussi; A. Buzzati-Travers and C.C. Lindegren; Ø. Winge. (p. 129) S. Fogel. (p. 271) I. Herskowitz. (p. 337) Map conceived by participants of 1970 Yeast Course. (p. 359) G. Fink and D. Botstein. (p. 415) F. Sherman; G. Fink; Postcard (front and back).

**Library of Congress Cataloging-in-Publication Data**

The Early days of Yeast genetics / edited by Michael N. Hall and Patrick Linder.
p. cm.
Includes bibliographical references and index.
ISBN 0-87969-378-9
1. Saccharomyces cerevisiae--Genetics. 2. Schizosaccharomyces pombe--Genetics. 3. Yeast fungi--Genetics--History. I. Hall, Michael N. II. Linder, Patrick.
QK623.S23E27 1993
589.2′33--dc20 93-4709
CIP



All Cold Spring Harbor Laboratory Press publications may be ordered directly from Cold Spring Harbor Laboratory Press, 10 Skyline Drive, Plainview, New York 11803-2500. Phone: 1-800-843-4388 in Continental U.S. and Canada. All other locations: (516) 349-1930. FAX: (516) 349-1946.

# Contents

## MITOCHONDRIA AND CYTOPLASMIC INHERITANCE



## MATING



## CELL CYCLE



## GENE STRUCTURE AND EXPRESSION

## MOLECULAR BIOLOGY



## INSTITUTIONS

# *Preface*

Yeast has recently joined maize, *Drosophila, Neurospora,* and *Escherichia coli,* among others, in the pantheon of experimental organisms. The rise to prominence covered a great distance in a relatively brief period. As is related in the ensuing chapters, yeast as an experimental genetics organism started out not too long ago and with the severe handicap of suspect genetic properties. Because the ascent was so rapid and recent, we are in the fortunate position that the history of yeast genetics can be recorded as viewed by the protagonists themselves. This is the purpose of this volume.

With a few exceptions, the authors were requested to describe how specific topics or lines of investigation in which they were involved developed within a context of influencing events and personalities. Each chapter was to convey a feel not only for the science, but also for the events and faces behind the actual published work. Thus, this book was intended to contain a series of scientific reminiscences much along the lines of *Phage and the Origins of Molecular Biology*, edited by J. Cairns, G.S. Stent, and J.D. Watson. Some chapters were edited more than others, but, in all cases, the editing was kept to a minimum; an intent of this collection of stories was also to capture the personalities of the authors.

To create the list of contributors, we had the authors pick themselves. We first made a short list of potential authors whose inclusion in such a volume could not be contested. The members of this skeleton list were then asked to contribute a chapter and to recommend additional contributors. Those whose names were submitted multiple times were then approached. There are, of course, some obvious omissions. Unfortunately, for either personal or health reasons, a few of those contacted declined to contribute.

Except for Urs Leupold's and Murdoch Mitchison's chapters on *Schizosaccharomyces pombe*, all chapters are about *Saccharomyces cerevisiae*. We apologize to those who believe that *S. pombe* is not adequately represented. We can offer only that this imbalance reflects the relative states of development of the two yeasts during the period covered. Genetic research on *S. pombe* started several years later than that on *S. cerevisiae*.

We thank, first and foremost, the authors. They are the ones who made this book, in more ways than one. A special tribute is owed to Bob Mortimer. Without him, neither this book nor the field of yeast genetics would be what they are. As his name keeps recurring throughout this volume, we need not mention his contributions to yeast genetics. On the book, he not only produced three chapters but also gave much valuable advice, in his usual unassuming manner. We also thank Oxford University Press and Annual Reviews Inc. (Palo Alto) for allowing us to reproduce the chapters by Boris Ephrussi (d.1979) and Herschel Roman (d.1989), respectively. Finally, we thank our colleagues at Cold Spring Harbor Laboratory Press, John Inglis, Nancy Ford, and in particular, Dorothy Brown and Joan Ebert, with whom collaborating was indeed a pleasure.

*Michael N. Hall*
*Patrick Linder*
*Basel, November, 1992*

# BEGINNINGS

R.K. Mortimer and D. Hawthorne
(Basel, Switzerland 1991)

M.S. Fox, H. Roman, and
U. Leupold (CSHL 1958)

B. Ephrussi (CSHL 1951)

H. Ephrussi-Taylor, B. Ephrussi, and L. Szilard (CSHL 1951)

A. Buzzati-Traverso and
C.C. Lindegren (CSHL 1951)

Ø. Winge
(date unknown)

Ø. Winge (1956)

C.C. Lindegren and G. Magni (Carbondale, Illinois 1956)

C.C. Lindegren (1952)

H. Roman
(University of Washington 1962)

B. Ephrussi and H. Roman (13 rue Pierre Curie 1953)

B. Ephrussi (date unknown)

# Øjvind Winge: Founder of Yeast Genetics

ROBERT K. MORTIMER
*Department of Molecular and Cell Biology*
*Division of Genetics*
*University of California, Berkeley, California 94720*

Øjvind Winge was born in Århus, Denmark, in 1886. He attended Copenhagen University and earned the mag. scient. degree in 1910 with a major in botany. He then did postgraduate work in the United States and at the Carlsberg Laboratory in Copenhagen. It was at the Carlsberg Laboratory that he started his genetic studies on *Humulus* (hops) and *Lebistes* (an aquarium fish), which he was to continue for most of his life. His doctoral thesis (Winge 1917) was a classic, *The Chromosomes: Their Numbers and General Importance*. Winge formulated in this study the hypothesis that interspecific diploid hybrids are sterile because of meiotic pairing difficulties and that this sterility can be overcome by an increase in ploidy to, for example, tetraploid. The higher ploidy cells would be fertile and, in effect, would be a new species. The existence of several natural polyploidy series of plants is consistent with this hypothesis.

In 1921, Winge was appointed to the Chair in Genetics at the Veterinary and Agricultural University in Copenhagen, and he taught genetics at this University for several years. In 1928, he published the *Textbook in Genetics* which was used for many years (Winge 1928). From 1929 to 1935, Winge taught an elementary genetics course, which was based on this textbook, at the University of Copenhagen. Much more detailed information on Winge's distinguished scientific career, which extended to several organisms and research topics, can be found in an excellent biography by Westergaard (1965). This present review, however, is restricted to Winge's research on yeast.

In 1933, Winge accepted the position of directorship of the Physiology Department at the Carlsberg Laboratory. Soon after he started in this new position, he initiated programs on the genetics of hops, barley, and yeast, all organisms of importance to the brewing industry. Winge already had considerable experience in plant genetics and, according to Westergaard, "He was unusually well qualified to tackle the problems of yeast genetics, because he had behind him 25 years of experience as a mycologist and more than 20 years of experience as a geneticist." The Carlsberg Laboratory, at the time of Winge's appointment, had a long tradition of research on yeast under the direction of Emil Christian Hansen and Albert Klöcker. These investigators were concerned mainly with the classification of natural and industrial yeast isolates and with studies of their life cycles. Winge's first publication on yeast (Winge and Hjort 1935) involved his attempts to recover the strains left at Carlsberg by these two investigators.

> With the purpose of taking up anew in the Carlsberg Laboratory the study of Saccharomycetes and certain other fungi an examination was

*The Early Days of Yeast Genetics*

> made, in 1933–1934, of all the original pure cultures left in this laboratory from the days of Emil Chr. Hansen and Alb. Klöcker. The cultures which were found to be living were freshened by new inoculations and examined for the power of spore formation.

These cultures had been stored in 10% saccharose solutions or in wort, some in the dark and some exposed to light. Winge was able to recover live cells from about 25% of these cultures, some as old as 46 years. The main problem in storage appears to have been the drying up of several of the cultures.

## THE EARLY PERIOD AND THE FIRST GENETIC ANALYSIS OF YEAST

Winge's next publication (Winge 1935) was particularly important because it represents the first attempt at genetic analysis in yeast.

> In beginning some studies on yeast in this laboratory, I found it necessary to try to elucidate the alternation between haploid and diploid cell generations of the Saccharomycetes of which our knowledge was very incomplete.

There was general disagreement at that time about the life cycle of yeasts and in particular whether cell conjugation (fertilization) was needed before spore formation. Some investigators argued that asci could be formed directly from haploid cells, i.e., parthenogenetically. Winge felt this to be unlikely.

> But how would it be possible that yeasts, in contrast to the other Ascomycetes, might be able to skip the process of fertilization and form spores in haploid asci? It certainly seems most improbable that the life cycle of the yeast should run contrary to the general concept of haplophase and diplophase.
>
> It is evident then that opinions differ greatly about the nuclear division in yeast, and that investigators who have worked extensively with Saccharomycetes do not take these yeast to be associated with any phasic change corresponding to haplophase and diplophase in other organisms.

To carry out his studies, Winge first developed special microchambers (operating chambers) to be used in isolation of individual spores and cells. He also designed a rotating stage and a special sterile box for his operations.

> Further, I have used an operating chamber constructed especially for isolation of asci or individual spores by means of a micromanipulator.
>
> It is sometimes desirable to turn the operating chamber around the optical axis of the microscope so as to make it more easy to introduce micropipettes and needles obliquely into the chamber.
>
> The mounting of the operating chamber takes place in a sterile operating box of my own construction, aimed at a greater assuredness of sterile manipulations than afforded by the Hansen box.

With this experimental setup, he started research on four strains of yeast, all of the genus *Saccharomyces*: *S. ellipsoideus* Hansen *forma Johannisberg*, *S. ellipsoideus* (Hansen), *S. validus* Hansen, and *S. marchalianus*. These strains originated from the collections of Hansen and Klöcker that Winge had revived earlier. He sporulated these strains on plaster blocks or on carrot slabs and then isolated individual spores or asci for observation. The germination and subsequent events were recorded by drawings or by photomicrographs; observations were made at various times after the spores were placed on germination media. Some of the most frequent observations are described below.

> The spores were usually found conjugating in pairs at germination, but several, in particular older spores germinated without conjugating, forming then some roundish cells which sometimes increased in number up to 10 or more, forming a dense little heap. As far as the occurrence of these two ways of germination is concerned, I can only confirm the findings of previous investigators. It was characteristic that the cell groups formed in the latter way consisted of smaller and more roundish cells than those which originated from pairs of conjugating spores. This condition is really the haploid vegetative form of the species, a characteristic feature of which is that the budding usually starts between the mother- and daughter-cells resulting in a characteristic short-shoot germination.
>
> It happens seldom, however, that any great number of cells of this type are produced by single-spore germination. The rule is that the haploid cells begin very soon to fuse in pairs, resulting in figure-8-formed diploid cells (zygotes). It is further characteristic that as soon as a zygote is formed it starts energetically to form diploid cells which are more elongated and considerably larger than haploid cells; also the manner of branching differs on the whole from the haploid growth in this, that the buds are most often set out distally, resulting in a long-shoot growth.
>
> The transition from haplophase to diplophase is subject to great variation, as illustrated in the drawing and microphotos, fertilization taking place sometimes at an early stage when the spore has formed only three daughter cells; in this instance the four cells conjugate two by two, giving rise to a characteristic double zygote, and this is the end of the haplont. Sometimes, when the haplont is multicellular the zygote production is accompanied by several fertilizations here and there in the haplont; this results in a considerable number of figure-8-formed zygotes which send out vigorous diploid cells that soon outgrow the haploid cells if there be any left.

One of the most frequently observed forms of germination and fertilization, to form the characteristic twin zygotes, occurred after two divisions.

> The zygotes have a most characteristic figure-8-shape with a half-way partition. In other words: a spore has conjugated with its daughter-daughter cell, and the two daughter cells conjugated at the same time.

> This formation of twin zygotes is exceedingly common and may be varied in several ways.

With this study, Winge established that *Saccharomyces* has a normal alternation between haploid and diploid cell phases and would be expected to behave genetically like other organisms.

Winge's next study, plus several following, was carried out in collaboration with O. Laustsen and involved the first tetrad analysis of yeast (Winge and Laustsen 1937). They chose a commercial Press Yeast, catalog no. 146, for this study and reasoned that it was likely to be diploid and heterozygous for many mutations that had accumulated in its continued propagation. A single clone was isolated that had arisen from a single cell in the original sample, and all subsequent studies were carried out on this sample. They realized that a suitable technique must be developed for the isolation and cultivation of all of the spores derived from individual asci.

> The junior author (Laustsen) succeeded in elaborating a technique which allows to isolate with complete certainty all the spores in an ascus and cultivate them.
>
> Notwithstanding the small size of the spore (hardly 3μ), this isolation of spores now proceeds so precisely that it only seldom happens that we are unable to isolate all the spores. With some training a person with the proper knack is able to isolate 70 spores a day.

They observed two types of spore germinations in the isolated spores from this diploid. Some spores germinated directly, without the formation of obvious zygotic forms, to produce elongated diploid cells that would then form asci on gypsum. The other spores germinated to produce haploid cells. In some of the haploid cell outgrowths, zygotes formed between pairs of cells and the clone eventually became mainly diploid. However, some clones appeared to remain as haploids and could not be induced to sporulate. Winge was concerned about how a diploid cell could arise from a haploid cell without cell fusion and proposed nuclear fusion as a possible explanation.

> The only reasonable explanation of the germination of ascospores with elongated cells will then be: that a *nuclear* fusion takes place between the mother-nucleus and the daughter-nucleus immediately at the germination of the spore.

In support of this hypothesis, Winge and Laustsen showed that some of the germinating spores were binucleate. They also collected all the apparent stable haploid cell lines that had been accumulated in the laboratory and paired these up in various combinations to determine if fertilization occurred: "In no instance, however, did this pairing inoculation give rise to ascus formation."

Winge and Laustsen then looked for segregation of genetically determined traits among the four spores of individual tetrads. The trait they chose to examine was "giant colony" formation. This involved making one to a few stab inoculations on wort-gelatin plates (8% lager-wort and 10% gelatin) and incubating at 15°C for at least 3 weeks. In one tetrad described in this study, each spore produced colonies with unique morphologies, but a given spore always produced the same type of colony. Winge concluded that colony morphology was under the control of more than one gene and that these genes were freely recombining.

Winge and Laustsen (1938, 1939a) next constructed interspecific hybrids by pairing individual spores from each species in a droplet of wort solution in the microchamber. In the first of these trials (baking yeast x *S. validus*), one in seven pairings yielded a zygote. The zygote developed into a yeast strain that had properties of both parents with respect to giant colony formation. Fourteen new yeast types were then produced by the same method from eight different crossings. Each spore-spore pairing was of course recognized to be able to produce a different hybrid if the parents were heterozygous for different traits. Both giant colony phenotype and fermentative abilities were determined for each hybrid. All fermentative abilities of one or both parents were expressed by the hybrid, and the authors concluded that these traits were controlled in a dominant fashion. Giant colony phenotypes were sometimes intermediate and sometimes widely different with respect to either parent. Asci were dissected from each hybrid and spore viabilities were determined. The hybrids fell into two categories with respect to this property. The percentage of germinations were in the range of 0–13% and 50–94%, and it was concluded that the former crosses were between different species, whereas the latter were within-species crosses. One of the new hybrids was found to have superior qualities for yeast production and was put into commercial use. Fermentative abilities were determined for the meiotic segregants, but the data were not presented. It was reported only that there was a considerable excess of fermentation-positive segregants.

## STUDIES ON *SACCHAROMYCODES LUDWIGII* AND THE DISCOVERY OF HETEROTHALLISM IN YEASTS

Winge and Laustsen (1939b) and Winge (1946) published two remarkable papers on the yeast *Saccharomycodes ludwigii*. This yeast, first characterized by Hansen 50 years earlier, was known to divide with a septum between mother cell and bud. It also was known to have unusual features of spore formation and germination. The spore pairs at each end of the ascus tended to adhere tightly to each other and, if left to germinate, usually formed a zygote from each spore pair. The authors dissected many tetrads and either allowed all four spores to germinate separately or separated one spore pair and left the other to form a zygote. If separated, one member of each pair divided continuously and the other stopped dividing after a few divisions. In addition, each spore pair differed in the type of cell produced, either short or long. Thus, each spore pair differed in mating type, normal versus lethal cell growth and normal versus elongated cell type. The conclusion was that *S. ludwigii* was heterothallic, and attempts were then made to mate the surviving spore clones in various combinations. However, no mating or spore formation occurred, and the authors concluded that mating type and lethality were linked. This type of segregation required that all three genes be near their respective centromeres and that at least normal/lethal and normal/elongated be on different chromosomes. Although one ascus was found that yielded four viable spores, the authors appeared not to have made any attempt to mate the cells in the spore clones from this ascus. If they had done so, heterothallic haploid cell lines would have been established in this organism several years before such lines were established in *Saccharomyces*. Later work in a strain of *S. ludwigii* that yielded four viable spores confirmed the findings of Winge and Laustsen (1939b) and in addition showed that 22 nutritional mutants that had been induced in haploid isolates of this strain all mapped at or very near several different centromeres (Yamazaki et al. 1976); 60 gene pairs were studied in 888 asci and only 5 tetratype asci were observed. This heterothallic yeast apparently has little or no recombination along its chromosomes, as originally suggested by the work of Winge and Laustsen (1939b).

## THE POSTWAR PERIOD: REACTION TO LINDEGREN AND THE COLLABORATION WITH CATHERINE ROBERTS

The two founders of yeast genetics, Øjvind Winge and Carl C. Lindegren, met apparently for the first time in July 1947. At that time, both investigators were presenting papers at the IVth International Microbiological Congress in Copenhagen. According to Winge and Roberts (1948), Lindegren presented a version of his cytogene hypothesis in which a dominant "cytogene" could be transmitted to the locus of the recessive allele, leading to a replacement of one gene with another by a mechanism that was not crossing-over or mutation but a "sort of contamination." Winge and Roberts obviously did not think much of this hypothesis. They wrote,

> As it appears, very far-fetched hypotheses are here involved, which, their correctness granted, would turn upside down our previous conceptions of genes and heredity, and which therefore cannot be accepted without sufficient proof. The first reaction must involuntarily be one of doubt as to whether the inheritance phenomenon occurring in the yeast fungi are completely distinct from those occurring in other organisms, and one may well question the correctness of the experimental data which form the basis for such conclusions.

In 1948, Winge started a very successful collaboration with Catherine Roberts (from Berkeley, California) that was to continue for about 15 years. Their first paper (Winge and Roberts 1948) describes the results of a cross they made between a strain of *Saccharomyces cerevisiae* (American Yeast Foam) purchased in 1940 from the Northwestern Yeast Company (Chicago) and *Saccharomyces chevalieri*. The former yeast was heterothallic and fermented maltose and galactose rapidly, whereas the latter yeast was homothallic and did not ferment maltose and fermented galactose only slowly. These authors showed by a series of genetic crosses that the *S. cerevisiae* strain carried three dominant genes for maltose fermentation (*M1*, *M2*, and *M3*) and that the difference between the two strains in galactose fermentation abilities segregated in a 2:2 fashion, indicating a single gene difference. We can conclude that the heterothallic strain used by Winge (*S. cerevisiae*, American Yeast Foam) is different from the strain used by Lindegren and Lindegren (1943). The former strain had three maltose genes segregating, whereas Lindegren's strain was heterozygous for a single maltose gene.

*Discovery of the Gene for Homothallism and the Use of Heterothallic Strains*

Winge and Roberts (1949) then described further experiments involving the above heterothallic x homothallic (*S. cerevisiae* x *S. chevalieri*) cross. They found that in each of 30 asci, two spore clones were able to mate and did not sporulate and two were unable to mate and did sporulate. The latter spores, upon germination, formed the characteristic twin zygotes described earlier by Winge (1935). They named the gene determining this genetic difference, *D*, a gene for diploidization. This gene, of course, is the now well-known homothallism gene *HO*. They concluded that two genes, one controlling mating type and the other diploidization, were segregating; their data showed that they were unlinked because they obtained 7 ditype segregations and 23 tetratype segregations between these two genes. They concluded that

> An interaction exists between the mating gene pair *A-a* and the gene pair *D-d* which is responsible for the presence or absence of diploidization at spore germination.

It is not clear why Winge did not start to work on heterothallic strains at this time because it was obvious that he recognized the value of such strains in genetic analyses. Regarding the Lindegrens' discovery of heterothallism (Lindegren and Lindegren 1943) in *Saccharomyces* he wrote (Winge 1949),

> Lindegren and Lindegren established the occurrence of heterothallism in certain Saccharomyces types, particularly in *S. cerevisiae*. They were able to demonstrate the occurrence of two mating types in single spore cultures of their strain of *S. cerevisiae*. Single spore cultures most often remained haploid, but when they were mixed together in pairs, there sometimes occurred a mating reaction, followed by the formation of zygotes, and it was clear that the mating took place according to the usual bipolar scheme.
>
> This observation is of great value, since it is possible in this way to use a single spore culture in a number of crossings. However, this method involves a risk, the risk that the haploid single spore cultures used may diploidize spontaneously, which sometimes happens, so that the mating experiment may result in a mixture of zygotes of hybrid origin and of homozygotic zygotes belonging to one or both of the parent types.

It is possible that Winge was concerned about this risk of diploidization of the parents, and this discouraged him from working with heterothallic strains. However, such problems arise only if the mating mixture is sporulated, as described by Lindegren and Lindegren (1943). Winge and Roberts (1950b) recognized the importance of the prototroph selection procedure for isolating diploids as described by Pomper and Burkholder (1949) as a means of minimizing this risk associated with heterothallic strains. Winge and Roberts also stated that they had isolated individual zygotes by micromanipulation as another method to minimize the risk of diploidization of the parents. Incorporation of even a few genetic markers into each parent, which is now standard procedure, would have made this risk negligible. Such markers were available because he had induced at least one biochemical marker, and others could have been obtained from Seymour Pomper or Sheldon Reaume. Despite this, Winge continued to use homothallic strains and spore-spore matings for most of the remainder of his career.

### *Attempts to Explain Gene Conversion by Conventional Means*

During the 1950s, Winge and Roberts published a series of papers that dealt with the phenomenon of gene conversion that Lindegren was promoting. They never accepted gene conversion as a real genetic process and argued, rather convincingly, that most irregular (non-Mendelian) segregations could be explained by conventional genetic means. Winge and Roberts (1950a) concentrated on the segregation of the three maltose genes and apparently felt obliged to answer many of the claims made by Lindegren and co-workers regarding irregular segregations or gene conversion.

> Especially because of the existence of the very sensational hypotheses of the LINDEGREN school which have brought a deplorable state of confusion into yeast genetics, we have deemed it necessary to undertake a more thorough study of the polymeric genes for maltose fermentation in yeasts in order to establish incontestably that only mendelian segregation occurs, and that 4:0 and 3:1 segregation in the ascus is due simply to polymery.

For these studies, Winge designed a new fermentometer to remove some of the ambiguity in scoring of fermentation phenotypes. Winge and Roberts also introduced the use of X-rays with the goal of producing mutations of the maltose genes. The study confirmed that three maltose genes were present in the *S. cerevisiae* strain (American Yeast Foam) and that each of these three genes segregated regularly in a 2:2 fashion. Winge produced a red adenine-deficient mutant in the heterothallic Yeast Foam strain and this was called B1. This gene was shown to segregate 2:2 in 35 asci. Winge and Roberts also were able to induce a new maltose gene *M4* and found a spontaneous mutation in the same gene as well. They irradiated diploids that were heterozygous for different maltose genes and found maltose-negative survivors at a frequency of about 1%, which they considered was due to mutation; however, it is most likely that they were inducing mitotic crossing-over or mitotic gene conversion. We know now that all of the maltose and other fermentation genes are on the ends of chromosome arms and that X-rays induce mitotic crossing-over and gene conversion efficiently (Nakai and Mortimer 1969). Another important technical step was taken in the research described in the Winge and Roberts (1950a) publication. Instead of sporulating the mating mixture in heterothallic crosses, which Winge considered to be risky, individual zygotes were isolated by micromanipulation. They concluded that this approach, or the prototrophic isolation procedure of Pomper and Burkholder (1949), was preferable to sporulating the mating mixture. In this study, they followed the segregation of the three maltose genes, the galactose gene, and the adenine-deficient red gene and saw only 2:2 segregations. They concluded:

> The results of these, as well as our preceding investigations, have established the fact that enzymatic characters in yeasts are inherited in a strictly mendelian fashion. We therefore reject the various explanatory hypotheses proposed by LINDEGREN and his colleagues, which all have the common assumption that inheritance in yeasts is governed by laws hitherto unknown in genetics.

Winge and Roberts (1950b) proposed that in some meioses, there was an extra mitosis to yield asci with eight nuclei. Random inclusion of four of these nuclei into four-spored tetrads would result in aberrant segregation ratios. They calculated that for a single heterozygous site, the percentages of 4:0, 3:1, 2:2, 1:3, and 0:4 ratios expected were 1.4, 22.9, 51.4, 22.9, and 1.4, and pointed out that if only 10% of the meioses were followed by a mitotic division, then 3:1 and 1:3 ratios would be expected in the 2% range, which is near the observed levels. They showed that a diploid that yielded 1 3:1 segregation and 12 2:2 segregations for a fermentation marker produced some five- and six-spored asci. This was considered to be evidence for supernumerary mitoses and support for their explanation of aberrant segregation ratios.

More convincing evidence that postmeiotic mitoses could explain at least some irregular segregations was published a few years later (Winge and Roberts 1954a,b). They constructed a hybrid that was heterozygous for the unlinked maltose and sucrose fermentation genes *M3* and *R2* and analyzed 37 tetrads from this hybrid; 31 tetrads segregated 2:2 for both markers and 6 segregated aberrantly for one or the other of the two markers. The strain they used was homozygous for the *D* (*HO*) gene, and this permitted them to analyze genetically the four spores in four of the six aberrant tetrads. Winge and Roberts found in each tetrad that three of the spores segregated 4:0 or 0:4 for the two markers, indicating homozygosity, but one spore clone was heterozygous for one or both markers. They proposed that postmeiotic mitosis had occurred and that in one of the spores, two nuclei had been included. This can explain the aberrant ratios and the heterozygosity of the markers in the spore clones. They then performed cytological studies and showed that in some asci, there was an extra mitosis to produce four pairs of spore nuclei. They also observed some binucleate spores. They argued that these observations supported their explanation of irregular segregations, although they were careful to point out that not all aberrant asci can be accounted for by this mechanism. In 1956, studies in *Neurospora* and *Saccharomyces* had established that gene conversion was a valid genetic process (Mitchell 1955; Roman 1957). However, as late as 1957, Winge continued to reject gene conversion as a genetic phenomenon (Winge and Roberts 1957).

### *Characterization of the Fermentation Genes in Different* Saccharomyces *Species*

Winge attended the 8th International Congress of Genetics, which was held in Stockholm in July 1948, and presented a major paper entitled "Inheritance of Enzymatic Characters in Yeasts" (Winge 1949). In this paper, he reviewed Lindegren's work on sugar fermentation genes and then presented research on the same subject carried out at the Carlsberg Laboratory. He then described the results of the cross *S. cerevisiae* x *S. chevalieri* which had been described in Winge and Roberts (1948). In this cross, they demonstrated the presence of three polymeric genes for maltose fermentation in *S. cerevisiae* as well as a dominant gene for galactose fermentation. Both strains were able to ferment sucrose, and thus the segregations of genes involved in this fermentation could not be studied. Winge then described a cross between *S. italicus* and *S. chevalieri* that was performed by Gilliland (1949) while he was at the Carlsberg Laboratory. The former strain was a nonfermenter of sucrose, and Winge and Roberts observed 4:0, 3:1, and 2:2 segregations for fermentation:non-fermentation of this sugar in the ratios 14:6:2. From this, they deduced that *S. chevalieri* carried at least three polymeric genes, *S1*, *S2*, and *S3*, for sucrose fermentation, which was later confirmed by appropriate backcrosses (Winge and Roberts 1952). Winge and Roberts then studied the third possible interspecific cross *S. cerevisiae* x *S. italicus* and determined that *S. cerevisiae* carried *S1* and *S2* and *S. italicus* carried *M1*. Thus, in this series of crosses and related backcrosses, these investigators identified three maltose genes, three sucrose genes, and a gene for galactose fermentation. These genes were distributed in the three strains as follows:

| | |
|---|---|
| *S. cerevisiae* | *S1 S2 s3 M1 M2 M3 G* |
| *S. chevalieri* | *S1 S2 S3 m1 m2 m3 g* |
| *S. italicus* | *s1 s2 s3 M1 m2 m3 G* |

These genes and related genes are still being actively studied by several groups. The *S. italicus* x *S. chevalieri* cross that Gilliland (1949) had analyzed gave segregation ratios for sucrose fermentation that were consistent with there being at least three polymeric sucrose genes in *S. chevalieri*. Winge and Roberts (1952) repeated the above cross between these two species and then performed a series of backcrosses to the sucrose-negative parent to identify each of the three *S1*, *S2*, and *S3* genes. All of these crosses were carried out by the spore-spore mating procedure developed by Winge and Laustsen (1937), and it is of interest to note the success of this method. Winge required that the two spores that were paired up form a zygote before it could be considered a successful mating. If either cell budded before mating, the mating was rejected, even if a zygote was eventually formed. With these criteria, only about one in six spore-spore pairings was successful (1/12, 2/12, 1/6, 4/28, 1/12, 2/24, 4/12, 4/12, 1/26, 4/12, 1/6, 3/12, 2/12, 2/12, 3/20). They observed in the various crosses some irregular segregations that they ascribed to mutation. In a cross heterozygous for *R2*, 13 asci were obtained that segregated 2:2 and 1 that segregated 1:3, which they said was "obviously due to mutation." They also found one 0:4 segregation of *R1* among 64 tetrads and stated that this "was undoubtedly due to the occurrence of a mutation in the hybrid." Presumably, they meant that this "mutation" occurred premeiotically. The same explanation was given for a 0:4 segregation of *M1* seen among 14 otherwise normal tetrads. It now seems more likely that these events were due to meiotic gene conversion (1:3) or mitotic gene conversion or crossing-over (0:4), rather than to mutation. The authors noted an obvious case of linkage in their crosses involving *M1* and *R1*. In 74 tetrads, only 2 tetratype asci were observed giving a map distance of 1.3 cM. This is the gene pair (*SUC1-MAL1*) that has been mapped near the telomere on the right arm of chromosome VII (Mortimer et al. 1989). Winge and Roberts (1953, 1955) then determined by genetic crosses that the *S. cerevisiae* strain American Yeast Foam carried the sucrose and raffinose fermentation gene *S2*, in addition to the maltose genes *M1*, *M2*, and *M3*. In these crosses, they also showed that *S3* and *M3* are tightly linked. This gene pair (*SUC3-MAL3*) has since been mapped near the telomere on the right arm of chromosome II (Mortimer et al. 1989).

Winge and Roberts (1958) published a review of the field of yeast genetics for the period 1935 up to about 1957. In this review, they discuss most of the work carried out at the Carlsberg Laboratory as well as research by the Lindegrens and other groups.

Several scientists spent time at the Carlsberg Laboratory, and Winge merits a lot of credit for this training activity. Urs Leupold was a student of Professor Hans Wanner in Zurich and was given permission to complete his doctoral studies with Winge in Copenhagen. Winge suggested that Leupold work on *Schizosaccharomyces pombe*, although there is no record that Winge worked on this yeast himself. Leupold developed the genetics of *S. pombe* while at Carlsberg, and this started another very active area of cell genetics (see Leupold, this volume). Giovanni Magni spent a year in Winge's laboratory and then returned to Italy to start a yeast genetics program. The academic situation in Italian genetics at that time was such that Magni felt it necessary to first work on *Drosophila* (he worked on the segregation distorter gene) before starting work on yeast. Hence, the start of his yeast program was delayed several years (G. Magni, pers. comm.). R.B. Gilliland worked out some of the genetics of sugar fermentation while at the Carlsberg. He then returned to the Guiness Brewing Company where he established a yeast genetics program. Other scientists who have worked with Winge are J. Wynants, J.L. Jinks, and C. Barry. Catherine Roberts spent almost 15 years working with Winge. She initially held a Sarah Berliner Research Fellowship granted by the American Association of University Women.

## SUMMARY AND COMMENTS

Winge made several very notable contributions to the field of yeast genetics. He developed the instrumentation necessary to carry out reliable ascus dissection and then first used this to clear up a major debate in the literature. He showed convincingly that yeast cells alternate between haplophase and diplophase just as had been described for many other organisms. He then did tetrad analysis for the first time on a commercial yeast. No clear segregations of colony morphology or fermentation phenotypes were observed because of polygenic control of these traits. He did show, in 1939, Mendelian segregation of three traits in another yeast, *S. ludwigii*, and also demonstrated that this yeast was heterothallic. He can thus be credited with these discoveries. He then made the very important discovery that the difference between homothallic and heterothallic strains was controlled by a single gene and this opened up the field of mating-type control. Finally, he characterized many of the fermentation genes present in type strains of *Saccharomyces* and showed that some of these genes were linked. This also opened up a major field of study. Winge made several major discoveries but, except for his work on the fermentation genes, he did not follow up most of them. According to Westergaard (1965),

> This was typical of Winge's method of work, especially in his younger years. He liked to skim the cream of a problem, but would then switch to something else, and leave the following-up to others.

Westergaard adds the following description of Winge:

> Winge was fundamentally a "lone wolf". He was actually a rather shy man, an introvert, to whom oral communication with others probably never came easily, and actual team work was alien to his personality. When facing a large audience Winge had the manners of a rather reserved aristocrat...

Winge was the recipient of several honors. He received honorary degrees from Stockholm Högskula and from Oxford University, and he was invited, in 1957, to the University of California, Berkeley, as Hitchcock Professor. He was a member of the Royal Danish Academy of Letters and Science and a foreign member of both the Royal Society of London and the U.S. National Academy of Sciences.

The two founders of yeast genetics, Øjvind Winge and Carl Lindegren, could not have been much different. One was a strict Mendelist and the other was challenging all of the rules of Mendelism. These two scientists seldom agreed on any subject, and because of such a controversial beginning, it is amazing that yeast genetics ever got started. Some of this controversy is dealt with in the biography of Winge prepared by Westergaard (1965).

> It is impossible to deal with the last part of Winge's yeast work without mentioning his running arguments with the American geneticist Carl Lindegren. In many ways these two brilliant scientists complemented each other. Lindegren: Imaginative, controversial, and often leaving a few loose ends to be tied up in his publications. Winge: Critical, solid, thorough and sound, and loosing [sic] no time in pointing out the weak points in Lindegren's arguments. This did not mean that Winge was always in the right and Lindegren always in the wrong. In most cases the truth was thrashed out between them and to Lindegren goes the undisputed credit of having discovered the mating type system in yeast and "gene conversion". Winge reluctantly accepted this last phenomenon-although he had a sound aversion against this obnoxious term...

Obviously anticipating this volume, Westergaard adds:

> When some day the history of yeast genetics will be written, it will no doubt be acknowledged that its rapid progress owes a great deal to the Winge-Lindegren discussions. It should also be recorded that despite their many arguments the two antagonists liked and respected each other, and both recognized in his opponent a scientist of the highest scientific integrity.

It is this writer's view that the arguments between Winge and Lindegren had more negative than positive effects on the progress of yeast genetics. This debate occurred during a very critical period in the development of modern biology when the role of DNA as the genetic material was being established and the structure and functions of this molecule were being determined. Most young investigators started work on bacteria and bacteriophages at that time and, had they considered yeast as an alternative, probably decided against this organism because of the controversy existing in the literature at that time. Both Spiegelman and Lederberg worked briefly on yeast during this tumultuous period in yeast genetics but then shifted to bacteria. *Neurospora* was a far more attractive alternative as an experimental organism at least until the early 1960s. It is significant that the gene conversion controversy was finally settled by an experiment carried out in *Neurosopora* by Mary Mitchell (Mitchell 1955). Only after this were yeast experiments given credence (Roman 1957).

Winge argued too long and too strongly that gene conversion did not exist, and this was unfortunate. He seemed to be completely convinced that all genetic phenomena must obey the rules of Morgan-Mendelism just as Lindegren was convinced that these same rules were mostly wrong. Winge observed some irregular segregations that were very likely due to gene conversion and he ascribed them to mutation. Lindegren observed irregular segregations also and ascribed them to gene conversion, when they mostly should have been attributed to polygenic inheritance. These two founders of yeast genetics never came to a resolution of their differences on this topic or on many of the other topics on which they differed. They nevertheless started a field of research that today is one of the most exciting in eukaryotic cell biology.

## ACKNOWLEDGMENTS

This article was written while I was a guest of the Department of Biochemistry at the Biocenter, University of Basel. I want to thank Mike Hall and the other members of the Biocenter for a very pleasant and rewarding stay. This article benefited from interviews with Giovanni Magni and Urs Leupold, who both spent time in Winge's laboratory, and I want to thank them for these interviews and the reprints and other materials they provided. Don Hawthorne made available to me an almost complete set of Winge's reprints, which were invaluable. Finally, several friends and associates have read the manuscript at various stages and have offered valuable suggestions.

## REFERENCES

Gilliland, R.B. 1949. A yeast hybrid heterozygous in four fermentation characters. *C.R. Trav. Lab. Carlsberg Ser. Physiol.* **24:** 347–356.

Lindegren, C.C. and G. Lindegren. 1943. A new method of hybridizing yeast. *Proc. Natl. Acad. Sci.* **29:** 306–308.

Mitchell, M. B. 1955. Aberrant recombination of pyridoxine mutants of *Neurospora*. *Proc. Natl. Acad. Sci.* **41:** 215–220.

Mortimer, R.K., D. Schild, R. Contopoulou, and J.A. Kans. 1989. Genetic map of *Saccharomyces cerevisiae*, edition 10. *Yeast* **5:** 321-403.

Nakai, S. and R.K. Mortimer. 1969. Studies of the genetic mechanism of radiation-induced mitotic segregation in yeast. *Mol. Gen. Genet.* **103:** 329–338.

Pomper, S. and P.R. Burkholder. 1949. Studies on the biochemical genetics of yeast. *Proc. Natl. Acad. Sci.* **35:** 456–464.

Roman, H. 1957. Studies of gene mutation in *Saccharomyces*. *Cold Spring Harbor Symp. Quant. Biol.* **21:** 175–185.

Westergaard, M. 1965. Øjvind Winge. *C.R. Trav. Lab. Carlsberg Ser. Physiol.* **34:** 1–24.

Winge, Ø. 1917. The chromosomes. Their numbers and general importance. *C.R. Trav. Lab. Carlsberg Ser. Physiol.* **17:** 131–275.
———. 1928. Arvelighedslaere paa eksperimentelt og cytologisk. *Grundlagen Kbh.* (pp. 301, 147 figs.).
———. 1935. On haplophase and diplophase in some Saccharomycetes. *C.R. Trav. Lab. Carlsberg Ser. Physiol.* **21:** 77–111.
———. 1946. The segregation in the ascus of *Saccharomycodes ludwigii. C.R. Trav. Lab. Carlsberg Ser. Physiol.* **24:** 223–226.
———. 1949. Inheritance of enzymatic characters in yeasts. (*Proc. 8th Int. Congr. Genet.*) *Hereditas* (suppl.) 520–529.
Winge, Ø. and Å. Hjort. 1935. On some Saccharomycetes and other fungi still alive in the pure culture of Emil Chr. Hansen and Alb. Klöcker. *C.R. Trav. Lab. Carlsberg Ser. Physiol.* **21:** 51–58.
Winge, Ø. and O. Laustsen. 1937. On two types of spore germination, and on genetic segregations in *Saccharomyces*, demonstrated through single-spore cultures. *C.R. Trav. Lab. Carlsberg Ser. Physiol.* **22:** 99–116.
———. 1938. Artificial species hybridization in yeast. *C.R. Trav. Lab. Carlsberg Ser. Physiol.* **22:** 235–244.
———. 1939a. On 14 new yeast types, produced by hybridization. *C.R. Trav. Lab. Carlsberg Ser. Physiol.* **22:** 337–352.
———. 1939b. *Saccharomycodes ludwigii* Hansen, a balanced heterozygote. *C.R. Trav. Lab. Carlsberg Ser. Physiol.* **22:** 357–374.
———. 1940. On a cytoplasmic effect of inbreeding of inbreeding in homozygous yeast. *C.R. Trav. Lab. Carlsberg Ser. Physiol.* **23:** 17–38.
Winge, Ø. and C. Roberts. 1948. Inheritance of enzymatic characters in yeasts, and the phenomenon of leng-term adaptation. *C.R. Trav. Lab. Carlsberg Ser. Physiol.* **24:** 263–315.
———. 1949. A gene for diploidization in yeasts. *C.R. Trav. Lab. Carlsberg Ser. Physiol.* **24:** 341–346.
———. 1950a. The polymeric genes for maltose fermentation in yeasts, and their mutability. *C.R. Trav. Lab. Carlsberg Ser. Physiol.* **25:** 35–83.
———. 1950b. Non Mendelian segregations from heterozygotic yeast asci. Nature **165:** 157.
———. 1952. The relation between the polymeric genes for maltose, raffinose, and sucrose fermentation in yeasts. *C.R. Trav. Lab. Carlsberg Ser. Physiol.* **25:** 141–171.
———. 1953. The genes for maltose and raffinose fermentation in *Saccharomyces cerevisiae,* strain yeast foam. *C.R. Trav. Lab. Carlsberg Ser. Physiol.* **25:** 241–251.
———. 1954a. Causes of deviations from 2:2 segregations in the tetrads from monohybrid yeast. *C.R. Trav. Lab. Carlsberg Ser. Physiol.* **25:** 283–329.
———. 1954b. On tetrad analyses apparently inconsistent with Mendelian Law. *Heredity* **8:** 295–304.
———. 1955. Identification of the maltase genes in some American haploid and European diploid yeasts. *C.R. Trav. Lab. Carlsberg Ser. Physiol.* **25:** 331–340.
———. 1957. Remarks on irregular segregations in *Saccharomyces. Genetica* **25:** 489–496.
———. 1958. Yeast genetics. In *Chemistry and biology of yeasts* (ed.A. H. Cooke), pp. 123–156. Academic Press, New York.
Yamazaki, T., Y. Ohara, and Y. Oshima. 1976. Rare occurrence of tetratype tetrads in *Saccharomycodes ludwigii. J. Bacteriol.* **125:** 461–466.

# Carl C. Lindegren: Iconoclastic Father of *Neurospora* and Yeast Genetics

ROBERT K. MORTIMER
*Department of Molecular and Cell Biology*
*Division of Genetics*
*University of California, Berkeley, California 94720*

Carl C. Lindegren was born in Ashland, Wisconsin in 1896. He was the first son of Swedish immigrants and had a sister, Ingeborg, and a brother, Erik. Carl was raised in the town of Rhinelander, Wisconsin, where his father opened and ran Lindey's Cleaners, a laundry and dry-cleaning shop; Carl's brother Erik still runs this business. The town was populated mostly with Germans, Swedes, and more established Americans (Yankees); the Swedes were at the bottom of this sociological pole. Carl attended the University of Wisconsin where he received B.S. and M.S. degrees in plant pathology and chemistry, the latter in the early 1920s. His college career was interrupted by time spent training as a pilot in the U.S. Army Air Corps during and after World War I. While flying with another pilot, who was doing aerobatics and who Carl had decided was out to kill both of them, he jumped out of the plane without a parachute and landed in a lake near Madison. The pilot landed without incident, but Carl spent 3 months in a hospital. He later married Ruth Thomson who had a daughter Claire. Claire later married Leonard Lerman, the Massachusetts Institute of Technology molecular biologist. The marriage with Ruth Thomson did not last, and Carl later married Gertrude Schiller, daughter of a prominent Chicago family. Gertrude became his devoted wife and scientific colleague until her death in 1976.

In the early 1920s, the mycologist Bernard O. Dodge was engaged in work on *Neurospora* at the New York Botanical Garden, and Dodge's friend, Thomas Hunt Morgan, at Columbia University, was in his second decade of work on *Drosophila* genetics. Nobel laureate Robert A. Millikan, who carried out the famous oil drop experiment that established the particulate nature of the electric charge, had become interested in genetics because of the apparent particulate nature of genes. As President of the Executive Council of the California Institute of Technology (Cal-Tech), he invited Morgan and his group to Cal-Tech where, in 1928, they opened the new Biology Division.

Carl and Gertrude had earlier moved to Pasadena where he was "convalescing from a long illness" (Lindegren 1973). He had enrolled as a part-time student in organic chemistry at Cal-Tech, planning to continue his graduate studies. He approached Morgan about the possibility of doing a graduate research project and was given the cultures of *Neurospora* that Dodge earlier had given to Morgan plus some publications on *Neurospora*. Carl soon worked out procedures for ascus dissection and discovered in these strains several naturally occurring morphological traits (e.g., crisp, peach, fluffy, and tan) that showed Mendelian segregation. He found linkage between some of these traits, and this enabled him to develop the first genetic map in *Neurospora*. His studies, which were continued after he moved to the University of Southern California (USC), permitted him to establish that crossing-over occurred at the four-strand stage and that only two of the four strands were involved at any one site (Lindegren 1936; Lindegren and Lindegren 1937, 1939). Gertrude (Jerry) Lindegren played a major and positive role in Carl's experiments throughout his career, while he was at Cal-Tech, USC, and later.

While Lindegren was a graduate student, George Beadle was also at Cal-Tech working with Sterling Emerson. Beadle and Emerson carried out important research on strand relationships of multiple exchanges using attached X chromosomes in *Drosophila*. Beadle was keenly aware of Lindegren's research on *Neurospora*. In his Nobel prize address (Beadle 1977), he writes:

> Dodge was an enthusiastic supporter of *Neurospora* as an organism for genetic work. It's even better than *Drosophila*, he insisted to Thomas Hunt Morgan, whose laboratory he often visited. He finally persuaded Morgan to take a collection of *Neurospora* cultures with him from Columbia to the new Biology Division of the California Institute of Technology, which he established in 1928.
>
> Shortly thereafter when Carl Lindegren came to Morgan's laboratory to become a graduate student, it was suggested that he should work on the genetics of *Neurospora* as a basis for his thesis. This was a fortunate choice, for Lindegren had an abundance of imagination, enthusiasm and energy and at the same time the advice of E.G. Anderson, C.B. Bridges, S. Emerson, A.H. Sturtevant and others at the Institute who at that time were actively interested in problems of crossing-over as a part of the mechanism of meiosis. In this favorable setting, Lindegren soon worked out much of the basic genetics of *Neurospora*. New characters were found and a good start was made toward mapping the chromosomes.
>
> Thus, Tatum and I realized that *Neurospora* was genetically an almost ideal organism for use in our new approach... . Thus encouraged, we obtained strains of *Neurospora crassa* from Lindegren and from Dodge.

After leaving Cal-Tech, Carl and Gertrude spent 6 years (1933–1939) at USC, where they continued their work on *Neurospora*. This was followed by 1 year at the University of Missouri in Lewis J. Stadler's laboratory, where they explored the use of *Neurospora* conidia for radiation-induced mutation studies. Lindegren's pioneering research on *Neurospora* is also discussed in a recent article by Perkins (1992). From about 1941 to 1948, the Lindegrens were at Washington University in St. Louis where he held the appointment of Research Professor in the Henry Shaw School of Botany. It was during this time that the Lindegrens began their research on yeast. We have not been able to determine the reasons for their change from *Neurospora* to yeast after such a promising start with *Neurospora*. Lindegren acknowledges, in the preface to his book (Lindegren 1949), the generous support of Anheuser-Busch Inc., and it seems likely that this was an important factor. It is clear, however, that the Lindegrens were fully committed to research on yeast. In research for this chapter, a long correspondence (1943–1948) was uncovered between Gertrude

Lindegren and Lynferd J. Wickerham, of the Northern Regional Research Laboratory in Peoria, Illinois, regarding various species of *Saccharomyces* with differing fermentative capabilities. The Lindegrens obtained many such strains from Wickerham and incorporated them into crosses. One of their prime interests was to obtain strains that were unable to ferment particular sugars, as well as corresponding fermentation-positive strains, and they soon had strains differing in their abilities to ferment a variety of sugars. Although they recognized multigenic inheritance of fermentative ability, they concentrated on deriving strains that differed in only a single gene for a given phenotype. This was a hallmark of their research for many years, and they explained that this was essential because their primary goal was to map the genes (Lindegren and Lindegren 1953):

> In this laboratory it has been the practice to select only a single stock capable of fermenting a specific sugar and to discard all other stocks. This policy has its origin in the fact that our primary objective has been to map the chromosomes, and in mapping it is difficult to deal with genes of the same phenotype. We have purposely built up stocks in which only a single gene controls the reaction to a specific metabolite (sugar, amino acid, B-vitamin, purine, or pyrimidine) and have discarded or stored stocks in which a repetition of the phenotype appeared. We have only recently broken this rule in the analysis of the inheritance of melezitose fermentation.

Nevertheless, Lindegren and his colleagues published a series of articles on the genetics of melezitose fermentation in which the assumption is made that only a single gene, albeit with many unusual alleles, controlled the fermentation of melezitose and four related sugars (see section below on Directed Mutation and the Galactose and Melezitose Genes).

The Lindegrens first reported heterothallism in *Saccharomyces* and also described a new method (mass mating) for hybridizing yeasts (Lindegren and Lindegren 1943a,b,c). Ø. Winge's method of hybridizing yeasts by spore-to-spore pairings and his almost exclusive use of homothallic strains precluded such a demonstration of heterothallism. Lindegren had obtained a heterothallic diploid, EM93, from Emil Mrak (University of California, Berkeley, and University of California, Davis) and had shown that stable haploid cultures of opposite mating types could be obtained by sporulation of this diploid. Mrak had isolated this diploid from some rotting figs in Merced, California in 1938. In the years between 1943 and 1949, **a** and α mating-type haploids from this diploid, EM93-1C(α) and EM93-3B(**a**), were sent by Lindegren to laboratories in Seattle (Roman, Douglas, and Hawthorne), Berkeley (Zirkle, Tobias, and Mortimer), Stanford (Tatum and Reaume), and Yale (Burkholder and Pomper). These two strains provided the principal material for the start of the American yeast genetics program (Mortimer and Johnston 1986). Lindegren deserves proper recognition for these major contributions to the field of yeast genetics. It seems appropriate to discuss the uses to which these strains were put during the first few years after Lindegren sent them out. Pomper and Burkholder (1949) reported the induction of several biochemical mutants and the subsequent characterization of these mutants. These two authors also confirmed the Lindegrens' report that heterothallism was under the control of a single gene and showed that several other genes segregated in a Mendelian fashion. They reported for the first time the isolation of diploid hybrids by a prototrophic selection procedure, pointing out that this method avoided the difficulties inherent in Lindegren's mass mating procedure (Winge and Roberts 1948), which was subject to diploidization and new rounds of sporulation and mating of the parents. Reaume and Tatum (1949) reported the induction, by nitrogen mustard exposure, of several nutritional mutants including the adenine-red mutant *ade1*. A strain carrying this mutation was sent to Lindegren, who demonstrated that *ade1* was centromere-linked and identified chromosome I. Reaume also obtained some mutants from Pomper (*trp1, met1,* and *ura1*) and incorporated these into crosses. At the time, I was working as a graduate student in biophysics at Berkeley with Cornelius Tobias, who also had obtained these same strains from Lindegren. Zirkle and Tobias (1953) had shown that diploid yeast cells (EM93) were much more resistant to inactivation by X-ray exposure than were haploid cells and had formulated a model which stated that haploids were killed by a recessive lethal event in one of several (20–64) sites, whereas diploids required a homozygous pair of such events in any one of these sites to be inactivated. I reasoned that a genetic approach was needed to test this model and this started my career in yeast genetics. I benefited from considerable help from the Tatum group at Stanford and, in particular, from Sheldon Reaume who provided me with genetically marked strains and advice on genetic manipulations. As discussed below, Roman, Hawthorne, and Douglas (Roman et al. 1951) started their genetic studies using the same two strains. Their initial experiments were designed to find alternative explanations for Lindegren's claims about gene conversion (see section below on Gene Conversion). The many contributions of this outstanding group are covered in other chapters in this volume.

In 1942, Herschel Roman had completed his Ph.D. degree with L.J. Stadler, of the University of Missouri, on studies of the B chromosome of corn and had moved to Seattle to join the Botany Department of the University of Washington. He soon discovered that Seattle was not a good place for research on corn and began to consider other organisms. In 1947, Roman invited Carl and Gertrude Lindegren to Seattle to give a series of lectures on yeast genetics to determine whether yeast might be a suitable alternative to corn. In a recent article on this period, Roman (1986) wrote, regarding the period that Lindegren was in Seattle:

> Some of his views were quite controversial, and we had many discussions during the two weeks or so that he was in Seattle: discussions in which alternative explanations were proposed to account for the same data. His unorthodox interpretations presented a challenge that was largely responsible for my choosing yeast as an experimental organism.

The lectures that Lindegren gave in Seattle were incorporated into the book *The Yeast Cell, Its Genetics and Cytology* (Lindegren 1949), which included a chapter describing the first genetic map of *Saccharomyces* as well as chapters on procedures and other aspects of yeast genetics and cytology. For many years, this was the only resource book for those wanting to start research on the genetics of yeast. As one who was introduced to yeast genetics by this book, I found it to be useful in some areas but also confusing and difficult to follow in others. Lindegren dedicated this book to his wife Gertrude

> ...in recognition of the major role that she played in its creation. By virtue of her impatient and energetic initiative, which demands immediate translation of thought into experiment, and her unflagging enthusiasm which has always proved equal to the discouraging task of testing each new alternative that experiment implacably imposes, she has succeeded in the extraordinarily difficult task of constructing the chromosome maps of *Saccharomyces*. This has graduated *Saccharomyces* from a promising organism for experimental breeding to a full-fledged membership in the *Drosophila*-maize-*Neurospora* hierarchy.

In 1948, the Lindegrens moved to the University of Southern Illinois at Carbondale, where they stayed for the remainder of their careers. From 1943 to 1973, more than 170 publications on various aspects of yeast genetics and cytology appeared from Washington University and from the Biological Research Laboratory at Carbondale, which Lindegren founded and directed.

The First International Conference on Yeast Genetics was hosted by Carl and Gertrude Lindegren and Maurice Ogur and was held in Carbondale on November 16–18, 1961. The meeting was arranged by Jack von Borstel and Carl Lindegren and was financed by the Committee for the Maintenance of Genetic Stocks, a standing committee of the Genetics Society of America. A total of 11 yeast geneticists were in attendance, five of whom were from Carbondale: Carl Lindegren, Gertrude Lindegren, Alvin Sarachek, Ernest Shult, and David Pittman. The outsiders were Robert Drysdale, Sy Fogel, Don Hawthorne, Giovanni Magni, Robert Mortimer, and Jack von Borstel (von Borstel 1962). It is of interest to note that the first *Neurospora* Conference was also held in 1961 and was attended by 92 *Neurospora* geneticists. Carl and Gertrude Lindegren were among the 53 yeast geneticists who attended the Second International Conference held in Gif-sur-Yvette in 1963 (J. Johnston, pers. comm.). The Third International Conference on Yeast Genetics was held in Seattle on September 13–14, 1965 (von Borstel 1966). This meeting, which was attended by 64 scientists, was organized by Herschel Roman to celebrate the opening of the new "J Wing" (built for the Genetics and Biochemistry Departments) of the University of Washington Medical School, and "to commemorate the distinguished contribution that Carl Lindegren has rendered to genetics in general and to the area of yeast genetics in particular."

In addition to his 1949 book, Lindegren published two other books: *The Cold War in Biology* (Lindegren 1966) and *The Theory and Practice of Natural Healing* (Lindegren 1981). I have recently learned, from Lindegren's granddaughter Averil Lerman that Lindegren had started another book entitled "*Theoretical and Speculative Cell Biology*." I have requested a copy of the manuscript of this book, but the revelation of its contents will have to wait for some future discussion of Lindegren's contributions. Ms. Lerman also told me that Lindegren was a friend of "Doc" in Steinbeck's "*Cannery Row*." Doc was Doc Ricketts who managed a small biological supply company in Monterey, California; Doc was acquainted with scientists and students at the Marine Biological Laboratory in Pacific Grove, California, as well as with all the artists and derelicts of the region. After a long career, which at times was distinguished and other times controversial, Carl Lindegren died in 1987. His wife and long-time collaborator, Gertrude, died in 1976.

Carl Lindegren's scientific career could be described as both unusual and tragic. Although he made some very important scientific contributions, he was largely rejected and ostracized by his scientific colleagues. He was associated, over a number of years, with some of the top biological scientists in the world but was not recognized as a leader himself. From the time he earned his Ph.D. degree in 1933, at the age of 37, until 1948, he was unable to find a stable academic position. At the age of 52, he accepted an appointment at Southern Illinois University, which at that time was primarily a teacher's college (Alvin Sarachek, pers. comm.).

Lindegren nevertheless developed, under these difficult circumstances, an active, productive, and devoted research group. He helped to establish a graduate program at Southern Illinois University and formed the Department of Microbiology. He trained a significant number of graduate students and postdoctoral fellows, and several of these have since gone on to very productive scientific careers. For many years, however, he kept his research group isolated from the larger yeast genetics and general genetics communities. Alvin Sarachek, one of Lindegren's former graduate students, and now a professor at Wichita State University, explains, "The exchange of people and information with other laboratories usual in science simply weren't promoted or considered essential." Further comments from Sarachek are presented below.

> The relatively few people who had opportunity to work directly with him have been intensely respectful and appreciative of the experience, and he returned them much support and loyalty. But he was not one to engender collegial affection outside that group.
>
> He was frequently in a hurry to make impact and not everything that emerged from the lab was experimentally well founded. Gertrude and his associates put a brake on much of this. But, unfortunately, enough shallow and arguable stuff was published to create uncertainties in many quarters about the laboratory's output, generally. He argued so fervently and for so long for gene conversion on inadequate grounds that his eventual proofs were probably ignored.
>
> Considering the singular quality of his early work, that lack of meticulousness may have been due, partially at least, to his limited physical and collegial resources, the drive of a uniquely imaginative and resourceful mind, and a felt need to move rapidly in hope of identification with uncovery of new and major principles.
>
> ...no comment on the Carbondale group would be complete without attention to Lindegren's truly special abilities to inspire his colleagues, and to create valid scientists out of the most improbable material.

Ernest Shult, now a Kansas Regents Distinguished Professor of Mathematics at Kansas State University, started out as an assistant in Lindegren's laboratory. He was encouraged by Lindegren to apply his considerable talents in mathematics to problems in genetics being worked on in the laboratory at that time. Yasuji Oshima, one of the world's leading yeast geneticists, spent 2 years (1963–1965) as a postdoc in Lindegren's laboratory. His research on the mating-type loci is discussed in another chapter in this volume.

Sol Spiegelman began his scientific career as a student of Lindegren soon after Lindegren had moved to Washington University and had begun his studies on yeast. Spiegelman and Lindegren began work on "adaptive enzyme synthesis" using sugar fermentative enzymes and associated genes in yeast as a model system. From experiments with the gene for melibiose fermentation, Spiegelman proposed that "once the enzyme is formed, further formation of enzyme molecules can proceed without intervention of the gene." This "plasmagene" model was later rejected by Spiegelman on the basis of additional experiments, but Lindegren modified the model into his "cytogene theory" and continued to promote this model (see section below on Cytogene Theory). This situation, as well as other aspects of Lindegren's research, are discussed in a recent book by Sapp (1987):

> Lindegren claimed that his experiments showed that the locus on the chromosomes, which he called the "chromogene", was simply a locus for the active element, the "cytogene", which he believed was capable of self-duplication in the cytoplasm independently of the "chromogene". This duality of the gene, he surmised, was corroborated by the work of Sonneborn on the *Kappa* substance in *Paramecium*.

Sapp adds,

> When Carl Lindegren broke from the Morgan school and began to support the inheritance of acquired characteristics, he (Lindegren 1949, Chapter 27) explicitly referred to the bias of indoctrination: "Recently in America there has been a recognition of the phenomenon discovered by the German workers and much discussion of cytoplasmic inheritance especially by Sonneborn, and it is noteworthy that his early training did not involve genetic indoctrination as a member or close associate of the Morgan school.
>
> In yeast genetics we encountered many examples of non-Mendelian phenomenon and in our early work I interpreted these as involving the hereditary transmission of autonomous entities. This was a rather difficult thing for me to do because I had been thoroughly indoctrinated in Gene Theory by my long association with Dr. Morgan and other members of his famous staff."
>
> Lindegren emerged as a staunch defender of the importance of cytoplasmic inheritance and the inheritance of acquired characteristics. By the 1960s, however, he was dismissed to the periphery of American genetics where he criticized what he called the anti-intellectual, atheoretic, and doctrinaire climate of American biology (Lindegren 1966).

It seems that Lindegren's apparent rejection of Morgan-Mendelism and his adoption of neo-Lamarckian reasoning in the interpretation of his findings (see section below on Directed Mutation and the Galactose and Melezitose

Genes) were the main reasons for his being rejected and dismissed by his colleagues. This is also discussed in a recent book by Brock (1990):

> Although Spiegelman himself soon abandoned the plasmagene model, Lindegren continued to insist on variants of it for his cytogene hypothesis (and even a version involving a kind of reverse transcription!). Spiegelman eventually became a respected contributor to molecular biology, but Lindegren's stubborn insistence on radical ideas unsupported by solid experimental evidence led to his eventual ostracism from the genetics community.

Brock very succinctly describes the basis of Lindegren's problems: a "stubborn insistence on radical ideas unsupported by solid experimental evidence." Lindegren's researches on gene conversion, adaptive enzyme synthesis, yeast cytology, yeast viruses, preferential segregation of chromosomes, strand relationships of multiple crossing-over, and other areas all are compromised by his continued promotion of untenable ideas. Many of his ideas initially were very stimulating, but these ideas soon lost their luster once they lacked the support of experimental results. Others, such as his notions about gene conversion, turned out to be correct even though they were based on incorrect interpretations of his data. To have radical ideas in science, as Lindegren did, is fine as long as these ideas are backed up by solid evidence and as long as alternative explanations are eliminated. Lindegren failed to provide the solid evidence to support his ideas, and he did not fairly eliminate counterproposals. Below I consider in more detail several of Lindegren's research topics.

## CHIASMA AND CHROMATID INTERFERENCE AND GENETIC RECOMBINATION

Lindegren, in his early *Neurospora* work, developed a partial genetic map of chromosome I and carried out a five-point cross involving the centromere and two markers on each side of the centromere. Centromere segregation was deduced from the order of spores in the linear ascus. From these crosses, Carl and Gertrude Lindegren established that crossing-over occurred at the four-strand stage and that only two of the four strands were involved in any one exchange. He deduced that second-division segregation of a gene resulted from crossing-over between this gene and the centromere and showed that each genetic site had a characteristic second-division segregation frequency (Lindegren 1936; Lindegren and Lindegren 1937, 1939). His studies opened up the field of *Neurospora* genetics and were a major reason Beadle and Tatum later chose this organism for their studies (Beadle 1977). These were and remain major contributions to the field of genetics.

The Lindegrens observed in the crosses multiple crossovers in individual asci and classified these according to the strands involved in adjacent double exchanges. For a given first interhomolog exchange, four possibilities exist for the second exchange. One of these is an exchange between the same two strands involved in the first exchange (two-strand double), one is between the two strands not involved in the first exchange (four-strand double), and the remaining two are between one involved and one uninvolved strand (three-strand double). If all four possible second exchanges are equally frequent (no chromatid interference), 2-strand:3-strand:4-strand double exchanges are expected in the ratio 1:2:1. The Lindegrens, however, observed an apparent nonrandom involvement of strands, or chromatid interference, in the direction of a large excess of two-strand doubles (Lindegren and Lindegren 1937, 1939, 1942). This chromatid interference was seen mostly for the regions spanning the centromere, where the evidence of centromere segregation depended entirely on the order of ascospore pairs in the linear ascus.

Several years later, Branch Howe (1956) and David Stadler (1956) independently showed that apparent two-strand doubles across the centromere in *Neurospora* are the result not of chromatid interference but of nuclear passing. Lindegren assumed that sister nuclei resulting from the second meiotic division were always side by side, one pair at one end of the ascus and the other pair at the opposite end. Postmeiotic mitoses were assumed to merely duplicate this pattern to generate the characteristic eight-spore linear ascus. A gene (*B/b*) that is very near a centromere would be expected to yield mostly *BB BB bb bb* asci, and departures from this pattern would be due to crossing-over between the gene and its centromere. However, nuclear "passing" can distort the arrangement of spores in the ascus and thus cause artifactual scoring of crossing-over and centromere segregation. This can account for the major part of the Lindegrens' excess of apparent two-strand doubles.

Lindegren and his co-workers also concluded that the same type of chromatid interference occurred in some of their yeast crosses (Desborough et al. 1960; Lindegren 1964), although work by Don Hawthorne and myself (Hawthorne and Mortimer 1960) and by the Carbondale group (Shult and Lindegren 1956a; Desborough and Shult 1962) failed to show such interference in most regions. Despite these and the many other studies in yeast and *Neurospora* that failed to show chromatid interference, Lindegren stated as late as 1964 (Lindegren 1964), in a discussion of different models of crossing-over, that

> The analysis of yeast hybrids has proved beyond question that in yeast, as in *Neurospora*, the frequency of 2-strand doubles exceeds the frequency of 3-strand doubles and that 4-strand doubles are the least frequent of the three categories. I shall propose a new explanation of crossing-over consistent with this distribution.

To have made such a statement, Lindegren seems to have ignored or to have been unaware of data from his own and other laboratories. He then went on to propose a "new" model of crossing-over. As might be expected, this model was not taken very seriously. A later study (Fogel et al. 1979), which was based on a large number of tetrads, was partially aimed at the question of whether or not there was an excess of two-strand double exchanges in yeast. The original Holliday (1964) model of genetic recombination predicted such an excess; we failed to find a significant departure from the frequency of two-strand doubles expected with no chromatid interference (256 2-strand:457 3-strand:200 4-strand). These findings, which were initially reported by Sy Fogel and me at the 1973 Aviemore (Scotland) meeting on Genetic Recombination, provided a basis for an essential feature of the Meselson-Radding model of genetic recombination (Meselson and Radding 1975).

## DIRECTED MUTATION AND THE GALACTOSE AND MELEZITOSE GENES

Lindegren and Pittman (1953) studied yeast strains that showed long-term adaptation to galactose fermentation (presumably these strains carried the *gal3* mutation) and found mutations to rapid fermentation following exposure of the cells to galactose. They concluded on the basis of fluctuation experiments that these mutations were induced by exposure of the cells to galactose and were not spontaneous mutations that had been selected by the experimental procedure:

> It is inferred that the induction of the mutation results from the impingement of galactose on the gene-surface thus modelling the gene-surface into a mirror-image of the substrate.

This argument is carried further in a later paper (Lindegren 1955a).

> There is a specific characteristic which is not consistent with the view that genes are composed of deoxyribonucleic acid. Evidence (Lindegren and Pittman 1953) concerning the gene controlling galactose-fermentation in *Saccharomyces* has been interpreted to indicate that genes may be plastic, since the substrate for a gene controlling the production of an adaptive enzyme is itself a mutagen controlling the mutation of the gene from a non-functional to a functional state.
>
> Rather than a rigid structure such as deoxyribonucleic acid, the genes themselves may be plastic materials attached to the deoxyribonucleic acid.
>
> The basic functions of deoxyribonucleic acid are to hold the plastic material of the gene in place and to transmit information from it to a functioning cell centre.

In their first report on the melezitose locus, Lindegren and Lindegren (1953) observed that this gene controlled the fermentation of up to five sugars. The ability to ferment the different sugars was subject to the adaptive synthesis of a single enzyme (Palleroni and Lindegren 1953) following exposure of the cells to one of these sugars. They observed stable alterations of the melezitose (*MZ*) gene to five different states, and these states were determined by the sugar used to cause adaptation and characterized by the combinations of these sugars fermented.

In a lengthy treatise (Lindegren 1957) entitled "The Integrated Cell" Lindegren argues for the autonomous role of the cytoplasm, the structural as opposed to informational role for DNA, the plasticity of the gene (which he said is not made up of DNA), and the directed mutation of this gene by its substrate.

> The gene is a plastic structure which can be remodelled by substrate and specifically remodelled by a specific substrate.

Most of Lindegren's conclusions about directed mutation are based on studies of the *MZ* locus (Hwang and Lindegren 1964; D.S. Hwang et al. 1964;

Lindegren and Lindegren 1953, 1956; Lindegren and Pittman 1958, 1959; Ouchi and Lindegren 1963). As an example, Lindegren et al. (1956a) showed that UV radiation caused specific "degradation" of the *MZ* locus so that only four or fewer sugars could be fermented. They argued:

> The MZ locus is a single gene with at least five different allele states. On this hypothesis, the radiation damage to the genetic material can be interpreted as an intragenic phenomenon. The totipotent allele is inferred to comprise a "Mirror-image" template surface of at least five different substrates. The minimal radiation damage to the totipotent allele involves a change in conformation which preserves the capacity to "fit" four of the substrates (turanose, maltose, sucrose, methyl-a-D-glucopyranoside and melezitose) but renders it incapable of fitting melezitose with sufficient precision to permit the latter to function as an inducer of enzyme.

In an article entitled "Directed Mutation in Yeast and Bacteria" Lindegren (1963b) developed a more advanced and detailed model for directed mutation involving a receptor as part of the gene.

> ...the gene (as distinguished from its fragment) must contain a heritable, mutable protein-component, the receptor (in addition to the "structural" DNA element) which can be imprinted by the substrate.

It is remarkable that these radical ideas were advanced at a time when evidence to the contrary was overwhelming and almost universally accepted.

Lindegren (1963c), in a paper entitled "Lamarckian Proteins," believed that Koshland's demonstration of a conformational change of a protein when it interacted with its substrate was strong support for his notions about genes and their directed mutation. The receptor hypothesis discussed above is developed fully in another article (Lindegren 1963a). This hypothesis was advanced as an alternative to the Central Dogma that in part states that the sequence of DNA of a gene determines the structure of the protein encoded by that gene. In this model, Lindegren recognizes the role of mRNA and admits to heritable changes in the DNA structural part as well as the protein part of his concept of the gene. Lindegren also found difficulties with the Jacob-Monod operon hypothesis and proposed his receptor model as an alternative.

> The receptor-hypothesis of the induction of gene-controlled adaptive enzymes proposes that the locus of a gene comprises a protein receptor and a DNA structural component. The structural component carries a segment of RNA which functions as messenger RNA. When certain inducers make contact with the receptor, they initiate an excitation which releases the messenger RNA from the structural component. The protein-component of the gene is hereditary, in the sense that the conformation, or coiling, of the protein is determined by the coiling of the receptor already at the locus, and on this hypothesis, mutations may occur either in the structural component or in the receptor.

It is interesting that many of the papers on directed mutation and neo-Lamarckian genetics are published with Lindegren as sole author. We do not know if this was by choice or if other members of his laboratory did not want to be associated with the advancement of these radical ideas.

## GENE CONVERSION

Lindegren (1949, 1953a,b) and Lindegren and Lindegren (1956) proposed that departures from the 2:2 segregation ratios expected from a heterozygous diploid were examples of the phenomenon of gene conversion in which the alleles can affect (contaminate) each other to change wild type to mutant or vice versa. On the basis of earlier studies of irregular segregations (gene conversion) in basidiomycetes and mosses, Winkler (1930) had proposed an alternative mechanism by which recombinant chromosomes could be generated. He assumed that gene conversion events, which were assumed to occur only at sites of heterozygosity, if coordinated along a chromosome, could generate recombinants that involved several loci. However, the elegant experiments of Stern (1932) demonstrated that crossing-over involved a physical exchange of large segments of the recombining chromosomes and that Winkler's hypothesis was wrong. Lindegren (1953b) correctly points out, however, that "Stern showed that Winkler was wrong in proposing that all recombination was the result of gene conversion; he did not prove that gene conversion did not occur."

At the time, revival of the notion that gene conversion could explain irregular segregation ratios was considered radical, and Roman et al. (1951) and Winge and Roberts (1954a,b, 1957) countered with more conventional genetic mechanisms such as polyploidy, multigenic control, or postmeiotic mitoses to explain these irregular ratios. These investigators carried out careful experiments to show that such processes occurred in Lindegren's strains and could account for at least some of the aberrant segregation ratios. A more detailed discussion of Winge's reaction to Lindegren's studies on gene conversion are included in the chapter on Winge in this volume. Many of Lindegren's studies of irregular segregation ratios were based on the segregation of what he thought was a single gene controlling the fermentation of the sugar α-methyl glucoside (Lindegren 1953b). Hawthorne (1956) obtained parental strains from Lindegren and showed, however, that the fermentation of this sugar was under the control of at least five genes. He established that fermentation-positive strains must carry *MGL1* + *MGL2*, *MGL3* + *MGL2*, or *MGL4* + *MAL1*. A diploid strain heterozygous for *MGL1*, *MGL2*, and *MGL3* would be expected to yield asci that segregate 2:2, 1:3, and 0:4 for fermentation-positive:fermentation-negative in ratios of 19:16:1. A diploid heterozygous for all five genes would be expected to segregate 19(4:0):352(3:1):799(2:2):120(1:3):6(0:4). (These ratios differ somewhat from those presented by Hawthorne. I am forced to conclude that either he or I or both of us made a mistake in this rather simple genetic exercise.) The types and frequencies of aberrant ratios that Lindegren saw varied in different crosses as would be expected if he were studying a trait under multigenic control. Lindegren later described a 1:3 segregation of the *ade1* gene and showed convincingly that this irregular ratio could not be explained by conventional means and that it was therefore a legitimate example of gene conversion (Lindegren 1955b; Lindegren et al. 1956b). However, Mitchell

(1955) and Roman (1957) had by now also shown that gene conversion (although they chose not to use this term) existed in both *Neurospora* and yeast, and Lindegren's valid example was mostly ignored. Although Lindegren was the first to draw attention to this phenomenon, he did so mainly on the basis of faulty interpretation of his data. When he did present a valid example, he was largely ignored. He nevertheless felt, and with some justification, that he and others should have been given more recognition than they received (see Lindegren's comments in Roman 1957; Lindegren 1958a). Later work in several laboratories has shown that gene conversion is an important and general phenomenon that is central to the process of genetic recombination (Holliday 1964; Fogel and Mortimer 1969; Meselson and Radding 1975; Fogel et al. 1979).

## TETRAPLOID GENETICS

In the early 1950s, several groups reported the isolation and genetic analyses of polyploid yeast. Roman et al. (1951) analyzed a diploid strain heterozygous for three genetic markers. Among 64 tetrads from this cross, one showed irregular segregation ratios for each of the three traits. All four spores were of **a**/α mating-type constitution and could be analyzed genetically to confirm the tetraploid origin of this ascus. The spontaneous occurrence of this tetraploid in an otherwise diploid culture was cited as one possible origin of the irregular segregation ratios that had been cited by Lindegren as evidence for gene conversion. Lindegren and Lindegren (1951b) also observed one ascus, from an otherwise diploid cross, that showed tetraploid segregation ratios. Two spores were **a**/α, one was **a**/**a**, and the other was α/α. The two mating diploid spores were crossed to each other and to haploids to produce tetraploid and triploid strains. The segregation of genes in this tetraploid was then studied genetically. The Lindegrens (Lindegren and Lindegren 1949) and Mundkur (1949) argued correctly that polyploidy can explain irregular ratios that are 4:0 and 3:1, but 1:3 and 0:4 ratios cannot be accounted for by this mechanism. The polyploid series constructed in this study was then used for a number of subsequent studies. Ogur et al. (1952) determined the amount of DNA/cell in this polyploid series, and the values presented have been the accepted standard for this ratio for most of the time since then. Other studies of polyploid segregation were carried out by Pomper et al. (1954), Roman et al. (1955), Leupold (1956a,b), and Mortimer (1958).

## GENETIC MAPS

The first genetic maps of *Saccharomyces* were published by the Lindegrens (Lindegren 1949; Lindegren and Lindegren 1951a), and additional mapping data appeared from this laboratory a few years later (Desborough and Lindegren 1959; Desborough et al. 1960; Lindegren et al. 1959), about the time that Don Hawthorne and I published our first map (Hawthorne and Mortimer 1960). Following the Carbondale meeting (von Borstel 1962), a uniform nomenclature of genes was established for the first time, and this involved an exchange of strains between my laboratory and that of the Lindegrens. The Carbondale group published additional mapping papers during the 1960s (Lindegren et al. 1962b; Y.L. Hwang et al. 1963, 1964; Hwang and Lindegren

1966; Shult et al. 1967) as did Don Hawthorne and I (Mortimer and Hawthorne 1966, 1973; Hawthorne and Mortimer 1968). The Lindegren maps usually contained extra "chromosomes," which were genes or groups of genes that were claimed to segregate preferentially with particular chromosomes, as well as some linkages that could not be repeated in other laboratories. On the basis of their studies, they correctly identified the centromere markers that define chromosomes I (*ade1*), II (*gal1*), III (*MAT*), V (*ura3*), X (*met3*), XI (*met14*), and XIII (*lys7*) and mapped several genes on these and other chromosomes.

## PREFERENTIAL SEGREGATION AND SITES OF AFFINITY

Mendel's law of independent assortment applies to pairs of genes on different chromosomes or to pairs of genes far apart on the same chromosome. This law has been shown to apply to a large number of different organisms and pairs of linkage groups within these organisms. In the early 1950s, departures from independent assortment were observed for certain pairs of genes on different chromosomes in corn and in mice. It was proposed that "sites of affinity" existed on nonhomologous chromosomes that led to the segregation of either parental or nonparental combinations of chromosomes to the same pole in meiosis I. If the sites of affinity correspond to the centromeres, then either all parental or all recombinant gametes are expected for genes at the corresponding centromeres.

Lindegren and co-workers reported examples of such preferential segregation in some of their yeast crosses (Lindegren and Shult 1956a; Shult and Lindegren 1957a,b). In one detailed publication (Shult et al. 1962), they described 25 independent PD, NPD, and T distributions out of a total of 1487 such distributions that showed a significant (<1% level) excess of NPD over PD tetrads. This number was considerably greater than the 7.4 expected by chance, and they interpreted this excess to be due to preferential segregation of nonhomologous chromosomes.

One possible source of apparent reverse linkage (excess of NPD tetrads) or of false positive linkage (excess of PD tetrads for unlinked genes) in tetrad analysis is differential viability of meiotic products with different combinations of parental genes. Consider the cross *A b* x *a B*, where *A*/*a* and *B*/*b* are on different chromosomes and the four possible gene combinations *A B*, *A b*, *a B*, and *a b* have relative spore viabilities of 1, 1, 1, and 0.5. The three ascus types PD, NPD, and T would have relative probabilities of having all four spores alive of 1:0.25:0.5. A sample of 120 asci that were originally distributed over these classes in the ratios 1:1:4 would be scored, among the complete tetrads, as 20PD:5NPD:40T. This would appear to be a significant linkage, yet this linkage would be artifactual. In an earlier study (R.K. Mortimer and S. Fogel, unpubl.), we observed a PD:NPD ratio for *trp1* and *pet1*, which are, respectively, near the centromeres of chromosomes IV and VII, of 80:33. This is a highly significant departure from the 1:1 ratio expected for independent assortment. Since this was part of a gene conversion study, only complete tetrads were analyzed. Because of this highly aberrant result, we then examined the tetrads that yielded only two or three viable spores, and when this was done, the PD:NPD ratio in the total sample did not depart significantly from 1:1. We then deduced that nearly all of the inviable spores carried *trp1* and *thr1* (linked to *pet1*). Presumably, the medium on which the tetrads were dissected and germinated was partially deficient in these amino acids.

We do not know if this could have been the basis for some of the aberrant ratios reported by Lindegren and co-workers as examples of preferential segregation. We believe this group analyzed only complete tetrads, but they rarely reported overall spore viabilities or fractions of complete tetrads (most yeast genetics groups still do not report such data). It is also our understanding that the tetrad dissection and analysis of the spore clones were carried out entirely by Gertrude Lindegren and her staff and that E. Shult, S. Desborough, C.C. Lindegren, and others, who analyzed the data, were not involved in this part of the experiments. Certainly, the initial results from the example cited above, involving chromosomes IV and VIII, if analyzed without question, could have been used to provide strong support for the phenomenon of preferential segregation.

## THE CYTOGENE THEORY

As mentioned above, Sol Spiegelman was a student of Lindegren at Washington University in St. Louis. There was considerable interest at that time in "adaptive enzyme formation" in which an enzymatic activity appears in response to the addition of substrate. Spiegelman and Lindegren proposed that the enzyme, once produced, could then increase in activity in response to substrate in an autonomous fashion without further need of the gene (Lindegren et al. 1944; Spiegelman et al. 1945). This was called the "plasmagene hypothesis," and these ideas were strongly influenced by the prevailing interest in cytoplasmic inheritance. Spiegelman soon abandoned this model but Lindegren continued to promote it or variations of it (the cytogene theory) for several more years (Lindegren 1946, 1949; Lindegren and Lindegren 1947a,b). He proposed that the dominant gene on a chromosome transmits a hypothetical cytogene to the cytoplasm and, if the corresponding substrate is present, enzyme production is started. For example, a *MEL* gene would transmit a corresponding *MEL* cytogene to the cytoplasm, which would then produce melibiase if melibiose was present. If the substrate is removed, the cytogene is recaptured by the chromosomal gene. In a heterozygote, a dominant cytogene could occasionally be transmitted to the locus of the recessive allele.

The experiments on which Lindegren based this cytogene theory involved a cross between *S. carlsbergensis* (Mel$^+$) and *S. cerevisiae* (Mel$^-$). Lindegren obtained six asci from this cross. Three asci segregated 4:0 for melibiose fermentation, two were 3:1, and one was 2:2, and Lindegren assumed, on the basis of these results, that *S. carlsbergensis* carried two nonallelic *MEL* genes. Later, however, he appears to have deserted this reasonable model and instead interpreted these results as evidence for his cytogene theory. Winge (1949) comments caustically about these experiments and their interpretation:

> Lindegren abandoned, however, his original interpretation and, in order to explain the inheritance behaviour, proposed a number of theories, which differed from all of those previously accepted in genetics, and the correctness of which we have not been able to verify through our own investigations.

> In 1945 Lindegren set forth his so-called cytogene hypothesis according to which an enzyme (in particular, melibiase and galactozymase) is primarily produced by the effect of a specific gene in a chromosome, but the ability to produce the enzyme is transmitted to the cytoplasm where a hypothetical cytogene takes over enzyme production, as long as the necessary sugar (melibiose or galactose) is present in the substratum. If the sugar is removed, enzyme production by the cytoplasm ceases, and the dominant gene in the chromosome collects practically all the cytogenic material. Lindegren would in this way attempt to show that enzymatic ability under certain conditions may be transmitted to the offspring through the cytoplasm, even if the gene itself is lacking in the chromosome.
>
> In 1947 he set forth, in addition, the hypothesis that in a heterozygote there can occur transmission of the dominant gene to the locus of the recessive allelomorph by contact with it.
>
> From our own investigations, we have concluded that these revolutionary hypotheses are superfluous.

Although wrong, the plasmagene theory and even the cytogene theory had considerable heuristic value at the time they were proposed. These theories envisioned the need for informational transfer from the nucleus to the cytoplasm and subsequent synthesis of protein independent of the gene.

## YEAST CYTOLOGY AND THE NUCLEAR VACUOLE

Lindegren devoted several sections in his book (Lindegren 1949) to yeast cytology. The experiments described were based primarily on earlier studies (Lindegren and Lindegren 1947a). He presented photomicrographs that he claimed showed the presence of rod-like chromosomes (six pairs) in the cell vacuole. These results were obtained using toluidine blue as a stain. Additional studies from Carbondale confirmed this view (Lindegren 1952; Lindegren and Rafalko 1950). Curiously, Townsend and Lindegren (1953) reported that rod-like structures in the vacuole could be generated by the stain peeling off the inner vacuole wall. Toluidine blue, when dried down on a slide, also forms such structures. Lindegren still argued that the structures that he saw in the vacuole were chromosomes because they were paired. The Feulgen-positive body adjacent to the vacuole, which Winge (1935) and others had identified as the nucleus, was called the centrosome. This radical view persisted in publications from Lindegren's laboratory until 1957 when Dan McClary and others at Carbondale published two papers that identified the nucleus as the Giemsa-staining body on the outside of the vacuole (McClary et al. 1957a,b). Sadly, Lindegren still continued to write about the nuclear vacuole for many years after these excellent reports appeared (Lindegren 1981).

## THE ZYMOPHAGE

In a short series of papers (Lindegren 1958b; Lindegren and Bang 1961, 1964; Lindegren et al. 1962a), Lindegren reported evidence for a yeast phage.

> We have recently discovered an infectious disease of yeast for which we have coined the name zymophage, to parallel the names bacteriophage and actinophage.

This phage was elusive.

> The same culture spread on different days may or may not show plaques.
>
> ...the plaques on yeast plates attain full size when they first appear (about 4 hours) and do not become larger as time goes on...
>
> Many plaques that appear early are soon overgrown, and colonies grow in the center of persistent plaques...

However, mixed inoculations of an "infected" and a normal strain sometimes failed to become turbid, and lysed cells were seen in infected areas on plates. Electron micrographs of "infected" cells sometimes showed electron-dense bodies that were virus-like in appearance.

It seems likely now that these "plaques" were artifactual. Droplets of condensation from the lid of the petri dish, bubbles, or nonwettable areas on the agar surface all can create temporary plaque-like patterns of growth. The electron-dense objects could have been the now well-known "killer" particles, and the failure of some mixed cultures to grow could have been due to the killer factor. These studies were not followed further, but it seems likely that, had they been pursued in the proper setting, they could have led to the identification of the killer-virus-like particle.

## THEORY, METHODS, POLEMICS, AND CRITIQUES

Several articles appeared from the Lindegren laboratory that presented theoretical arguments or were critiques of other studies. Other articles presented new methods, some of which are still in use. The theoretical articles fall mainly into two categories: those that were written by Ernest Shult and Carl Lindegren and those that were written by Lindegren alone. Shult and Lindegren (1956a) published an important paper on tetrad analysis theory that uses a novel mathematical approach. Shult and Lindegren (1955a,b, 1956b, 1957a,b,c) also published several other articles of less impact that deal with genetic interference, preferential segregation of chromosomes, and methods of ordering genes using tetrad analysis data. Lindegren also published a large number of short papers that deal with a variety of other topics. For example, he challenged Lederberg's Nobel prize address for his failure to give proper recognition to Lindegren and others for their demonstration of the role of substrate in changing the gene (Lindegren 1960). A particularly interesting and revealing paper (Lindegren and Shult 1956b), which sums up many of the arguments presented in the above papers, and in fact in many of Lindegren's papers, appears to be a challenge to most of the basic tenets of genetics, such as (1) independent assortment of centromeres, (2) stability of the gene in heterozygous state, (3) randomness of crossing-over at the four-strand stage, (4) positive chiasmata interference as a general finding, (5) direct relationship between crossing-over and genetic recombination, and (6) independence of crossing-over events on different chromosomes. They claim to have presented evidence that questions all of these assumptions, and some of these topics are discussed in the earlier sections of this article.

Lindegren and his co-workers should be given credit for development of some important experimental protocols. Ogur et al. (1954, 1957) described a convenient plate method for detecting respiratory-deficient clones that has been used effectively for several years. Lindegren et al. (1965) described a procedure that is still in use for producing yeast mutations using ethylmethanesulfonate as a mutagen. Lindegren and Lindegren (1951a) described a procedure that employed only seven different test mixtures to identify any of up to 28 nutritional requirements. Their scheme is still used by *Neurospora* geneticists to identify the requirements of new mutants (D. Perkins, pers. comm.).

## EPILOGUE

I have presented here a brief biographical sketch of Carl Lindegren and have discussed several areas of research in which he was involved during his scientific career. In nearly all of these areas, he presented evidence that he interpreted in ways that were far from the then mainstream of thinking in cell biology and genetics. He challenged the concept of the stability of the gene in heterozygous condition in his gene conversion studies. He argued for the presence of the chromosomes in the cell vacuole, contrary to the findings of many years of cytological research. He argued for nonrandomness in the strand involvement in multiple crossing-over and for nonrandom disjunction of nonhomologous chromosomes, which is a direct challenge to Mendel's laws. Most importantly, he challenged the Central Dogma which states that the gene is made up of DNA, which in turn determines the sequence of a protein. He believed that the informational part of the gene was a protein and that this protein could be changed by directed mutation. This makes a consistent story only if one accepts the assertion of Sapp that Lindegren broke with Morgan-Mendelism and adopted a neo-Lamarckian philosophy. He then tried to show that the basic principles of Morgan-Mendelism were wrong and in addition argued for directed mutation and the inheritance of acquired characteristics as an alternative way of interpreting genetic data. Lindegren nevertheless made some great scientific contributions. These include pioneering the field of *Neurospora* genetics and, along with Winge, the field of yeast genetics, and the discovery and distribution of heterothallic yeast strains. His development of procedures for genetic analysis helped to make possible yeast genetics as we know it today.

## ACKNOWLEDGMENTS

This paper was written while I was a guest of the Department of Biochemistry at the Biocenter, University of Basel. I want to thank Mike Hall and the other members of the Biocenter for making my stay a very pleasant and rewarding one. This paper benefited from interviews with David Stadler, David Perkins, and Caryl Roman, and I want to thank them for these interviews and the reprints and other materials they provided for my use. Some parts of this article were based on an earlier study done in collaboration with John Johnston. I want to thank Alvin Sarachek, J. Bhattacharjee, and Roberto Palleroni for their thoughtful and informative letters about their experiences at Carbondale. After an early draft of this article was written, I was contacted by Averil Lindegren Lerman, Lindegren's granddaughter. She provided me with valuable insights into Lindegren's early years. Finally, several friends and associates have read the manuscript at various stages and have offered valuable suggestions.

## REFERENCES

Beadle, G.W. 1977. Genes and chemical reactions in *Neurospora*. In *Nobel Lecture in Molecular Biology, 1933-1975*. Elsevier, Amsterdam.

Brock, T.D. 1990. *The emergence of bacterial genetics*. Cold Spring Harbor Laboratory Press, Cold Spring Harbor, New York.

Desborough, S. and G. Lindegren. 1959. Chromosome mapping of linkage data from *Saccharomyces* by tetrad data. *Genetica* **30:** 346–383.

Desborough, S. and E.E. Shult. 1962. An assay for chromosomal and chromatid interference in chromosome V of *Saccharomyces. Genetica* **33:** 69–78.

Desborough, S., E.E. Shult, T. Yoshida, and C.C. Lindegren. 1960. Interference patterns in family Y-1 of *Saccharomyces. Genetics* **45:** 1467–1480.

Fogel, S. and R.K. Mortimer. 1969. Informational transfer in meiotic gene conversion. *Proc. Natl. Acad. Sci.* **62:** 96–103.

Fogel, S., R.K. Mortimer, K. Lusnak, and F. Tavares. 1979. Meiotic gene conversion: A signal of the basic recombination event in yeast. *Cold Spring Harbor Symp. Quant. Biol.* **43:** 1325–1341.

Hawthorne, D.C. 1956. The genetics of alpha-methyl-glucoside fermentation in *Saccharomyces. Heredity* **12:** 273–284.

Hawthorne, D.C. and R.K. Mortimer. 1960. Chromosome mapping in *Saccharomyces*: Centromere linked genes. *Genetics* **45:** 1085–1110.

———. 1968. Genetic mapping of nonsense suppressors in yeast. *Genetics* **60:** 735–742.

Holliday, R. 1964. A mechanism for gene conversion in fungi. *Genet. Res.* **5:** 282–304.

Howe, H.B. 1956. Crossing over and nuclear passing in *Neurospora crassa. Genetics* **41:** 610–622.

Hwang, D.S. and C.C. Lindegren. 1964. Palatinose element of the receptor of the melezitose locus in *Saccharomyces. Nature* **203:** 791–792.

Hwang, D.S., C.C. Lindegren, J.K. Bhattacharjee, and A. Roshanmanesh. 1964. The genetic integrity of upgraded and downgraded alleles of the melezitose locus in *Saccharomyces. Can. J. Genet. Cytol.* **6:** 414–418.

Hwang, Y.L. and G. Lindegren. 1966. Sites of affinity and linear arrangement of genes on chromosome V of *Saccharomyces. Can. J. Genet. Cytol.***8:** 677–694.

Hwang, Y.L., G. Lindegren, and C.C. Lindegren. 1963. Mapping the eleventh centromere in *Saccharomyces. Can. J. Genet. Cytol.* **5:** 290–298.

———. 1964. The twelfth chromosome of *Saccharomyces. Can. J. Genet. Cytol.***6:** 373–380.

Leupold, U. 1956a. Tetraploid inheritance in *Saccharomyces. J. Genet.* **54:** 411–426.

———. 1956b. Tetrad analysis of segregation in auto tetraploids. *J. Genet.* **54:** 427–439.

Lindegren, C.C. 1936. A six-point map of the sex chromosome of *Neurospora crassa. J. Genet.* **32:** 243–256.

———. 1945. Mendelian and cytoplasmic inheritance in yeasts. *Ann. Mo. Bot. Gard.* **32:** 107–123.

———. 1946. A new gene theory and an explanation of the phenomenon of dominance to Mendelian segregation of the cytogene. *Proc. Natl. Acad. Sci.* **32:** 68–70.

———. 1949. *The yeast cell, its genetics and cytology*. Educational Publishers, St. Louis, Misouri.

———. 1952. The structure of the yeast cell. *Symp. Soc. Exp. Biol.* **6:** 277–289.

———. 1953a. Concepts of gene-structure and gene-action derived from tetrad analysis of *Saccharomyces. Experientia* **9:** 75–80.

———. 1953b. Gene conversion in *Saccharomyces*. *J. Genet.* **51:** 625–637..
———. 1955a. Is the gene a prime mover? *Nature* **176:** 1244–1245.
———. 1955b. Non-Mendelian segregation in a single tetrad of *Saccharomyces* ascribed to gene conversion. *Science* **121:** 605–607.
———. 1957. The integrated cell. *Cytologia* **22:** 415–441.
———. 1958a. Priority in gene conversion. *Experientia* **14:** 444.
———. 1958b. Respiratory deficiency and the zymophage: Two factors capable of causing degeneration in brewery yeasts. In *Proceedings of the American Society of Brewery Chemists 1958 Meeting*, pp. 86–91. ASBC, St. Paul, Minnesota.
———. 1960. Genetical theory. *Science* **131:** 1569.
———. 1963a. The receptor-hypothesis of induction of gene-controlled adaptive enzymes. *J. Theor. Biol.* **5:** 192–210.
———. 1963b. Directed mutations in yeast and bacteria. *Bull. Res. Counc. Israel* (11-A)**4:** 363–368.
———. 1963c. Lamarckian proteins. *Nature* **198:** 1224.
———. 1964. A new theory to explain crossing-over between genes on chromosomes. *Nature* **204:** 322–324.
———. 1966. *Cold war in biology*. Planarian, Ann Arbor, Michigan.
———. 1973. Reminiscences of B.O. Dodge and the beginnings of *Neurospora* genetics. *Neurospora Newsl.* **20:** 13–14.
———. 1981. *The theory and practice of natural healing*. Vantage, New York.
Lindegren, C.C. and Y.N. Bang. 1961. The zymophage. *Antonie Leeuwenhoek* **27:** 1–18.
———. 1964. Origin from mitochondria of the double-walled viral membranes in *Saccharomyces. Nature* **203:** 431–432.
Lindegren, C.C. and G. Lindegren. 1937. Non-random crossing over in *Neurospora. J. Heredity* **28:** 105–113.
———. 1939. Non-random crossing over in the second chromosome of *Neurospora crassa. Genetics* **24:** 1–7.
———. 1942. Locally specific patterns of chromatid and chromosome interference in *Neurospora. Genetics* **27:** 1–24.
———. 1943a. A new method of hybridizing yeast. *Proc. Natl. Acad. Sci.* **29:** 306–308.
———. 1943b. Environmental and genetic variations in yield and colony size of commercial yeast. *Ann. Mo. Bot. Gard.* **30:** 71–82.
———. 1943c. Segregation, mutation and copulation in *Saccharomyces cerevisiae. Ann. Mo. Bot. Gard.* **30:** 453–468.
———. 1947a. The cytogene theory. *Cold Spring Harbor Symp. Quant. Biol.* **11:** 115–129.
———. 1947b. Depletion mutations in Saccharomyces. *Proc. Natl. Acad. Sci.* **33:** 314–318.
———. 1949. Unusual gene-controlled combinations of carbohydrate fermentations in yeast hybrids. *Proc. Natl. Acad. Sci.* **35:** 23–27.
———. 1951a. Linkage relationships in *Saccharomyces* of genes controlling the fermentation of carbohydrates and the synthesis of vitamins, amino acids and nucleic acid components. *Indian Phytopathol.* **4:** 11–20.
———. 1951b. Tetraploid *Saccharomyces. J. Gen. Microbiol.* **5:** 885–893.
———. 1953. The genetics of melezitose fermentation in *Saccharomyces. Genetica* **26:** 430–444.
———. 1956. Effect of local chromosomal environment upon the genotype. *Nature* **178:** 796–797.
Lindegren, C.C. and D.D. Pittman. 1953. The induction in a *Saccharomyces* sp. of the gene-mutation controlling utilization of galactose by exposure to galactose. *J. Gen. Micobiol.* **9:** 494–511.
———. 1958. Effect of environment (radiation, substrate and allelic genes) on the melezitose gene in *Saccharomyces. Nature* **182:** 272–272.
———. 1959. The "all-or-none" nature of X-ray induced backmutation at the melezitose locus in *Saccharomyces. Genetica* **30:** 169–189.
Lindegren, C.C. and M. Rafalko. 1950. The structure of the nucleus of *Saccharomyces bayanus. Exp. Cell Res.* **1:** 169–187.
Lindegren, C.C. and E.E. Shult (1956a). Nonrandom assortment of centromeres with implications regarding random assortment of chromosomes. *Experientia* **12:** 177.

———. 1956b. The interpretation of genetical data. *Experientia* **13:** 48–51.
Lindegren, C.C., Y.N. Bang, and T. Hirano. 1962a. Progress report on the zymophage. *Trans. N.Y. Acad. Sci.* **24:** 540–566.
Lindegren, C.C., D.D. Pittman, and B. Ranganathan. 1956a. Orderly degradation of the MZ locus by ultraviolet radiation and its regeneration by contact with substrate. *Cytologia* (suppl.) 42–50.
Lindegren, C.C., S. Spiegelman, and G. Lindegren. 1944. Mendelian inheritance of adaptive enzymes in yeast. *Proc. Natl. Acad. Sci.* **30:** 346–352.
Lindegren, C.C., G. Lindegren, E.E. Shult, and S. Desborough. 1959. Chromosome maps of *Saccharomyces. Nature* **183:** 800–802.
Lindegren, C.C., G. Lindegren, E. Shult, and Y.L. Hwang. 1962b. Centromeres, sites of affinity and gene loci on the chromosomes of *Saccharomyces. Nature* **194:** 260–265.
Lindegren, C.C., G. Lindegren, R.B. Drysdale, J.P Hughes, and A. Bernes-Pomales. 1956b. Genetical analysis of the clones from a single tetrad of *Saccharomyces* showing non-Mendelian segregation. *Genetica* **28:** 1–24.
Lindegren, G., Y.L. Hwang, Y. Oshima, and C.C. Lindegren. 1965. Genetical mutants induced by ethyl methanesulfonate in *Saccharomyces. Can. J. Genet. Cytol.* **7:** 491–499.
McClary, D.O., M.A. Williams, and C.C. Lindegren. 1957a. Nuclear changes in the life cycle of *Saccharomyces. J. Bacteriol.* **73:** 754–757.
McClary, D.O., M.A. Williams, M. Ogur, and C.C. Lindegren. 1957b. Chromosome counts in a polyploid series of *Saccharomyces. J. Bacteriol.* **73:** 360–364.
Meselson, M.S. and C. M. Radding. 1975. A general model of genetic recombination. *Proc. Natl. Acad. Sci.* **72:** 358–361.
Mitchell, M.B. 1955. Aberrant recombination of pyridoxine mutants of *Neurospora. Proc. Natl. Acad. Sci.* **41:** 215–220.
Mortimer, R.K. 1958. Radiobiological and genetic studies on a polyploid series (haploid to hexaploid) of *Saccharomyces cerevisiae. Radiat. Res.* **9:** 312–326.
Mortimer, R.K. and D.C. Hawthorne. 1966. Genetic mapping in *Saccharomyces. Genetics* **53:** 165–173.
———. 1973. Genetic mapping in *Saccharomyces.* IV. Mapping of temperature-sensitive genes and use of disomic strains in localizing genes. *Genetics* **74:** 33–54.
Mortimer, R.K. and J.R. Johnston. 1986. Genealogy of principal strains of the Yeast Genetic Stock Center. *Genetics* **113:** 35–43.
Mundkur, B.D. 1949. Evidence excluding mutations, polysomy, and polyploidy as possible causes of non-Mendelian segregation in *Saccharomyces. Ann. Mo. Bot. Gard.* **36:** 259–281.
Ogur, M., G. Lindegren, and C.C. Lindegren. 1954. A simple screening test for genetic studies of respiration deficiency in yeast. *J. Bacteriol.* **68:** 391–392.
Ogur, M., R.C. St. John, and S. Nagai. 1957. Tetrazolium overlay technique for population studies of respiratory deficiency in yeast. *Science* **125:** 928–929.
Ogur, M., S. Minckler, G. Lindegren, and C.C. Lindegren. 1952. The nucleic acids in a polyploid series of *Saccharomyces. Arch. Biochem. Biophys.* **40:** 175–184.
Ouchi, S. and C.C. Lindegren. 1963. Genic interaction in *Saccharomyces. Can J. Genet. Cytol.* **5:** 257–267.
Palleroni, N.J. and C.C. Lindegren. 1953. A single adaptive enzyme in *Saccharomyces* elicited by several related substrates. *J. Bacteriol.* **65:** 122–130.
Perkins, D.D. 1992. *Neurospora*: The organism behind the molecular revolution. *Genetics* **130:** 687–701.
Pomper, S. and P.R. Burkholder. 1949. Studies on the biochemical genetics of yeast. *Proc. Natl. Acad. Sci.* **35:** 456–464.
Pomper, S., K.M. Daniels, and D.W. McKee. 1954. Genetic analysis of polyploid yeast. *Genetics* **39:** 343–355.
Reaume, S.E. and E.L. Tatum. 1949. Spontaneous and nitrogen mustard-induced nutritional deficiencies in *Saccharomyces cerevisiae. Arch. Biochem. Biophys.* **22:** 331–338.
Roman, H. 1957. Studies of gene mutation in *Saccharomyces. Cold Spring Harbor Symp. Quant. Biol.* **21:** 175–185.
———. 1986. The early days of yeast genetics: A personal narrative. *Annu. Rev. Genet.* **20:** 1–12.
Roman, H., D.C. Hawthorne, and H.C. Douglas. 1951. Polyploidy in yeast and its bear-

ing on the occurrence of irregular genetic ratios. *Proc. Natl. Acad. Sci.* **37:** 79–84.
Roman, H., M.M. Phillips, and S.M. Sands. 1955. Studies of polyploid *Saccharomyces*. I. Tetraploid segregation. *Genetics* **40:** 546–561.
Sapp, J. 1987. *Beyond the gene. Cytoplasmic inheritance and the struggle for authority in genetics.* Oxford University Press, England.
Shult, E.E. and C.C. Lindegren. 1955a. The determination of the arrangement of genes from tetrad data. *Cytologia* **20:** 291–295.
———. 1955b. The hypothesis of chromosomal interference. *Nature* **175:** 507.
———. 1956a. A general theory of crossing over. *J. Genet.* **54:** 343–357.
———. 1956b. Mapping methods in tetrad analysis. I. Provisional arrangement and ordering of loci preliminary to map construction by analysis of tetrad distribution. *Genetica* **28:** 165–176.
———. 1957a. Direct effect of preferential segregation on the origin of polysomy. *Nature* **179:** 683.
———. 1957b. Orthoorientation: A new tool for genetical analysis. *Genetica* **29:** 58–82.
———. 1957c. The localized crossover and a new hypothesis of chromosomal interference. *Experientia* **13** 393.
Shult, E.E., S. Desborough, and C.C. Lindegren. 1962. Preferential segregation in *Saccharomyces. Genet. Res.* **3:** 196–209.
Shult, E.E., G. Lindegren, and C.C. Lindegren. 1967. Hybrid specific linkage relations in *Saccharomyces. Can. J. Genet. Cytol.* **9:** 723–759.
Spiegelman, S., C.C. Lindegren, and G. Lindegren. 1945. Maintenance and increase of a genetic character by a substrate-cytoplasmic interaction in the absence of the specific gene. *Proc. Natl. Acad. Sci.* **31:** 95–102.
Stadler, D.R. 1956. Double crossing over in *Neurospora. Genetics* **41:** 623–630.
Stern, C. 1932. Über die Konversiontheorie. *Biol. Zentralbl.* **52:** 367–379.
Townsend, G.F. and C.C. Lindegren . 1953. Structures in the yeast cell revealed in wet mounts. *Cytologia* **18:** 183–201.
von Borstel, R.C. 1962. Yeast Genetics Conference. *Science* **142:** 1594.
———. 1966. Yeast genetics. *Science* **152:** 1287–1288.
Winge, Ø. 1935. On haplophase and diplophase in some Saccharomycetes. *C.R. Trav. Lab. Carlsberg Ser. Physiol.* **21:** 77–112.
Winge, Ø. 1949. Inheritance of enzymatic characters in yeasts. (*Proc. Int. Congr. Genet.*) *Hereditas* (suppl.), pp. 520–529.
Winge, Ø. and C. Roberts. 1948. Inheritance of enzymatic characters in yeast, and the phenomenon of long-term adaptation. *C.R. Trav. Lab. Carlsberg Ser Physiol.* **24:** 263–315.
———. 1954a. Causes of deviation from 2:2 segregation in tetrads of monohybrid yeasts. *C.R. Trav. Lab. Carlsberg Ser. Physiol.* **25:** 285–329.
———. 1954b. On tetrad analyses apparently inconsistent with Mendelian Law. *Heredity* **8:** 295–304.
———. 1957. Remarks on irregular segregation in *Saccharomyces. Genetica* **28:** 489–496.
Winkler, H. 1930. *Die Konversion der Gene.* Gustav Fischer, Jena, Germany.
Zirkle, R.E. and C.A. Tobias. 1953. Effects of ploidy and linear energy transfer on radiobiological survival curves. *Arch. Biochem. Biophys.* **47:** 282–306.

# Nucleo-cytoplasmic Relations in Micro-organisms: Their Bearing on Cell Heredity and Differentiation—First Lecture

BORIS EPHRUSSI

Reprinted from *Nucleo-cytoplasmic Relations in Micro-organisms: Their Bearing on Cell Heredity and Differentiation* by Boris Ephrussi (1953) by permission of Oxford University Press.

# I

*'Inheritance must be looked at as merely a form of growth'*
(DARWIN, *Variation in Animals and Plants*)

## I. GENERAL INTRODUCTION

THE most decisive step in the development of biology into a scientific discipline was the formulation of the Cell Theory. Its original content is summed up in this sentence of Virchow, the founder of cell pathology: '*Every animal appears as a sum of vital units*, each of which bears in itself the complete characteristics of life.'[39] The emphasis in the quoted sentence is on the cell as a vital unit, and this unit, in Virchow's time, was regarded as the *ultimate* unit of life.

You know that among all the characteristics of life, none of which taken singly permits one to distinguish the living from the non-living,[30] the most astonishing and the most profoundly significant one is the faculty of identical perpetuation, resulting in the permanence of the multitude of living forms.

With the demonstration of cell division, and with the accumulating evidence that a cell can arise only from a pre-existing cell, the fact of identical perpetuation received the beginnings of an explanation.

A series of brilliant discoveries soon came to lay the foundation of the modern science of heredity: the discovery of the fundamental role of the nucleus in the life of the cell, and of the individuality and permanence of its chromosomes; the discovery of the nature of fertilization and notably of the fact that the only link between successive generations are single-celled gametes, one of which most often consists of barely more than a nucleus. This led to the notion of genetic determinants and pointed the way to the chromosome theory of heredity. Following the rediscovery of Mendel's laws of transmission of hereditary traits in sexual reproduction, which proved the discreteness of the hereditary material, more refined genetical studies established the complete similarity in the behaviour of groups of the hypothetical genetic determinants (the Mendelian genes) and the visible chromosomes. The gene theory, almost in its present shape, was born: you can see how its development shifted attention, on the one hand, from the subtle identical perpetuation involved in any sort of growth and multiplication to that particular form of identical perpetuation which strikes the eye as soon as we turn to sexual reproduction; and, on the other, from the cell as a whole to one of its constituents.

These shifts of attention were, of course, dictated in part by opportunity. Singly, the fact of identical perpetuation, in cell division for example, does not provide proof of the existence of genetic determinants. The demonstration of their existence requires, as you know, the occurrence of genetic variation in the first place, and of the possibility of crossing individuals carrying different forms of the same determinants in the second. Thus, for the purpose of analysis, attention was focused on sexual reproduction.

The demonstration of the existence of genetic determinants was, of course, a tremendous achievement in the analysis of the cell. Taken together with the countless proofs that Mendelian genes control an amazing variety of functions, and, above all, with the fact, so clearly shown by Muller in a paper,[29] in many ways prophetic, written in 1926, that the genes possess the unique, fundamental property of identical reduplication, which would otherwise have had to be ascribed to the cell as a whole, it led to the notion of the nucleus as the governing body of the cell, while the simultaneous absence of evidence for an equally universal cytoplasmic heredity incited geneticists to look at the rest of the cell as a by-product of gene activity. Moreover, the ability of the genes to vary and, when they vary (mutate), to reproduce themselves in their new form, confers on these cell elements, as Muller has so convincingly pointed out, the properties of the building blocks required by the process of evolution.

Thus, the cell, robbed of its noblest prerogative, was no longer the ultimate unit of life. This title was now conferred on the genes, subcellular elements, of which the cell nucleus contained many thousands and, more precisely, like Noah's ark, two of each kind.

However, the nuclear genes did not remain long the sole possessors of the title. Viruses were soon shown to be endowed with the same marvellous property of identical or *covariant* reproduction. Although viruses were soon obtained in crystalline form, the question was raised whether they were living or not living. The question obviously has no answer, because life cannot be characterized by a single criterion: no more by the criterion of self-reproduction than by any other. As Szent-Györgyi half-jokingly said, under the auspices of this same Lectureship: 'Everyone knows this much Biology—that one rabbit could never reproduce itself, and if life is characterized by self-reproduction, one rabbit could not be called alive at all, and one rabbit is no rabbit, and only two rabbits are one rabbit.'[37]

I am quoting Szent-Györgyi at this point because his remark is a sound protest against judging life by the sole criterion of self-reproduction. But the argument has a flaw, which illustrates what I have mentioned already: the excessive concentration of attention on the self-reproduction involved in sexual transmission of characters from parent to offspring. *The rabbit began by being a rabbit's egg*; the rabbit was gradually evolved in its increasing and exquisite complexity, through the numerous divisions of a single cell. Even if, starting with a microscopic mass of rabbit protoplasm, we had ended up simply with a pound of exactly the same stuff, we would have had to conclude that the material we started with was endowed with the property of identical reproduction. We could have ascribed this to nuclear genes. *But the development of a rabbit obviously involves much more than making more rabbit protoplasm. It involves both variation of descendants of the initial cell, resulting in a branching lineage of different cell types, and the inheritance of these variations.* The inheritance of the differences between somatic cells, *cell heredity*, is claimed here not only because the mass of each of the differentiated cell types increases during the development and subsequent growth of an individual, but above all because it has been shown by experiments in tissue culture that each cell type can 'breed true' for a practically indefinite time.

Here we have identical perpetuation again, but, in this case, the inherited differences can hardly be ascribed to nuclear genes, for the different cell types which make up a rabbit are all derived from the egg cell by equational mitosis: they must therefore all possess the same genotype. How then can differentiation be explained? To what are the apparently permanent differences between somatic cells due?

*Unless development involves a rather unlikely process of orderly and directed gene mutation, the differential must have its seat in the cytoplasm.*

The cytoplasm of the egg is indeed heterogeneous, and different cell lines take their origin in portions of cytoplasm of different composition. But it must be emphasized again that, using Medawar's words, 'the mere sharing-out of ancestral cytoplasm among daughter cells and then in turn among their progeny is not a sufficient explanation of cellular inheritance, since the number of claimants to the legacy increases in the extreme case exponentially while the legacy itself is fixed'.[28] *No, if the cytoplasm causes differentiation, it must be endowed with the power of perpetuation of cell-type.*

Consideration of the phenomenon of differentiation thus leads us to a seemingly simple alternative, but the experimental evidence, at least at first sight, lends no support to either of its terms. On the one hand, the outstanding role of the nucleus in the life of the cell is established beyond doubt, and it is natural to expect it to play a decisive part in cell differentiation. In fact, the earliest theories of differentiation postulated the sorting out of nuclear determinants in the course of development. However, the equational character of nuclear division has been proven by the functional equivalence of the nuclei in the early stages of development, and geneticists know no means of inducing specific gene mutations. This seems to point to the cytoplasm as the seat of the causes of the persistent changes involved in differentiation. But, on the other hand, genetic experiments have revealed cytoplasmic heredity only in exceptional cases. As S. Wright pointed out, the chief objection to the cytoplasmic theory of differentiation 'is that it ascribes enormous importance in cell lineages to a process which is only rarely responsible for differences between germ cells, at least within a species'.[47]

Obviously, what is required is more than deductions from the behaviour of germ cells; what is needed is direct genetic analysis of somatic cells, for the assumed functional equivalence of irreversibly differentiated somatic cells, however plausible, is only an hypothesis. Crosses between such cells being impossible, only nuclear transplantation from one somatic cell to another, or grafting of fragments of cytoplasm, could provide the required information; such experiments however will have to await the development of adequate technical devices.* In the meantime, the

* Nuclear transplantation has been successfully accomplished in amoebae by Comandon and de Fonbrune[4] and by Lorch and Danielli.[23] A recent paper by Briggs and King[1] gives a first account of successful transplantation of nuclei from frog blastulae cells into enucleated frog eggs. Since the nuclei of blastula cells are, according to the authors, almost as small as those of differentiated cells of older embryos, it may be hoped that experiments performed with the help of the technique of Briggs and King will, in the near future, yield most interesting information.

closest approximation to the evidence we would like to have is provided by the study of lower forms which propagate by vegetative reproduction and possess no isolated germ line. Studies of some micro-organisms have brought to light some facts of great interest on the respective roles of the cytoplasm and the nucleus, and in the following discussion I intend to present the evidence which appears to me of particular significance in connexion with the problems of cell differentiation and cell heredity. You will see that these studies confirm the view that the cytoplasm, like the genes, is endowed with genetic continuity. The genes are therefore no longer to be regarded as the sole cell-constituents endowed with this property. Moreover, these studies will reveal to us some nucleo-cytoplasmic relations which will permit us to return to a more general consideration of embryonic development. And this, you will see, will lead us to restore to the cell a great part of its lost prestige.

## 2. VISIBLE CYTOPLASMIC ELEMENTS ENDOWED WITH GENETIC CONTINUITY

Since we are going to be concerned to a large extent with the cytoplasm, let me begin with the following remark. If the cytoplasm has genetic properties of its own, these properties may or may not be linked with particulate elements. The existence of particles endowed with genetic continuity can be suggested by simple observation. The existence of autonomous cytoplasmic properties not correlated with the presence of particulate elements can be suggested only by experiment. The same is true for the rigorous demonstration, not simply the suggestion, of the genetic continuity of particulate elements. Since our knowledge of cells in general, and of micro-organisms in particular, was, to begin with, based solely on observation, it is not surprising that first to be discovered was the existence of cytoplasmic particles apparently endowed

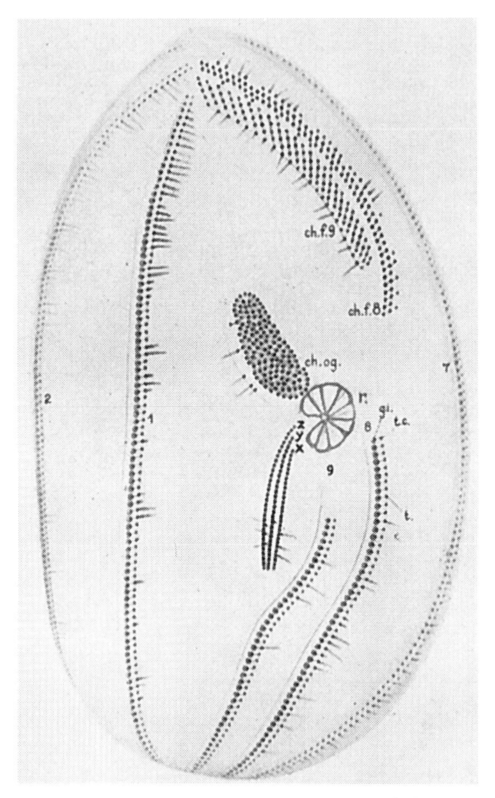


FIG. 1. The system of kinetosomes of the tomite of *Gymnodinioides inkystans* (from Chatton and Lwoff, 1935, see Ref. 2).

with genetic continuity. The experimental proof of this property of some of the particles came later. Genetic experimentation, which is the only means of demonstrating those genetic properties of the cytoplasm that are not linked to visible particles, was developed still later: therefore, the discovery of properties probably belonging to this category came only recently.*

I do not intend to trace in these lectures the history of these different notions. Nevertheless, for reasons of convenience, I will present them in the indicated order.

Simple observation long ago suggested to cytologists that all sorts of cells contain in their cytoplasm a variety of particles endowed with genetic continuity. To this class of intracellular elements were considered to belong centrioles, which, in most animal and some plant cells, can be seen at the poles of the mitotic spindles; blepharoplasts and kinetosomes, which can be found at the base of cilia and flagella of many micro-organisms and of ciliated and flagellate cells of Metazoa; the kinetoplasts of Trypanosomes; mitochondria and chondriosomes, ubiquitous elements apparently present in any living cell, whether of animal or plant origin; plastids of several sort (chloroplasts, leucoplasts), assuming various functions in plant cells and in some Flagellates. Genetic continuity was ascribed to all these cytoplasmic elements because they have never been seen to arise *de novo*, that is, otherwise than from pre-existing elements of the same sort. Indeed, some cytological pictures suggest that elements of this kind multiply, like the cells themselves, by direct division. Just to give you some examples, I have selected a few figures. Fig. 1 represents the amazingly complex and constant system of kinetosomes underlying the ciliary system of the protozoon *Gymnodinioides inkystans*

* There is no possibility of distinguishing, on the basis of crosses, between cytoplasmic inheritance and some other form of extrachromosomal heredity. However, the involvement of the cytoplasm appears to me most probable in view of its highly organized state.

in the masterly description of Chatton and Lwoff;[2] and Fig. 2 shows how, in one of the stages of the complicated life-cycle of this animal, one row of kinetosomes, through division, gives rise to another row of similar particles. I can only mention here that the kinetosomes are polyvalent elements which give rise to a variety of structures (in *Trichomonas*, for example, the same kinetosome can give rise to a flagellum, a fibre, a parabasal body

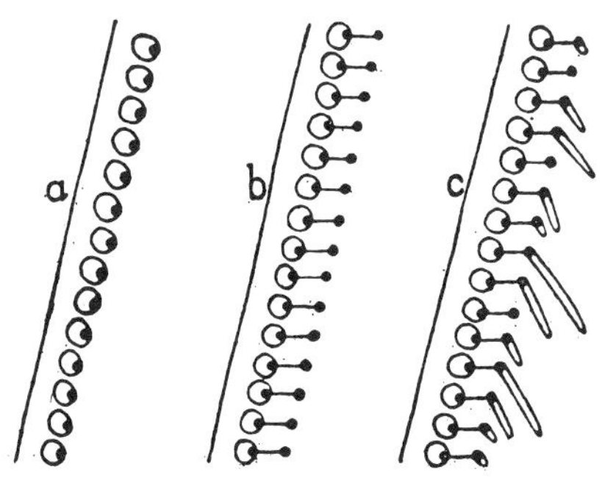


FIG. 2. Division of kinetosomes of *Gymnodinioides inkystans*. The normal row of kinetosomes (*a*) divides, giving rise to trichocystosomes (*b*), which later produce trychocysts (*c*) (from Chatton and Lwoff, 1935, see Ref. 2).

and an axostyle),* and which play an important part in the morphogenesis of Protista.[14, 25, 40] Although the claim of genetic continuity of the kinetosomes is based solely on the interpretation of the cytological pictures, one must admit that the evidence, as far as it goes, is extremely impressive.

Fig. 3, due to another very fine observer, Fauré-Fremiet, shows on the left mitochondria of the ciliated protozoan *Spirostomum ambiguum* in what is believed to be the process of division.[13] The case of mitochondria, as will appear to you later, is particularly interesting in several connexions. At this point I will only mention that genetic continuity of these elements, as well as their role as bearers

* The kind of structure to which a kinetosome gives rise depends, according to Lwoff,[25] on the quality of the particular region of cortical cytoplasm in which it is embedded.

of hereditary properties, was postulated many years ago by Meves, on the basis of cytological observations. Unfortunately, attempts at proving the point experimentally have thus far not been conclusive.

Experiments have on the contrary provided convincing proof of the genetic continuity of the kinetoplasts of Trypanosomes (Fig. 4). These disk-shaped bodies, which

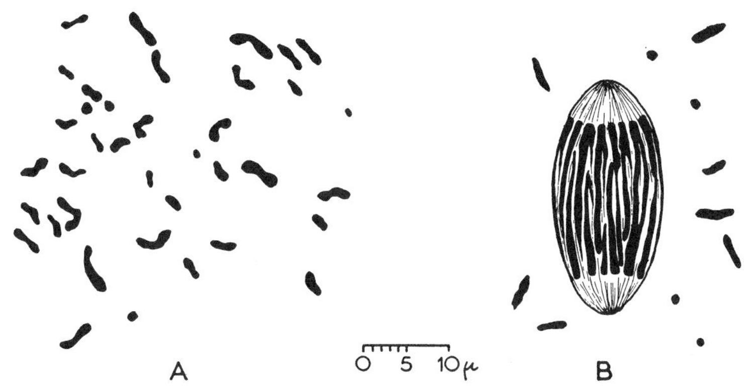


FIG. 3. A. Division of mitochondria of *Spirostomum ambiguum*. B. Simultaneous division of mitochondria and micronucleus of *Urostyla grandis* (redrawn from Fauré-Fremiet, 1910, see Ref. 13).

can be observed at the bases of the flagella, just below the kinetosomes, are curious structures which, like the nuclei, take the Feulgen stain,[24] characteristic of desoxyribonucleic acids, and are reproduced by simple division, which occurs prior to the cytoplasmic fission and about simultaneously with the division of the kinetosome (Fig. 4 A and B). It was observed many years ago by Werbitzki[41] that when Trypanosomes are grown in the presence of certain acridine dyes, the division of the kinetoplasts is blocked, while that of all the other structures proceeds undisturbed (Fig. 4 C). As a result, fission of organisms in the presence of these dyes gives rise to unequal pairs of animals, one of which is devoid of the kinetoplast (Fig. 4 D). This loss of the kinetoplast is irreversible. It is thus evident that these cell elements are endowed with genetic continuity. Aside from this experimentally induced loss of the kinetoplast, spontaneous loss no doubt occurs from time to time: such must be the origin of akinetoplastic races of Trypanosomes which occur as parasites in certain animals.[15, 16]

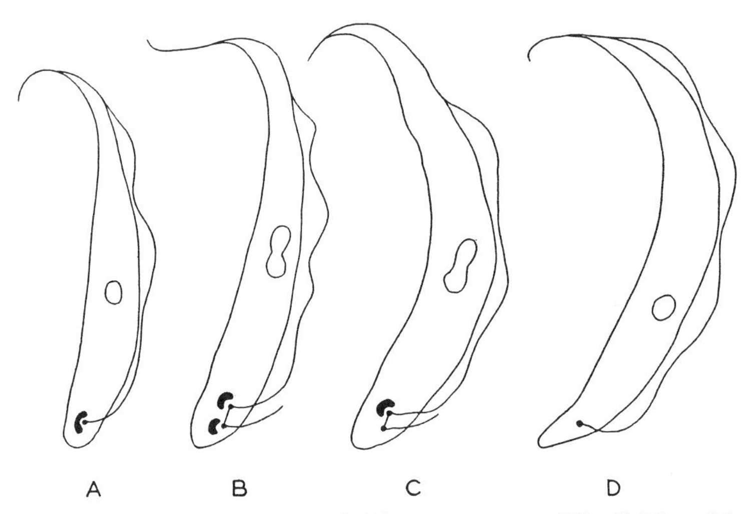


FIG. 4. Effect of acriflavine on a dividing trypanosome. The division of the kinetoplast of a normal Trypanosome (A) is shown in B. C shows the absence of division of the kinetoplast in the presence of acriflavine. D, the resulting akinetoplastic individual (redrawn from Lwoff, 1949, see Ref. 24).

The physiological function of the kinetoplast is not known. Since animals can survive its loss, it is evident that either it plays no essential part in metabolism, or that organisms devoid of these elements have some alternative metabolic pathways, which permit them, so to say, to by-pass the function of the kinetoplast.

That the latter interpretation is the correct one is made rather probable by the observations of Robertson on the flagellate *Bodo caudatus*.[32] This organism also possesses a kinetoplast, shown in Fig. 5 A, which either degenerates, or fails to divide in the presence of acriflavine (Fig. 5 D), giving rise to akinetoplastic individuals like the one shown in Fig. 5 B. Robertson did not succeed, however, in establishing a permanently akinetoplastic clone. It thus appears probable that the kinetoplast is involved in an essential function and that, in the species studied by Robertson, there is no alternative metabolic pathway which permits compensation for the loss of the kinetoplast.

A very similar situation exists in the case of the chloroplasts of certain Flagellates. These cell elements also

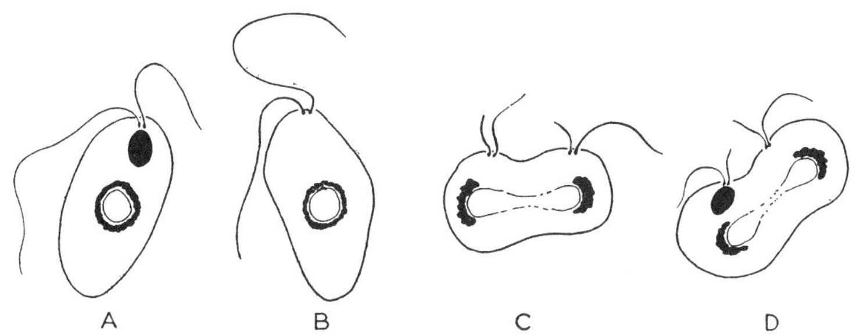


FIG. 5. The origin of akinetoplastic *Bodo caudatus*. A, normal type; B and C, akinetoplastic individual and its division; D, division of a normal *Bodo* in acriflavine producing one normal and one akinetoplastic individual (redrawn from Robertson, 1929, see Ref. 32).

multiply by direct division, but they do enjoy a certain autonomy with respect to the rest of the cell: the rhythm of their multiplication does not necessarily follow that of the organism as a whole. Thus, for example, when *Euglena mesnili* is cultivated in the dark, the division of the chloroplasts is slowed down and this process results in the decrease of the average number of chloroplasts per organism, as shown in Fig. 6. Eventually, organisms arise which are devoid of chloroplasts. This loss of the chloroplasts is irreversible. However, no achloroplastic clone could be established because the organisms with no chloroplasts are apparently incapable of prolonged multiplication. The loss of chloroplasts can also be induced in this species by growing it in the presence of streptomycin, but achloroplastic organisms obtained in this way cannot multiply either. It is thus clear that the loss of the chloroplast is accompanied in this case by the loss of an essential

12 VISIBLE CYTOPLASMIC ELEMENTS

function, and it can be shown that it is not the lack of photosynthesis which is the cause of the death of the organisms devoid of chloroplasts.[26]

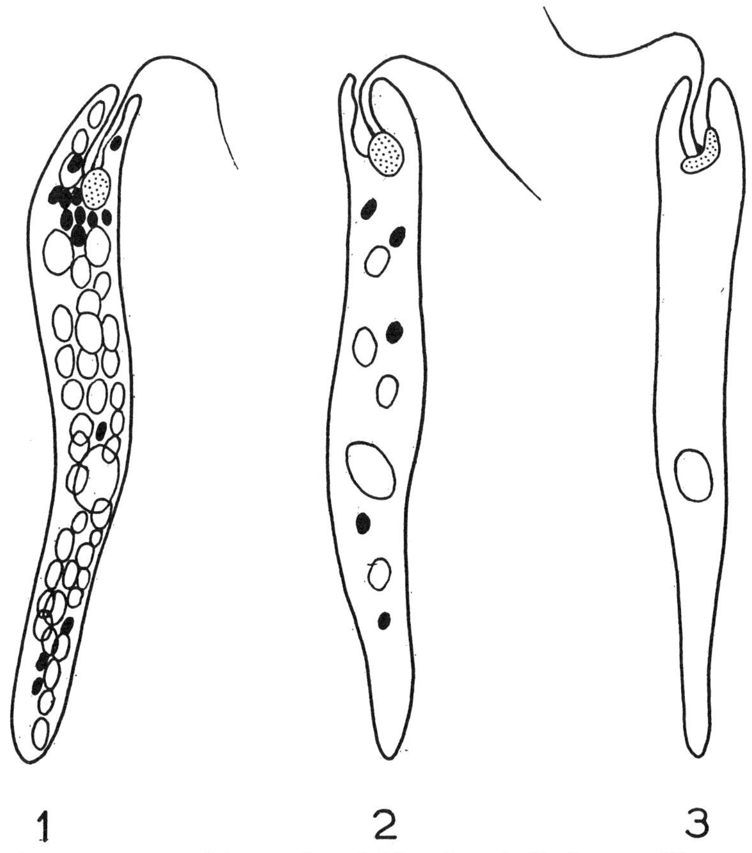


FIG. 6. Decrease of the number of chloroplasts in *Euglena mesnili* grown in the dark. 1, *Euglena* grown in light; 2, same after a short time of growth in the dark; 3, achloroplastic individual (redrawn from Lwoff, 1949, see Ref. 24).

Treatment of *Euglena gracilis* with streptomycin, although it also results in the development of permanently 'bleached' organisms, gives somewhat different results.[31] The bleaching is the result of blocked chlorophyll synthesis, that is, of the alteration of a function of the chloroplast.

But the chloroplast does not seem to disappear altogether: its other functions probably remain unaltered and this permits the continuous reproduction of bleached organisms.[26]

In the light of these results the chloroplasts appear to be complex structures, possibly comprising independent components in charge of several different functions and endowed with genetic continuity.[26]

The case of chloroplasts, apart from providing an example of autonomous cytoplasmic elements endowed with genetic continuity, is particularly instructive because it shows to what extent the demonstration of such cell elements depends on the ability of organisms to survive the loss or alteration of the elements in question. The importance of this point will appear to you later, when we discuss the intriguing problem of the rarity of proofs of cytoplasmic heredity, which is in such striking contrast with the considerable number of self-reproducing cytoplasmic elements.

## 3. CYTOPLASMIC HEREDITY IN YEASTS

From Flagellates we now turn our attention to yeasts, organisms which, I am sure, all of you know and appreciate. Yeasts have served Science well in the past, thanks mainly to the discoveries of the great Pasteur. You will be amused to hear that Pasteur undertook the study of yeasts for patriotic reasons. 'I was inspired in these investigations by our misfortunes', says Pasteur in the Introduction to his classic book *Études sur la Bière*. 'I undertook them directly after the 1870 war and continued them relentlessly ever since, with the resolution of carrying them far enough to stamp with a lasting progress an industry in which Germany is superior to us.' I am afraid that in spite of Pasteur's efforts, German beer remained much better than French beer. Meanwhile, however, Pasteur's studies on yeast have laid the foundation of modern biochemistry.

Ever since Pasteur's day yeast has been one of the most frequent objects of biochemical studies and I doubt that any other organism is as well known today from the biochemical point of view. Genetical investigation of yeasts, on the contrary, began only very recently, mainly because their life-cycle was not understood. It is known today, thanks to the works of Kruis and Satava[17] and of Winge,[42, 44] and I will begin by giving you a brief account of the life-cycle of baker's yeast (Fig. 7).

A culture of baker's yeast (*Saccharomyces cerevisiae*) is composed of oval cells which are diploid and multiply by budding, a process in which a small protoplasmic bulb grows out of one side of the cell. The nucleus divides and one of the daughter nuclei migrates into the bud, which is then walled off from the mother cell before it reaches the adult size. This vegetative reproduction can go on for very prolonged periods of time and sexual reproduction does not intervene usually until the conditions for vegetative multiplication become unfavourable. The cell wall then apparently thickens and, within the at first undivided cytoplasm, the nucleus undergoes two meiotic divisions, resulting in the formation of four haploid nuclei. The cytoplasm then condenses around each of the four nuclei to form four spherical masses: we have now what is called an ascus with its four haploid ascospores which can function as gametes.

The equivalent of fertilization sets in when the asci are placed in favourable conditions. It most frequently takes one of the following two courses (Fig. 7). The ascospores swell by absorption of water and either fuse to form what is called a *spore zygote* (Fig. 7. 1) or germinate to produce a few round haploid cells which fuse in pairs, forming a *cell zygote*. Cell zygotes can be formed by fusion of two cells produced by the same or by two different ascospores (Fig. 7. 2 and Fig. 7. 3). In any case, the result is a diploid

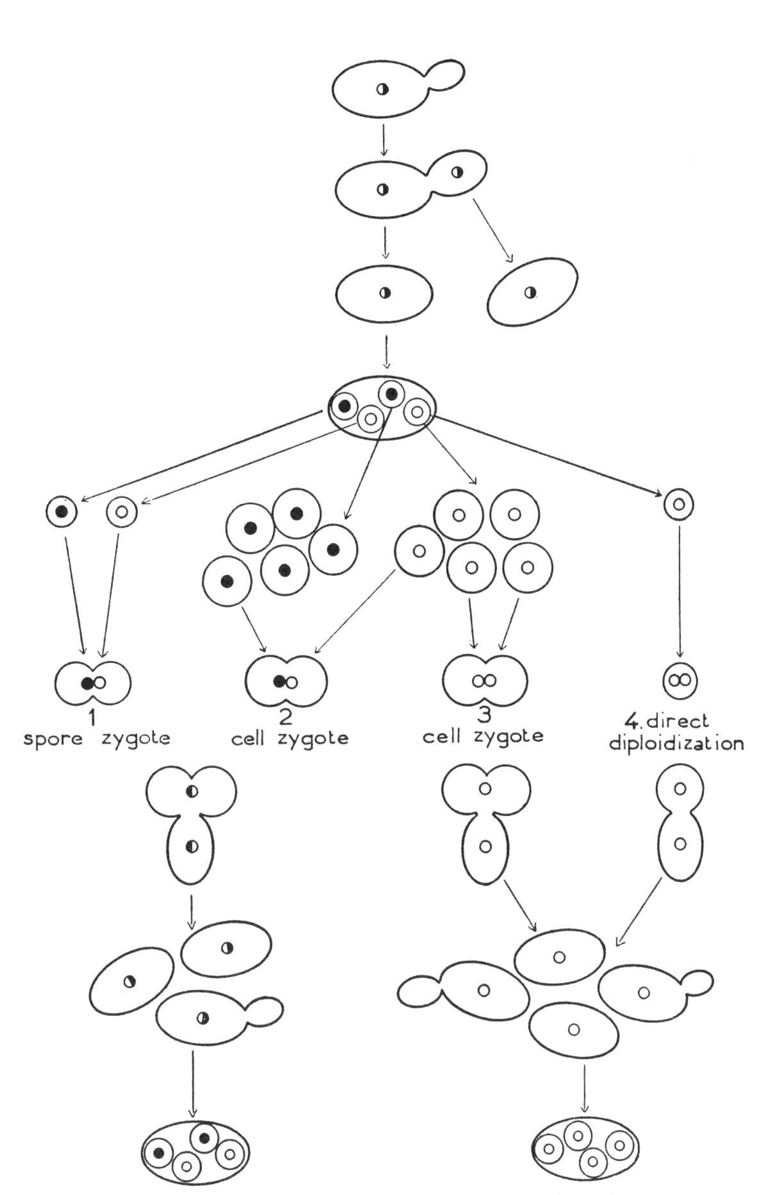


FIG. 7. The life-cycle of *Saccharomyces*. The figure is to be read downwards. In the upper part of the figure a diploid (oval) cell is shown in the process of budding. Immediately below is shown an ascus containing four haploid (round) ascospores. Further down can be seen the four ways of diploidization which lead back to the diplophase.

zygote, which soon produces oval diploid cells by budding. In very rare cases this result is achieved in a more direct way called *direct diploidization* (Fig. 7. 4). The spore nucleus divides and the two daughter nuclei immediately fuse to form a diploid nucleus: the spore then germinates, giving rise to a diploid cell.[43, 45]

All these methods of diploidization occur only in yeast species which we call *homothallic*, that is in species in which all spores are of the same physiological type, so that any spore can fuse with any other. In other yeasts, however, as shown by experiments of Lindegren,[20] each ascus contains two spores of one mating type (+) and two of the other (−).

In these so-called *heterothallic* yeasts, copulations normally take place only between spores or cells of opposite mating types. Imagine that in Fig. 7 the black and white nuclei represent nuclei or genes of opposite mating type. It is then seen that, of the three frequent ways of diploidization shown in Fig. 7, only 1 and 2 are open to heterothallic yeasts, unless, of course, mutation of the mating-type genes takes place.

You will notice that in the species I am talking about the haploid phase is confined to a very small part of the cycle. However, the artificial isolation of single ascospores here permits the establishment of relatively stable haploid cell lines. Using two such strains of opposite mating type, one can at a chosen moment obtain fusion by mixing the two cultures. One thus obtains a new diplophase, the cells of which are capable of sporulation.[21]

This brief description of the life-cycle of baker's yeast must be supplemented by a few comments on the transmission of hereditary traits. When two haploid strains differing by a pair of alternative morphological or biochemical characters are crossed, the transmission, in the enormous majority of cases, is Mendelian: it follows the same pattern as the transmission of mating type already illustrated in Fig. 7. The black and white circles can be taken to represent either the alternative mating type genes or any other pair of genes. The nuclei of the two haploid strains having fused in the zygote, the diploid cells formed by the zygote are heterozygous and exhibit the dominant trait. At meiosis the genes responsible for these traits segregate: if the alternative characters of the two crossed strains are due to a single gene difference, each ascus contains two spores of one type and two spores of the alternative type. When the distinguishing characters are controlled by more than one gene pair, the segregations are somewhat more complex.[18, 19, 46] I will not give you now more precise information, since I will have to return to this point later; all I want to say at this moment is that, in any case, we know exactly what segregation is expected to occur if a character is controlled by two, three, or more genes, whatever the interactions between these genes. As a corollary we also know what behaviour of pairs of characters is incompatible with the Mendelian mode of transmission, that is, with inheritance through nuclear genes.

I will now turn to cases in which such abnormal, non-Mendelian behaviour was observed.

The first of these cases was described by Winge and Laustsen.[45] It concerns the germinative power of the spores of the homothallic variety *ellipsoideus* of *S. cerevisiae*. In the original strain 68 per cent. of the spores germinate and produce viable clones. However, when the new diploid strains formed from single ascospores are studied, it is found that the germinative power of the next spore-generation depends on the mode of origin of the particular diploid line. This dependence is shown in the comparison between diploids formed within a clone derived from a single ascospore by cell zygote formation on the one hand, and by direct diploidization on the other (Fig. 8). Although in both cases fusion is between identical nuclei and therefore leads to the formation of identical homozygous diploids, the first of these produces roughly 40 per cent. of viable spores, the second 0 to 6 per cent. Fig. 9 shows that the loss of the germinative power of these spores is permanent and cannot be restored by the intervention of another method of diploidization. Taken together, these facts indubitably point to the operation of a non-chromosomal

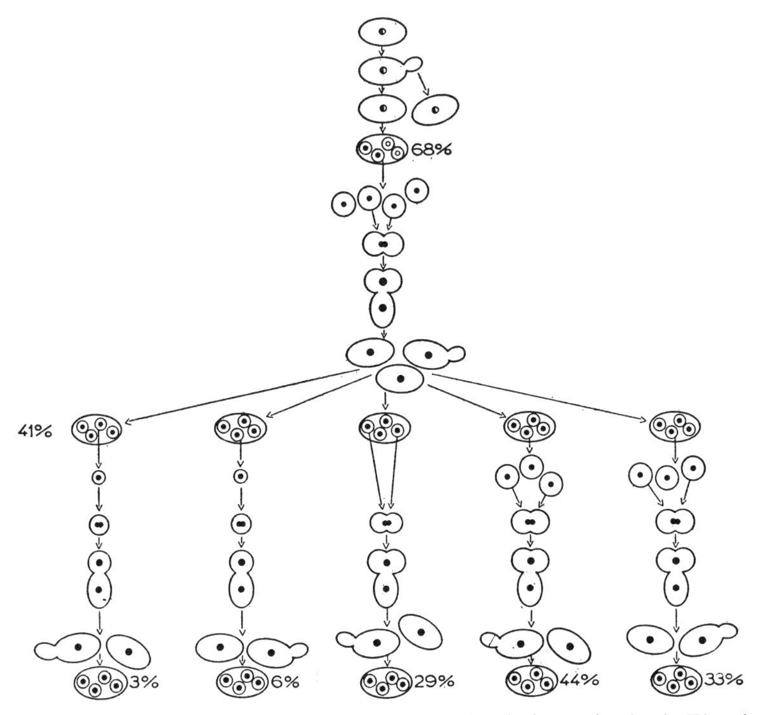


FIG. 8. Effect of different modes of diploidization (schematized as in Fig. 7) on the germinative power of ascospores (after Winge and Laustsen, 1940, see Ref. 45).

hereditary mechanism. Winge and Laustsen therefore suggested that the difference between these diploids may reside not in their nuclei, but in their chondriosome content. Assume that the chondriosomes divide once per cell cycle, shortly after the nuclear division. Since, in direct diploidization, the two nuclei resulting from mitosis of the spore nucleus immediately fuse to form a diploid nucleus, the division of the chondriosomes may be

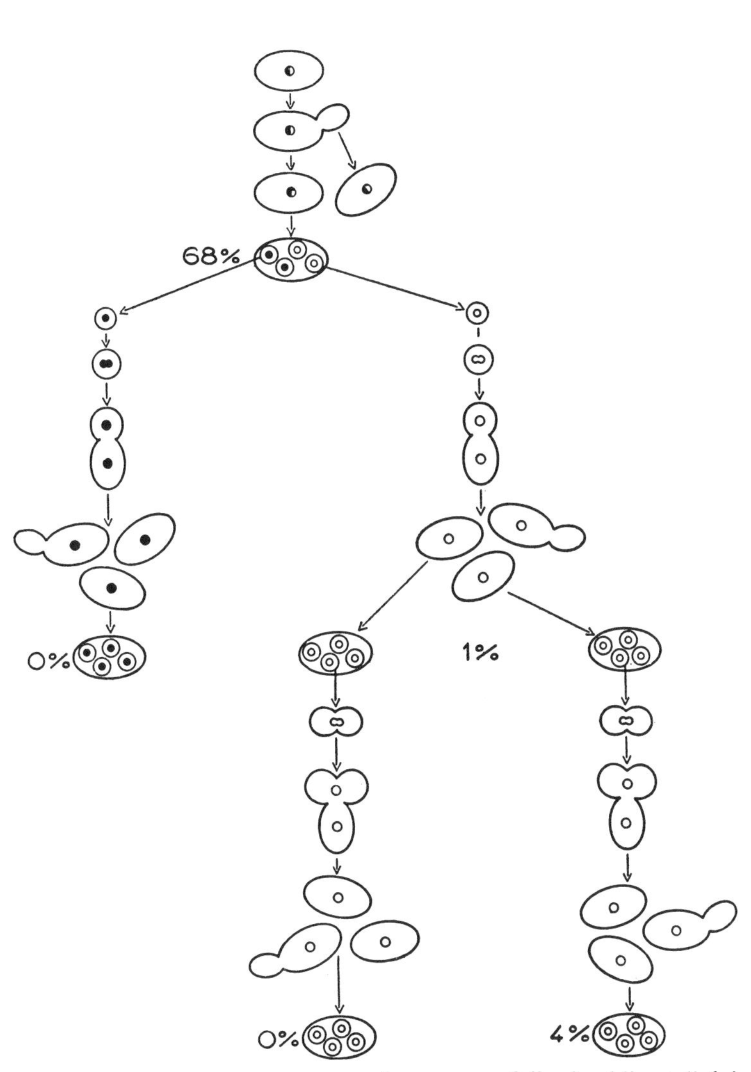


FIG. 9. Loss of germinative power of ascospores following 'direct diploidization' (after Winge and Laustsen, 1940, see Ref. 45).

suspended: their number will be unchanged in the diploids produced by direct diploidization. In cell zygotes, on the contrary, it is doubled, since in this case each of the haploid cells contributes its chondriosome content (Fig. 10). Winge and Laustsen thus are led to ascribe to the chondriosomes of the yeast cell the quality of autonomous cytoplasmic elements, the multiplication of which is closely correlated with nuclear division.

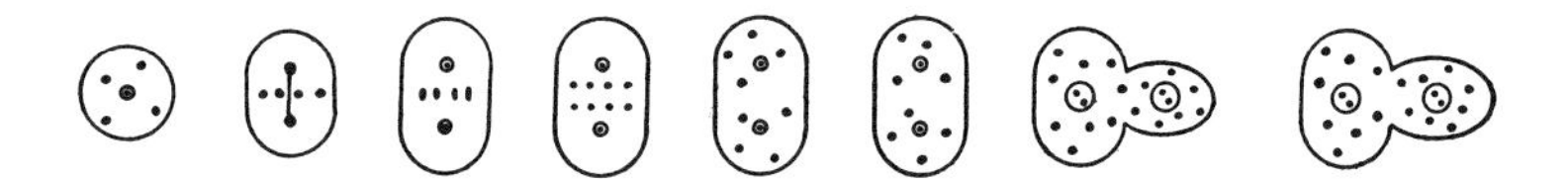

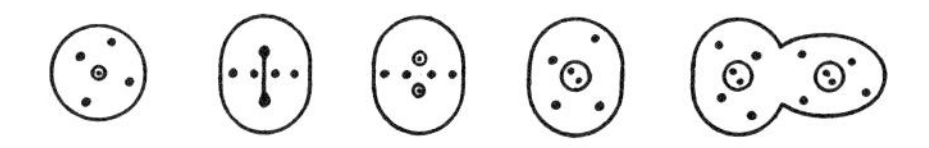

FIG. 10. Hypothetical explanation of the origin of different chondriosome contents in diploid yeasts produced by 'cell-zygote' formation (upper row) and by 'direct diploidization' (lower row) (after Winge and Laustsen, 1940, see Ref. 45).

This interpretation is, of course, hypothetical, but it is not entirely unfounded: you will remember what has been said earlier concerning the genetic continuity of mitochondria (cf. Fig. 3 B). Moreover, as you will presently see, this idea recently gained some support in the discovery of another case of cytoplasmic heredity in yeast, the study of which has kept several of my co-workers and myself busy for the last six years. Fortunately, it does not take that long to tell the story, and I hope I can give you its essence in the remainder of this lecture hour.

When a culture of baker's yeast, whether diploid or haploid, is plated, that is, spread on the surface of a solid nutrient medium in a Petri dish, each of the cells gives rise in the course of the next few days to a colony. The great majority of these colonies are of very nearly identical size, but one usually finds also a very small number—say 1

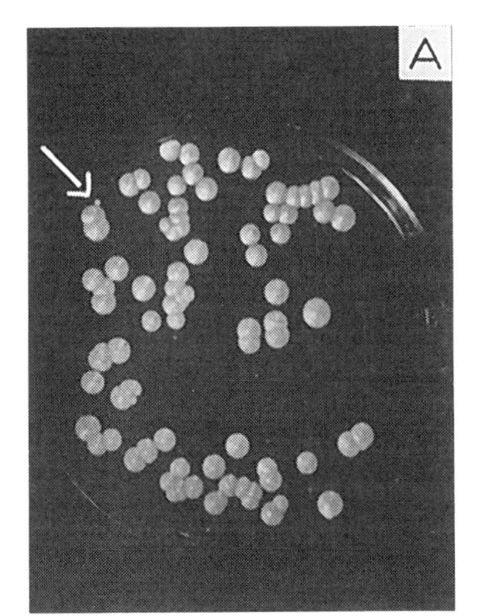


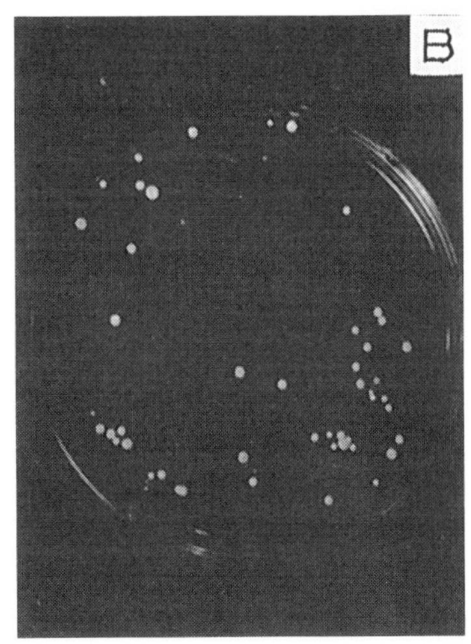


FIG. 11. Colonies formed by baker's yeast on a solid medium. A, colonies of a normal yeast, showing one small colony (arrow); B, colonies formed by the same yeast grown prior to plating in the presence of acriflavine.

or 2 per cent.—of distinctly smaller colonies: their diameter is only one-third or half of the diameter of the bigger colonies (Fig. 11 A). These facts suggest that the population of cells which was plated was heterogeneous and that it may be possible to purify it by taking cells from either the big or the small colonies only. The results of such a selection show, however, that cells from the big colonies, when suspended in liquid and replated on nutrient agar, again and again produce the two types of colonies, while the cells from small colonies give rise to small colonies only. No selection within the big colonies can change this result: obviously these colonies are composed of cells which, in the course of their vegetative reproduction, constantly give rise both to cells similar to themselves and to cells of a different type, characterized by the small size of the colonies they form. These latter cells, on the contrary, apparently never give rise to cells different from themselves: they represent a stable type which may be regarded as the result of a hereditary variation, that is, of a mutation. Although the individual cells of the two types can hardly be distinguished by their size, I will speak of 'big' and 'little' cells, and, in order to emphasize the origin of the latter in the course of vegetative reproduction, I will call them 'vegetative littles' or, alternatively, 'vegetative mutants'.[9]

The statement that the 'littles' represent a stable cell type is based on the fact that many strains of littles have been kept by serial passages over thousands of cell generations without reverting to the original (normal) type. It must be pointed out, however, that when cells of a number of small colonies, formed on plates of a *normal* yeast, are isolated and sub-cultured, it is found that some of the clones are rather quickly invaded by normal cells which, under the usual culture conditions, enjoy a considerable selective advantage. Although this points to the possible existence of 'reversible littles', the origin of the normal cells in these clones remains uncertain. Most yeast colonies are initiated by groups of several cells rather than by single cells, and it is possible that some of the 'reversions' are due to the inclusion, in the midst of a small colony, of a few temporarily inhibited normal cells.[9, 12]

Before we go on to the study of the genetic nature of the littles, a few words should be said about the reasons which cause the mutant type of yeast to grow more slowly than the normal yeast. It must be said first of all that this difference is observed only when the two cell types grow in the presence of oxygen; in its absence, on the contrary, their rates of growth are similar. As you certainly know, yeast can derive the energy required for its growth from the utilization of sugar in two ways: by respiration, when oxygen is present, or by fermentation when oxygen is absent. You certainly know also that the utilization of sugar is much more efficient when the respiratory path is used: for a given amount of glucose, five times more living matter is formed in aerobiosis than in anaerobiosis. The fact that the mutant yeast grows more slowly than normal yeast in aerobiosis suggests that something is wrong with its respiratory mechanism and that its fermentative mechanisms are normal.[38] This can be confirmed directly by measuring the respiratory quotient of the two cell types: this quotient is found to be of the order of 1 in normal yeast and near infinity in the mutant yeast: in other words, the respiration of the mutant yeast is nearly abolished.[33]

The causes of the respiratory deficiency of the mutant yeast have been thoroughly investigated[34, 35] and it has been established that it is due to the loss of the ability to synthesize a whole series of respiratory enzymes. Among these, the loss of two enzymes of the cytochrome system —succinic dehydrogenase and cytochrome oxidase—is particularly interesting because the presence of these enzymes in normal cells is accompanied by the occurrence in the spectrum of these cells of two characteristic bands (cytochromes *a* and *b*). These bands are absent in the mutant yeast. The difference between two cultures of yeast, one normal, one mutant, can thus be directly established by the examination of their spectra (Fig. 12). Thus the experimenter, in order to establish the normal or mutant character of a yeast strain, has at his disposal several means: the observation of the colony size, the observation of the spectrum of the yeast, and the determination of the respiratory quotient. Still a fourth means

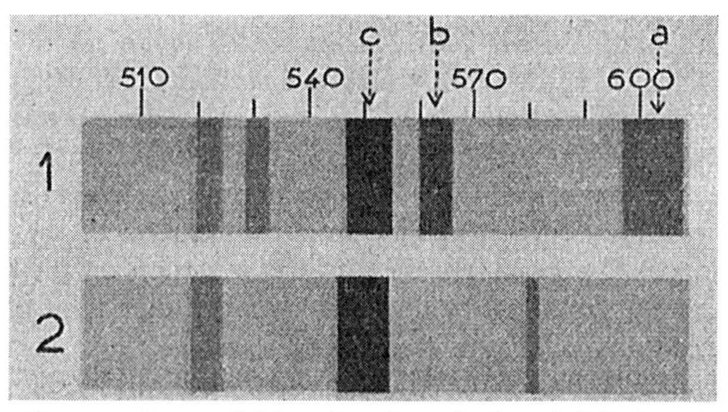


FIG. 12. Spectra of normal (above) and respiration-deficient mutant yeast (below).

is the so-called Nadi reaction—a colour reaction characteristic of indophenoloxidase, which is probably identical with cytochrome oxidase. Indophenoloxidase is present in normal yeast and absent in the mutant yeast: the Nadi reaction performed on a suspension of normal cells gives a deep blue colour, and gives no colour when performed on the mutant yeast.

Slonimski[34] has performed a very thorough comparative study of the physiological properties of normal yeast and of vegetative littles. In brief, the most prominent characteristics of the mutant cells are: (*a*) the total absence of cyanide-sensitive respiration, and the presence of cyanide-resistant respiration, and of aerobic fermentation; (*b*) the almost complete abolition of endogenous metabolism; (*c*) a fermentation which is equally active in the presence and in the absence of oxygen, and which, compared with the aerobic fermentation of normal yeast-cells, is much more sensitive to monoiodoacetic acid and much less to fluoride.

These metabolic characteristics of the mutants are the results of profound enzymatic changes in the cells as a consequence of the mutation. Apart from lacking the enzymes mentioned (cytochromes *a* and *b*, cytochrome oxidase, and succinic dehydrogenase), the mutants are apparently totally deficient in several other enzymes (cytochrome *e*, reduced coenzyme I-cytochrome *c* reductase, α-glycerophosphate-dehydrogenase) and contain reduced amounts of malic dehydrogenase linked to coenzyme I. The contents of alcohol-dehydrogenase and of cytochrome *c* are, on the contrary, higher than in normal cells. There seems to be no difference between normal and mutant yeasts in their content of catalase and of lactic dehydrogenase. Lastly, the mutant cells seem to contain two compounds not found in normal cells: cytochrome $a_1$ and malic cytochrome *c* reductase, independent of coenzyme I.

The modification of the enzymatic constitution of the cell is, according to Slonimski, an effect of the mutation on the protein moiety of the haemin-containing enzymes, rather than on their prosthetic groups.

Study of fractions of yeast homogenates obtained by differential centrifugation showed that the mutation causes the disappearance only of enzymes linked to particulate material of the cell. The observed differences in the lyoenzymes are of an entirely different order of magnitude and are regarded as a consequence of the cytochrome-oxidase deficiency.

Apart from providing us with practical means of distinguishing mutant cells from normal cells, this investigation of the biochemical differences between the two cell types gives a rather striking picture: the simultaneous loss of a series of different enzymes is not what would be expected to be the result of a gene mutation, for gene mutations usually result in single enzyme deficiencies.

A similar impression, that we are not dealing with genic mutants in the case of vegetative littles, is gained from the consideration of the frequency of the formation of cells of this type. The spontaneous mutation rate, that is, the frequency of the spontaneous occurrence of vegetative littles in the course of vegetative reproduction of baker's yeast, is of the order of 2 per 1,000; that is, approximately two out of a thousand buds formed by normal or big cells give rise to 'vegetative mutants'. This is a mutation-rate much higher than that of the usual gene mutations. Although the process of mutation is thus frequent and unidirectional, yeast populations usually do not transform into populations of mutants because the high mutation rate is compensated by the selective advantage of the normal cells, that is, their more rapid multiplication. A total conversion of a population of normal yeast into a population of mutants is however possible if, for example, the yeast is grown in the presence of certain acridines.[12] These dyes appear to be very highly active and specific mutagenic agents: they enormously increase the frequency of the mutations without appreciably affecting the selection process. Thus, a culture of baker's yeast, grown for 48 hours in the presence of acriflavine, consists almost exclusively of mutant cells which on plating will give rise to dwarf colonies (Fig. 11 B). The examination of their spectra and of their respiration clearly shows that all these colonies are composed of the same type of mutant cells, which we have called 'vegetative littles'.

You will, of course, have noticed that since the percentage of mutant cells in a population is the result of two forces acting in opposite directions, viz. mutation and selection, a similar result, that is, the replacement of normal cells by mutant cells, would also be achieved if the acridine deprived the normal cells of their selective advantage without affecting at all the mutation rate. In fact, many modifications of populations of micro-organisms have found their explanation in such terms of modified selection. The following experiments[7] show, however, that the action of acridines on yeast is of a different sort. Single cells from a normal yeast strain are isolated with the help of a micromanipulator in droplets of culture medium containing a low concentration of euflavine. The budding of the cells is directly watched under the microscope and the successive daughter cells are removed from the mother cell, in the order of their formation, and transferred to droplets of normal medium. Similarly, second-generation

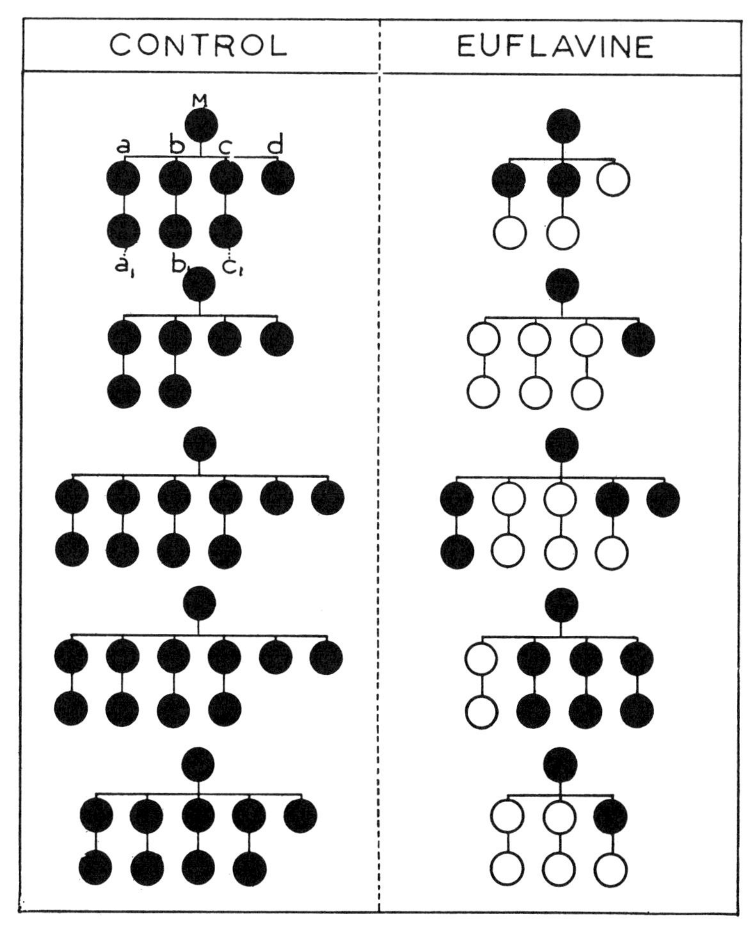


FIG. 13. Descendants of isolated yeast cells proliferating in the presence or absence of euflavine. M, mother cell; a, b, c, &c., successive buds formed by the mother cell; $a_1$, $b_1$, $c_1$, &c., successive second-generation buds. Black and white circles indicate normal and mutant cells, respectively.

buds are separated from each of the daughter cells. Finally, at the end of the experiment, the mother cell itself is transferred to normal medium.

Similar isolations are performed in the absence of euflavine. They constitute the controls.

After the isolated cells have multiplied for 24 hours, they are transferred to bigger volumes of culture medium. After 2 or 3 days of growth, the characteristics of each of the clones can be established by subjecting them to the Nadi reaction. Fig. 13 gives the results of one such experiment. It can be seen that, whereas none of the five control mother cells gave rise to mutant buds, more than one-half of the buds produced by the five mother cells multiplying in the acridine were mutants. Serial transfers of these clones and repeated tests showed that the mutant character was irreversible. Since the design of the experiment is such as to offer no possibility for selection (apart from cell mortality which has been found to be practically nil), it is clear that the acridine has a formidable mutagenic effect.

Using a different technique it has been shown that in the presence of purified euflavine, mutation rates close, if not equal, to 1 can be obtained:[27] practically every bud formed in the presence of the dye is a mutant bud.*

It must be added that the acridines produce their mutagenic effect only on actively multiplying cells; in the absence of cell multiplication there is no mutation.[9]

In these experiments on the isolation of single cells, the diagnostic test employed (Nadi reaction) is applied to the clone as a whole. Since a mixture of normal and mutant cells (up to 80 per cent. of the latter) gives a positive Nadi reaction, the clones classified as 'normal' may have contained a considerable proportion of

* The mutation rate is defined as the ratio $N_p/(N_g+N_p)$, where $N_g$ and $N_p$ are, respectively, the numbers of normal and mutant buds formed by a certain number of normal cells, that is, as the probability of a bud taken at random to be mutant. The justification of this definition has been given by Marcovich.[27]

Mutation rates, thus defined, approach the value of 1 in concentrations of purified euflavine which have practically no toxic effect, as measured by the lengthening of their generation time.[27]

mutants. That this is often the case can be shown by plating out the very young clones formed by the isolated buds. Many clones then give rise to various proportions of normal and small colonies. The variability of this proportion suggests that the mutants appear at various times in the history of the clones or, in other words, that the high mutation rate induced by the acridine is maintained after the transfer of the cells into normal medium. Study of pedigrees by the last-mentioned technique (plating) indubitably shows that the

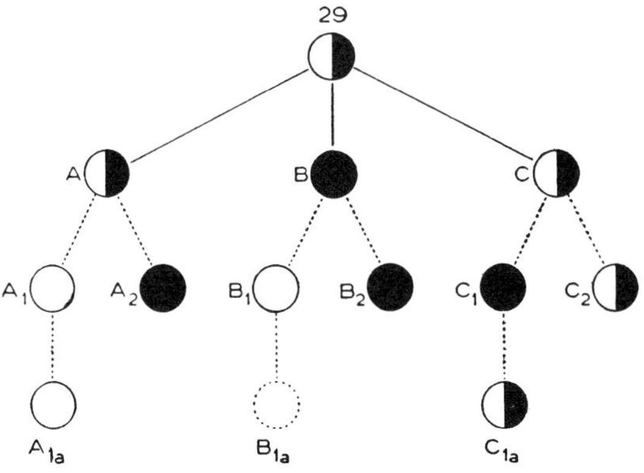


FIG. 14. Pedigree showing the instability induced in a yeast cell by treatment with euflavine. Black, white, and half-black circles indicate cells which have given rise to normal, mutant, and mixed clones respectively.

acridines can induce in the cells an *unstable state* which manifests itself by a long-lasting high mutation rate: the acridine-treated cell and some of its offspring, placed in normal medium, go on producing normal and mutant buds alternately over several generations. Fig. 14 gives an example of such an induced instability. Consider in particular cell C. This cell, initiated in acridine, has produced first a normal cell ($C_1$), then cell $C_2$ which was initiated in the normal medium, but which gave rise to a mixed clone of normal and mutant cells; thereafter cell C produced a clone which, when plated, developed into a mixture of big, small, and scalloped (see below, p. 34) colonies. In this case a high mutation rate was indubitably preserved over numerous cell generations.[8]

The picture of the mutation process, as it appears in the experiments I am referring to, is very different from that of induced gene mutations. As you certainly know, gene mutations have never been observed to occur with such high rates; also, no mutagenic substance has ever been discovered, the action of which results in the appearance of invariably the same mutation. It therefore appears again very improbable that the mutants of yeast, which we are concerned with, are the results of genic mutations.

However, the two facts which I have quoted so far as suggesting that the respiratory mutants of yeast might not be due to gene mutations are, at best, presumptions. Only the usual technique of genetics, that is crossing, can permit more definite conclusions.

Let us see, then, how the mutant character behaves in sexual reproduction. Here, unfortunately, the possibilities of experimentation are limited to one type of cross. Sporulation of yeast is an aerobic process. Diploid respiratory mutants are unable to respire and therefore do not sporulate. Consequently it is impossible to establish whether the mutant character is maintained through meiosis. The only possible cross is between a normal yeast and a vegetative mutant of opposite mating type. Each of these strains carries, of course, some Mendelian 'markers': their presence permits one to ascertain that the nuclear behaviour of the hybrid is normal, for the markers show a 2:2 segregation in each ascus.

The yeast used in this work[10] was 'Yeast Foam', an American strain of diploid *Saccharomyces cerevisiae*. From an ascus of this yeast two spores were isolated which gave rise to two haploid strains of opposite mating type. One (276/3 *b*) is adenine independent (*A*), thiamine dependent (*t*) and of mating type —. The other (276/3 *d*) is both adenine and thiamine independent (*AT*) and of mating type +. Both strains are normal with respect to respiratory mechanism ('big').

Among the colonies formed by the first of these strains on an agar plate, a red colony was found. It is the origin of a strain of red yeast (276/3 *br*) which, like the yeast clone within which it arose, is haploid, thiamine dependent and of mating type —. However, contrary to the latter, it is adenine dependent (*a*). Both the adenine dependence and the red pigment are effects of the same recessive *a*.

The red yeast, like the strain of origin, possesses a normal respiration. Like any other strain of yeast, it frequently gives rise to vegetative littles, the frequency of which is increased by growth in the presence of acridines: strain 1*A*-2 involved in the cross to be described was obtained from such treatment. An interesting and technically convenient feature of the vegetative mutants of the red yeast is that, although they contain the recessive gene *a*, they form white colonies when grown in the usual manner on media containing 2–3 per cent. glucose.

All crosses to be described now were made between pairs of haploid yeast cultures of which one is *AT*+ and the other *at*−. Every spore isolated from the asci of the hybrid was tested for these markers. All asci showed a 2:2 segregation of the mating type genes. A 2:2 segregation for genes *A*/*a* was observed in 129 out of 134 asci. A similar segregation for the pair *T*/*t* was observed only in 120 out of 134 asci. This rather poor result is ascribed to imperfection of the technique of testing. The important point is, however, that all the asci contained both *A* and *a* and *T* and *t* spores; in other words, they were 'legitimate' hybrid asci.

The results of the cross between normal yeast and a 'vegetative little' are summarized in Fig. 15. The fusion of the two cell types results in the formation of diploid cells which are normal with respect to respiration. These diploid cells are able to sporulate and the genetic analysis of the asci is therefore possible, by the isolation of single ascospores. Such analysis shows that all four spores of the hybrid asci give rise to clones with normal respiration. In other words, while the genetic markers undergo perfectly normal segregation, the mutant character we are interested in, that is the respiratory deficiency, vanishes on crossing and does not reappear in the spore progeny.[10]

If the respiratory deficiency were the result of the mutation of a single gene, it should reappear, like the genetic markers, in two of the four spores of each ascus. Since this is not the case, we may conclude that the respiratory deficiency of the vegetative littles is not caused by the mutation of a single gene. No conclusion beyond this can, however, be drawn, since it must be kept in mind that if the respiratory deficiency were due to the simultaneous mutation of several genes to their recessive allelomorphs, the mutant character would reappear only in a fraction of

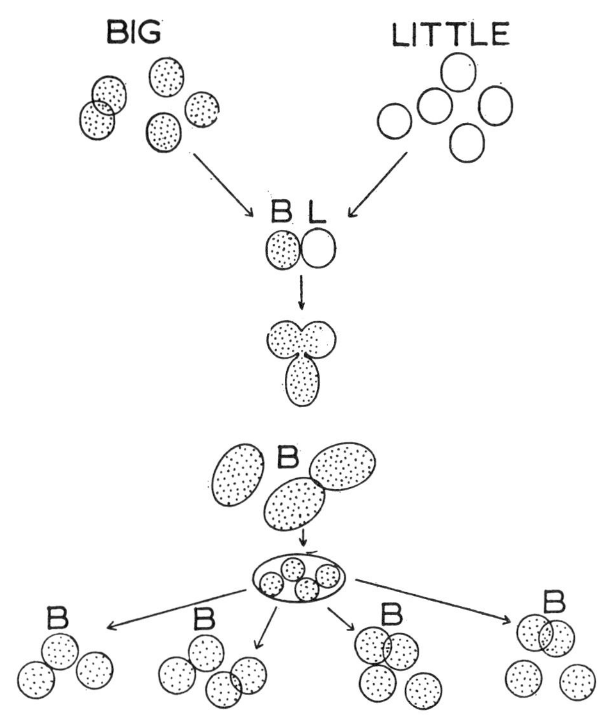


FIG. 15. Results of a cross between a normal strain and a 'vegetative mutant' of yeast. The cytoplasm of the normal cells is indicated by stippling.

asci, the proportion of which would depend on the number of genes involved: the greater the number of recessive genes involved, the smaller the proportion of asci containing mutant spores should be. Just to give you an example: if the respiratory deficiency were due to the simultaneous presence of four recessive genes, it would reappear only in some 12 per cent. of the asci. The fact that the mutant character was not observed to reappear in the spore progeny of the hybrid may therefore be due simply to the necessarily small number of asci which it is possible to analyse in practice, and the possibility must be recognized that different results might have been obtained if a much greater number of asci had been studied. This difficulty can obviously be removed by the use of forced labour, but fortunately less drastic methods can achieve the same result: one can make use of the well-known technique of backcrossing, that is, of repeatedly crossing the spores formed by the hybrid with the mutant parent. In this way the genes of the mutant parent accumulate in the hybrids of the successive generations. Consequently, if the mutant character required the simultaneous presence of several recessives to manifest itself, the proportion of asci containing mutant spores should rapidly increase in the course of the successive backcrosses.

This method has indeed been used[10] to follow up the results of the first generation which I have stated earlier. Five successive backcrosses have been made, but the result expected on the basis of the polygenic hypothesis was not achieved.* A few mutant spores did occur in the course of the backcrosses, but their frequency was not greater than the frequency of spontaneous mutations, and calculations have shown that, in order to interpret the experimental results in terms of Mendelian genes, one would have to assume that the manifestation of the mutant character under consideration depends on the simultaneous presence of at least a dozen recessive genes.[10] This does not, of course, prove that the genic interpretation is wrong: it merely makes it very unlikely. As I have pointed out earlier, the mutation rate we would have to assume to account for the frequency of the spontaneous occurrence of vegetative mutants if they were due to changes of single genes is unusually high; but still higher, and in fact

* The actual number of asci analysed in the first cross and in the five successive backcrosses are, respectively: 31, 23, 18, 20, 57, and 6, making a total of 620 spores. Among these, five were mutant.

incredibly high, mutation rates of individual genes have to be assumed on the polygenic hypothesis. Thus, the results of the crosses support the presumptions quoted earlier against the interpretation of the respiratory deficiency in terms of mutations of Mendelian genes and point to the intervention of an extrachromosomal factor. That is why I some years ago[5] adopted, as a working hypothesis, an interpretation which assumes that *the vegetative littles are due to a cytoplasmic, rather than a nuclear mutation.** This interpretation postulates that the synthesis of the respiratory enzymes, missing in the mutants, is dependent upon the presence of a cytoplasmic factor, which is self-reproducing and particulate. If it is further assumed that this factor is distributed at random between the mother cell and the bud when the latter is formed, and that the average number of the hypothetical particles per cell of the normal strain is rather low, say of the order of 10, it is easy to see how some buds happen to contain no particles at the moment they are walled off from the mother cell: such buds, containing no particles and consequently unable to synthesize the respiratory enzymes, are, in other words, the result of a loss mutation. The action of the acridines can be accounted for by supposing that these substances electively affect the cytoplasmic particles in a manner analogous to that in which they act on the kinetoplasts of Trypanosomes, that is, either by destroying them or by preventing their multiplication. (One could, of course, assume that the cytoplasmic particles are not lost in the mutants, but that they have mutated to an inactive condition. No test permits one at present to distinguish between loss and inactivation.)

If a suspension of normal yeast is plated on a solid medium containing at least one part in 300,000 of acriflavine, all the cells give rise to small colonies, composed of respiration-deficient mutants.

* Lindegren and Lindegren[22] have observed very similar results with another strain of red yeast. Their interpretation is, however, different from the one proposed below.

If the concentration of acriflavine is slightly lower (one part in 500,000), all the colonies are big, but have a characteristic irregular contour. Similar 'scalloped' colonies develop when a yeast culture, or isolated cells (like those in the pedigrees described on p. 28), are grown for a short time in liquid medium containing mutagenic acridines, and are then plated on normal nutrient agar.

When a 'scalloped' colony is dissociated and replated, it gives rise to a mixture of normal and small colonies. Since normal cells possess a distinct selective advantage, the formation of scalloped colonies is ascribed to the unstable state of the cell from which the colony originated (see above, p. 28). This cell gives rise alternately to similar unstable cells, to normal cells, and to vegetative mutants, as in the pedigrees described on p. 28.

The simplest interpretation of the induced unstable state is that it is due to the diminished number of cytoplasmic particles, owing to destruction of the latter by the acridine. Obviously, the smaller the number of particles per cell, the higher the probability of their loss, that is, of the mutation.[8]

This interpretation has the advantage of offering also a simple explanation of the results of the cross between a normal yeast and a vegetative little. Since the cross results in the mixture of the cytoplasms of the cells undergoing fusion, the diploid cells contain the cytoplasmic factor supplied by the normal parent. Furthermore, since, in the process of sporulation, the spores are simply cut out of the cytoplasm of the ascus-forming cell, all spores contain some particles, and the mutant character of the mutant parent, caused by the total absence of particles, does not reappear in the asci.*

Can the hypothetical particles be identified with any known visible elements of the yeast cell? In normal yeast

* In view of this behaviour in crosses of the character under consideration, the opinion has been expressed that the terms 'mutation' and 'mutant' should be avoided in speaking of the vegetative littles. It has been suggested that, instead, one might speak of 'differentiation' and 'phenocopy'. However, the use of such terms appears to me undesirable, for the mechanisms of the phenomena for the description of which they have been invented are unknown, and it is precisely my purpose to suggest that differentiation may be due to discontinuous and irreversible changes in the heredity of cells, that is, to mutations.

the enzymes lacking in the mutant are linked to particulate material separable by differential centrifugation.[34,35] The enzyme-carrying particles, sometimes called macrosomes, can probably be identified with mitochondria. The idea therefore naturally suggests itself that the hypothetical particles, postulated on the ground of genetic experiments, are in fact mitochondria, and that the mutation which results in the formation of vegetative littles consists of the loss of mitochondria.[6,35] Microscopic observation of normal yeast cells on which has been performed the Nadi reaction, characteristic of indophenoloxidase (one of the enzymes missing in the littles), shows that the colour is concentrated in a small number of cytoplasmic granules. Each cell contains many granules, but only some of them show the blue colour. The cytoplasm of the mutant cells also contains such visible granules, but none of them are coloured. It is tempting to assume that the blue granules of the normal cells *are* the particles postulated on genetic grounds. It is particularly tempting because, if proved, it would explain, without resort to other factors, the simultaneous disappearance of the several different enzymes present in normal cells and absent in the mutants. It must be admitted though that, however tempting, such conclusions are not justified at the present stage. First of all, there is at present no proof that the indophenol blue is actually produced in, rather than adsorbed on, the visible granules. Secondly, were this proven, one would be entitled to say that the visible granules carry the indophenoloxidase, but the identity of the enzyme-carrying elements with the hypothetical gene-like particles would still require demonstration, which so far has been beyond our reach. *The only conclusion we may draw today with a high probability of being right is that the normal yeast and the vegetative mutants differ by the presence in the former and the absence in the latter of cytoplasmic units endowed with genetic continuity and required for the synthesis of certain respiratory enzymes.*

Sturtevant has suggested in private discussion a different interpretation which likens our 'vegetative mutations' to the variegations observed in *Drosophila*. In Sturtevant's opinion the latter are due to phenotypic inactivity of definite genes, unstable in the somatic line (the instability being caused by a translocation into the neighbourhood of heterochromatin, for example), but necessarily restored to normal activity during meiosis.

The inability of 'vegetative mutants' to sporulate has thus far not permitted this ingenious hypothesis to be subjected to experimental test.

While a certain number of most intriguing problems thus remain momentarily unanswerable, we may turn to others and, in particular, to the question of the relation between the postulated cytoplasmic genetic units of the yeast cell and its nucleus. Are these units totally independent of the nucleus in their reproduction and their function?

Although all the experiments thus far described favour the view that they are completely autonomous, you will see in a moment that such *autonomy* can never be safely claimed, by contrast to *lack of autonomy* (that is dependence) which can be directly proved. However, you will see also that the demonstration of the dependence of any cell element on the nucleus is entirely subject to the accident of discovery of a nuclear constitution which interferes with the normal multiplication or the normal functioning of the element in question. Consequently, an apparent autonomy can always be ascribed to the fact that the proper nuclear constitution has not yet been discovered.

All this will become clearer to you after I have told you the results of some other observations. The experiments I have described so far were all performed on an American strain of yeast, called Yeast Foam; in the experiments I am going to describe now,[3] a strain of French baker's yeast, belonging to the same species and called B-II, was used. I do not want to imply anything about national characteristics, but you will see that the behaviour of the French strain is somewhat more fanciful. Like the American strain, it produces some vegetative mutants in the course of its vegetative reproduction, but in addition it produces in its asci many mutant spores, that is, spores which give rise to clones entirely composed of cells which present the same biochemical characteristics as the vegetative littles you are now familiar with. In spite of this similarity, I am going to designate them by a different term; I will speak of 'segregational mutants': you will soon see that this distinctive is justified.

The germinative power of the ascospores of this French strain of yeast is very poor: only 4 per cent. of all the ascospores germinate to produce viable clones. It is therefore impossible to determine in what ratios the mutant spores occur in the asci. It was found, however, that the frequency of mutant spores among all the viable spores was close to 50 per cent. This suggested that the cells of strains B-II might be heterozygous for a recessive Mendelian gene controlling the character of the spores. This was worth investigating because, if this were the case, it would demonstrate that the respiratory deficiency can be caused by the mutation of a nuclear gene as well as by the loss of cytoplasmic particles; and this in turn might indicate a relation between these two sorts of cell constituents.

These expectations were borne out by crosses between three haploid strains derived from the French yeast. The precise relationship between these strains is shown in Fig. 16. From two asci formed by the original strain B-II, two ascospores of opposite mating type were isolated. One of these was normal and gave rise to a clone of haploid cells (B 15) normal with respect to respiration; the other was a 'segregational mutant' and produced a clone (B 26) of mutant, respiration-deficient haploid cells. From the former (B 15) a 'vegetative mutant' (B 15p4) was isolated. In Fig. 16 the mutant cells are indicated by a smaller size.

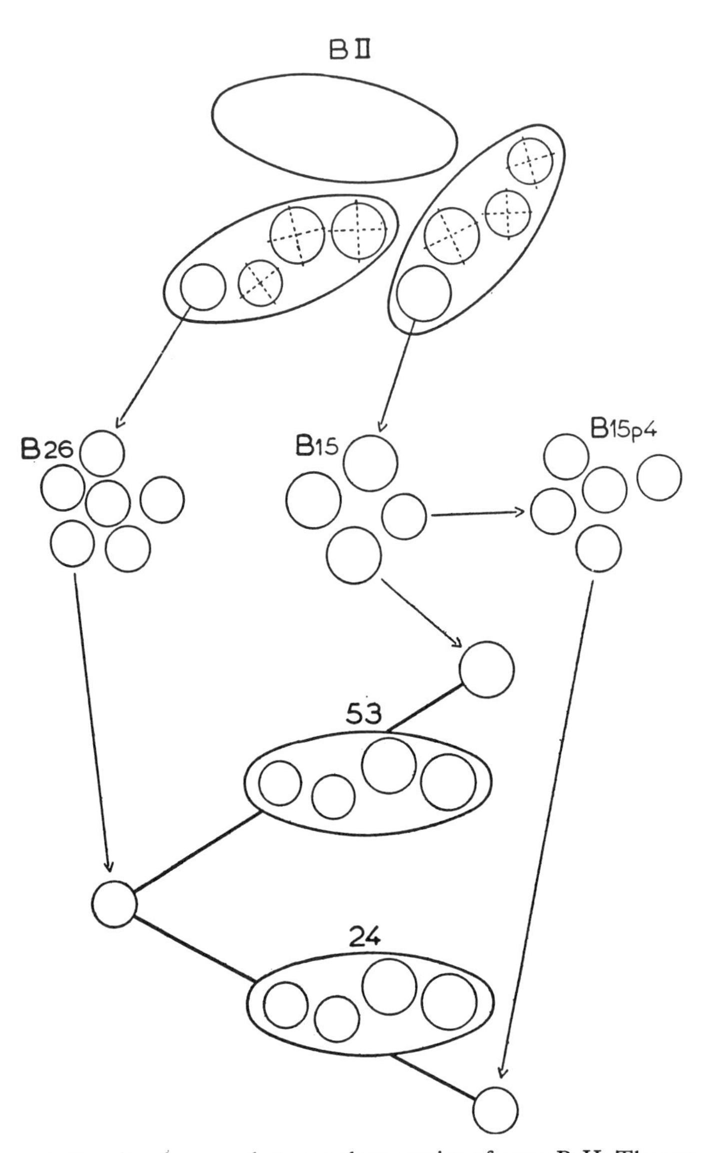


FIG. 16. Results of crosses between three strains of yeast B-II. The upper part of the figure shows the origin of the three haploid strains. B 15 is a yeast with normal respiration, B 15p4 and B 26 are respectively clones of 'vegetative' and 'segregational' littles. The mutant phenotype is indicated by the small size of the cells.

Between these three strains two crosses are possible: the 'segregational mutant', of mating type –, can be crossed with each of the other two strains, which are both of + mating type.

The cross between the segregational mutant (B 26) and the normal strain (B 15) results in the formation of diploid cells with normal respiration. These cells are therefore able to sporulate. The analysis of the asci showed that each of them contained two normal and two mutant spores. This result confirms the hypothesis that the difference between 'normal' and 'segregational mutant' is genic, and the clear unifactorial 2:2 ratio found in the asci shows that these two strains are differentiated by a single gene pair. Since the diploid cells formed as a result of the cross are normal, it is clear that the normal strain B 15 carries a dominant gene, say *R*, and that the respiratory deficiency of the segregational mutant B 26 is due to the presence of the recessive allelomorph *r*.

The second cross provided us with results much more difficult to explain in Mendelian terms. Following the cross of the same 'segregational mutant' B 26 with the 'vegetative mutant' (B 15p4), diploid cells are formed which possess a normal respiration. Analysis of the asci formed by these diploids showed that these asci too each contain two normal and two mutant ascospores. The reconstitution of a normal diploid cell by the fusion of two haploid mutants can easily be explained if it is assumed that each of the strains crossed owes its mutant character to the mutation of a different, non allelic, gene; but the sporulation of such a doubly heterozygous hybrid should, you remember, result in segregation ratios more complex than the typically unifactorial 2:2 ratio actually observed.

If the two crossed strains of littles owed their character to the presence of two different non-allelic recessives, the expected segregations would be 0:4, 1:3, and 2:2 in different asci. If the two genes in question were not linked, the proportion of asci of the 1:3 type would depend on the distance of the genes from the centromeres, the rest of the asci being equally divided between the 0:4 and the 2:2 types. If the genes were linked, the frequency of 2:2 asci should be higher than that of asci of the 0:4 type.

While the experimental results are thus difficult to explain if only genic differences between the three strains are postulated, they are completely and easily accounted for by an interpretation postulating the intervention of both genic and cytoplasmic factors.

Assume that in this yeast, as in the American one (Yeast Foam), there is no genic difference between the normal haploid B 15 and the vegetative mutant B 15p4 derived from it, and, that the vegetative mutant is the result of a loss-mutation, the loss involving the cytoplasmic particles necessary for the synthesis of respiratory enzymes. Assume, on the other hand, that the segregational mutant does contain the cytoplasmic particles, but that it carries a recessive gene in the presence of which the cytoplasmic granules are physiologically inactive. Fig. 17, in which the inactive particles are represented by dots and the active ones by rods, shows in its centre the expected results of the two crosses I described (cross 53: B 26×B 15 and cross 24: B 26×B 15p4). You can see that on our assumptions both crosses should result in the formation of diploids with normal respiration because in both cases the diploids formed carry the dominant gene *R* and contain the cytoplasmic particles. They are both normal biochemically because in each of them the cytoplasmic particles are activated by the dominant gene; and they both produce asci showing a 2:2 segregation because they both are heterozygous for a single gene pair *R*/*r*: after meiosis the cytoplasmic particles remain active only in the two spores carrying the dominant *R*; in the two others, carrying the recessive *r*, the particles become, on the contrary, physiologically inactive.

The fact that segregational mutants manifest the respiratory deficiency although they contain the cytoplasmic particles endowed with genetic continuity proves that the genetic units and the enzymes with which we are concerned are distinct entities. On the other hand, the question whether the enzyme-carrying par-

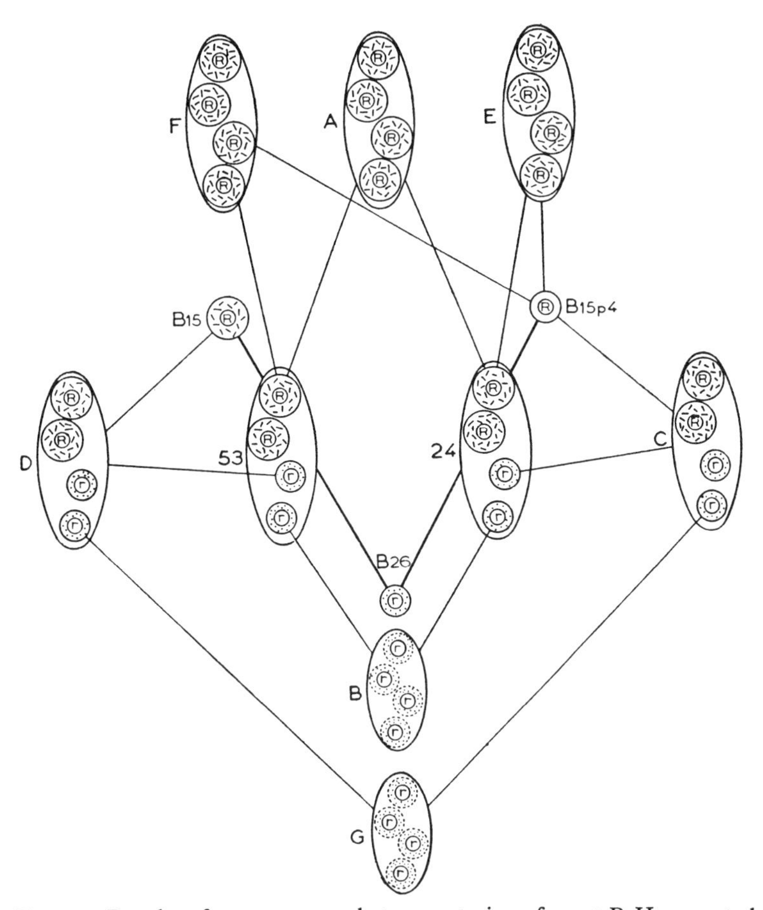


FIG. 17. Results of seven crosses between strains of yeast B-II, expected on the basis of the interpretation given on p. 40. In the middle of the figure are represented the same two crosses as in Fig. 16. The size of the asci indicates the normal (big) or mutant (little) character of the diploid hybrids. The size of the ascospores symbolizes their normal or mutant character. *R* and *r* are two allelic nuclear genes. Active cytoplasmic particles are represented as rods, inactive particles as dots.

ticles sedimentable by centrifugation and the particles postulated on genetic grounds are one and the same thing remains at present unanswered. The assumption that they are identical is simplest in the sense that it offers a simple mechanical explanation for the simultaneous disappearance from the mutant cells of several enzymes known to be linked to sedimentable cell components.

The suggested interpretation thus fully accounts for the observed results. The correctness of its postulates can be further checked by a number of crosses. The expected results of seven of these are given in Fig. 17 (A–G) and in the 2nd and 4th columns of Table I.

TABLE I

| | *Nadi reaction of diploid hybrid* | | *Segregation* | | |
|---|---|---|---|---|---|
| *Cross** | *Expected* | *Observed* | *Expected* | *Observed* | *Nature of spores in the aberrant asci*† |
| 53 | + | + | 2:2 | 31 (2:2) | .. |
| 24 | + | + | 2:2 | 15 (2:2) | .. |
| A | + | + | 4:0 | 9 (4:0) | .. |
| B | — | — | 0:4 | no asci | .. |
| C | + | + | 2:2 | 18 (2:2) | .. |
| D | + | + | 2:2 | 20 (2:2) | .. |
| | | | | 1 (0:4) | (vm vm sm dm) |
| E | + | + | 4:0 | 11 (4:0) | .. |
| | | | | 3 (3:1) | 3 (vm) |
| F | + | + | 4:0 | 19 (4:0) | .. |
| | | | | 1 (3:1) | (vm) |
| | | | | 2 (1:3) | 6 (vm) |
| | | | | 5 (0:4) | 20 (vm) |
| G | — | — | 0:4 | no asci | .. |

* The designation of the crosses corresponds to that of Fig. 17.
† vm : vegetative mutant; sm : segregational mutant; dm : dual mutant.

Let us first consider what the character of the diploid hybrids of these different crosses should be according to our scheme.

Crosses A, C, D, E, and F should lead to the formation of diploids carrying the dominant gene *R* either in homozygous or heterozygous form. Since, on the other hand, all of them involve at least one strain bringing in the cytoplasmic particles, all diploid hybrids of these crosses should possess normal respiratory characteristics.

Crosses B and G, on the contrary, each involve two segregational mutants, that is cells containing inactive cytoplasmic particles and the recessive gene *r*. The diploids formed in these two crosses should therefore be respiration-deficient.

Examination of the data of columns 2 and 3 in Table I shows that these expectations are fulfilled.

Let us now turn to the segregations expected to occur in the asci of the seven crosses, and begin with crosses A and F which are particularly important, for they are tests of the correctness of our assumption that the normal strain (B 15) and the vegetative mutant (B 15p4) derived from it, are genically identical, both carrying the dominant gene *R*. Because these two strains are of the same mating type, we were unable to verify this assumption by crossing them. We can now bridge the gap in the demonstration: as a result of gene recombination which has taken place during spore formation in the $F_1$ hybrid (53) between normal (B 15) and segregational little (B 26), many of the normal spores (that is of the spores carrying gene *R* inherited from the normal parent) are now of mating type −. These spores can now be backcrossed to the vegetative mutant parent (B 15p4). Cross F is such a backcross. If the vegetative mutant contains, as postulated, gene *R*, this cross must result in asci containing four normal spores (4:0 segregation).

Cross A is important for the same reason. It is a cross between two normal clones derived from ascospores of the two $F_1$ crosses respectively. Each of them is supposed to have inherited the same *R* allele. Cross A is therefore expected to result in asci containing only normal spores.

A similar 4:0 ratio is expected in cross E which is a backcross of an $F_1$ normal from cross 24 to the vegetative mutant parent of presumably identical genotype.

Crosses C and D are backcrosses of $F_1$ littles from the two crosses 53 and 24 to their respective parents presumed to carry gene *R*. (In the first case this parent is a vegetative mutant, in the second it is normal.) Both crosses are therefore expected to give 2:2 segregations.

Lastly, crosses B and G should result in asci containing four mutant spores each (0:4 segregations), since they are combinations of segregational mutants.

These are the expectations.

Comparison of columns 4 and 5 of Table I, which give the expected and observed segregations in the asci formed by the different hybrids, reveals a situation which is somewhat more complex than anticipated. Two of the hybrids (B and G) did not sporulate: this should not surprise us, however, since, as stated earlier, diploid vegetative mutants with deficient respiration are unable to sporulate. Obviously, diploid segregational mutants cannot sporulate either. More serious is the fact that in several of the other crosses, in addition to the expected segregations, some unexpected ratios were observed.

Does this mean that the hypothesis we formulated is wrong? Let me reassure you: you will see very shortly that unexpected results, as so often happens in science, are as valuable in the demonstration of the correctness of our hypothesis as expected ones: they are the exceptions which prove the rule.

You will have noticed that all the asci which I may call 'exceptional' contain an *excess* of mutant ascospores. Let us determine to what type the mutant spores of the exceptional asci belong. Following our hypothesis, this can be done in the manner shown in Fig. 18. The clone of cells derived from the ascospores to be tested (marked in the figure by a question mark) is crossed to two tester strains: one of them is a strain of vegetative mutants, the other a strain of segregational mutants. If the tested strain is a strain of vegetative littles, the first cross will result in the formation of respiration-deficient diploid cells; the second in the formation of normal diploids. If the tested strain is a strain of segregational mutants, the results of the two crosses will be reversed. Finally, there obviously is a third possibility: the tested strain could be what I shall call a 'dual mutant', the cells carrying neither dominant gene, nor cytoplasmic particles. Such a 'dual mutant' should, according to our hypothesis, form respiration-deficient diploids in both test crosses.

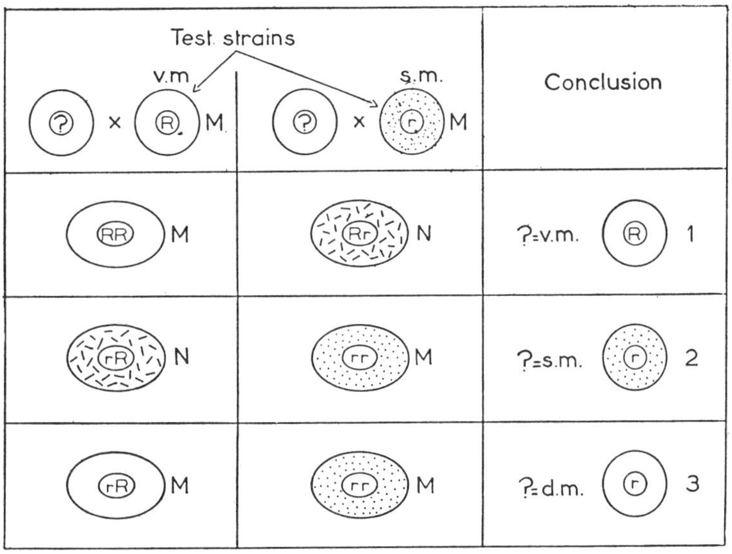

FIG. 18. Diagram indicating the method of testing mutant ascospores found in 'aberrant' asci. N, normal phenotype; M, mutant phenotype; vm, vegetative mutant; sm, segregational mutant; dm, dual mutant. Other symbols as in Fig. 17.

The results of this test, applied to all mutant spores of the exceptional asci, are given in the last column of Table I. These results clearly show that the genic segregation was in agreement with expectation in all the exceptional asci. In other words, the exceptional phenotypic ratios were in every case due to the loss of the cytoplasmic particles.

Summing up, the study of the two strains of baker's yeast shows that *the synthesis of respiratory enzymes by these organisms requires the simultaneous presence of a cytoplasmic factor and of a dominant nuclear gene.*[3] The cytoplasmic factor appears, in the light of the experiments I have described, to be dependent on the nucleus in its function, and independent of it in its reproduction.

This statement requires, however, a qualification. While the functional dependence of the cytoplasmic particles on the nucleus appears to be established beyond doubt, their apparently complete autonomy in reproduction is deduced from purely negative evidence. We have seen how misleading this kind of evidence can be: the cytoplasmic particles appeared to be completely autonomous in their function until the investigations of strain B-II were made. Similarly, their independence with respect to the nucleus in reproduction may be only apparent.

That this may be the case is actually suggested by observations[11] on strains which present, in populations in equilibrium, very high proportions of vegetative mutants. It appears that this characteristic is due to a high spontaneous mutation rate and is controlled by a recessive gene. A cross between a normal strain and one with a high mutation rate results in a 2:2 segregation of the character. Since the mutation itself seems to consist in the loss of cytoplasmic particles, the role of the mutability-gene may be interpreted as increasing the probability of the loss.

This situation can be accounted for by assuming that the mutability-gene controls some intracellular condition which limits the average content of particles per cell to a low number. This result could be achieved if, for example, the recessive gene caused a slow multiplication of the cytoplasmic particles.

That the gene does not affect the intrinsic properties of the particles is proved by the fact that hybrids produced by the cross of a mutable strain with a vegetative little, isolated from a 'stable' strain, produces spore progeny which again segregates into 'highly mutable' and 'stable'. In other words, the cytoplasmic particles of a 'highly mutable' yeast again show normal behaviour when placed in the presence of the dominant gene.

*Long-term adaptation to galactose fermentation.* Winge and Roberts[46] have described in *Saccharomyces Chevalieri* strains differing in the ability to ferment galactose. This difference has been shown to be based on the presence in the fermenters of a dominant gene *G* and of its recessive allele $g_s$ in the slowly adapting strains.

Study of the phenomenon of 'long-term adaptation' to galactose fermentation in a strain of the latter sort has led Spiegelman and co-workers[36] to the conclusion that it involves the induction, in a small proportion of 'positives' exposed to the substrate, of a cytoplasmically transmitted enzyme-forming system, distinct from the enzyme itself. This cytoplasmic factor is maintained during growth in the presence of galactose, but rapidly and suddenly disappears when the 'positives' are placed in glucose-medium. The loss is, however, not permanent, and a certain proportion of negative cells remain capable of producing the enzyme-forming system again when placed in the presence of galactose.

The kinetics of the reversion from positive to negative has been studied by bud analysis of the sort which led to the demonstration of the mutagenic action of euflavine. The pedigrees obtained show a great similarity with those described above. The authors conclude that enzyme formation is dependent on a cytoplasmic particulate enzyme-forming system, randomly distributed between mother cell and bud.

By subjecting galactose-adapted cells to euflavine treatment, Spiegelman and co-workers observed vegetative respiration-deficient mutants which contain galactozymase and transmit the enzyme-forming capacity to their progeny. The reversion to negatives follows in these vegetative mutants the same course as in the cells with normal respiration.

It is concluded that the particles responsible for galactozymase formation are distinct from those involved in the synthesis of respiratory enzymes.

## REFERENCES

1. BRIGGS, R., and KING, T. J. Transplantation of living nuclei from blastula cells into enucleated frogs' eggs. *Proc. Nat. Acad. Sci.* 1952, **38**, 455–63.
2. CHATTON, E., and LWOFF, A. Les Ciliés apostomes. I. Aperçu historique et général; étude monographique des genres et des espèces. *Arch. Zool. exp. et gén.* 1935, **77**, 1–453.
3. CHEN, S. Y., EPHRUSSI, B., and HOTTINGUER, H. Nature génétique des mutants à déficience respiratoire de la souche B-II de la levure de boulangerie. *Heredity*, 1950, **4**, 337–51.

4. COMMANDON, J., and DE FONBRUNE, P. Greffe nucléaire totale, simple ou multiple chez une Amibe. *C.R. Soc. Biol.* 1939, **130**, 744–8.
5. EPHRUSSI, B. Action de l'acriflavine sur les levures. In *Unités biologiques douées de continuité génétique*. Édition du Centre Nat. Rech. Sci., Paris, 1949, 165–80.
6. —— Remarks on cell heredity. In *Genetics in the 20th Century*. Macmillan, New York, 1951, 241–62.
7. —— and HOTTINGUER, H. Direct demonstration of the mutagenic action of euflavine on baker's yeast. *Nature*, 1950, **166**, 956.
8. —— —— On an unstable cell state in yeast. *Cold Spring Harbor Symp. Quant. Biol.* 1951, **16**, 75–84.
9. —— —— and CHIMENES, A. M. Action de l'acriflavine sur les levures. I. La mutation 'petite colonie'. *Ann. Inst. Pasteur*, 1949, **76**, 351–64.
10. —— —— and TAVLITZKI, J. Action de l'acriflavine sur les levures. II. Étude génétique du mutant 'petite colonie'. *Ann. Inst. Pasteur*, 1949, **76**, 419–50.
11. —— —— and LEUPOLD, U. Unpublished.
12. —— L'HERITIER, PH., and HOTTINGUER, H. Action de l'acriflavine sur les levures. VI. Analyse quantitative de la transformation des populations. *Ann. Inst. Pasteur*, 1949, **77**, 64–83.
13. FAURE-FRÉMIET, E. La continuité des mitochondries à travers les générations cellulaires et le rôle de ces éléments. *Anat. Anz.* 1910, **36**, 186–91.
14. —— Les mécanismes de la morphogénèse chez les Ciliés. *Folia Biotheoretica*, 1948, **3**, 25–28.
15. HOARE, C. A. Recent studies on the kinetoplast in relation to heritable variation in Trypanosomes. *J.R. Micr. Soc.* 1940, **60**, 26–35.
16. —— Races of *Trypanosoma evansi* produced by mutation. *Trans. R. Soc. Trop. Med. Hyg.* 1950, **43**, 360–1.
17. KRUIS, K., and ŠATAVA, J. O vývoji a klíčeni spŏr jakož i sexualitě kvasinek. *Nakl. C. Akad.* Praha, 1918.
18. LINDEGREN, C. C. Chromosome maps of *Saccharomyces*. *Proc. 8th Int. Congr. Genetics.* 1948, 338–55.
19. —— *The Yeast Cell, its Genetics and Cytology*. Educational Publishers, Saint Louis, 1949.
20. —— and LINDEGREN, G. Selecting, inbreeding, recombining and hybridizing commercial yeasts. *J. Bact.* 1943, **46**, 405–19.
21. —— —— A new method for hybridizing yeast. *Proc. Nat. Acad. Sci.* 1943, **29**, 306–8.
22. —— —— Depletion mutation in *Saccharomyces*. *Proc. Nat. Acad. Sci.* 1947, **33**, 314–18.
23. LORCH, I. J., and DANIELLI, J. F. Transplantation of nuclei from cell to cell. *Nature*, 1950, **166**, 329–30.
24. LWOFF, A. Les organites doués de continuité génétique chez les Protisdes. In *Unités biologiques douées de continuité génétique*, Édition du Centre Nat. de la Rech. Sci., Paris, 1949, 7–23.

25. LWOFF, A. *Problems of Morphogenesis in Ciliates.* Wiley and Sons, New York, 1950.
26. —— La synthèse de l'amidon chez les Leucophytes et la valeur morphologique du réseau de Volkonsky. *New Phytologist,* 1950, **49**, 72–80.
27. MARCOVICH, H. Action de l'acriflavine sur les levures. VIII. Détermination du composant actif et étude de l'euflavine. *Ann. Inst. Pasteur,* 1951, **81**, 452–68.
28. MEDAWAR, P. B. Cellular inheritance and transformation. *Biol. Rev.* 1947, **22**, 360–89.
29. MULLER, H. J. The gene as the basis of life. *Proc. 4th Int. Congr. Plant. Sci.* Ithaca, 1929, **1**, 897–921.
30. PIRIE, N. W. The meaninglessness of the terms of Life and Living. In *Perspectives in Biochemistry,* Cambridge University Press, 1937, 11–22.
31. PROVASOLI, L., HUTNER, S. H., and SCHATZ, A. Streptomycin-induced chlorophyll-less races of *Euglena. Proc. Soc. Exp. Biol.* 1948, **69**, 279–82.
32. ROBERTSON, M. The action of acriflavine upon *Bodo caudatus. Parasitology,* 1929, **21**, 375–416.
33. SLONIMSKI, P. Action de l'acriflavine sur les levures. IV. Mode d'utilisation du glucose par les mutants 'petite colonie'. *Ann. Inst. Pasteur,* 1949, **76**, 510–30.
34. —— Recherches sur la formation des enzymes respiratoires chez la levure. Thesis, Faculté des Sciences, Paris, 1952.
35. —— and EPHRUSSI, B. Action de l'acriflavine sur les levures. V. Le système des cytochromes des mutants 'petite colonie'. *Ann. Inst. Pasteur,* 1949, **77**, 47–63.
36. SPIEGELMAN, S., DE LORENZO, W. F., and CAMPBELL, A. M. A single-cell analysis of the transmission of enzyme-forming capacity in Yeast. *Proc. Nat. Acad. Sci.* 1951, **37**, 513–524.
37. SZENT-GYÖRGYI, A. *Nature of life.* Academic Press, New York, 1948.
38. TAVLITZKI, J. Action de l'acriflavine sur les levures. III. Étude de la croissance des mutants 'petite colonie'. *Ann. Inst. Pasteur,* 1949, **76**, 497–509.
39. VIRCHOW, R. *Die Cellular-pathologie in ihrer Begründung auf physiologische und pathologische Gewebelehre,* Berlin, 1858.
40. WEISZ, P. A general mechanism of differentiation based on morphogenetic studies in Ciliates. *Amer. Nat.* 1951, **85**, 293–311.
41. WERBITZKI, F. W. Über blepharoblastlose Trypanosomen. *Zblt. f. Bact.* 1910, **53**, 303–315.
42. WINGE, O. On haplophase and diplophase in some *Saccharomycetes. C.R. Lab. Carlsberg,* 1935, **21**, 77–111.
43. —— and LAUSTSEN, O. On two types of spore germination, and on genetic segregations in *Saccharomyces,* demonstrated through single spore cultures. *C.R. Lab. Carlsberg,* 1937, **22**, 99–117.
44. —— —— Artificial species hybridization in yeast. *C.R. Lab. Carlsberg,* 1938, **22**, 235–44.

45. WINGE, O, and LAUSTSEN, O. On a cytoplasmic effect of inbreeding in homozygous yeast. *C.R. Lab. Carlsberg*, 1940, **23**, 17–40.
46. —— and ROBERTS, C. Inheritance of enzymatic characters in yeasts, and the phenomenon of long term adaptation. *C.R. Lab. Carlsberg*, 1948, **24**, 263–315.
47. WRIGHT, S. The physiology of the gene. *Physiol. Rev.* 1941, **21**, 487–527.

# The Early Days of Yeast Genetics: A Personal Narrative

HERSCHEL ROMAN

*Ann. Rev. Genet. 1986. 20:1–12*


# THE EARLY DAYS OF YEAST GENETICS: A PERSONAL NARRATIVE[1]

*Herschel Roman*

Department of Genetics, SK-50, University of Washington, Seattle, Washington 98195

---

CONTENTS



## THE FIRST YEAST GENETICS LABORATORIES

The first studies of the genetics of yeast were made by Ö. Winge in 1935 at the Carlsberg Laboratory in Copenhagen, when he demonstrated the alternation of haplophase and diplophase in the yeasts that he examined (38). A series of publications from the same laboratory followed, which described the inheritance of colony morphology and, later, the inheritance of fermentative

[1]The term "yeast" is used in this review to mean *Saccharomyces cerevisiae*. This is done for convenience; I do not mean to disparage the important work of investigators of other yeasts, particularly *Schizosaccharomyces pombe*.



0066-4197/86/1215-0001$02.00

variants (40). A second laboratory that achieved prominence was directed by Carl C. Lindegren, first at Washington University, St. Louis, and later at the University of Southern Illinois, Carbondale. Lindegren came to yeast from *Neurospora,* with which he had pioneered the development of tetrad analysis in the Ascomycetes. A third laboratory that played an important role in the history of yeast genetics was that of Boris Ephrussi, who worked in Paris and later in Gif-sur-Yvette, some 30 kilometers south of Paris. Ephrussi's major contribution in yeast research was the discovery of cytoplasmic inheritance; he, together with his collaborators, notably Piotr Slonimski, provided the basis for the flourishing field of mitochondrial genetics. The fourth yeast laboratory was established in the late 1940s at the University of Washington, Seattle, under the joint direction of Howard C. Douglas and myself, with the able participation of Donald C. Hawthorne, an exceptional graduate student.

## PERSONAL PREHISTORY AND WHY YEAST WAS CHOSEN

I was a graduate student at the University of Missouri when the first experiments of the Winge laboratory came to my attention. The principle organisms for genetic research at that time were *Drosophila* and corn *(Zea mays).* Ascomycetes *(Neurospora),* protozoa *(Paramecium),* bacteria *(Escherichia coli),* and the bacteriophages were to become important as experimental organisms shortly thereafter. I was extremely fortunate to join one of the nation's preeminent groups in genetics, which was under the leadership of Lewis J. Stadler. In addition to Stadler, the group included Barbara McClintock, Ernest R. Sears (wheat cytogenetics), George F. Sprague, Joseph O'Mara, Fred Uber (biophysics of ultraviolet irradiation of corn pollen), and, more peripherally, Daniel Mazia (general physiology). Spencer Brown, Seymour Fogel, John Laughnan, James Cameron, Irwin Herskowitz, Frances Clark (Beard), and Katherine Mills (DeBoer) were fellow graduate students. Bentley Glass was also a member of the group, holding a position at nearby Stephens College until he left for Johns Hopkins University. The group was not formally organized but was held together by common intellectual interests and by the cementing influence of Stadler. The individuals were dispersed among several departments.

In 1942, with a Ph.D. in hand, I accepted a position at the University of Washington, intending to continue research with corn. I had not investigated climatic conditions in Seattle and soon found that corn could not be matured under these conditions. The greenhouses were quite inadequate until new ones were added in 1950. In the interim, except for time given to World War II, I grew corn for three summers (1946–1948) at the California Institute of

Technology, supported by a Gosney Fellowship, and for one summer (1949) in eastern Washington on rented acreage.

I became convinced that it was time to change organisms if I were to remain in Seattle. In 1947 we invited Carl Lindegren to come to Seattle to present his views on the usefulness of yeast for experimental work in genetics. Some of his views were quite controversial, and we had many discussions during the two weeks or so that he was in Seattle: discussions in which alternative explanations were proposed to account for the same data. His unorthodox interpretations presented a challenge that was largely responsible for my choosing yeast as an experimental organism. Also the interest in yeast of my colleague Howard C. Douglas, a microbial physiologist, was important in encouraging me to reach this decision. Douglas later played a leading role in unraveling the elements of control in the metabolism of galactose (3).

Yeast appeared attractive for its own intrinsic properties. It was a simple, single-celled eukaryotic organism that could be handled in much the same way as bacteria. It had a short life cycle, about 1½ hours at 30°C during exponential growth. It had long been a favorite research tool of biochemists, and therefore much was known of its biochemistry. Unlike many Ascomycetes used experimentally, *Saccharomyces cerevisiae* had a stable diplophase and haplophase. It turned out that it had a normal meiosis as well as a stable mitosis. It seemed therefore a model eukaryotic organism, simple by comparison with others then in use, and it proved to be very useful in genetic research.

This review covers the early years of controversy and discovery in yeast genetics, principally in *Saccharomyces cerevisiae* and its near relatives. I do not include the findings in *Schizosaccharomyces pombe,* an organism whose potential for genetic study was first demonstrated by Urs Leupold while he was a guest at the Carlsberg Laboratory, in Winge's department. Leupold, now at the University of Bern, Switzerland, has over the years trained most of the investigators working with *Schizosaccharomyces*. The results he and his followers have obtained parallel and complement those achieved with *Saccharomyces cerevisiae*.

I also do not include more recent findings in *Saccharomyces cerevisiae,* because these are adequately reviewed in three volumes published in 1981 (36), 1981 (32), and 1982 (33). Studies of those aspects of yeast genetics common to all eukaryotic organisms (notably those dealing with gene regulation and gene mutation) are also not included, except for those investigations in which yeast was a significant contributor.

## THE DISCOVERY OF HETEROTHALLISM

Today over 500 yeast genes have been mapped among 16 or 17 chromosomes (the number hasn't been settled to everyone's satisfaction). Many of the genes have known functions, e.g. affecting macromolecular synthesis and specific points in the pathway of synthesis of amino acids, fatty acids, etc, and controlling the various steps that provide energy in metabolic processes. By contrast, when Winge began his studies with yeast, colony shape and structure were used as criteria that demonstrated inheritance. Of course, colony morphology was recognized as being multigenic, but intensive inbreeding produced segregants among the haploid progeny of a hybrid in a ratio of approximately 2:2. Later Winge and his collaborators, principally Catherine Roberts, studied the inheritance of the fermentation of a number of sugars, which also gave multigenic ratios. It soon became clear that although several genes governed the fermentation of a given sugar, these genes could be sorted out so that simple 2:2 ratios were obtained. The finding of several genes affecting the same character was the first of its kind in any organism, and the genes were categorized as "polymeric." This term was applied to genes controlling fermentations of raffinose, sucrose, and maltose (40).

Winge and his collaborators made all their crosses by spore-to-spore matings. This was a laborious technique, and it often failed for reasons that will become apparent. In 1943 an important technical step was taken when the Lindegrens found heterothallism in *S. cerevisiae* (16, 17). There were two mating types, **a** and $\alpha$, and a cross could be achieved only if the two parental strains were of opposite mating type: i.e. **a** cells mixed with $\alpha$ cells would mate, whereas **a** cells mixed with **a** cells, and $\alpha$ cells mixed with $\alpha$ cells, would not. The advantages of the technique were twofold: the haploid cultures arising from each spore could be diagnosed for mating type by testing with known **a** and $\alpha$ strains, and they could then be crossed with strains of the opposite mating type bearing the desired markers. The Lindegren technique had the advantage that the parents of the cross were not consumed in the process of mating; therefore, the identical cross could be repeated. Haploid cultures could thus be retained in storage for later use as required.

At first, following the discovery of mating types, a cross was accomplished by a mass mating of the haploid cultures of opposite mating types, followed by analysis of the asci produced by the population of independently formed diploids. When it was realized that heterogeneity in the cultures could arise as a consequence of vegetative growth (as we shall see), crosses were made by the mass culture technique, followed by the isolation of individual colonies, which originated from a single diploid cell. A sample of these colonies was grown to stationary phase and was then sporulated. By this technique the composition of the parents of the cross could be ascertained, or if something unusual had happened to give parents of an unexpected genotype, the dissections would reveal such aberrations, and the results of the cross could be disregarded.

## THE CONTROVERSY OF GENE CONVERSION

The Lindegren laboratory was responsible for a second observation, and it was one that resulted in considerable controversy—a controversy in which several laboratories became embroiled. This resulted from the finding that irregular ratios (such as 3:1 and 1:3, which differ from the 2:2 ratio expected from Mendelian segregation) were obtained quite regularly with frequencies of about 1% for one or another of the markers introduced in heterozygous condition in an individual hybrid (15). This departure from Mendelian segregation, first noticed by Winkler (41), who coined the term "gene conversion," led to efforts in other laboratories to find more orthodox explanations for the irregular segregations. In fact, in the early stages of yeast genetics the arguments became quite vituperative. Geneticists accustomed to the results obtained with more conventional organisms such as *Drosophila* and maize (in which irregular segregations would have been virtually impossible to detect because the four products of an individual meiosis could not be isolated and analyzed) had doubts that yeast was stable enough in its inheritance to be useful for genetic experimentation.

Our laboratory in Seattle soon confirmed the unorthodox findings of the Lindegren laboratory. However, we were skeptical of gene conversion as the interpretation and, on the assumption that yeast followed the rules of Mendelian inheritance, sought other explanations to account for the irregular segregations. One of the possibilities we considered was that polyploidy could account for most, perhaps all, of the irregular segregations obtained by the two laboratories. Evidence for polyploidy was indeed found among the asci dissected by D. C. Hawthorne in our laboratory (27), and similar observations were reported by the Lindegrens in the same year (18).

The finding of heterogeneity in cultures grown from single spores was the next landmark in yeast genetics (29). It had been assumed, as was the case in another Ascomycete, *Neurospora,* that mitotic divisions regularly produced haploid nuclei in heterothallic strains. In our laboratory, we observed that frequent transfers of a spore culture resulted in the loss of mating ability and that such cultures were able to sporulate. Dissections of the asci showed that the culture had become diploidized by virtue of mutation at the mating ability to the opposite mating type. The observation of mutation had been made also by Ahmad, as mentioned in Catcheside (2). The **a**$\alpha$ diploid cells soon dominated the culture because they divided more rapidly than the **a** or $\alpha$ haploids from which they arose.

The diploid cells were larger than haploids and could easily be distinguished and isolated from the surrounding haploid cells. It became evident that the larger cells were of two types: **a**$\alpha$ and **aa** if they arose in the haploid **a** culture, or **a**$\alpha$ and $\alpha\alpha$ if they arose from $\alpha$ haploids. Neither **aa** nor $\alpha\alpha$ cells could be induced to sporulate, but both were capable of mating with cells of opposite mating type, either haploid or diploid. Thus, **aa**$\alpha\alpha$ tetraploids, and triploids of two types—**aa**$\alpha$ and **a**$\alpha\alpha$—could be produced. Also, aneuploid spores could in turn be obtained from triploid, and to some extent from tetraploid, cultures, owing to occasional nondisjunction during meiosis in the latter. Thus, the necessary elements contributing to irregular segregation could be obtained from the products of polyploid cells.

It soon became clear that polyploidy was not the sole answer to the problem of irregular segregation. As methods were developed that selected for several mutant alleles at the same locus the opportunity presented itself for crosses between alleles of independent origin. Mary Mitchell (20) crossed two strains that carried two different recessive alleles at the pyridoxine locus in *Neurospora* and found 2:6 segregations of wild-type: recessive phenotypes in addition to the predominant $4^+:4^-$ type. (The 2:6 segregation in the eight-spored *Neurospora* ascus is equivalent to a 1:3 segregation in the four-spored ascus of yeast.) Cultures from the spores that were of the recessive phenotype were in a 2:4 ratio for one or the other of the two alleles. The wild-type spore cultures exhibited a new phenotype and could not have arisen as a result of mutation, polyploidy, or any other known mechanism. Mitchell also demonstrated that crossing over between markers bracketing the pyridoxine locus occurred with a much higher than expected frequency.

I made a similar finding in mitotic cells, which was published the following year (25). Previously it had been shown that *ade*2 strains, which produced a red pigment owing to the accumulation of an intermediate in the pathway of adenine synthesis, produced cells that gave rise to white colonies. The white cells had a growth advantage over the reds; therefore there was a strong selection for whites in the red culture. This was particularly true when the culture was grown in a rich medium, such as one containing yeast extract and peptone, with glucose as the carbon source. In synthetic media the two types of cells grew equally well. The reason for this difference is not obvious even today.

An analysis of the white mutants showed that six loci were involved, any one of which could block the production of the red pigment. Mutations at five of the loci interrupted the adenine pathway prior to the block that produced the red pigment. A sixth mutation, *ade*3, affected both the histidine and adenine pathways and was shown to be due to a defect in the tetrahydrofolate cycle (14). The white mutants were therefore double mutants, carrying both the original *ade*2 mutation as well as another *ade* mutation at any of six different loci. Thus, large numbers of mutations could readily be obtained in each of the six loci and could be separated from *ade*2 by crosses with wild-type strains. As in *Neurospora*, crosses between strains carrying alleles of different origin but belonging to the same locus produced wild-type alleles with relatively high frequency (compared to mutation) in mitotic division. Wild-type alleles were also produced in meiosis, where 1:3 segregations occurred as frequently as 1%. The alleles that produced the wild types were designated "heteroalleles," and diploids constructed such that they were homozygous for the same recessive allele were designated "homoallelic" diploids (28). The latter produce wild types with a mutational frequency clearly lower than that obtained from heteroallelic diploids.

The phenomenon disclosed by these techniques resulted in some controversy concerning nomenclature. The term "nonreciprocal recombination," which reflected the 1:3 segregation (as opposed to "reciprocal" recombination, which was characterized by the 2:2 segregation), was proposed to replace the term "gene conversion." Other terms such as "intragenic recombination" and "interallelic recombination" were also used, to avoid the connotation of one gene converting its allele by a mutational process, as the term "gene conversion" suggested. It was shown, in fact, that the three recessive alleles in the 1:3 segregation were identical (7, 25). These terms were short-lived, and "gene conversion" became, and is today, the term applied to cases of irregular segregation not attributable to orthodox causes (such as polyploidy).

The unequivocal demonstration that gene conversion really existed, and neither was due to faulty observation nor could be explained as the result of more orthodox causes, gave yeast a certain respectability among geneticists. An additional reason for its acceptance was the fact that many more markers began to become available, and attempts at mapping them revealed linkage between genes and between a gene and its centromere. The most successful of these efforts were those of D. C. Hawthorne and R. K. Mortimer, who teamed together to provide genetic maps that became the standard in the field [The latest version has been assembled by Mortimer & Schild (21)].

## OTHER CONTROVERSIES AND PLEASANT SURPRISES

Another controversy had arisen in the meantime, concerning the location of the nucleus and the number of chromosomes it contained. The chromosome number was variously given as from 1 to 7, the number in some cases coming after information of genes linked to different centromeres became available. Thus, genetic evidence for chromosome number seemed to precede cytological observation rather than the reverse; the chromosome number increased as the number of known genes increased and the linkage groups became more complete.

There was an unsuspected consequence of chromosome number in relation to DNA content. As the chromosome number increased and reliable DNA measurements of the haploid nucleus were made it became evident that each chromosome was on the average considerably shorter than that of *Escherichia coli,* and only about the length of five T4 phage chromosomes. This made the yeast chromosome amenable to physical separation and measurement (23).

As for the location of the nucleus within the cell, there was an early suggestion that the vacuole, especially evident in stationary-phase cells, was the nucleus, and chromosome counts of twelve in the diploid nucleus were reported. An alternative view was that this structure was indeed a vacuole and that the chromosome-containing nucleus was a separate unit, stainable by traditional stains and regular in size. This view proved to be correct; the nucleus was quite small and was most clearly seen with the electron microscope.

The nonsense codons, UAA, UAG, and UGA, were first identified in bacteria and bacteriophage as stop codons that interrupted protein elongation prematurely if a mutation producing them occurred within a gene. A second mutation in a tRNA gene can change the specificity of its codon-recognition site, with the result that the nonsense triplet can now be read and an amino acid inserted to restore protein elongation and yield a functional product. The same three nonsense codons were discovered in yeast and thereby gave additional support to the hypothesis of the universality of the genetic code (10, 11, 30).

Moreover, a much more complex story emerged from the yeast studies. For a given nonsense codon there were several phenotypic classes of supersuppressors (as they came to be called) arising not only from the particular amino acid inserted but also from the efficiency of the translation of the nonsense codon. Thus, some supersuppressors restored function to approximate the wild type of some genes and not of others, whereas other supersuppressors had another range of restoration. The interpretation was much like that in *E. coli* and T4 phage, but differed in the number of tRNA suppressors found. As an example, there are eight tRNA loci that specify tyrosine as the inserted amino acid. These loci are located on different chromosomes although they are alike in their composition and in their function. The questions of why eight are needed where one would suffice and of how these genes retain their homology have not yet been answered.

## CYTOPLASMIC INHERITANCE; MITOCHONDRIAL GENETICS

Quite a different aspect of yeast genetics was developed under the auspices of Boris Ephrussi in France. His principal collaborator was his student, later his successor, Piotr Slonimski. Ephrussi discovered the yeast mutant phenotype "petite." Physiologically the petite phenotype was due to the absence of certain respiratory enzymes. As a result of this deficiency, the mutant made less effective use of glucose and grew into a smaller colony (therefore the name). Crosses between the petite and the normal strain resulted in the loss of the petite phenotype in the hybrid and in its failure to be transmitted to the spore progeny. Further experiments showed that the petite mutant was due to the loss of a cytoplasmic factor.

The existence of a petite mutant that showed a 2:2 segregation in crosses with the wild type later demonstrated that there were also nuclear mutations that could produce the petite phenotype. A third class of petites, the "suppressive" petites, was found in crosses with wild-type cultures from which high percentages of petites (often rising above 90%) were found in contrast to the "neutral" petites, which produced virtually no petites in the same crosses. The studies of cytoplasmic inheritance at this stage were summarized by Ephrussi (5).

An important technique developed by Wilkie (37) made possible the identification of the loci on mitochondrial DNA, alterations of which produced the cytoplasmic petite. Wilkie and his colleagues found that resistance to antibiotic markers was caused by mutations in the mitochondrial DNA, and thereby they provided a method for the study of the genetics of cytoplasmic inheritance. Slonimski and his colleagues used this technique to expand rapidly our knowledge of mitochondrial DNA. Their work has been summarized in a Cold Spring Harbor Monograph (34; see also 4).

## HOMOTHALLISM vs HETEROTHALLISM

Homothallic strains had been used extensively by Winge and his collaborators. One such strain was *Saccharomyces chevalieri,* which formed diploid cells from single spores. Winge & Roberts (39) crossed *S. chevalieri* with *S. cerevisiae* by spore-to-spore matings. Diploids obtained from these crosses and the spores of an ascus were of two types: two were homothallic, like *S. chevalieri,* and two were heterothallic, like *S. cerevisiae*. The two heterothallics could be either of mating type **a,** of mating type $\alpha$, or of both mating types. Winge & Roberts inferred that there was a gene for homothallism (designated *D* for diploidization, now given the designation *HO* for homothallism), segregating independently from the genes specifying mating type. They further inferred that homothallic strains carried either the **a** or the $\alpha$ allele. Later Hawthorne (8) and Oeser (22) demonstrated that early in the growth of the homothallic culture there was a switch of some cells to the opposite mating type and subsequent fusion of the switched cells with the original cells. Once the **a**$\alpha$ diploid condition was established, *D* had no further effect. Hawthorne also proposed that what had been regarded as the mating-type locus was actually bipartite, containing both the **a** and $\alpha$ cistrons.

He based his interpretation on his finding that a deletion, seemingly terminating within the mating-type locus, converted an $\alpha$ mating type to an **a** mating type (9). The remarkable ramifications of further developments in the mating-type story are documented in *The Molecular Biology of the Yeast* Saccharomyces: *Life Cycle and Inheritance* (12).

## THEORIES OF THE BASIS OF GENE CONVERSION

Gene conversion was an established fact, but its mechanism remained an enigma until Holliday (13) suggested the then novel idea of strand transfer between DNA duplexes to form symmetrical heteroduplexes. If a mutant allele fell within the heteroduplex opposite a wild-type allele, a correction mechanism would excise one or the other of the two alleles to bring about complementarity. If correction occurred in the same sense in both heteroduplexes, either a $1^{+}:3^{-}$ or $3^{+}:1^{-}$ ratio would result. The cross-configuration resulting from the exchange of segments would be cleaved in both strands so that the two duplexes could separate, or the two strands not involved in the transfer would be cut and rejoined with their homologous segments, again for the purpose of freeing the duplexes. The effect of the former would be to leave flanking markers in the same relative position as before, and the effect of the latter would be to recombine flanking markers.

The proposal made by Holliday has been altered considerably since its introduction to take into account the rarity of ratios that are obtained experimentally but are not predicted from the original proposal (19, 35; also 7 for pertinent segregational data). Recent evidence (26) suggests that conversion can be temporally separated from crossing-over in mitosis. Conversion in G1, in which there are two DNA duplexes for each chromosome in the diploid, is followed by crossing-over in G2, after DNA replication has increased the number of duplexes to four for each chromosome. There is, therefore, a correlation between conversion and crossing-over. The previous hypotheses do not predict this temporal separation. The significance of the separation is not yet clear (see also 31).

## THE CELL DIVISION CYCLE

Finally, the analysis of the mitotic cell cycle by obtaining temperature-sensitive mutations that affected the cycle was originally begun by Hartwell and his colleagues (24). They obtained mutations affecting several steps in the cycle, such as those affecting different stages of G1, budding, DNA synthesis, those in G2, and those in the different stages of cell division until division was completed. The cytological basis for some of these steps was worked out by Byers and his collaborators (1). A similar analysis of mutations affecting meiosis and spore formation is given by Esposito & Klapholz (6).

## APOLOGIES AND PROSPECTS

I apologize again for leaving out a large number of investigators who contributed importantly to various aspects of yeast genetics, such as the effects of radiation. As I said in the introductory remarks, this is a personal narrative of the beginnings and early development of yeast genetics and represents my view of the important landmarks during the early period. Later developments, especially those dealing with the molecular aspects of yeast, have also been omitted. References are given so that the reader can become acquainted with the more modern facts about the genetics and molecular biology of yeast.

Yeast today has achieved a gratifying popularity as a simple eukaryotic organism for investigating basic problems in genetics and molecular biology. It likewise occupies an important position in modern biotechnology. The increase in the number of investigators working with yeast, from the few in the early days to upwards of a thousand today, attests to its value as an experimental organism.

ACKNOWLEDGMENTS

The research of my laboratory was supported by the National Institutes of Health, AI 00328 and GM 27949.

### *Literature Cited*

1. Byers, B. 1981. Cytology of the yeast life cycle. See Ref. 32, pp. 59–96
2. Catcheside, D. G. 1949. *The Genetics of Micro-Organisms*, London: Pitman. 223 pp.
3. Douglas, H. C., Hawthorne, D. C. 1964. Enzymatic expression and genetic linkage of genes controlling galactose utilization in *Saccharomyces*. *Genetics* 49:837–44
4. Dujon, B. 1981. Mitochondrial genetics and functions. See Ref. 32, pp. 505–635
5. Ephrussi, B. 1953. *Nucleo-Cytoplasmic Relations in Microorganisms*. Oxford, England: Clarendon
6. Esposito, R. E., Klapholz, S. 1981. Meiosis and ascospore development. See Ref. 32, pp. 211–87
7. Fogel, S., Mortimer, R. K., Lusnak, K. 1981. Mechanisms of meiotic gene conversion, or "wanderings on a foreign strand." See Ref. 32, pp. 289–339
8. Hawthorne, D. C. 1963. Directed mutation of the mating type alleles as an explanation of homothallism in yeast. *Proc. XI Intern. Congr. Genet.* 1:34–35
9. Hawthorne, D. C. 1963. A deletion in yeast and its bearing on the structure of the mating type locus. *Genetics* 48:1727–29
10. Hawthorne, D. C. 1976. UGA Mutations and UGA suppressors in yeast. *Biochimie* 58:179–82
11. Hawthorne, D. C., Mortimer, R. K. 1963. Supersuppressors in yeast. *Genetics* 48:617–20
12. Herskowitz, I., Oshima, Y. Control of cell type in *Saccharomyces cerevisiae:* mating type and mating-type interconversion. 1981. See Ref. 32, pp. 181–209
13. Holliday, R. 1984. A mechanism for gene conversion in fungi. *Genet. Res.* 5:282–304
14. Jones, E. W., Fink, G. R. 1982. Regulation of amino acid and nucleotide biosynthesis in yeast. See Ref. 33, pp. 181–299
15. Lindegren, C. C. 1953. Gene conversion in *Saccharomyces*. *J. Genet.* 51:625–37
16. Lindegren, C. C., Lindegren, G. 1943. Segregation, mutation, and copulation in *Saccharomyces cerevisiae*. *Ann. Mo. Bot. Gard.* 30:453–69
17. Lindegren, C. C., Lindegren, G. 1943. A new method for hybridizing yeast. *Proc. Natl. Acad. Sci. USA* 29:306–8
18. Lindegren, C. C., Lindegren, G. 1951.

Tetraploid *Saccharomyces*. *J. Gen. Microbiol.* 5:885–93
19. Meselson, M. S., Radding, C. M. 1975. A general model for genetic recombination. *Proc. Natl. Acad. Sci. USA* 72:358–61
20. Mitchell, M. B. 1955. Aberrant recombination of pyridoxine mutants of *Neurospora*. *Proc. Natl. Acad. Sci. USA* 41:215–20
21. Mortimer, R. K., Schild, D. 1985. Genetic map of *Saccharomyces cerevisiae*. *Microbiol. Rev.* 49:181–212
22. Oeser, H. 1962. Genetische Untersuchungen über das Paarungstypverhalten bei Saccharomyces und die Maltose-gene einiger untergaringer Bierhefen. *Arch. Mikrobiol.* 44:47–74
23. Petes, D. D., Fangman, W. L. 1972. Sedimentation properties of yeast chromosomal DNA. *Proc. Natl. Acad. Sci. USA* 69:1188–91
24. Pringle, J. R., Hartwell, L. H. 1981. The *Saccharomyces cerevisiae* cell cycle. See Ref. 32, pp. 97–142
25. Roman, H. 1956. Studies of gene mutation in *Saccharomyces*. *Cold Spring Harbor Symp. Quant. Biol.* 21:175–85
26. Roman, H., Fabre, F. 1983. Gene conversion and associated reciprocal recombination are separable events in vegetative cells of *Saccharomyces cerevisiae*. *Proc. Natl. Acad. Sci. USA* 80:6912–16
27. Roman, H., Hawthorne, D. C., Douglas, H. C. 1951. Polyploidy in yeast and its bearing on the occurrence of irregular genetic ratios. *Proc. Natl. Acad. Sci. USA* 37:79–84
28. Roman, H., Jacob, F. 1958. A comparison of spontaneous and ultraviolet-induced allelic recombination with reference to the recombination of outside markers. *Cold Spring Harbor Symp. Quant. Biol.* 23:155–60
29. Roman, H., Sands, S. M. 1953. Heterogeneity of clones of Saccharomyces derived from haploid ascospores. *Proc. Natl. Acad. Sci. USA* 39:171–79
30. Sherman, F. 1982. Suppression in the yeast *Saccharomyces cerevisiae*. See Ref. 33, pp. 463–86
31. Sherman, F., Roman, H. 1963. Evidence for two types of allelic recombination in yeast. *Genetics* 48:255–61
32. Strathern, J. N., Jones, E. W., Broach, J. R., eds. 1981. *The Molecular Biology of the Yeast* Saccharomyces: *Life Cycle and Inheritance*. Cold Spring Harbor, New York: Cold Spring Harbor Lab. 751 pp.
33. Strathern, J. N., Jones, E. W., Broach, J. R., eds. 1982. *The Molecular Biology of the Yeast* Saccharomyces: *Metabolism and Gene Expression*. Cold Spring Harbor, New York: Cold Spring Harbor Lab. 680 pp.
34. Slonimski, P., Borst, P., Attardi, G., eds. 1982. *Mitochondrial Genes*. Cold Spring Harbor, New York: Cold Spring Harbor Lab. 500 pp.
35. Szostak, J. W., Orr-Weaver, T. L., Rothstein, R. J., Stahl, F. W. 1983. The double-strand-break repair model for recombination. *Cell* 33:25–35
36. von Wettstein, D., Stenderup, A., Kielland-Brandt, M., Friis, J., eds. 1981. *Molecular Genetics in Yeast. Alfred Benzon Symposium 16*. Copenhagen: Munksgaard. 443 pp.
37. Wilkie, D. 1970. Analysis of mitochondrial drug resistance in *Saccharomyces cerevisiae*. *Symp. Soc. Exp. Biol.* 24:71–83. London: Cambridge Univ. Press
38. Winge, Ö. 1935. On haplophase and diplophase in some Saccharomycetes. *C. R. Trav. Lab. Carlsberg Ser. Physiol.* 21:77–109
39. Winge, Ö., Roberts, C. 1949. A gene for diploidization in yeasts. *C. R. Trav. Lab. Carlsberg Ser. Physiol.* 24:341–46
40. Winge, Ö., Roberts, C. 1952. The relation between the polymeric genes for maltose, raffinose, and sucrose fermentation in yeasts. *C. R. Trav. Lab. Carlsberg Ser. Physiol.* 25:141–73
41. Winkler, H. 1930. *Die Konversion der Gene*. Jena: Fischer

# *Saccharomyces* Studies 1950–1960

DONALD C. HAWTHORNE
*Department of Genetics*
*University of Washington*
*Seattle, Washington 98195*

When Carl Lindegren came to Seattle as a Walker-Ames Lecturer in 1947, I was still an undergraduate, a sophomore, and had not yet taken a course in bacteriology. Regrettably, Howard Douglas, assigned as my advisor when I chose bacteriology as my major course of studies, did not recommend that I attend Lindegren's graduate seminar series. As a junior in 1948, I took my first bacteriology course, one that was in the curriculum for the pre-dental students. I also took a course that year in genetics taught by Herschel Roman and then his one advanced course, cytogenetics, which had an accompanying laboratory section limited by the number of microscopes available to only eight students. Since microscopy had been a hobby of mine since my high school days when I had bought a second-hand Leitz microscope with proceeds from my yard-care route, I was often the first in the laboratory to display the chromosomes in the day's preparation, and this was duly noted by Roman, who directed the laboratory session himself.

The following year when Douglas and Roman were looking for a student to initiate their yeast genetics project "*Saccharomyces* Studies," I was chosen. It was December 1949 when I was given a transcript of Lindegren's lectures, a micromanipulator still in its packing case, and two haploid yeast strains, 93-3B *a f g* and 93-1C α *F G*. In the week or so before the Christmas holidays, I read the transcript, set up the de Fonbrune micromanipulator in a little room hidden behind the herbarium, had a dissection chamber made up to my specifications, and prepared the special media: Lindegren's presporulation medium, gypsum slants, and galactose fermentation tubes with gas traps.

I was told by Roman and Douglas to cross the haploid stocks, sporulate the diploids, dissect some asci, score the spore clones, and see if I could find a tetrad with a case of "gene conversion." With the start of classes that winter quarter, I began Lindegren's protocol: Half-milliliter suspensions of the freshly grown haploid parents were put into 1 ml of fresh YEP broth, the tube went on the shaker, and within a couple hours, there were aggregates of mating cells. I showed restraint and waited until the next morning before putting a few drops of the mating mixture over the presporulation slant. From a 2-day growth on the presporulation medium, a slurry of cells, about 1 ml, was layered on the top portion of the gypsum slant; 1 ml of dilute acetic acid (5%) was then pipetted onto the bottom half of the gypsum slant. I waited 2 days and sure enough found lots of asci.

The next step did not go as smoothly. According to the transcript protocol, the dissection of the asci was to be made on agar, with the ascus wall disrupted by "gently" rolling the ascus around under the flat surface of the perpendicular needle. I got nowhere with this procedure. I persisted for about half an hour, proceeding from a gentle rolling to rough treatment, only to bury the ascus in the agar. I realized that in order to disrupt the ascus wall, the dissection would have to be made on glass. The protocol presented by Øjvind Winge called for dissection on glass but with two needles: one very slender and wimpy to be placed over the ascus, lying between spores, and a stouter one to bring pressure against the first and force it down to the glass. Well I had no way of handling two needles and not enough nerve to ask for a second micromanipulator, so I tried to adapt the needles made to Lindegren's specifications, a perpendicular needle presenting a flat end about 20 µm across, to dissecting on glass.

The sporulating preparation had four-spored asci in three shapes: linear, pyramidal, and lozenge. I tried rolling the pyramidal- and lozenge-shaped asci under the flat end of the needle, only to crush a spore before disrupting the wall. With the linear asci, I tried another tactic. I noticed that many of the linear asci had a bit of ascus wall extending beyond the compact array of four spores. Thus, I found that I could bring up the edge of the needle over just the extended envelope and grind it off against the glass. Occasionally, the end spore would pop out, but usually the four spores stayed in the sleeve of the envelope. The next step was to bring the edge of the needle up against the spore at the closed end and carefully pinch a bit. It took only a few trials to get the right touch to squeeze out one spore at a time.

The dissection procedure seems rather laborious nowadays. The moist chamber held two 25-mm square coverslips. The inner coverslip had a big drop of water with the asci. A linear ascus at the periphery of the drop was dragged free of the water for dissection. When the envelope had been ruptured and a spore freed, a droplet of water was brought to the spore to allow pickup with surface tension to the needle. The spore was transferred to one of four drops of YEP agar on the outer coverslip. With the transfer of all four spores, the moist chamber was removed from the microscope, and the coverslip with the agar drops and isolated spores was placed on a deep-well culture slide and sealed with parawax. With 2 or 3 days at 30°C, the colonies from the spores could be picked. The spore viability was about 80% and I was saving asci with three surviving spore clones as well as the complete tetrads. The manipulations might have contributed to the spore lethality, but the dissections where a spore was crushed or even where the spore order was lost were aborted. In a week or so, I was dissecting three linear asci an hour, keeping the spores in order. This rate meant I was analyzing about ten tetrads per week. This was a part-time project since I was still completing my senior year curriculum.

The ability to isolate the spores in order presented a new problem. I was not at all sure I was going to find any asci showing gene conversion; but while I was dissecting asci and scoring tetrads, I could check to see if the linear spore array would be of use in mapping centromeres. With every week's scoring, I looked for a predominant noncrossover array. Lindegren's lecture transcript had a linkage map with four centromeres marked by *ad* (*adenine*), *g* (*galactose*), *pb* (*p-aminobenzoic acid*), *a*/α (*mating type*). In my cross, I was particularly following galactose fermentation since "*g*" was more tightly centromere-linked than "α." After 3 or 4 weeks, I had found no predominance of + + – – arrays for *galactose*, *flocculence*, or *mating type*. However, I did not know if I should be expecting a first-division segregation array as in *Neurospora crassa*. Linear asci in a yeast called *Saccharomycodes ludwigii* had been analyzed by Winge and Laustsen (1939) to the extent that non-sister nuclei were found in each half of the ascus as a consequence of parallel spindles at meiosis II.

When I had 80 asci scored with no predominant spore array to indicate the nuclear assortment, I figured it was time to do the tetrad analyses. It had been obvious that there was not linkage for any pair of the three genes, but the analyses gave pretty much 1 (parental ditype):4 (tetratype):1 (nonparental ditype) ratios for all three combinations of the genes. Something was wrong! I was expecting an excess of ditype tetrads for the *galactose* versus *mating-type* combination, the two centromere-linked genes. I was disappointed, but I wrote it up for my undergraduate research report. Douglas generously gave me an "A" and I graduated.

After a week's break, I was back working full time for the summer, still dissecting linear asci while looking for an ascus with a "gene conversion." I found it within a week, on Saturday morning, June 24, 1950. While scoring a batch of tetrads, I found that the four spore clones from ascus number 88 had not mated with either the *a* or α testers; three of the spore clones were fermenting galactose and one was not, and all four spore clones were scored as flocculent. I was lucky; professors in those days did not show up in the laboratory on Saturdays so I had until Monday to figure out for myself what was going on.

I went to a model airplane meeting at the Sand Point Naval Air Station that afternoon. After a couple of hours with the buzz of the modeler's gas engines in my ears, I figured that the spore clones could be *a*/α diploids and might be able to sporulate. I hurried back to the laboratory to look at the cells in the synthetic liquid medium used for the flocculence assay. The cells from the four spore clones of ascus 88 were larger and less tightly clumped than the cells from *F* segregants of sister tetrads. This was encouraging enough for me to inoculate presporulation slants. On Monday, I had the four spore clones on gypsum slants before I had to tell Professors Roman and Douglas about my discovery of a "gene conversion" ascus.

I continued with the dissections of the original cross for several more days, but by Wednesday, the four spore clones from ascus 88 had in turn given spores on gypsum. Dissections were started at once with these preparations, and in 3 weeks, I had scored three tetrads from each. As expected, all spore clones of ascus 88 were heterozygous for the mating-type alleles and for *F/f*. Two of the spore clones were heterozygous for *G/g* and the third galactose fermenter was *G/G*. Thus, as far as we could tell with these three markers, ascus 88 was derived from a tetraploid meiosis.

June 24, 1950, was a fateful day for me in another way; it was the day we got news of the invasion of South Korea. I was in the Naval Reserve and with the increasing American involvement, I could foresee that I was likely to be called up for active duty. I therefore pursued the analyses of the diploid spore clones from the tetraploid ascus earnestly and met my deadline, which was imposed by a Naval Reserve training cruise scheduled for the first half of August. When I got back from the training cruise, a letter to report to active duty was waiting. I had 4 days to pack up and take my belongings home. The laboratory had been tidied up prior to the training cruise.

I reported for active duty on August 25, 1950, and within 1 week, I was assigned to the crew of the LST 1083 "mothballed" at Tongue Point, the naval facility on the Columbia River a few miles from Astoria, Oregon. It took 3 weeks to get the ship seaworthy so that we could take it up the Washington coast and then into Puget Sound to the shipyard at Bremerton. We were there for another 2 weeks for further overhaul and outfittings. While there, my shipmates were impressed when a letter from Herschel Roman to the ship's captain managed to get me a workday liberty to go to Seattle to advise my successor, Marcia McDowell, on the intricacies of ascus dissection.

After 4 years of college, I didn't mind the call up to active duty in the Navy. I was the ship's leading electronics technician and had a reasonably satisfying job. The 13-month stint would prove to be a disciplined adventure for me. From Bremerton, our squadron of six refurbished LSTs sailed to San Diego, where we had 2 months of training in beaching, etc. On December 20, 1950, we headed for Korea with brief stops at Pearl Harbor, Hawaii, and Yokosuka, Japan. The LST 1083 operated in Korean or Japanese waters from late January to early June in 1951. We had seemingly mundane tasks: transferring POWs from Pusan down to the island of Koje Do where a prison camp had been set up and transporting engineering companies, first from the Pusan area up to Inchon and then from Okinawa to a spot up river from Kunsan where an airfield was to be built. We did all that was asked of us without ever coming under enemy fire. On June 8, 1951, we headed back to the United States; our destination was San Diego.

With the announcement of regulations on the demobilization of the reservists, in the spring of 1951, I could plan on being discharged October 11, 1951 after having completed a year's extension of my enlistment. I wrote to Herschel Roman to let him know. Just before we departed from Yokosuka, I received a letter (see facing page) from him saying that I would be expected for the fall quarter at the University of Washington. He also said that a paper on the tetraploid ascus had been published (Roman et al. 1951).

Our ship, following the great circle from Yokosuka, arrived at San Diego on June 28, 1951. I had 2 weeks leave to get home and to stop by Seattle to see Professors Roman and Douglas. I put in an application for graduate school and got a registration appointment set for as late as possible, October 2, 1951, hoping that I might be able to get discharged a week or so ahead of time. No such luck, but I did get a week's leave which enabled me to register and attend the first 3 days of classes. I rejoined the ship at San Francisco and then reported to the separation center at Treasure Island. I was released the 16th of October and was back in time for classes on the 17th. Meanwhile, Herschel Roman had sat in on the biochemistry lectures and took notes for me on the four lectures I missed.

When I got back into the laboratory, I was eager to continue the investigation of the linear spore array. We had received several strains with nutritional markers: 1914 *hi an* (*his1 trp2*) from Caroline Raut, 62P *tr ur* (*trp1 ura1*), and 67P *me* (*met1*) from Seymour Pomper. I set about to construct a diploid heterozygous for the genes of all these nutritional markers plus the three fermentation markers, galactose, melibiose, and sucrose. To make the crosses, I paired the parental cells with the micromanipulator. While working through the preliminary crosses, I also attempted to cross two homothallic strains (from Howard Douglas' collection) that showed a high percentage of linear asci. Spore pairings with strain 245 (*S. fragilis*) and 279 (at that time dubbed the "Volmer strain" but currently called *S. douglasii*) failed to give any matings in

UNIVERSITY OF WASHINGTON
DEPARTMENT OF BOTANY
SEATTLE 5

May 24, 1951

Donald C. Hawthorne, ET-2, 386-94-12
USS LST 1083
c/o F.P.O.
San Francisco, California

Dear Don:

Your very welcome letter arrived yesterday with the news that you will be here in the next couple of months. I was glad to see that you are finding life in the Navy an interesting experience but I was even more glad that you will return to the University in October. Even if you should arrive too late for Fall registration, it will be possible to make some credit arrangement that will avoid loss of time in your work towards a degree.

You are now an author. I am sending under separate cover a reprint of a paper that had as its germ the findings in ascus 88 (referred to as ascus 55 in the paper). The ascus turned out to have a much greater significance than we thought at the time of its discovery and you will find that the paper has blossomed out into an explanation of Lindegren's irregular ratios. My delay in writing you is partly due to my desire to wait until the reprints were available so that I could send you one and partly to the very bad habit I have of not clearing my desk.

The yeast work is progressing rather slowly. The mating-type picture is considerably more complex than we originally thought. We are getting, for example, spore clones that mate within the clone to give spores and the latter exhibit again, in many cases, the specific mating types that went into the original cross. In other words, these promiscuous mating types probably have both the a and alpha alleles and the question arises as to whether or not the promiscuous mating type depends on a certain ratio of these two alleles in the vegetative cells. We have also found some curious things in connection with the cytochrome deficient strains and it looks at present as if we have a strain that lacks cytochrome a and has both b and c present, in contrast with the petite strain which lacked both a and b. The genetic significance of these strains is still to be worked out.

I hope that you will be able to take charge of the yeast dissection work on your return. You will have two capable people to help you out. Stanley Sands has learned how to dissect and is doing a good job. Marcia McDowell has been doing most of the dissecting and is familiar with all of the techniques that are involved in making the various tests. With your good news at hand, I am looking forward to a very productive year. I am going to order another micromanipulator so there won't be any shortage of tools for the job.

Yours,

Herschel Roman

Herschel Roman

25 trials. The motivation for these trials was a photograph of a linear ascus from a hybrid of *S. cerevisiae* and *S. fragilis* with two round spores and two kidney-shaped spores: p17-9 in *The Yeast Cell: Its Genetics and Cytology* (Lindegren 1949). I also tried with *S. cerevisiae* as a parent, and eight spore pairings with *S. fragilis* failed, but there was one successful mating in the first set of four spore pairings with the Volmer strain. With this cross, there would be no spore morphology to follow, but I was hoping to enhance the percentage of linear asci while bringing in a maltose-negative allele into the breeding stock. The hybrid did give mostly linear asci but no viable spores from the dissection of 12 asci. By this time, I had constructed the multi-marked diploid that I wanted, so this infertility of the 279/*S. cerevisiae* hybrid was merely noted and not reexamined for some 25 years.

In the winter quarter of 1952, I started the dissection of a diploid that gave 40% linear asci and had a spore viability close to 90%. It took only a couple of weeks to see that I had found a marker giving a predominant tetrad array, + – + –. It was one of the two tryptophan mutations, but fortunately I could get a complete scoring of the tetrads, i.e., 2:2 segregations, on tryptophan-less medium supplemented with anthranilic acid. The tryptophan mutation giving the + – + – array was designated *trp1*, and the mutation satisfied by anthranilic acid was designated *trp2*.

By the summer of 1952, I had scored 74 complete tetrads, and the + – + – array for *TRP1* versus *trp1* occurred in about 95% of them. I was pretty sure that non-sister nuclei were alternating in the linear ascus, but there was no confirming evidence with *mating type* or the other markers. *Mating type* gave the + – + – array in a little more than 40% of the tetrads, and the other markers showed more or less random distributions for the + + – –, + – – +, – + + –, and + – + – arrays. At Roman's urging, I set out to get an independent determination of centromere linkage for *trp1* by making a tetrad analysis of a tetraploid with the genotype *trp1/trp1/TRP1/TRP1*. It took a little more than a year to construct that tetraploid and to get enough tetrads scored to show a centromere linkage of 2.5 map units for *trp1* (Hawthorne 1955a).

Following the discovery of the tetraploid ascus, in the summer of 1950, Herschel Roman, together with technicians Stanley Sands and Marcia McDowell, commenced the examination of the large, putative diploid, cells that appeared in the haploid cultures after prolonged cultivation. From both *a* and α haploid strains, they isolated large-celled clones lacking mating potential but capable of sporulation. Upon dissection and tetrad analysis, they were shown to be diploids heterozygous for *a*/α; thus, they could have arisen following a mating-type mutation. The large-celled clones retaining their original mating potential were unable to sporulate; however, they were intercrossed to give hybrids that did sporulate and when analyzed showed tetraploid segregations (Roman and Sands 1953; Roman et al. 1955).

I was assigned an ancillary project: to determine what role "illegitimate matings" might have in the formation of the diploid cells retaining their mating potential. It took a while to construct sets of strains with the nutritional markers *his1 trp2* and *ura1 met1* so I could select for rare matings on a minimal medium and not be concerned about back mutations to prototrophy in one of the parents. The *a* x *a* crosses gave rare zygotic clones that lacked mating potential but were able to sporulate. The dissection of asci from a sample of nine clones gave four viable spores per ascus and 2:2 segregations for *a* versus α. On the other hand, the α x α crosses gave higher frequencies of zygotic clones, about half of which retained their mating potential and did not sporulate. Seventeen zygotic clones lacking mating potential but able to sporulate were dissected: Ten gave four viable spores per ascus and 2:2 segregations for *a*/α and seven gave two viable spores per ascus that were always of α mating type. A lethal mutation of *a* was our hypothesis. It was another 10 years before the availability of markers distal to *mating type, thr4,* and *MAL2* made it possible to characterize the lethal event as a deletion of some 30 map units (Hawthorne 1963).

A limited number of α mating-type diploids derived from "illegitimate" α x α matings were used as a parent in the construction of tetraploids for the *trp1*-centromere linkage study. In these few instances, both α alleles appeared to be equivalent in that the tetraploid segregations for α/α/*a*/*a* were as expected; in particular, there were tetrads with four nonmating spores.

In the academic year 1952–1953, Herschel Roman obtained a Guggenheim Fellowship to work with Boris Ephrussi at the University of Paris. While there, he was involved in the discovery and description of "suppressive *petite*" strains (Ephrussi et al. 1955). Roman sent us two haploid strains from the Laboratoire de Génétique, 276/3br α *GAL SUC MAL ade2* and B15 *a gal1 gal2 SUC MAL,* along with his letters of encouragement.

The above two strains and two strains just received from Carl Lindegren, 1428 α *GAL SUC mal* α*mg* and 1426A *a gal1 SUC MAL* α*MG*, brought into our stocks enough polymeric and complementary fermentation genes to keep me busy for a couple of years in the sorting out of six maltose genes, five sucrose genes, and four α-methyl-glucoside genes (Hawthorne 1955b). The *ade2* mutation was a welcome acquisition to the breeding stock, but the red pigmentation was not exploited until Roman's return. Of immediate interest was the *gal1* mutation; with the tetrad analysis of the *GAL1/gal1* diploids, I again saw the predominance of the + – + – array in the linear asci.

With the acquisition of the *gal1* stock, Howard Douglas commenced his characterization of the genes controlling galactose utilization. Together with Frances Condie, he showed that the *gal2* cell was likely defective in a permease for galactose import, whereas *gal1* was postulated to be defective in an enzyme required for the conversion of galactose to glucose-6-phosphate (Douglas and Condie 1954). Studies on the kinetics of galactose adaptation correlated with those on gene dosage for *GAL1* or *GAL2* in sets of tetraploid yeasts (*GAL1/gal1/gal1/gal1*; *GAL1/GAL1/gal1/gal1*; etc.), and these data were used as a thesis project by Nels Nelson. After making whole-cell (Warburg) assays with both the *GAL1* and *GAL2* series, Nelson reexamined the *GAL1* series with cell-free preparations, testing for galacto-kinase (Nelson and Douglas 1963).

I wonder if I contributed to the 8-year gestation for this paper. After leaving Seattle, I sent off *gal1* and *gal2* stocks to Hugette de Robichon-Szulmajster, who characterized the *gal1* strain as lacking the galacto-kinase but having the galacto-transferase and the galacto-epimerase after induction by galactose (de Robichon-Szulmajster 1958).

In August of 1953, 2 years into my graduate studies, I presented a paper at the annual meeting of the American Society of Microbiology in San Francisco. David Perkins was the chairman of the session where I talked about the *TRP1/trp1* + – + – spore arrays in the linear asci of *Saccharomyces*. He brought to my attention the observations of Rizet and Engelmann (1949) on a single obligatory crossover between the centromere and the mating-type locus in *Podospora anserina*. Happily, I could respond with a little tetrad data for *gal1-trp1* that I had held back for just this sort of a comment.

It was at this meeting that I also met Carl Lindegren. He presented a paper on gene conversion in yeast in which α-methyl-glucoside fermentation was the character providing the most extensive data for irregular segregations. I spoke up in the discussion, stating that I too had been following the fermentation of α-methyl-glucoside but was interpreting the data as the segregation of both complementary and polymeric genes. Afterward, Lindegren graciously invited me to dinner at the congress hotel. I accepted, flattered but with some reluctance (my comrades were going off to explore the "Barbary Coast"). I was surprised to find I was dining just with Carl, not even Gertrude Lindegren joined us. Throughout the whole dinner, I was more or less interrogated on my findings that α-methyl-glucoside fermentation involved at least three genes; *αMG1 αMG2 αMG3*, *αMG1 αMG2 αmg2*, and *αmg1 αMG2 αMG3* were the fermenter genotypes, whereas any combination with *αmg2* gave a nonfermenter. Since I was polite and not at all argumentative, I did not convince him that this situation could apply to his observations (see facing page) (Lindegren 1953).

With the San Francisco meeting behind me, I had 2 more years of graduate school. My thesis project was to make a start on a linkage map for *Saccharomyces cerevisiae*. At this point, I had five centromeres marked by *trp1-gal3*, *his2*, *gal1*, *mating type*, and *film* (pellicle formation scored with the fermentation assays). Within 1 year, we obtained another centromere marker, *ade1*, in strain 99R from Edward Tatum. In addition to the *trp1-gal3* linkage, two other cases of gene-to-gene linkage that held up were *his1-trp2* and *gal1-MEL*.

In parallel with my tetrad analyses, Roman, Sands, and McDowell were establishing linkage groups with disomic stocks originating from triploid dissections. Four disomic stocks that had been backcrossed to purity were marked by *mating type*, *his1-trp2*, *ura1-met1*, and *SUC2*. Moreover, we knew that *trp1* and *gal2* were not located on these four disomic chromosomes. The few test crosses with the later acquisitions, *gal1*, *his2*, and *ade2*, were also negative. Thus, we could define at least six linkage groups by using two different approaches.

After Roman returned from Paris in September 1953, these triploid-aneuploid spore-disomic stock studies languished, since a new problem now held most of his attention. To obtain diploid clones, the red *ade2* stock had been put through a regimen of prolonged cultivation, i.e., six consecutive aerobic cultures, and then plated. What stood out was the relatively high frequency (5–10%) of large white colonies. These colonies were not *petites* and they were still adenine-dependent. When backcrossed to adenine prototrophs, they gave diploids that yielded tetrads with digenic ratios, 2:2, 1:3, and 0:4, for the adenine requirement. Roman obtained from these white colonies five adenine mutations distinguished by complementation tests (Roman 1956). Independent isolates of alleles within a complementation category could in turn be distinguished by demonstrating "heteroallelic recombination" when the diploid with the pair of alleles was put through sporulation (Roman 1957).

In September of 1955, I moved on to the California Institute of Technology at Pasadena with a National Science Foundation postdoctoral fellowship. I

SOUTHERN ILLINOIS UNIVERSITY
CARBONDALE, ILLINOIS

August 25, 1953

Dr. Donald Hawthorne
Department of Bacteriology
University of Washington
Seattle, Washington

Dear Don:

I am enclosing a marked copy of the article on gene conversion in Saccharomyces. The data seem to indicate that the three factor analysis is not involved. We have set up a table indicating the kinds of offspring which one would expect under the hypothesis which you outlined in our discussion. With three factors involved, three of the eight genotypes would be fermenters and five would be non fermenters. If each genotype is indicated by a number from 1 to 8, the triply heterozygous hybrid would produce 12 kinds of asci with the frequencies indicated. A total of 36 kinds of matings would be possible between the 8 different genotypes to produce asci with the frequencies indicated. Chi-square analysis indicates that the asci from family XV might fall into the triply heterozygous type but asci from families XI, XII and XVII are completely unexplicable on your hypothesis. The data from families VI, X, IV and XXVIII show a variance from expectation greater than would normally occur according to Chi-square analysis. Only those families with small numbers of individuals could be fitted into your scheme. I would be very much interested to hear more about the possibility of your hypothesis applying to this phenomenon. I realize how important it is to check every possibility in the advancement of such a revolutionary idea, but I do not believe that the scheme which you have proposed will work.

Hoping to hear from you soon, I am

Sincerely yours,

Carl C. Lindegren, Director
Biological Research Laboratory

CCL:gb

Encl.

had expected to work with Sterling Emerson, but he decided that 1955–1956 would be a good year to take a leave and serve as an advisor for the Division of Biology and Medicine of the Atomic Energy Commission, Washington, D.C. In his stead, George Beadle consented to act as my sponsor. Professor Beadle was the director of the Division of Biology and not involved with a research program at that time, so my contacts with him were the occasional departmental social events.

During the first week or so at Cal-Tech, I gave an evening seminar in a series aimed at the geneticists in the Biology Division. I talked on my multigenic explanation of α-methyl-glucoside fermentation in yeast and Lindegren's gene conversion interpretation of similar irregular ratios. I can recall a question from Max Delbrück: "Why don't you work with isogenic strains?" The only reply that I could muster was that I hadn't been isolating mutations but was working with variants found in nature.

It was just after this talk that my training at Cal-Tech began; I was invited to take part in a rather ritualized afternoon coffee break at the "Greasy Spoon." Usually it was Alfred Sturtevant who would summon me; he would stand in the doorway of my laboratory (Emerson's lab) in Kirchhoff Hall and stamp his foot if I had not seen him. (I never scheduled ascus dissections for mid-afternoon.) Norm Horowitz and Ed Lewis would be waiting at the stairway, and together we would go down to the ground floor and across the campus to the coffee shop. Sometimes Ray Owen would join us there, and during their stay at Cal-Tech, a couple of visitors, Dan Lewis and Howard Gest, were "regulars" at these sessions. These sessions would last from 30 to 45 minutes and were generally devoted to science. Sometimes a journal article was brought up, but most often it was the latest results. I got the floor at least every other week, and so after my accumulated observations from my graduate work had been exhausted, I was hard-pressed to provide new results. However, I recall one observation that intrigued them. To identify the maltose and sucrose genes I had found, I had been crossing my heterothallic strains to the homothallic tester strains from Øjvind Winge. The procedure of pairing four spores from the ascus of one parent with four spores from an ascus of the other gave, in one case, three successful matings. The tetrad analyses of the three hybrids gave 2:2 segregations for the homothallism gene *D* but also in each case the appearance of both *a* and α among the heterothallic spore clones. Thus, I could deduce that the homothallic strain was heterozygous for the mating-type alleles and that when these alleles were segregated away from *D*, they were stable.

In the fall of 1955, I was asked to contribute a paper to a special volume of the *Comptes Rendus des travaux des Laboratoire Carlsberg Série Physiologigue* to honor Ø. Winge on his 70th birthday. I worked over a chapter on the genes for galactose fermentation from my thesis (Hawthorne 1956). It was not very polished; it escaped Herschel Roman's red pencil.

In the spring of 1956, I got a telephone call from Bob Mortimer at Berkeley. He said that he had been up to Seattle and had been advised by Herschel Roman to contact me. Then he started talking about the exotic nutritional mutants that he had isolated, arginine, glutamine, isoleucine, lysine, etc. I had to interrupt to ask if he would like a little help in mapping a few of them. Well he said "Sure!" and the next week he came down to Pasadena for a day's visit and brought along a dozen mutants, all isolated in S288C. I had already been working with the white adenine mutants from Roman, but even with the doubling of the number of markers being followed in the crosses, linkages were hard to find. It took us almost 4 years before we could tie down enough genes linked to the centromeres to make a worthwhile map (Hawthorne and Mortimer 1960).

Herschel Roman received a Fulbright research award for the academic year 1956–1957 to work in Ephrussi's laboratory, which had relocated to Gif-sur-Yvette about 25 km south of Paris. I decided that I would try to go there as well. I had no success with my original project to find chromosome rearrangements that could be exploited for linkage studies in *Saccharomyces*. Therefore, when I sought a renewal of the National Science Foundation fellowship, I put in a different proposal to investigate suppressor genes in yeast. Suppressors of *trp1* had been isolated at Seattle (Parks and Douglas 1957), and I had made a couple of surprising observations while working with them: (1) The selection of a suppressor for *trp1* in a sucrose nonfermenter stock was accompanied by the restoration of sucrose fermentation and (2) crosses with some *trp1* suppressors would give a diploid in which the suppressor was expressed yet the suppressor was not recovered in the spores. Unfortunately, I had no explanation for these observations and thus I did not present a scheme that could be tested. I did not get a second year of the NSF fellowship.

I turned to the National Institutes of Health and worked up a proposal to follow through meiosis the hypothetical particles involved in adaptation to galactose, and perhaps melibiose and maltose. The kinetics of adaptation and deadaptation to galactose with *g-s* (*gal3*) strains inspired the postulation of substrate-dependent autocatalytic enzyme-forming particles (Spiegelman et al. 1950). Our own experience with *gal3* strains indicated that the adaptations to the fermentation of other sugars were also affected by this mutation. Both Norm Horowitz and Boris Ephrussi were pleased with my new proposal since they were very skeptical of the "plasmagene" hypothesis. From the NIH, I obtained a 2-year fellowship to go to France.

In January of 1957, I joined Ephrussi's group in the new Laboratoire de Génétique Physiologique du CNRS at Gif-sur-Yvette. I found that there were almost as many American visitors as French staff. Besides Herschel Roman, there were Dave Bonner and Kenneth Raper as visiting faculty and Roger Milkman, Bruce Bonner, and myself as postdocs. Resident researchers included Harriet Ephrussi-Taylor, Piotr Slonimski, and Yoshio Yotsuyanagi, and as candidates at various stages on the road to the Doctorate d'Etat were George Prevost, Pierre Galzy, Nicole Prudhomme, and Janine Beisson.

The laboratories had a small canteen that provided two sittings for lunch so there developed a tendency to eat together and then come back to the laboratory for a "Nes-Cafe." The Ephrussi's ate at home but would join us for coffee at the laboratory. This was an opportunity for Boris to ask what was new with the research projects. That accomplished, he would tell a story or an innocuous joke; no "dirty jokes" but he wasn't above telling ethnic jokes. I recall one appropriate for this volume that he told for the benefit of Professor Raper who had gotten NIH funding to come to Paris to find again some long lost myxomycetes, originally found in the Jardin des Plantes in 1890 but never seen again. So Professor Ephrussi told this story about Dutchmen and Frenchmen, supposedly set in 1950.

> A brash American graduate student (who must have been somewhat conversant with French) while touring the Jardin des Plantes wandered

> beyond the confines of the little zoo and came upon an antiquated laboratory where he found an old professor peering through a top fine-adjustment microscope. Sizing him up as a fellow microbiologist, he proudly introduced himself as a student of C.B. van Niel. "*Qui?*" was the professor's response. C.B. van Niel, he was a student of Kluyver. "*Qui?*" again from the professor. A.J. Kluyver, he was a student of Stelling-Dekker. "*Qui?*" N.M. Stelling-Dekker, she was a student of M.W. Beijerinck. "*Ah! Beijerinck, lui je l'ai très bien connu.*"

I was delighted with this story since I had met both C.B. van Niel and A.J. Kluyver. In 1950, van Niel was in Seattle for a week to give a series of seminars and I had him for 1 hour at the micromanipulator dissecting a linear ascus. In 1954, Kluyver visited Seattle for a day and I was allotted 10 minutes to tell him what I was doing about mapping the yeast chromosomes. I wanted to ask Ephrussi the identity of the American student, but I did not dare in case the story was a complete fabrication. I was thinking that if the date for this encounter was actually 10 years or so earlier, it could have been R.Y. Stanier.

The experiments to examine the particulate nature of enzyme-forming systems did not get very far. The presence of the sugars throughout the sporulation regimen markedly inhibited spore formation. With the *GAL3/gal3* diploid, the few four-spored asci dissected gave four rapid fermenters per ascus. However, this was not necessarily a demonstration for the persistence of a particle or even the galactose enzymes through meiosis, since the spores upon dissection had been placed on galactose agar drops so several tens of thousands cells had accumulated in the presence of galactose before the fermentation assays.

I resumed the linkage studies and with Ephrussi's encouragement, I included the nuclear *pet1* mutation from B-II, Boulangerie-II (Chen et al. 1950). The *pet1* mutation was found to be centromere-linked, proximal to the centromere marker *arg4*. Together with the centromere and the distal markers *thr1* and *CUP1*, these markers delimited four intervals where we could follow multiple exchanges and thus we could comment on crossing over and interference in the paper on centromere linkage (Hawthorne and Mortimer 1960).

One more accomplishment of my stay in France was the finalization of a paper on α-methylglucoside fermentation. This was realized under the watchful eye of Herschel Roman, with a coffee at a sidewalk table at La Coupole, Blvd. Montparnasse. For a month or so in the late spring of 1957, we would leave the laboratory in Gif about 3:30 to drive to Paris. (The Romans lived in a pension in the Montparnasse neighborhood and at that hour he could find a parking spot close by.) These revision sessions sometimes lasted an hour just to fine tune one or two paragraphs. The paper (Hawthorne 1958) came out 5 years after Lindegren's paper in the *Journal of Genetics* (Lindegren 1953) and so it appeared well past the time for arguments on the existence of "gene conversion."

I am sure Roman had returned to Ephrussi's laboratory in 1956–1957 with the intention of making a more detailed characterization of the suppressive *petites* as a primary project. However, Ephrussi had set aside this project for a postdoc, Roger Milkman, a novice in microbiological techniques, and this evidently exposed some incompatibilities in the philosophy on operations and procedures for the triad Ephrussi, Milkman, and Roman. By the time I showed up at Gif, Roman was no longer involved in the suppressive *petite* studies and had turned full time to studying mitotic recombination (Roman 1958). Hence, it had been Roman's custom to return to Paris at mid-afternoon long before I went along with him in May. He had been going to the Pasteur Institute to join François Jacob in experiments on the effects of UV irradiation on the induction of "heteroallelic recombination" at the *iso1* locus in yeast (Roman and Jacob 1959).

In September of 1958, I returned to Seattle. I had accepted an offer from Herschel Roman to join him in the Botany Department as a research instructor. Since I had NIH support, my only teaching responsibility was a cytogenetics laboratory course, given once a year in the spring quarter. For a while, I gave some help to Roman in the analyses of white/red-sectored colonies arising from mitotic recombination in the *ADE6 SUC1/ade6 MAL1 ade2/ade2* system. But since he had adequate hands with his technician, Agnes Towe, and a student helper, Beth Jones, I turned to writing up the centromere linkage studies.

In January of 1959, I went down to Berkeley for a week to get together with Bob Mortimer so that we could combine our contributions and rework them into a first draft for the mapping paper. It took another year of rewriting to get the paper by our internal critics and submit it to *Genetics* (Hawthorne and Mortimer 1960). By the time this was accomplished, we had found several new centromeres and gene-to-gene linkages so we were making plans for the next map.

My visit to Berkeley gave impetus to our adopting their techniques: ascus dissolution with snail gut juice and spore isolation on agar slabs. To cope with the fivefold increase in asci dissected, we moved over to replica plating for all the diagnostic tests. The changes were reinforced when Fred Sherman came up from Berkeley for a postdoc fellowship in Seattle. With Roman, he examined the kinetics of heteroallelic recombination in an *iso1-1/iso1-2* diploid during sporulation (Sherman and Roman 1963).

In making plans for my next cross to look at multiple exchanges and interference, I decided to contend with Max Delbrück's remark concerning isogenic strains. I figured the easiest way to get an isogenic cross would be to start with a haploid carrying 10 to 12 different nutritional markers and then start selecting for back mutations sequentially along two different lines. Once I got all the requirements covered by the two multi-revertant lines, I could cross them and select for a mating-type switch on a minimal plate. This would give me a diploid isogenic except for the various back mutations.

I was frustrated in building the isogenic stock in this fashion. To my dismay, I found that selecting for adenine prototrophy would give prototrophy to arginine, isoleucine, lysine, and tyrosine. The same sort of multi-prototrophic response was found for revertants selected on arginine. I abandoned my plans for an isogenic cross and took up the investigation of the suppressor genes, "supersuppressors," acting upon several different phenotypes.

While I had systematically selected for adenine (*ade2-1*) revertants and then for arginine (*arg4-2*) revertants, I heard from Bob Mortimer that he had stumbled upon a tryptophan (*trp5-2*) revertant that carried a suppressor capable of acting upon several other nutritional markers. When we compared the alleles suppressed, we figured we might be working with the same suppressors. Intercrosses with the first three suppressor stocks gave us one case of allelism, i.e., two different suppressor loci (Hawthorne and Mortimer 1963).

Although the supersuppressors were restoring to wild type many mutant phenotypes, they proved to be allele-specific, with about one quarter of the alleles tested responding to the suppressor. It was clear that the suppressible alleles formed a class with a common defect, but what sort of a defect could be remedied by the product of a single gene? Could it be a protein-folding problem that might be remedied by a change in the molarity of the cytosol? Could it be a mutation in the promoter that might affect RNA polymerase binding? Stop codons and nonsense mutations had not yet been invented in *Escherichia coli*, so we did not even think of them.

We set out to test for growth on hypertonic diagnostic media strains with suppressible alleles and as controls, strains with nonsuppressible alleles. Happily, we got a nice clear-cut result: None of the strains with the suppressible alleles responded to the osmotically enhanced media, but about one fifth of the nonsuppressible alleles definitely responded. With further testing, these "osmotic-remedial" mutants were generally found to be temperature sensitive with regard to prototrophy. With Jørgen Friis, we plotted molarity-temperature growth curves for representative mutants (Hawthorne and Friis 1964).

Another screen was applied to our mutant collection: Intra-allelic complementation. The suppressible alleles were noncomplementing; however, the osmotic-remedial alleles were nearly always complementing if intra-allelic complementation was observed at all for the locus in question.

It seemed that everybody was getting on the suppressor bandwagon. Down in Berkeley, Tom Manney was screening for suppressor susceptibility of the *trp5* alleles he was mapping by X-ray-induced recombination. He was finding suppressible alleles located throughout his fine-structure map of the *trp5* locus. Again the suppressible alleles were in the noncomplementing class in the intra-allelic complementation tests, with one exception for a suppressible allele mapping at the end of the map (Manney 1964). Back in Seattle, Satya Kakar, Herschel Roman's first Ph.D. student, encountered allele-specific suppressors during the course of his thesis project. He was looking at spontaneous mitotic revertants to isoleucine prototrophy in diploids heteroallelic for combinations of *iso1* alleles at four sites and found polarity for the homozygosity of the distal flanking marker, *trp2*. On this basis, the four sites were ordered with respect to the centromere, *his1*, and *trp2* (Kakar 1963a). Some spontaneous revertants in the *iso1-1*/*iso1-2* diploid were found to be due to an allele-specific suppressor acting on *iso1-1*. Several isolates of a suppressor acting on both *iso1-1* and *iso1-2* were also found. A search made for suppressors of *iso1-2* in UV-irradiated haploids gave allele-specific suppressors as well as a suppressor acting upon both *iso1-1* and *iso1-2*. Three tests for allelism indicated identity with the prior isolates of the "nonspecific" suppressor (Kakar 1963b). In retrospect, this nonspecific suppressor could well be one of the omnipotent suppressors acting upon UAA, UAG, and UGA nonsense alleles, since *iso1-1* and *iso1-2* have been characterized as ocher and amber alleles (Hawthorne and Leupold 1974; Liebman et al. 1976).

Meanwhile, Howard Douglas and his students down in microbiology were isolating and characterizing mutants involved in galactose fermentation. A mutant designated *gal4*, obtained from Y. Oshima (1959), was found to lack the three inducible enzymes galacto-kinase, galacto-transferase, and galacto-epimerase. The next mutation isolated was *gal5*, phosphoglucomutase (Douglas 1961). Then a mutation, *i-gal* (*gal80*), for the constitutive synthesis of the galactose enzymes was described by Douglas and Pelroy (1963). After Douglas had found mutations specifically for the galacto-transferase, *gal7*, and the galacto-epimerase, *gal10*, I got involved. The *gal1*, *gal7*, and *gal10* genes were tightly linked, with recombination more frequent via gene conversion than a reciprocal recombination (Douglas and Hawthorne 1964).

In September, 1959, the Genetics Department budded off from the Botany Department at the University of Washington. Besides Herschel Roman as chairman, the Botany Department contributed Dave Stadler, Norm Eaton, and myself. The other faculty were joint appointments with Medicine, Stan Gartler and Arno Motulsky, and Microbiology, Howard Douglas. As the Genetics Department, we attracted, even in our formative years, a number of visitors wishing to work with yeast: Sy Fogel, David Wilkie Jørgen Friis, Toshio Takahashi, Alberta Herman, Fritz Zimmermann, and Robin Holliday. I hope they were content with their experiences in Seattle and have warm memories of friends in Seattle and of Herschel Roman in particular.

## ACKNOWLEDGMENT

Most of the studies cited were supported by the National Institutes of Health, U.S. Public Health Service, grant E-328 "*Saccharomyces* Studies."

## REFERENCES

Chen, S., B. Ephrussi, and H. Hottinguer. 1950. Nature génétique des mutants à deficience respiratoire de la souche B-II de la levure de Boulangerie. *Heredity* **4:** 337–351.

de Robichon-Szulmajster, H. 1958. Induction of enzymes of the galactose pathway in mutants of *Saccharomyces cerevisiae. Science* **127:** 28–29.

Douglas, H.C. 1961. A mutation in *Saccharomyces* that affects phosphoglucomutase activity and galactose utilization. *Biochim. Biophys. Acta* **52:** 209–211.

Douglas, H.C. and F. Condie. 1954. The genetic control of galactose utilization in *Saccharomyces. J. Bacteriol.* **68:** 662–670.

Douglas, H.C. and D.C. Hawthorne. 1964. Enzymatic expression and genetic linkage of genes controlling galactose utilization in *Saccharomyces. Genetics* **49:** 837–844.

Douglas, H.C. and G. Pelroy. 1963. A gene controlling inducibility of the galactose pathway enzymes in *Saccharomyces. Biochim. Biophys. Acta* **68:** 155–156.

Ephrussi, B., H. Hottinguer, and H. Roman. 1955. Supressiveness: A new factor in the genetic determinism of the synthesis of respiratory enzymes in yeast. *Proc. Natl. Acad. Sci.* **41:** 1065–1071.

Hawthorne, D.C. 1955a. The use of linear asci for chromosome mapping in *Saccharomyces. Genetics* **40:** 511–518.

———. 1955b. "Chromosome mapping in *Saccharomyces.*" Ph.D. thesis, University of Washington, Seattle.

———. 1956. The genetics of galactose fermentation in *Saccharomyces* hybrids. *C.R. Trav. Lab. Carlsberg Ser. Physiol.* **26:** 149–160.

———. 1958. The genetics of α-methyl-glucoside fermentation in *Saccharomyces. Heredity* **12:** 273–284.

———. 1963. A deletion in yeast and its bearing on the structure of the mating type locus. *Genetics* **48:** 1727–1729.

Hawthorne, D.C. and J. Friis. 1964. Osmotic-remedial mutants. A new classification for nutritional mutants in yeast. *Genetics* **50:** 829–839.

Hawthorne, D.C. and U. Leupold. 1974. Suppressor mutations in yeast. *Curr. Top. Microbiol. Immunol.* **64:** 1–47.

Hawthorne, D.C. and R.K. Mortimer. 1960. Chromosome mapping in *Saccharomyces*: Centromere-linked genes. *Genetics* **45:** 1085–1110.

———. 1963. Super-suppressors in yeast. *Genetics* **48:** 617–620.
Kakar, S.N. 1963a. Allelic recombination and its relation to recombination of outside markers in yeast. *Genetics* **48:** 957–966.
———. 1963b. Suppressor mutations for the isoleucine locus in *Saccharomyces. Genetics* **48:** 967–979.
Liebman, S.W., F. Sherman, and J.W. Stewart. 1976. Isolation and characterization of amber suppressors in yeast. *Genetics* **82:** 251–272.
Lindegren, C.C. 1949. *The yeast cell: Its genetics and cytology*. Educational Publishers, St. Louis, Missouri.
———. 1953. Gene conversion in *Saccharomyces. J. Genet.* **51:** 625–637.
Manney, T.R. 1964. Action of a super-suppressor in yeast in relation to allelic mapping and complementation. *Genetics* **50:** 109–121.
Nelson, N.M. and H.C. Douglas. 1963. Gene dosage and galactose utilization by *Saccharomyces* tetraploids. *Genetics* **48:** 1585–1591.
Oshima, Y. 1959. Genetic studies in the fermentation of D-galactose by *Saccharomyces* yeast. *J. Ferment. Technol.* **37:** 419–425.
Parks, L.W. and H.C. Douglas. 1957. A genetic and biochemical analysis of mutants to tryptophan independence in *Saccharomyces. Genetics* **42:** 283–288.
Rizet, G. and C. Englemann. 1949. Contribution à l'étude génétique d'un ascomycete tétrasporé: *Podospora anserina. Rev. Cytol. Biol. Veg.* **11:** 202–301.
Roman, H. 1956. A system selective for mutations affecting the synthesis of adenine in yeast. *C.R. Trav. Lab. Carlsberg Ser. Physiol.* **26:** 299–314.
———. 1957. Studies of gene mutation in *Saccharomyces. Cold Spring Harbor Symp. Quant. Biol.* **21:** 175–183.
———. 1958. Sur les recombinaisons non réciproques chez *Saccharomyces cerevisiae* et sur les problèmes posés par ces phénomènes. *Ann. Genet.* **1:** 11–17.
Roman, H. and F. Jacob. 1959. A comparison of spontaneous and ultraviolet-induced allelic recombination with reference to the recombination of outside markers. *Cold Spring Harbor Symp. Quant. Biol.* **23:** 155–160.
Roman, H. and S.M. Sands. 1953. Heterogeneity of clones of *Saccharomyces* derived from haploid ascospores. *Proc. Natl. Acad. Sci.* **39:** 171–179.
Roman, H., D.C. Hawthorne, and H.C. Douglas. 1951. Polyploidy in yeast and its bearing on the occurrence of irregular genetic ratios. *Proc. Natl. Acad. Sci.* **37:** 79–84.
Roman, H., M.M. Phillips, and S.M. Sands. 1955. Studies of polyploid *Saccharomyces*. I. Tetraploid segregations. *Genetics* **40:** 546–561.
Sherman, F. and H. Roman. 1963. Evidence for two types of allelic recombination in yeast. *Genetics* **48:** 255–261.
Spiegelman, S., R.R. Sussman, and E. Pinska. 1950. On the cytoplasmic nature of "long-term adaptation" in yeast. *Proc. Natl. Acad. Sci.* **36:** 591–606.
Winge, Ø. and O. Laustsen. 1939. *Saccharomycodes ludwigii* Hansen, a balanced heterozygote. *C.R. Trav. Lab. Carlsberg Ser. Physiol.* **22:** 357–370.

# The Origin of *Schizosaccharomyces pombe* Genetics

URS LEUPOLD
*Niederweid*
*3088 Oberbütschel, Switzerland*

The present volume on the early days of yeast genetics is essentially devoted to the budding yeast *Saccharomyces cerevisiae*. Space limitations have allowed only Murdoch Mitchison, the founder of the cell biology, and myself, who started the genetics of *Schizosaccharomyces pombe*, to contribute chapters that concentrate on fission yeast. The result is that only the very beginnings of modern research on this yeast are treated and that most of the subsequent early developments in this important field must remain untold.

It was therefore with some hesitation that I consented to contribute the following brief description of my own love story in which, after some disappointments with a few other organisms, I finally chose *S. pombe* as the yeast of my life. Like any good love story, it limits itself to that first happy period of love that finally leads to marriage. The ensuing period which is always more difficult and serious but correspondingly also more interesting—not the least because it leads to children—is cunningly left out. I gladly leave it to Murdoch's and to my own scientific children and grandchildren, those that are now the leaders of present-day research on the cellular and molecular biology and genetics of *S. pombe*, to produce a similar book on the historical development of modern research on fission yeast alone, even if it will take some 15 years from now until they are ready to do so. After all, fission yeast genetics started 15 years later than the genetics of budding yeast. For the time being, interested readers will have to turn to a recent book, *The Molecular Biology of the Fission Yeast*, edited by A. Nasim, P. Young, and B.F. Johnson (1989). Although it lacks the personal and human touch of a book on the historical development of a research field written by the direct participants, it does give a good survey on the present state of fission yeast research—a field that has grown enormously since the early beginnings described here by Murdoch Mitchison and myself.

I have repeatedly been asked why I took up *S. pombe* as an organism for genetic study. The essence of the answer to this question can be stated in one single sentence: Øjvind Winge himself proposed it to me as a potentially useful organism for genetic work when I visited him for the first time in the fall of 1946 for 3 months.

I was then a young botany student only 23 years old but as a pupil of Hans Wanner and Ernst Hadorn, my teachers in general botany and zoology at the University of Zürich, I already had developed a deep interest in genetics. My ideas concerning a specific project in yeast genetics were still very vague. I simply hoped to be able to do some work on the recombination of genes in a yeast, and it was entirely Winge's merit to propose *S. pombe* for this purpose.

*The Early Days of Yeast Genetics*
 0-87969-378-9/93 $5 + .00

To my knowledge, he never had made any attempt to work with fission yeast himself. I do not remember with certainty his detailed comments on the virtues of the organism when he recommended it to me as a promising object for genetic studies. But as a geneticist, he was undoubtedly attracted not only by the regular formation of complete four-spored asci, but also by the linear arrangement of the four spores within the ascus. This promised the possibility of determining gene-centromere distances directly from the frequency of allele segregation, reflecting second-division segregations by the mixed order of the two parental allele types within the linear array of the four spores, a prediction that in fact turned out to be correct, since allele segregations giving exclusive first-division-type tetrads with an unmixed spore order were found.

It was my botany teacher Hans Wanner, a young plant physiologist who had recently become the head of the Institute of General Botany at the University of Zürich, who proposed the brilliant idea that I do my thesis work with Øjvind Winge in Denmark. Hans Wanner soon saw that my talents as a plant physiologist were rather unpromising, although a small piece of work in which I cooperated as his assistant did in fact lead to a publication by Wanner and myself, "Ueber die longitudinale Verteilung der Saccharaseaktivität in der Wurzelspitze," of *Vicia faba*, to be precise, which appeared in 1947 in the *Berichte der Schweizerischen Botanischen Gesellschaft*.

Having originally been trained as a plant cytogeneticist during his thesis work guided by his own botany teacher and predecessor as director of the Institute of General Botany, Alfred Ernst, Wanner was sufficiently interested in genetics to permit me to start a thesis project that met my own genetic interests. He first proposed that I repeat and carry further an earlier piece of work, published by H. Greis in 1941, to try to obtain monosexual mutants from the hermaphroditic fungus *Sordaria fimicola*, a beast growing on horse dung and producing both male and female organs in its wild-type form. When X-ray treatments failed to produce any mutants among 2580 survivors, each of which was individually and carefully tested, I gave up both collecting horse dung and proceeding with the project. Instead, I gladly accepted Wanner's proposal to go to Winge's laboratory to learn fungal genetics from a leading and experienced specialist in the field and to carry out my Ph.D. thesis in yeast genetics.

This magnanimous proposal of Hans Wanner provided me both with the yeast of my life and with the wife of my life, and the few months I spent in Copenhagen in 1946 kept me busy getting acquainted on a more intimate basis with both. I fell immediately in love with Rita Lavard, who was to become my wife during my second stay in Denmark in 1948, when I met her on the very first day of my arrival in Copenhagen in 1946. But the *S. pombe* strain of the Carlsberg collection that Winge gave me to study was rather disappointing. It turned out to be homothallic but quite infertile, and the germination rate of its spores did not exceed 50%. But at least I learned some of the basic methods, including micromanipulation, that were used in Winge's laboratory at that time.

It was only during the following year, 1947, after my return to Zürich that I found the true fission yeast strain of my life, i.e., *S. pombe* str. *liquefaciens*, which I obtained from the Centraalbureau voor Schimmelcultures in Delft, The Netherlands. This culture, which had originally been isolated by A. Osterwalder in Switzerland in 1924, turned out to be a mixture of three strains. Two were homothallic, one of very high fertility and one of only medium fertility and giving up to 90% and 40% spores, respectively (later called $h^{90}$ and $h^{40}$). The third was heterothallic and almost sterile when tested in pairwise combinations of the first three isolates obtained (because they were of the same heterothallic mating type, later called $h^+$, but this I only realized the following year in Copenhagen when I was able to isolate heterothallic strains of the opposite mating type, $h^-$, from the same culture). Unfortunately, my university studies in Zürich left me little time to proceed with these experiments in 1947, since I still had to pass an examination in the last of my side subjects, zoology, and it was only after my return to the Carlsberg Laboratory in the spring of 1948 for about a year that I was able to take up the work with the *liquefaciens* strain with full force.

From then on, my thesis work literally exploded, giving new results every week, and it was the fission yeast itself that took over the guidance of my work. The results are known and are described in my thesis which appeared in 1950 in the Comptes rendus du Laboratoire de Carlsberg (Leupold 1950). This was the happiest research period of my life. Winge gave me every freedom to work on my own and was quite satisfied to receive a personal report on the progress of my work every few weeks or even months only. But he was certainly interested in my progress reports and gave me the satisfaction a few years later of calling it "a beautiful piece of work." What contributed to the happiness of this research period was not only the freedom that I enjoyed in the friendly and helpful atmosphere of the laboratory of Winge and his coworkers, but also the complete lack of any pressure from a large international community of yeast geneticists, all interested in related topics and threatening to beat a lonesome and less well equipped research student in arriving at similar results, a situation very different from that of a Ph.D. student nowadays in the field of the molecular genetics and cell biology of yeasts. All I had to read in the field of budding yeast genetics were the two dozen papers written early on by Øjvind Winge and later on by Carl Lindegren and their coworkers between 1935 and 1949, a task that was easily overcome and did not need any computer search in the late 1940s! Now, every week brings an average of one to two new papers on fission yeast alone.

What followed after my second stay in Copenhagen in 1948/1949 were a few months in Zürich from the spring to the fall of 1949 when I wrote my thesis and passed my Ph.D. examination in botany, my main subject. After graduation, I spent 2 years, from the fall of 1949 to the fall of 1951, with Norman Horowitz at the California Institute of Technology in Pasadena, during which we contributed to the support of George Beadle's "One Gene–One Enzyme" hypothesis with an analysis of temperature-sensitive mutants isolated in *Escherichia coli*. I then spent 1 year, from the fall of 1951 to the fall of 1952, in Paris in the laboratory of Boris Ephrussi who asked me to introduce genetic markers from American strains into his French strains of *S. cerevisiae*, an undertaking that led to all sorts of aberrant segregations due to polygeny, polyploidy, and possibly even—horribile dictu!—to gene conversion. This experience certainly reinforced my wish to return as soon as possible to my work with *S. pombe* str. *liquefaciens*, which, with its monolithic origin, promised to provide a much higher degree of homogeneity in the genetic background of the mutants studied than *S. cerevisiae* does even up to the present day.

After returning to Hans Wanner's Botany Institute at the University of Zürich in the fall of 1952, I first completed the work on polyploid segregation in budding yeast, which I had started in Paris, before I went back again to the work with my beloved fission yeast strain *liquefaciens* of *S. pombe* to which I remained faithful for the rest of my life. My work on *S. pombe* continued first until 1963 at the University of Zürich, where I became an extraordinary Professor of Microbiology and head of the beginning of an independent Institute of Microbiology in 1961, and then from the spring of 1963 to the fall of 1986 at the University of Bern as a full Professor of Microbiology and (until 1980) head of a new Institute of General Microbiology. Since my retirement in the fall of 1986, I have enjoyed the life of a modest but happy farmer and *pombe* worker in Oberbütschel near Bern. Here, I have regained all the freedom for my work that I once had as a Ph.D. student in Winge's laboratory in Copenhagen.

## REFERENCES

Leupold, U. 1950. Die Vererbung von Homothallie und Heterothallie bei *Schizosaccharomyces pombe. C.R. Trav. Lab. Carlsberg. Ser. Physiol.* **24:** 381–480.

Nasim, A., P. Young, and B.F. Johnson, eds. 1989. *The molecular biology of the fission yeast*. Academic Press, New York.

# RECOMBINATION, GENE CONVERSION, MUTATION, AND REPAIR

S. Fogel (CSHL 1981)

A. Nasim, R.K. Mortimer, G. Magni, and F. Sherman (Osaka, Japan 1968)

C.A. Tobias (1983)

R.C. von Borstel (Louvain-la-Neuve, Belgium 1980)

# The Salad Days of Yeast Genetics and Meiotic Gene Conversion

SEYMOUR FOGEL
*University of California, College of Natural Resources*
*Department of Plant Biology, and California Agricultural Experiment Station*
*Berkeley, California 94720*

Conceived as a running narrative, the present account concerns the origin and development of my yeast program which began during 1958 at the University of Washington, following an invitation by Herschel Roman. I first met Herschel in 1941–1942 at the University of Missouri in Columbia. We were both graduate students of the late Louis J. Stadler. Herschel was about to complete his Ph.D. while John Laughnan and I were about to begin. Very likely, Herschel's invitation was prompted by the fact that we both found it difficult to pursue our maize research programs. Seattle's short growing season necessitated planting the experimental crops east of the Cascade Mountains, whereas vandalism and inadequate greenhouse facilities posed major obstacles to maize genetics in the New York area. I am under the impression that I was the first postdoctoral fellow in Seattle's training grant. This document is based purely on personal recollections and should not be construed as an accurate history. Mostly, it attempts to communicate the fellowship and collegiality that marked those early salad days when the worldwide yeast community comprised about 20 researchers.

During the fall of 1957, my scientific horizon was bleak indeed. My experimental garden for growing maize at Brooklyn College had been paved over to generate much needed parking space. In addition, my studies on X-ray-induced multiple alleles at the *dp* locus in *Drosophila melanogaster* had generated a significant disagreement with Herman J. Muller, the Nobel laureate, then located at the University of Indiana. This public interchange took place at the 1959 Genetics Society meeting in New York, where I reported that restorations to the wild-type as well as extreme *dp* alleles could be obtained as rare events from a hybrid between *dp-vortex/dp-oblique* x *dp*. Elof Carlson confirmed our findings a few years later. I had also pursued similar approaches with maize where, in collaboration with Louis J. Stadler, my Ph.D. mentor, I attempted to identify "recombinants" in addition to the reciprocal products of crossing-over between alleles at the *R* locus. Unlike the *Drosophila* study, in which I had utilized the closely linked markers *ed* and *cl* that flanked the central *dp* locus, the maize situation was severely limited by the unavailability of closely linked outside markers. I undertook to remedy this shortcoming by introducing the proximal marker "golden" and the distal cytological Knob 10. It remained for Marjorie Emmerling to consummate this work. Thus, in a screen of about 100,000 seedlings from the cross $R^g/r^{ch}$ x $r^g$, I identified several green seedlings that arose from kernels with colorless aleurone. Conceivably, these might represent mutational losses of either the seed color component from $R^g$ or the plant color component from $r^{ch}$. Alternatively, they might be taken as one of the possible reciprocal recombination products if crossing-over between the alleles had occurred. At best, such investigations were as tedious as they were laborious and slow. Melvin Green's successful study on crossing-over between lozenge alleles provided continuous impetus.

Equally slow, although fascinating, were the simultaneous studies involving highly mutable *R* alleles in maize whose behavior mimicked various aspects of Barbara McClintock's *AC,DS* system. These were spontaneous mutants found among the controls of a UV-induced mutation trial. Answers to these persistent, challenging questions and others are perhaps better understood if some previous history is provided. It is a pleasure here to record my deep appreciation for the encouragement and numerous suggestions accorded me by Barbara McClintock and Marcus Rhoades during this stressful and difficult period. Ruth Sager and Evelyn Witkin offered prudent counsel and much needed support. From time to time, Ruth visited my Brooklyn maize garden and Evelyn introduced me to the magic of bacterial and viral genetics. Thus, what were my scientific origins? How and where did I begin a career as a scientist? How did I go from maize to yeast?

I recall my first year in high school biology. I was sent to the principal's office to be disciplined because I was thought to be either impudent, disruptive, or probably dyslectic. In one lesson, we had studied the phases of cell division during which we committed to memory the chromosomal dance routine—interphase, prophase, metaphase, anaphase, telophase, and around again to interphase. Of course, at age 14, this was nothing more than mere verbal fluidity beyond our intellectual comprehension. Yet, the notion of a cycle was at least dimly perceived. It seemed clear that if the mitotic process continued cycle after cycle, organisms would grow to enormous proportions. Grow they did, but the limits to normal growth were self-evident. The next set of lesson plans centered on the endocrine system—an organ recital with hormonal contrapuntal accompaniments. To my naive mind, there was a causal connection between the action of the hormones and the initiation of a mitotic cycle. I thus posed this in the form of a question to my teacher. Summarily, I was sent to the principal's office. Of course, the major outlines concerning normal and neoplastic growth in relation to endocrine function are current foci in molecular biology.

During the winter quarter of 1958–1959, I undertook a program of becoming familiar with the yeast system as it was known in Herschel Roman's laboratory, then a part of the University of Washington's Botany Department. Herschel's office was flanked by two small rooms. One housed the microscopes and micromanipulators, and the other was inhabited by Don Hawthorne and a newly arrived postdoc freshly minted by Robert K. Mortimer's Donner laboratory in Berkeley, California. This was Fred Sherman. The subsequent, extensive exploitations of the yeast system by these two perceptive individuals are sufficiently well known to require no further comment here. The laboratory also housed Agnes Towe, Roman's research associate and general laboratory manager. Perhaps more than any other single individual, she instructed me in the rudiments of microbiological procedure then in vogue for handling yeast cultures. Satya Kakar was then a senior graduate student.

Later, he became a prominent educator at Haryana University in India. Satya completed the *dramatis personae* of North America's premier yeast establishment.

Under Agnes Towe's tutilege, I soon learned the essentials—dilution plating, toothpick tranfers, replica plating, drop overlay, complementation testing, zygote isolation, preparation of master plates for replica printing via velveteen transfer, etc. However, when I became curious about the various types of media that were utilized in these operations, I was led to a kitchen located on the floor above the laboratory. There I became acquainted with a singularly impressive human dynamo. A complete solo performer—weighing, stirring, cooking, autoclaving, pouring hot media into previously sterilized petri plates etc.—a complete medium production facility all under the cheerful, industrious control of a young undergraduate chemistry major who had only the faintest notions of what became of her labors. After luring her into the research laboratory and introducing her to the science of genetics, Elizabeth W. Jones was well on the way to being permanently hooked into the yeast enterprise.

I told Beth Jones that if her genetical interests continued, I would invite her to work in my New York laboratory over the summer when she graduated. Needless to say, it was a highly productive summer during which Beth lived both as a member of my family and as a young colleague during the day in the laboratory. Her interactions with Don Hurst were notable. Don quickly recognized her extraordinary analytical capabilities and her determined tenacity in problem solving. In his own inimitable way, he posed a new problem for her consideration each day of the summer. Subsequently, Beth returned to Seattle where she became a graduate student and completed her Ph.D. centered on the fine structure of *ade3* with Roman. Her ongoing and comprehensive contributions to yeast genetics and its molecular and cell biology are widely recognized at the international level. Currently a full professor at Carnegie-Mellon, she shoulders a wide variety of professional chores in addition to a vigorous teaching and research program.

By 1960, my family (Mary Jo Scott and son Miles Bradley) and I again found ourselves enroute to the west coast. On this occasion, I took a leave of absence from Brooklyn College to substitute for David Perkins who was about to enjoy a sabbatical leave. Ordinarily, he gave the undergraduate Genetics Course at Stanford. Since I had been teaching a similar course at Brooklyn for some 12 years, presenting only a lecture course at Stanford without the added effort required for organizing or supervising laboratory sections, I found abundant time to take advantage of the generosity advanced by Charles Yanofsky, who was then dissecting the tryptophan synthetase system in *Escherichia coli.* Along with learning how to process large volumes of *E. coli* cultures, isolating the desired protein, subsequently digesting it, and subjecting the digests to peptide analysis, I enjoyed frequent interactions with Yanofsky's talented postdocs, e.g., Ulf Henning, and several others. I recall also that Edward Penhoet (now professor of biochemistry at Berkeley and chief executive officer of Chiron Inc.) was among the outstanding students of the large Stanford class.

Incidentally, the Stanford undergraduates contributed meaningfully to my own professional development as an educator. From them I learned, under circumstances embarrassing to me, that proctoring an examination by a faculty member, especially a final examination, was tantamount to rudeness as well as insulting. This was my introduction to the honor system and since that memorable day, I regarded proctored exams as an abomination devoutly to be avoided. This experience served me well over the ensuing decades. It redefined for me the nature of the student-faculty relationship—a mutual, reciprocal learning experience wherein the teacher could learn at least as much from the student.

I acquired numerous insights and approaches from the Yanofsky group, particularly when the PI, postdocs, graduate students, and selected undergraduates jouneyed off northward for Saturday afternoon luncheons at a well-known Redwood City hofbrau that featured such delicacies as corned beef and pastrami, fare strange to the western palate. The initial Stanford experience was enhanced considerably by two individuals, namely, David Perkins, who provided me with a spacious comfortable office that encompassed a magnificent reference library and additionally a baby stroller distinguished by an overly long handle. The function of this seemingly ungainly apparatus became immediately evident when I took my infant son Miles Bradley for his regular afternoon stroll. My wife Mary Jo Scott and I celebrated his first birthday at the fountain and entrance to Stanford's medical school. Twenty years later, I returned to Stanford for retraining in molecular genetics and recombinant DNA protocols.

Another major contribution to my survival was provided by Stuart Brody, then a Yanofsky graduate student (now professor of biology at the University of California, San Diego), who also functioned as my teaching assistant. Surprisingly, I did not understand a teaching assistant's functions. Having been seasoned at Queens and Brooklyn Colleges, both autonomous institutions of the New York municipality, I was accustomed to performing all duties associated with the teaching mission without the benefit of assistance of any sort. It was not until 1968 when I came to the University of California at Berkeley, as chairman of genetics, that I was reintroduced to the teaching and research assistant functions as executed by graduate students. As I indicated earlier, these associations were richly rewarding as well as illuminating since they invariably provided insights into the personal doubts, needs, aspirations, and professional goals that graduate students maintained.

Finally, at Stanford, I was assigned to a laboratory area for my personal use. This was the former laboratory space utilized by George Beadle and Edward Tatum. It was a long narrow room with a single bare bulb and a rough plank across the distal end that reached to the rough foundation stones of the enclave. Yeast work was impossible there since the atmosphere was saturated with a legacy of *Neurospora*.

When I returned to Brooklyn College, I discovered that I had a new colleague, Donald D. Hurst. Trained in cilliate genetics with David Nanney, Hurst revealed himself to be a brilliant, indefatigable scientist and colleague over the next 10 years. His sense of geographical placement was astounding. Whenever we arrived at an airport, he could easily pilot a rental car to our destination without reference to a map or guide of any kind. Similarly, he could play exhibition chess blindfolded and defeat a room full of skilled opponents seemingly without effort. His ability to assimilate vast amounts of data and to recall the pedigrees of unique strains rivaled Louis J. Stadler's prodigious gifts in the corresponding arenas. Furthermore, and miraculously,

I had received a promotion to the rank of associate professor with a much needed raise in salary for I was now the head of a newly formed family. Our second son, Tracy, arrived in 1963.

The subsequent sections concern the development of yeast genetics at Brooklyn College, where laboratory-based research was an exception to the institution's primary mission as a teaching institution. Literary criticism, scholarly library research, painting, musical composition, classical studies in Latin and Greek, personality and psychological development, and other humanistically inclined disciplines such as philosophy were politically correct for the period, but hands-on science was virtually unknown. Equipment, facilities, or even space for scientific research could not be found under any circumstances. This grievous situation was ameliorated by the thoughtful consideration, cooperative understanding, and even allocation of funds by the college's distinguished president Harry D. Gidionse and his perceptive "chief of staff" Arthur Hilliary. Together, they permitted us to liberate a rarely used ladies restroom and remodel it into a functional laboratory. This was achieved by the vigorous cooperation of the college's skilled mechanics—masons, carpenters, electricians, plumbers, painters, etc.—who sensed that they were contributing an added, new dimension to the college's mission. In subsequent years, visitors frequently admired the laboratory's tiled walls and elegant terrazzo floors polished to a high shine.

A few years later, the Biology Department allowed us to recruit two new yeast people—Carl Beam from Yale, a specialist in radiobiology relative to the yeast budding cycle, and Norman Eaton from the University of Washington. Norman had accumulated considerable expertise in yeast enzymology and its genetic control and regulation. In short order, they were outfitted with adjacent laboratory space equivalent to our own. We shared dishwashing facilities, autoclaves, incubators, refrigerators, etc., and some talented graduate students who completed their Ph.D.s under our combined watchful eyes. Included in this pioneer group were Judith Wildenberg, Julie Silver, Nasim Kahn, Timothy Perper, Richard Needleman, Ann Mazlin, Isreal Benathan, and Michael Waxman. Fritz Zimmerman and David Wilkie joined our group at a later date.

Apart from the graduate students, we also attracted a coterie of unusual, intellectually eager, hungry, talented undergraduates. Among the most outstanding were Michael S. Esposito, Rochelle Easton, Wylie Burke, and Jay Tischfield. Mike devoted himself in a characteristically methodical and ecumenical fashion to isolating pure suspensions of spores, zygotes, or unbudded cells via a plethora of jury-rigged countercurrent distribution plumbing systems. Some arrangements were transcendental in the extreme. Rochelle, or Shelly, later married to Mike Esposito, concentrated on preparing an endless series of master plates for replica plating the following day.

Because they exhibited the imagination and curiosity rarely encountered among undergraduates, they soon found themselves sold into slavery on Seattle's pirate ship of graduate study captained by Herschel Roman. Wylie Burke, a professional ballet dancer, was a far-eastern studies major at Brooklyn who had the audacity to attain a perfect score on the general biology examinations. As a reward, Don and I shanghaied her into becoming a member of our growing laboratory family. But before long, she too was banished to Seattle as a graduate student on the assumption that parochial New Yorkers would flourish in proportion to the square of the distance that they were removed from their birthplaces. After completing a Ph.D. at Seattle, Wylie completed an M.D. degree there and eventually emerged as head of Human Genetics at Seattle. Jay Tischfield, a brilliant undergraduate distinguished by his ATP level, was shipped off to Frank Ruddle at Yale where he completed a memorable Ph.D. thesis. Now he is chair of the Human Genetics group at the University of Indiana. By this time, Roman had made his prophetically wise decision. He declined the attractive offer for a high-level appointment at the California Institute of Technology, perhaps as chairman, and established the first Department of Genetics in a liberal arts matrix at the University of Washington. For the most part, genetics departments in the United States were typically associated with Colleges of Agriculture as plant and animal breeding units. Accordingly, most failed to emerge as broadly constituted academic disciplines.

It was against this background that I became acutely aware of the power of tetrad analysis, perhaps mostly from the subtle suggestions emanating from the constitutively taciturn Don Hawthorne. During my early days in Seattle, Agnes Towe patiently taught me the conventional routine of ascus dissection. After 3 weeks of extremely discouraging and fruitless labors, I concluded, despite Øjvind Winge's assertion to the contrary, that finding, isolating, dissecting, and recovering a fully viable tetrad was beyond my abilities. Poorly sporulating cultures with few complete asci were commonplace in Seattle. Once an ascus was sighted within the hanging droplet carried on a coverslip, it was crushed against the coverslip with a glass needle carried in the hydraulically driven micromanipulator. Each spore was then transferred to a droplet of nutrient agar on the same coverslip. The entire assembly was incubated in a moist chamber. To recover a fully viable, complete ascus was a rare event, and I was convinced that yeast genetics based in tetrad analysis was impossible for me. In sum, 3 weeks of intensive activity with the micromanipulator yielded little in the way of meaningful results.

Maize cytogenetics and *Drosophila* salivary analysis, despite their inherent limitations, were relatively simple. It was during this depressingly stressful period that Don Hawthorne came to my rescue. One day, he invited me to peer through his microscope. A rather large slab of agar was mounted on a large glass slide inverted over a moist chamber—no fragile coverslip. The field of view was populated with nothing but four-spored asci, liberated, in some mystical way, each from its parental ascus cell wall. Immediately after lunch, I prevailed upon Don to share with me how he achieved this miracle, and I was all the more excited after he convinced me that every dissected ascus would yield four viable ascosporal colonies upon incubation. Given the same ascal suspension, I rapidly discovered that tetrad dissection no longer posed any serious difficulty. Finally, Don showed me a tiny volumetric flask that contained the filter-sterilized contents of a few snail or slug crops. He offered me a bacteriological loopful of this magical juice and very soon thereafter I was able to locate, dissect, and recover fully viable tetrads even from poorly sporulated cultures.

When I returned to Brooklyn, I vowed to make ascus dissection a readily available technique. In this connection, I was fortunate to be introduced to a master machinist and toolmaker, Desiré Light. He had an extensive background as an instrumentation specialist in the torpedo section of the French navy and worked as a professional troubleshooter on punch-card-programmed knitting machines. He fabricated a large mechanical stage and a manipulator that embodied our numerous changes superimposed on a prototype device originated by Brower. The resulting product was smaller, simpler, and readily affixed to the large mechanical stage that, in turn, was bolted to the microscope fork. This arrangement obviated the need for heavy vibration-dampened tables. The unit has been in continuous use and to date is completely functional. As our program expanded, aided by a grant from the National Institutes of Health, we sought to acquire another manipulator. In this instance, we were fortunate to locate Jack Harris, a skilled aeronautical engineer and talented scientific toolmaker. Previously, he had designed and fabricated numerous manipulators for the microsurgery group at New York University headed by Professor Kopac, who continued the pioneering intracellular microsurgical explorations inaugurated by Professor Chambers and his group. Harris fabricated a special stage according to my sketches. The X and Y drives were actuated by spring-loaded levers, a clutch, and a pair of Geneva mechanisms. The arrangement allowed us to move the stage in any direction and then return to the optical center of any previously given position—an essential requirement for our projected study on the recovery and analysis of tetrads that contained a prototrophic spore. William Scherr and Son of Brooklyn, New York, also improved on the Harris apparatus and devised a special turntable with an extra heavy rim, a device extremely useful for streaking out large numbers of cultures. The last of our modified mechanical dissection devices were produced by Wallace D. Lawrence of Precision Machine, in Hayward, California. Mr. Lawrence retired a few years ago. The ultimate, motor-driven, computer-controlled, dedicated micromanipulator was fashioned by Carl Singer of Singer Instruments, Roadwater, England. The instrument incorporates much of our experience in isolating yeast cells in various stages of the life cycle. The firm also supplies preformed dissection needles of excellent quality.

Finally, it is appropriate to note that while at Ron Davis's laboratory, where I learned the rudiments of hands-on recombinant DNA technology, I conveyed to most of the graduate students and postdoctoral fellows the elements of ascus dissection, zygote isolation, pedigree analysis, etc. The need for readily available additional manipulation equipment led me to design and construct a simplified version of the manipulator and stage that bypassed the requirement for highly accurate precision ground screws encased in finely lapped split-nuts—overall a rugged but precise arrangement. Instead, starting with a commercially available rack and pinion mechanical stage actuated by a coaxial *XY* control, I installed a series of spring-loaded detent stop mechanisms along the respective *X* and *Y* axes. Given the stage, a drill press, and the standard commercial detents, virtually any shop could fabricate the stage at a nominal cost. A connection clamp that conjoined the manipulator to the microscope and simultaneously brought the stage and manipulator into a working alignment completed the assembly. Working drawings for manufacturing the detented stage, the connector clamp, the micromanipulator, and the dissection plate holder have been distributed worldwide to scores of yeast researchers, at zero cost since 1979. I note with approval that since 1992, Carl Zeiss Inc. has marketed a version of this detented stage coupled to a version of the Sherman manipulator.

Exactly how we came to the procedure for isolating tetrads that manifested the outcome of a "recombinant event" between heteroalleles flanked by linked outside markers can be recorded. On a visit to Yale University, where I presented a seminar at the invitation of Norman Giles and Mary Case concerning our preliminary studies of gene conversion in yeast, my former student Ben Zion Dorfman prevailed upon me to demonstrate ascus dissection to Gerald R. Fink, a young, exceptionally brilliant graduate student. With a view toward recruiting him at Brooklyn, I invited Fink to drive back to New York with me and join my family for dinner featuring freshly caught bluefish. At some point, Fink asked whether or not it might be possible to identify conversional events by some selective tactic. His query precipitated a stream of trials aimed at recovering tetrads containing a budding spore after they had been challanged by incubation on an appropriate selective medium. Thus, in diploids carrying a pair of distinguishable heteroalleles at *his1* and the flanking markers *hom3* and *arg6*, we could scan digested, sporulated cultures for asci that carried three normal spores and a "recombinant" spore that had budded off at least one cell. The wall characteristics of the mother spore and the bud were readily apparent. Our *opus magnus* was provoked by Fink's fertile query.

Initially, Don Hurst and I restricted our efforts to the analysis of mitotic events in heteroallelic diploids. However, Herschel Roman soon suggested during a visit to the Brooklyn laboratory that we would do well to pursue this effort and to supplement it with a tetrad analysis approach. At the time, Herschel was regarded as a visionary guru. Accordingly, one of our undergraduates had constructed a conspicuous, publicly displayed poster that carried the message "Roman is never wrong." Some time after Roman departed, we found that he had modified the message by replacing "never" with "always." It is entirely appropriate to note here that Herschel was scientifically generous, in the extreme, and were it not for his unflagging enthusiasm and encouragement, yeast genetics would probably not have enjoyed the spectacular achievements and growth that it manifests today. Persistent, tireless, and with constant devotion, he along with Allan Bevan, Don Williamson, and Brian Cox from England, Piotr Slonimski from France, and Giovanni Magni from Italy shaped the development of the international yeast community. Today, yeast is a mainstay of the biotechnology industry, and the yeast community includes about 2500 individuals. In a single generation, the field expanded 100-fold.

Don Hurst and I convened the first conclave of yeast researchers, who happened to be attending a summer meeting of the Genetics Society of America (ca. 1960 or 1961). Among the attendees were Robert C. von Borstel then at the Oak Ridge Laboratory, Urs Leupold from Switzerland, and several distraught individuals from the Lindegren Carbondale establishment—David Pittman, Ernie Shult, Maurice Ogur (my former Brooklyn College research partner), and perhaps a few others. We resolved the anxiety of the Carbondale group by rapidly imbibing all of their opened and unopened liquid refreshments that had their historical origins via intimate association with our cherished organism, *Saccharomyces cerevisiae*. The gift of fruit and the spirits that enveloped them soon led to a resolution that set the next meeting at Carbondale, Illinois.

The first international yeast meeting at Carbondale, Illinois, chaired in a masterful manner by Giovanni Magni, was graciously and cordially hosted by the entire Lindegren group. Don Hawthorne and Robert Mortimer held forth on the nature of linkage, its detection, and estimation via tetrad analysis. Key among the various agreements was an extensive plan to exchange various strains with a view toward generating a uniform nomenclature adaptable to all yeast laboratories. Unfortunately, Roman did not attend this historic attempt to resolve interlaboratory differences. In turn, these differences tended to perpetuate the widely held, simmering undercurrent of perceived disagreement concerning the validity and generality of yeast data. Most general geneticists took the position that when yeast researchers presented a consistent, understandable body of data, their reports would then receive appropriate attention and approbation.

We now consider the substantive findings reported in our analysis of selected "recombinant" tetrads referred to in previous paragraphs (see Fogel and Hurst 1967). We considered that critical evaluation of the reciprocal and nonreciprocal aspects of recombination would require a genetic system with the following attributes: (1) All meiotic products must be available as viable, isolatable tetrads, (2) closely linked markers flanking the locus within which recombination is being studied must be present (the outside markers allow assessment of chromosome and chromatid involvement in crossing-over), and (3) a substantial sample size of recombinant tetrads must be recoverable.

Three diploids satisfying the above criteria were synthesized and, in all, we isolated and analyzed some 1109 recombinant tetrads. In virtually all instances, the prototrophic spore could be recognized by the presence of its associated budded derivatives. Placed at the head of the dissected spore array, as predicted, it invariably generated a wild-type ascosporal colony as regards the histidine requirement. The resultant data were then analyzed, first according to the classical theory of multiple exchanges which supposed that the recombinant wild-type spore arose exclusively via a reciprocal exchange mechanism, and second in terms of newer schemes proposed by Robin Holliday in 1964 and by Whitehouse and Hastings in 1965. Distinctive to such models was a postulated DNA dissociation cycle, hybrid DNA or heteroduplex formation with mismatched base pairs at the included mutant sites, and finally, enzymatic correction of the mispairings or heterologies. At the time, such notions were not warmly received. Generally speaking, gene conversion, at this time, was regarded as a biological oddity peculiar to yeasts and other lower eukaryotes such as mosses and liverworts. Today, it is a well-documented phenomenon in numerous life-forms.

The following are our principal findings: (1) Among the recombinant tetrads, only a minority or 10% arose consequent to a reciprocal exchange between the parental alleles, and among these chromosomes, interference was evident in distally marked regions. (2) 90% of the large sample represented nonreciprocal events at either the proximal or distal mutant sites. Thus, allelic recombination was, for the most part, a nonreciprocal recombinational event distinct from crossing-over. (3) One especially striking observation was that proximal allele convertants were, on average, six to ten times more frequent than distal allele convertants. A detailed study of unselected tetrads marked by three or four heterozygous sites within the *arg4* locus led to similar findings regarding polarity. These later studies by Fogel et al. (1971) and also Fogel and Mortimer (1971) reported that the high-conversion polarity region was situated close to the transcription initiation site. In the main, this conclusion was facilitated by exploiting Mortimer's complementing nonsense mutants localized in the low conversion frequency region of the polarity gradient and close to a terminus in his genetic fine-structure map (see also Hurst et al. 1972).

Four distinct regions were defined by five heterozygous sites situated on the right arm of the fifth linkage group, and the distributions of single and multiple crossovers were analyzed in detail, particularly with respect to chromosome and chromatid interference. The data derived from the 1100 *his1* "recombinant" tetrads were compared to the corresponding values obtained from 3030 unselected control asci. Among the recombinant tetrads, it was clear that the prototrophic strand was involved in about 95% of the associated exchanges that were positioned within the immediate proximal or distal intervals and only ramdomly in the exchanges localized in a more distant marked interval. Nevertheless, 1:2:1 ratios were obtained for two-, three-, and four-strand double exchanges in both the control and selected tetrads. Hence, although no evidence of chromatid interference was apparent, the frequency of double exchanges and strand involvement among control and recombinant tetrads was, respectively, 36:74:29 versus 33:68:28; however, the recombinant sample size was only 1081 compared to the control 3030 sample, a threefold difference (see also Mortimer and Fogel 1974).

Our first graduate student Judith Wildenberg conducted a thesis investigation utilizing the same diploid strains. She successfully identified and analyzed rare budding prototrophic cells among initially unbudded vegetative mitotic cells after X-irradition and challenge on the selective histidine-deficient medium. When she brought the news of our findings to Harvard where she was about to initiate a postdoctoral fellowship with Matthew Meselson, he responded with a request to examine our data in the form of the original laboratory notebooks. Of course, we were deeply pleased by this display of interest by a famous, acclaimed geneticist who had with Franklin Stahl established a major milestone in the recombination field. Within 1 week, Meselson arrived. He cornered himself quietly and inconspicuously with all our primary records over several days much in the manner of a highly scrupulous, certified accountant. After scrutinizing the diagrams and records for each ascus, the suspense was broken when he announced that he was fully convinced. Then he proceeded to suggest that we publish the entire body of data in the absence of any hypothesis. After attempting to satisfy this criterion, we found the task overly onerous. Moreover, in our hands, it led to a manuscript that was barely understandable. Thus, we generated a half-chiasma model based on the Holliday and Whitehouse and Hastings schema rooted in the behavior of DNA molecules and their interactions to yield heteroduplex DNA that could undergo enzymatic correction of mismatches via subsequent excision-repair reactions. In all, we had clearly established a firm, growing database for the conclusion that interallelic recombination was fundamentally and most frequently a nonreciprocal process distinct from crossing-over, although in later studies with Mortimer, we proposed that gene conversion was the prelude to reciprocal recombination. By an order of magnitude calculation, we demonstrated that conversion-associated exchange was sufficient by itself to account for the entire length of the genetic map. Supportive interactions and correspondence with Franklin Stahl, Robin Holliday, H.L.K. Whitehouse, Charles Radding, Miroslav Radman, and Henry Sobell were helpful and stimulating. Commentary and criticism from Norton Zinder and Joshua Lederberg were also prized for similar reasons.

During 1967–1968, I spent my sabbatical leave in Mortimer's laboratory housed in the Department of Biophysics headed by Professor Cornelius Tobias. He and Mortimer provided the essential impetus for the development of the first simple micromanipulator that was manufactured in the machine shop of the Lawrence Berkeley Laboratory by Mr. Brower.

Over the year, Mortimer and I developed a close working relationship that lasted for some 15 years, at which time we each decided, quite independently, that progress in understanding recombination via genetical inquiry had become increasingly difficult. Consequently, we turned to learn the tactics of molecular biology and recombinant DNA. Mortimer went off to John Carbon's laboratory and I again returned to the Stanford campus, this time in the Department of Biochemistry attached to the Ronald Davis laboratory group. Dan Stinchcomb, Thomas St. John, Stuart Scherer, Kevin Struhl, and Jasper Rine were most helpful, and cooperative. Collectively, they provided the impetus for converting me into a molecular geneticist. Thus, I was reborn. The next section concerns the Berkeley experience.

During the Berkeley period, Mortimer and I published a series of papers that centered on defining more precisely the various parameters and characteristics of the conversion process. Several contributions may be cited in this context. In our 1968 paper (Fogel and Mortimer 1968), we called attention to the seemingly paradoxical situation that reasonably consistent fine-structure maps could be constructed on the basis of a metric that was essentially an index of nonreciprocal recombination or gene conversion. A principal consideration was related to the understanding that selective procedures commonly used to study gene conversion failed to detect all intragenic events. Thus, we undertook a comprehensive survey of all intragenic events in a wholly unselected population of meiotic tetrads. We studied some 1600 tetrads from three different heteroallelic diploids that collectively harbored four different *arg4* alleles. Then, by utilizing Mortimer's complementing nonsense mutants that were localized to the low end of the *arg4* polarity gradient, we established that the high-frequency conversion end of the gradient coincided with the transcription initiation site. Our work led to the definition that "gene conversion involves replacement of the genetic information in the relevant DNA segment with information that is identical to that carried in the corresponding segment of the homologous non-sister chromatid. This process is designated as *informational transfer*."

The ordering of mutants within *arg4* was established by two independent tactics, namely, the X-ray mapping procedure devised by Manney and Mortimer, wherein an X-ray map unit was defined as one prototroph per million cells per roentgen and this corresponded to about 150–200 nucleotides. In addition, ordering of the mutant sites relative to the centromere was derived by identifying those asci in which a reciprocal exchange had occurred between the alleles. Among the principal findings were verification of *parity*, i.e., the number of 3+:1– asci and 1+:3– asci; conversions in either direction were equally probable. We identified single-site conversions of either the proximal or distal allele, reciprocal recombinants, and a novel category of symmetrical *double-site conversions*. Again, as in our previous work, for all tested allele pairs, reciprocal recombinants accounted for only a minority of the exceptional segregations. Thus, of 163 bona fide conversions, only 14 (or about 10%) arose via reciprocal recombination. For the alleles within *arg4*, an approximate linear relationship was found between the frequency of reciprocal recombinants and the length of the intragenic distance as estimated from the mitotic X-ray map. We underscored the finding that the magnitude of the symmetrical double-site conversion component, later felicitously designated as *coconversions* by David Stadler, was a linear function of the physical distance between the mutant sites. Widely separated alleles behaved as essentially independent sites. But, with closely spaced mutant alleles, the conversion of one site was often associated with conversion of the second site on the same DNA strand. The modal length of the replaced segment of information was estimated to extend over a physical length of several hundred nucleotides. Furthermore, as an extension of our conversion studies involving nonsense mutants of the ocher and amber varieties, we concluded that informational transfer occurred with virtually complete fidelity.

Testing several thousand tetrads required considerable labor. During this extensive testing period, Mortimer and I devised a variety of agar holders that led ultimately to the plate dissection technique. We poured dissection agar plates on a spirit-leveled plate glass platform. This yielded uniformly thick, nearly optically flat, agar. Asci were replica-plated directly from the dissection plate to a series of diagnostic plates for each segregating marker. For allele identification, we utilized suppressibility, complementation, or UV-induced reversion response, rather than homoallelic or heteroallelic reversion rates. Frequently, these and mating-type tests could be achieved by spraying plates with suspensions of the appropriate tester strains. Such innovations, particularly those that preserved the geometrical integrity of the ascosporal colonies, paved the way for our extensive analyses of postmeiotic segregation, reported first for yeast in *ade8-14* and *ade8-18* by Michael Esposito. The special significance of aberrant 4:4 segregations was defined in 1962 by Yoshiaki Kitani, then at Columbia University.

During much of this period, Ruth Lerner provided superlative technical assistance. Almost invariably she anticipated our needs for media and other supplies long before we requisitioned them. My stay at the Donner laboratory was enhanced by interactions with Mortimer's graduate student, Michael Resnick, who later came to spend some time at my Brooklyn laboratory after he completed his doctoral degree. Resnick had devised a double-strand break mechanism to account for gene conversion, and during the afternoon tea break, we would often take turns at rearranging the colored tapes to model the various exceptional tetrads. I was also intrigued by Resnick's ingenious adaptation of Howard Mel's STAFLO apparatus, a stable laminar flow system that included a density gradient in which cells could be separated according to their density, size, or charge. With the device, across which a voltage could be applied, Resnick could prepare spore suspensions that were essentially pure. During the following year, with machine shop assistance from the talented Jack Harris, I constructed a modified STAFLO system for isolating unbudded, synchronous cells. However, the system required a full-time, skilled technician, a luxury we could not afford at the time given our modest NIH research award. Later, we achieved this goal with zonal rotors or medium manipulations.

Immediately adjacent to Mortimer's laboratory was the office of an extraordinary biophysicist, Hal Anger. At the time, he was perfecting his now famous camera widely used in various radioactive scan procedures. Over the years, this same camera was key to diagnosing correctly several of my incapacitating major medical difficulties, including recent femoral popliteal artery replacements, a bovine graft to facilitate dialysis treatments, and finally, a cadaveric renal transplant in 1987.

Once again, I returned to home base in New York. Soon thereafter, during 1968, Everett Dempster, the chairman of the Berkeley Genetics Department, initiated a correspondence that eventually culminated in offering me a full professorship if I would agree to come aboard as chairperson. At long last, I was to be located on a campus where research was on a par with teaching. In fact, 85% of my research appointment was in the Agricultural Experiment Station and the remainder was classified as instructional or administrative.

1969 marks my official appointment at Berkeley. I was mandatorily retired in 1990 at age 70, although I am now provided with adequate space and facilities until 1992. This will allow me to complete studies in progress that are supported by NIEHS and NIH. Soon after arriving at Berkeley, I found that Alec Keith, a young assistant professor, and I shared many interests. Along with Resnick and Mortimer, he had previously isolated several mutants that affected membrane behavior and lipid biosynthesis. His expertise in nuclear magnetic resonance and the synthesis of electron-spin resonance reagents provided novel biophysical approaches to central cell biology issues. However, Keith failed to attain the coveted tenure rank, although he outpublished the collective efforts of his detractors. I was deeply disappointed by this grossly unfortunate event. Keith went on to become a highly creative phamaceutical entrepreneur. As a consequence, I inherited his beginning graduate students—Susan Henry, now chair of genetics at Carnegie-Mellon, Bernadine Wisneiski, professor of microbiology, at the University of California, Los Angeles, and Martin Bard who focused on the mechanism of nystatin resistance and the behavior of mutants affecting sterol biosynthesis.

At Berkeley, I enjoyed a protracted 20-year interaction with a succession of excellent graduate students, postdoctoral fellows, and two consistently supportive and dependable research associates. Among the Ph.D. students were Nina Halos, Carol Lax, Carole Cramer, David Radin, Dan Maloney, Beth Rockmill, Elizabeth Dowling, Marsha Williamson, and Lorraine Spector. The postdoctoral fellows included Robert Roth, John Cummings, Douglas Campbell, Amar Klar, Anne Delk, Lisa Sena, Charlotte Paquin, Ann Blecl, Steve Whittaker, John Game, Ayyamperumal Jeyaprakash, Kathy Atkinson, Barbara Kramer, and Wilfred Kramer. Each of them pursued a problem that impinged, in a general sense, on the recombination area. Soon, many established fruitful, independently funded programs.

Special recognition is appropriate for Karin Lusnack, a research associate who joined me in 1970. Much of her contributions are evident in our 1979 Cold Spring Harbor Symposium review (Fogel et al. 1979). Juliet W. Welch, who pioneered our studies on copper resistance, gene amplification, and recombination within the tandemly iterated *CUP1* locus, also merits equal recognition. Dan Maloney currently functions as my principal executive associate and his contributions are ongoing indeed. These last three persons are long-term, current mainstays in my Berkeley yeast laboratory. To say the very least, their contributions have been extensive and germinal.

The 35-year venture into yeast genetics and its molecular biology comprises an experience marked by a full measure of zest and excitement. It generated for me many prized international friendships, particularly those growing out of 16 International Conferences on Yeast Genetics and Molecular Biology. Early on, Roman invited me to function as chair of the U.S. Finance Committee to aid the International Yeast Genetics Conference. Every second year, private and public contributions allowed us to underwrite partially the travel or subsistence of some 20–30 young scientists. Their enthusiastic participation continues to play a highly significant role in the growth of yeast genetics and its molecular biology. Special recombination conferences at Aviemore, Nethybridge, and Cold Spring Harbor were milestone events.

Among these historical gatherings, I recall the meeting at Edinburgh. As I moved to the microphone to present my paper, a bagpiper entered and played "Happy Birthday." Equally memorable was the Pisa meeting in Italy at the nearby seashore resort hotel. The scene was comparable to a Fellini movie. On stage, in a dilapidated cinema, were Piotr Slonimski and David Wilkie engaged in heated dispute over some aspect of omega function. They were adroitly restrained by Roman and Magni. Finally, there was the unique conference at Chalk River, Canada. Called upon to offer thanks to our hosts, I remarked "When we came to Canada we saw billboards that exhorted us to DRINK CANADA DRY. We tried and failed." Such were the salad days of yeast genetics. A comprehensive, updated critical review concerning recombination in yeast was recently published (Petes et al. 1991).

## REFERENCES


Fogel S. and D.D. Hurst. 1967. Meiotic gene conversion in yeast tetrads and the theory of recombination. *Genetics* **57:** 455–481.

Fogel, S. and R.K. Mortimer. 1968. Informational transfer in meiotic gene conversion. *Proc. Natl. Acad. Sci.* **62:** 96–103.

———. 1971. Recombination in yeast. *Annu. Rev. Genet.* **5:** 219–236.

Fogel, S., D.D. Hurst, and R.K. Mortimer. 1971. Gene conversion in unselected tetrads from multipoint crosses. *Stadler Genet. Symp.* **2:** 88–110.

Fogel, S., R. Mortimer, K. Lusnak, and F. Tavares. 1979. Meiotic gene conversion: A signal of the basic recombination event in yeast. *Cold Spring Harbor Symp. Quant. Biol.* **43:** 1325–1341.

Hurst, D.D., S. Fogel, and R.K. Mortimer. 1972. Conversion-associated recombination in yeast. *Proc. Natl. Acad. Sci.* **69:** 101–105.

Mortimer, R.K. and S. Fogel. 1974. Genetical interference and gene conversion. In *Mechanisms in recombination* (ed. R. Grell), pp. 263–275. Plenum Press, New York.

Petes, T.D., R.E. Malone, and L.S. Symington. 1991. Recombination in yeast. In *The molecular and cellular biology of the yeast* Saccharomyces: *Genomic dynamics, protein synthesis, and energetics* (ed. J.R. Broach et al.), pp. 407–521. Cold Spring Harbor Laboratory Press, Cold Spring Harbor, New York.

# My Road to Repair in Yeast: The Importance of Being Ignorant

ROBERT H. HAYNES[1]
*Editor-in-Chief and President, Annual Reviews*
*Palo Alto, California 94306*

*The best of men have but a portion of good in them—a kind of spiritual yeast in their frames, which creates the ferment of existence.*

John Keats

## THE PORNOGRAPHY OF AUTOBIOGRAPHY

In 1876, Charles Darwin wrote a brief autobiography intended only as a private reminiscence for his children and grandchildren. However, an attenuated version of this work, edited by his son Francis, was published in 1887, 5 years after the author's death. Francis Darwin had wanted to publish the entire text, but other members of the family strenuously opposed this idea. Charles' wife Emma and their daughter Henrietta argued that the passages relating to his religious beliefs were very crude and that their publication would damage his reputation. An unbowdlerized text, fully restored and edited by Darwin's granddaughter, Nora Barlow, was not published until 1958, 99 years after the first appearance of the *Origin of Species*.

I think it is unfortunate, but hardly surprising, that many scientists even today seem to consider autobiography akin to autoeroticism, an unseemly pleasure, more properly indulged in private. However, as the present memoir amply demonstrates, I have been unable and unwilling to resist the temptation to talk about myself in print. In fact, I am delighted to have this opportunity to describe how I became involved with yeast genetics and molecular biology, even at the risk of damaging whatever reputation I may have acquired.

I regret that I have never kept a diary, especially since my "forgettory" is now more robust than my memory. Thus, what follows is by no means a scholarly, and quite possibly not even a very accurate, account of what motivated me to become a scientist, then to enter physics, and finally to embark on my 40-year migration from theoretical nuclear physics to DNA repair, mutagenesis, mitotic recombination, and deoxyribonucleotide (dNTP) metabolism in yeast. That my route from physics to molecular genetics happened to be by way of blood rheology, radiology, and mitotic chromosome dynamics is rather bizarre but, as I hope to show, not entirely inexplicable in light of the first steps I was able to take as a graduate student in science.

Successful and expanding scientific specialties, such as contemporary yeast genetics, seem inevitably to differentiate into a multitude of solitudes with dif-

[1]Distinguished Research Professor Emeritus, York University, Toronto, Ontario, Canada.

ferent immediate interests and domestic dialects. Therefore, to provide a frame of reference, and to set the stage for what follows, I shall begin with a synopsis of the biological significance of DNA repair and allied phenomena as I presently see it. This very general picture is, of course, distilled not only from my own research, but, more extensively, from the work of many people over many years; any interpretive errors or general misconceptions are my own. I then proceed, after a few autobiographical remarks, with a narrative account of the origin and development of my work on cellular responses to DNA-damaging agents, primarily in the yeast *Saccharomyces cerevisiae* and the bacterium *Escherichia coli*.

I have been deliberately biased in my selection of references as this paper is intended to be a personalized and impressionistic essay, rather than a formal review. In 1989, while on sabbatical leave at the Wissenschaftskolleg in Berlin, I began work on such a review, a historically and philosophically oriented monograph on the molecular basis of genetic stability and change. However, the excitements and allures of Berlin in the months preceding the Fall of the Wall were for me not conducive to scholarship of the monastic kind. Thus, the book remains but another gleam in my eye.

## THE IMPORTANCE OF DNA REPAIR FOR GENETIC STABILITY AND CHANGE

Natural selection is an immediate, or proximal, cause of evolutionary change, as observed at any level of biological organization. However, the ultimate *physical* sources of the diversification of organisms lie not in selection, but rather in the generation of heritable variation among them. Heredity is a conservative process. It is a manifestation of the stability of the genetic material and of its accurate replication, transmission, and utilization from one generation to the next. On the other hand, variation is a subversive process. It is a manifestation of many different sorts and degrees of change in the semantic content, arrangement, and expression of genes and genomes. The molecular basis of heredity (genetic stability) and variation (genetic change), and the surprisingly intimate relation between these superficially conflicting phenomena, is rooted in the complementary structure of double-helical DNA and the biochemical systems responsible for its replication, repair, recombination, and rearrangement.

In 1935, Max Delbrück suggested that genetic stability and change could be explained physically in terms of the Polanyi-Wigner theory of molecular fluctuations (Timoféeff-Ressovsky et al. 1935). The general assumption was that these were primarily physical rather than biochemical properties of genes. The genetic control of cellular sensitivities to mutagens was yet to be discovered. On this basis, spontaneous mutations were considered to arise from quantum-statistical fluctuations in the isomeric states of the genetic molecules. If these molecules were assumed to have unusually stable structures, such mutations would be rare events, as was actually observed in living cells. Induced mutations were thought to be inevitable physicochemical consequences of the interactions of ionizing or far ultraviolet (UV-C) radiations with the genes, as envisioned in the classical target theory (Timoféeff-Ressovsky and Zimmer 1947).

Chemical mutagens were unknown in 1935, and the rarity of mutations therefore seemed to be consistent also with the low ambient levels of mutagenic radiations in most environments. However, views began to change as a result of accumulating evidence (e.g., the discovery of mutator genes) that mutagenesis had important cellular, as well as physicochemical, components. Indeed, in a remarkably prophetic way, Muller (1954) argued as early as 1952 that mutation is the result of "some biochemical disorganization in which processes *normally tending to hold mutation frequencies in check are to some extent interfered with*" (my italics). Thus, *potential* mutation frequencies could be much higher than actual frequencies: The observed levels would then be *net* values, rather than absolute values.

The genetic material is now known to be composed of ordinary molecules, not endowed with any peculiar kind of physicochemical stability, nor is it protectively sequestered within cells from the "hurly-burly" of normal metabolism. DNA is subject to many types of spontaneous structural degradation, as would be expected (on purely chemical grounds) in warm, aqueous environments. In addition, cells are exposed naturally to many mutagenic agents of both endogenous and exogenous origin. Finally, the potential error rate of nonenzymatic DNA synthesis is high, on the order of $10^{-2}$ per base pair replicated, whereas observed error rates during normal replication are remarkably low, about $10^{-10}$ to $10^{-8}$ per base pair replicated.

If the various sources of DNA structural decay, damage, and replicational error had free rein, neither the informational integrity of DNA nor cell viability could be maintained. The well-regulated metabolism of living cells would collapse from what might be called "genetic meltdown." That such collapse does not occur arises from the fact that arrayed against these various destabilizing pressures, there exists an amazing battery of coordinated biochemical processes that actively maintain genetic stability and viability throughout the cell cycle. These stabilizing mechanisms include (1) those that promote high levels of replicational fidelity during normal semiconservative DNA replication, for example, 3′-exonucleolytic proofreading by DNA polymerases and methylation-instructed mismatch correction; (2) those that repair, or bypass, potentially lethal or mutagenic damage in DNA; and (3) those that chemically protect DNA by neutralizing or detoxifying mutagenic molecules of both endogenous and exogenous origin (for reviews, see Hurst and Nasim 1984; Friedberg 1985; Kirkwood et al. 1986; Friedberg et al. 1991).

In addition to these extensively studied processes, certain alterations in deoxyribonucleotide pools can provoke the entire range of genetic effects normally associated with exposure of cells to physical and chemical mutagens (Kunz 1982). Genetic loci known to control various modes of DNA repair in yeast and other organisms have pleiotropic effects on cellular responses to dNTP pool imbalances. Similarly, studies with various DNA polymerases have shown that the fidelity of DNA synthesis in vitro depends on the relative concentrations of the dNTPs in the reaction mixture; finally, certain mutator phenotypes arise from genetic defects in enzymes required for pyrimidine nucleotide biosynthesis. Thus, regulation of dNTP pools is yet another metabolic control system that promotes gene and chromosomal stability, as well as cell viability (Haynes 1985; Haynes and Kunz 1988; MacPhee et al. 1988).

For thermodynamic reasons, if none other, these remarkable error-correction and repair mechanisms do not function with perfect accuracy. Even though many mutations are deleterious, natural selection cannot drive mutation rates to zero and thereby eliminate the continuing production of new genetic variation. At the level of phenotypic evolution, the opposing processes of genetic stability and change emerge as complementary, rather than antagonistic, phenomena.

The possibilities for repair inherent in the informational redundancy of complementary base-pairing may account, at least in part, for the ubiquity of double-stranded DNA as the genetic material of contemporary cells. The processes promoting genetic stability are of fundamental importance for the integrity of living systems and so it seems likely that they arose very early in cellular, or even precellular, evolution. Their importance for contemporary organisms is well-attested by the extraordinary sensitivity to mutagens of strains of yeast deficient in DNA repair: The LD37 dose of germicidal UV light for mutants lacking the three major repair processes known to exist in *S. cerevisiae* corresponds to the formation of only one or two UV-induced pyrimidine dimers per genome. I cannot imagine how organisms totally deficient in DNA repair could arise and flourish in nature.

The DNA and protein synthesizing machinery of the cell is a remarkable example of a highly reliable, dynamic system built from vulnerable and not fully reliable parts. Many different genetic loci are involved in the biochemical stabilization of the genetic material, both directly and indirectly. The actual number of such loci in any organism is not known, but it is likely to be rather large; in yeast, more than 100 loci of this kind have been identified so far (cf. Haynes and Kunz 1981; Friedberg 1988; Friedberg et al. 1991).

In engineering practice, if high-fidelity performance is to be achieved with equipment of poor intrinsic precision, many checking and quality assurance procedures must be built into the system. For optimum economy, the energy cost of these procedures should be just sufficient to reduce the overall error rate to a tolerable level (Dancoff and Quastler 1953). This "principle of maximum error" is exemplified in the genetic machinery of cells. The evolution of long genetic messages has been made possible, in part, by the fact that they encode extensive instructions for their own correction. The energy cost clearly is not prohibitive, and the residual error rate is consistent with the genetic integrity of normal organisms and most of their progeny. Natural selection appears to have fashioned all major aspects of DNA metabolism in such a way as to ameliorate the effects of "genetic noise" and thereby to minimize mortality and mutability. However, in view of the existence of "error-prone" processes that simultaneously promote viability and generate mutations, it seems that viability takes precedence over fidelity in the economy of cells (Witkin 1969).

Discovery of the close relationship between DNA repair and recombination has had a significant impact on current thinking about genetic variation and evolution. For example, I am inclined to think that the adaptive significance of recombination may lie more in its ability to provide a mechanism for repairing damaged DNA than to produce heritable variation upon which natural selection *might* act favorably in the future (for a wide-ranging review of these revisionist ideas, see Bernstein and Bernstein 1991). Similarly, the primary benefit of inducible error-prone repair systems (e.g., the SOS response in *E. coli*), like that of the damage-inducible responses to alkylation and oxidation, may be to protect against the immediate toxic effects of the inducing agents, rather than to expand genetic variation. Nevertheless, should appropriate new mutations be produced through such mechanisms, they could be recruited by selection and thereby reduce the probability of population extinction in the face of the noxious material that induced the response in the first place (Echols 1982). Thus, the essential raw materials that make evolution possible, mutational and recombinational variation, may be adventitious by-products of processes initially selected to maintain replicational fidelity and cell viability and to ameliorate the deleterious effects of unavoidable genetic noise. As so well expressed by Reanney et al. (1983), "the variety of life is a surface phenomenon, superimposed on mechanisms of genetic homeostasis by the very noise those mechanisms were designed to combat."

The discovery of excision repair (Setlow and Carrier 1964) played a seminal role in the formulation of this current picture of the mechanisms underlying genetic stability and change, a picture based more on biochemical dynamics than on molecular statics. Elucidation of these fundamental biochemical processes has had a major impact on research in biology and medicine (Friedberg and Hanawalt 1988). In addition to improving our understanding of mutation and recombination, important insights into the etiology of certain genetic diseases and cancer, and even new ideas on speciation, aging, and the biological significance of sex, have been inspired by work in this area. Even though I remain troubled by the teleology implicit in the very idea of "repair" in the world of molecules, the DNA damage-repair hypothesis has proven its worth in deepening our insight into the macromolecular basis of life.

I had the good fortune to be associated with some of the radiobiological research from which ideas about the possibility and mechanisms of DNA repair emerged (Haynes 1964b; for a historical review, see Kimball 1987). My work on the lethal effects of radiation in *S. cerevisiae* and *E. coli* was initiated in 1960 in collaboration with Robert B. Uretz and his student, the late David Freifelder, at the University of Chicago. In 1962, I began similar experiments using the "radiomimetic" alkylating agent, nitrogen mustard (HN2), with W.R. Inch, an old friend from my student days at the University of Western Ontario. It was to prove significant for later recognition of the "generality" of excision repair that from the beginning, we adopted a comparative approach to these problems and worked with different organisms and examined the effects of five different inactivating agents (ionizing and UV radiations, acridine-sensitized visible light, heat at 60°C, and HN2). This being said, I shall now turn the clock back to about 1941 in order to explain how I became interested in physics and biology and, more importantly, how my pursuit of these interests led me to the University of Chicago in 1958.

## FROM THEOLOGY TO SCIENCE: SOME AUTOBIOGRAPHICAL ANECDOTES

"I want to be a bio-chemical-physicist" was my reply as a precocious preteen to avuncular inquiries about what I wanted to be when I grew up. Unfortunately, this answer always stimulated further questions about just what such people might do to earn a living, and why should I want to do whatever it was they did? I had no good answers for these questions and so I would blandly explain that it was a profession I had invented for myself because I was interested equally in biology, chemistry, and physics. It seemed that the only way out of my trilemma was to combine, somehow, all three subjects. In retrospect, it amazes me that I have in fact become something of a "bio-chemical-physicist."

My father was a charming, somewhat feckless salesman but a very well-informed and enthusiastic amateur scientist. He introduced me to chemistry (in our basement) and to astronomy (in our backyard) in the small Ontario town of Port Colborne, where we happened to be living during my primary school years. In contrast, my mother was a frugal, practical, church-going woman and a lifelong worrier over me and my father. However, she had an insouciant sense of humor. She frequently said that scientists were all a bit silly; she thought that Einstein's hair and Darwin's beard made them look funny. She also told me that Darwin must have been a very wicked man to claim that we came from monkeys [*sic*]. She urged me to be sensible, to study the Bible, to excel in school, and to become a doctor. Thus, my parents were very different people, with apparently few interests in common, other than their only child.

In the aftermath of the great depression, my mother's attitudes seemed wise and realistic especially since my father never had had an opportunity to trade his scientific knowledge for a paycheck. Thus, I worked hard in school, read voraciously, and studied the Bible with great interest, especially the book of *Genesis* and the *Revelation of St. John the Divine*. As the first and last chapters of scripture, it seemed to me that the subjects treated therein, the origin and fate of the universe, must be the most profoundly important things to try to understand in the course of one's life. Unfortunately however, in contrast to what I read in science, the biblical accounts of these matters struck me as being fantastic, unattestable, and contrary even to my youthful knowledge of the natural world. Ironically, study of my mother's Bible made me very serious indeed about my father's science.

My interest in science, and my fascination with the historical warfare between science and religion, grew stronger in high school (at the Brantford Collegiate Institute) as a result of my eclectic reading habits, which embraced popular accounts of recent scientific discoveries, as well as a few books on dogmatic theology and Christian apologetics. I also read successive chapters of the Bible every night as I had become a member of the British Scripture Union, an organization that provided small, explanatory guidebooks to accompany such regular reading. As a result of these disjunctive influences, I became afflicted with a form of "intellectual schizophrenia," possibly similar to that experienced by children raised in Communist countries who talked Marxism at school and in party meetings, but common sense with trusted and intimate friends. The psychic tension generated by my inability to compartmentalize science and religion in my thinking ultimately intensified my commitment to the pursuit of science and gave rise to my disdain for all dogmatic ideologies, whether secular or sacred.

Perhaps the most unusual high school influence came from my study of classical Greek. There was only one other pupil in my school taking Greek at that time; thus, it was inevitable that at each class meeting I would be called upon to translate and be quizzed on nuances of Attic and Homeric grammar. This discipline meant that Greek homework was a regular task of highest priority during my third and fourth years of high school. The more I studied Greek (and Latin), the more I became interested in classical culture. I soon discovered the Greek atomists, in particular Epicurus. The teachings of the "friendly philosopher" of the Garden provided a potent antidote against the uncompromising doctrines and bloody imagery of the Christian church. My encounter with Epicureanism also predisposed me to see as much "chance" as "necessity" in the world, a predilection that remains strong with me even today (Haynes 1987).

I did take seriously my mother's advice to go to medical school. I entered the University of Western Ontario in 1949, and at the end of my freshman year, I was accepted for the medical program. However, just a few weeks before I was to begin my premedical year, I changed course abruptly and entered physics and mathematics instead. My philosophical interests, aided and abetted by reading popular accounts of modern physics and cosmology, especially those by the British astrophysicist, Sir Arthur Stanley Eddington, convinced me that I simply could not go through life, in whatever occupation, without gaining at least some understanding of what I considered then (and now) to be the two most profound human attempts to penetrate the mysteries of the universe: the theories of relativity and quantum mechanics. My mother was more than slightly distressed by my sudden change in career plans, but my father provided quiet encouragement. Rather naughtily I countered my mother's worries about my economic future by telling her that I did not care about money; if I could not get a job as a scientist, I could always sell shoes, as my father once did. Actually, there was more truth than bravado in this remark. I feel I have been extraordinarily fortunate to be able to indulge my interest in science *and* to earn a good living in the bargain.

## FROM NUCLEAR PHYSICS TO BIOPHYSICS

I graduated in 1953 and immediately began work in theoretical nuclear physics at McGill University. It was there that I became seriously interested in cell biology and genetics. Like other young physicists of the day, my curiosity was aroused by reading Erwin Schrödinger's (1944) book, *What Is Life?*, and also F. Dessauer's (1954) *Quantenbiologie*. I became imbued with the fantasy that sufficiently clever physicists could, through further developments in the quantum mechanics of molecular interactions, solve the deep problems of biological reproduction and heredity: Biochemistry was messy and could be safely ignored. I was by no means alone among physicists in spouting such arrogant nonsense.

I had read some biology in high school, but my interest in it waned from lack of attention during my undergraduate years. In the older British/Canadian university tradition for honors students, virtually my entire undergraduate program was devoted to physics and mathematics. I had no "liberal" education in the American sense. Thus it was from Schrödinger that I first learned of genes, mutations, and the remarkable behavior of chromosomes in mitosis and meiosis. Schrödinger devoted an entire chapter of *What Is Life?* to Delbrück's purely physical theory of genetic stability and change in which the first point he made was "We may safely assert that there is no alternative to the molecular explanation of the hereditary substance. The physical aspect leaves no other possibility to account for its permanence. If the Delbrück picture should fail, we would have to give up further attempts." I was also surprised to read that X-rays could increase mutation rates in *Drosophila*. However, what really astounded me was the conjecture, based on a straightforward application of the classical target theory to X-ray mutagenesis data, that genes must be orders of magnitude smaller than chromosomes, perhaps an "association" (*Atomverband* in Delbrück's terminology) of no more than 1000 atoms. The implications of this conclusion for biology seemed truly profound. Schrödinger wrote, "the dislocation of just a few atoms within the group of 'governing atoms' of the germ cell suffices to bring about a well-defined change in the large-scale hereditary characteristics of the organism." Even though Delbrück's estimates of "gene size" were later shown to be wrong, Schrödinger's book did convince me that from a purely physical standpoint, the stability of genes is more remarkable, and of much more fundamental significance, than their mutability. Indeed, understanding stability (of anything) is logically prior to understanding change. For example, as Dirac noted in 1930, it was quite hopeless to try to understand the mechanisms of atomic and molecular change until the reasons for the stability of atoms and molecules were elucidated through the invention of quantum mechanics. On the basis of classical electrodynamics, the Rutherford-Bohr atom could not possibly be a stable structure—the negatively charged electrons would simply spiral into collision with the nucleus.

My curiosity about the "physical basis of life" led me to abandon my master's thesis work at McGill, which was very tedious mathematically and not going well in any case. Thus, I changed course again and applied for graduate study in quantum chemistry with C.A. Coulson at Oxford. I intended, with youthful naivety, to take a "bottom up" approach to cell division and mutation through quantum chemistry. My application was accepted and Coulson arranged for me to join him at Wadham College. Unfortunately, the National Research Council of Canada would not allow me to use my graduate fellowship for study abroad. Thus, I decided to enter biophysics under the supervision of Alan C. Burton, back at my alma mater, in London, Ontario.

Burton was a very British, very classical physicist who had come to biology by way of human physiology. He enjoyed a substantial international reputation and was soon to serve as president of the American Physiological Society and the Biophysical Society (for an autobiographical sketch, see Burton 1975). Among students he was famous for his playful scientific imagination, his distaste for specialization, his admiration for Sir Francis Bacon, and his very bushy eyebrows.

I shall never forget my first interview with him. I announced that I wanted to find out how radiation causes mutations. I asserted that through such work we might learn something fundamental about "the quantum physics of life." Burton's great eyebrows jumped in surprise at my innocent chutzpah, but his voice was patient as he replied, "Yes, yes, you may be right, but surely you know that here we work on the circulation!" He went on to explain that he wanted me to "look into" the rheology of human blood. "What's rheology?" I asked. "That's your problem," he replied, or words to that effect. I soon learned that rheology entailed study of the (non-Newtonian) flow properties of particulate suspensions such as blood. To be fair, however, Burton's request was by no means whimsical; as always, he had good reasons for suggesting somewhat offbeat research projects (for those curious about these reasons, see Haynes and Burton 1959).

Soon after I joined Burton's group, Herbert Jehle, an émigré theoretical physicist who had been a student of Schrödinger's in Berlin, visited our department. He gave what was for me an exciting seminar on his calculations of specific attractive forces (London–van der Waals interactions) between identical macromolecules (Jehle et al. 1959). Assuming the effectiveness of such forces for DNA in vivo, he proposed a "conservative" replication scheme for the genetic material. The fact that he seemed able to pull such an important rabbit out of a mathematical hat (no messy biochemistry), *and* that he had been associated with Schrödinger himself, buttressed my hopes for a new *Quantenbiologie* as the royal road to understanding genetics at the molecular level. I tried to follow Jehle's lead in theorizing about the forces involved in gene replication, but Burton wisely steered me back to the laboratory and to blood flow.

In those far-off days, most students did as they were told by their supervisors. Thus, I proceeded dutifully, but without much enthusiasm, to complete a Ph.D. thesis (Haynes 1957) on the rheology of blood. I was in no sense prepared academically for this work because, as luck would have it, I had never studied fluid dynamics beyond Poiseuille's law and I knew very little about blood or circulatory systems. Thus, my thesis work, like most of my subsequent research, was neither prejudiced nor informed by much previous knowledge of the subject. However, I eventually published several papers based on my thesis, the first of which was presented, before I graduated, at the First (March 1957) National Biophysics Conference in Columbus, Ohio (Haynes and Burton 1958). It was at this meeting that I met several members of the Yale University Department of Biophysics who were to become particularly close colleagues in later years. The most significant of these was Philip C. Hanawalt who was then a graduate student in Richard B. Setlow's laboratory. However, it was some years later, in a scruffy Oxford pub, during the 1964 International Congress of Photobiology, that our lifelong friendship was established. Soon thereafter, in California, we began our continuing collaboration in research, writing and speculating on the mechanisms and evolutionary significance of DNA repair.

## FROM RHEOLOGY TO RADIOLOGY

While working on blood flow, I kept up my interest, derived from Schrödinger and Dessauer, in the biological effects of radiation. Fortunately, Burton allowed me to accept a part-time job as a physicist in the Ontario Cancer Foundation's Clinic at Victoria Hospital, just across the street from the Medical School. I hoped to gain some practical experience there working with radiation in a biomedical context.

It was at Victoria Hospital that Ivan H. Smith pioneered the use of cobalt-60 γ-rays in cancer therapy. I was assigned the laborious task of calculating, by numerical integration, the radiation doses at the centers of deep-seated tumors in patients undergoing "rotation," rather than fixed beam, radiotherapy. To speed up this work, I developed a novel mathematical approach that greatly simplified these calculations. Thus, my first published paper was in the area of therapeutic radiology (Haynes and Froese 1957). To my surprise, Smith was sufficiently impressed to send me, together with my co-worker, W.R. Inch, to the 1956 International Congress of Radiology in Mexico City to give a paper on this new technique. At the Congress, I encountered Joseph Rotblat, a well-known nuclear physicist, and long-time secretary of the Pugwash movement, as well as others, including Paul Howard-Flanders, who were active in radiobiological research.

The following year, presumably on the strength of my work in radiological physics, I was awarded a postdoctoral fellowship by the British Empire Cancer Campaign to work with Rotblat at St. Bartholomew's Hospital Medical College in London. At "Barts," I was introduced to the London "radiobiology club," then dominated intellectually by the late L.H. Gray at Mount Vernon Hospital. However, another event had occurred during my first year with Burton that, in retrospect, set me off unwittingly on the road to repair in yeast.

## FROM BLOOD FLOW TO CHROMOSOME FLOW

Schrödinger's book also had stimulated my curiosity about the nature of the forces involved in the anaphase movement of chromosomes in mitosis. Thus, a few months after joining Burton's group, I made a point of seeing one of Warren H. Lewis' (Wistar Institute) time-lapse motion picture films of dividing rat fibroblasts in tissue culture. The apparent rotatory motion of the metaphase chromosomes (the so-called metaphase dance) and the relative rapidity of their anaphase separation, in what I assumed to be a rather viscous cytoplasmic medium, caught my eye; perhaps *here*, at a much more fundamental level than blood flow, was a really important problem in biorheology. A few rough calculations, based on Stokes' law, convinced me that the chromosomes could not be moved easily *through* the cytoplasm but rather might be flowing *with* it. I conjectured that the formation of the spindle itself, the metaphase configuration, the metaphase dance, and the anaphase separation of the chromosomes were all manifestations of an organized pattern of opposing, co-axial protoplasmic streams (giving rise to a "smoke-ring" vortex at the metaphase plate) inside the cells. This idea appealed to Burton's instincts as a classical physicist. We presented this theory at the 1955 meeting of the Society of General Physiologists at Woods Hole, Massachusetts (Burton and Haynes 1955). The talk stimulated a good deal of comment, not all of it favorable. However, we were sufficiently encouraged by some of the experts present to write up a full paper that we submitted to the *Journal of Cellular and Comparative Physiology*. After the usual editorial vicissitudes, it was accepted, but more than a year was to elapse, and I was already in England, before I received page proofs.

As a result of my interest in the physics of chromosome movement, Burton arranged a visit for me with his old friend Raymond E. Zirkle, at the University of Chicago, after the Woods Hole meeting. Zirkle was a prominent radiation worker who had served as Principal Biologist on the Manhattan Project and was then Chairman of the Committee on Biophysics at the University. He was an authority on the physical aspects of radiobiological action and also on the mechanisms of chromosome movement. Together with his former students, Robert B. Uretz and Robert P. Perry, and the distinguished histologist, William Bloom, Zirkle had developed various microbeam irradiation and time-lapse photomicrographic techniques as tools for probing the cytological changes exhibited by dividing cells in tissue culture (Zirkle 1957).

Zirkle asked me to give a seminar describing my ideas on mitosis. Bloom was gently skeptical of the theory of protoplasmic streaming. After my talk, he showed me a few of the many time-lapse films of abnormal chromosome movements in newt fibroblasts subject to microbeam irradiation that had been made by the Chicago group. I was particularly impressed with the phenomenon called false anaphase. This occurred in metaphase cells in which a small localized spot (about 2 μm in diameter) of cytoplasm, remote from the chromosomes, was irradiated with UV light. About 20 minutes after irradiation, the spindle seemed to have melted away; at any rate, it could no longer be detected in either phase contrast or polarization microscopy. As the spindle disappeared, the chromosomes first moved in an apparently haphazard way but then rearranged themselves into an orderly "quasirosette" configuration with all of the kinetochores clustered around a common center. Several hours later, the original quasirosette resolved itself into two "daughter" quasirosettes that moved apart from each other. This was followed by cytoplasmic constriction and the formation of two "daughter cells" with normally appearing, but highly aneuploid "interphase nuclei." I could see no way to explain such movements on the basis of my opposing protoplasmic streams. Clearly, chromosome dynamics was a far more complex problem than I had imagined. However, I was excited by the prospect that many new facts could be discovered about the nature of chromosome movement through the use of microbeam techniques.

## FROM "LONDON IN THE BUSH" TO "LONDON IN THE FOG"

I returned to Burton's laboratory in London, Ontario, and finished my degree with an increasing respect for the abilities of living cells to confound the simplistic ideas of brash young physicists seeking new fields to conquer. Thus, I set off in 1957 for my postdoctoral year in foggy old London with the sickening feeling that I had in press a paper on chromosome movement that was probably wrong. When the page proofs finally arrived at Barts I felt I had no choice but to withdraw the article from publication and to forget about the forces involved in chromosome movements. I did this without consulting Burton. When he discovered that I had "chickened out" so late in the day, he was not amused, especially since we had agreed that even if the protoplasmic streaming theory was wrong, our measurements and calculations might be useful to others. Today I feel that Burton's judgment again was right; I should not have withdrawn the paper so precipitously. In science, even incorrect theories, if well-defined and carefully thought out, can be helpful in stimulating new and better ideas along the slow, asymptotic path toward understanding.

While in London, I heard Francis Crick deliver a celebrated lecture in which he promulgated his "central dogma" of molecular biology. Despite the brilliance and cogency of his presentation, I was upset by the implications of his ideas; I might have to learn something about biochemistry after all if I was to understand the physical basis of life. Perhaps to be a physicist was not enough. Perhaps I had been misled by the romantic promise of Germanic *Quantenbiologie.* I was further depressed over the prospects for physicists moving into biology by the arguments of the brilliant, but controversial, medical physicist, P.R.J. Burch of the University of Leeds. I was collaborating with him, and G.W. Dolphin at Barts, on some new calculations in radiation physics. Burch insisted (correctly) that the target theory, as it was then formulated, could provide only the most superficial insight into the mechanisms of radiobiological action. And I was not much reassured, in light of Crick's lecture, by those other colleagues who argued that DNA was not the primary "target" for radiobiological effects in cells. I decided that a trip to Paris was necessary to relieve my funk and would provide an opportunity to meet Raymond Latarjet at the Institut du Radium.

Before leaving for Paris, someone suggested that I look up Latarjet's early work on the radiobiological "ploidy effect" in yeast. His data cheered me up because they seemed to lend credence to the target theory (Latarjet and Ephrussi 1949). At that time, I did not know of the more recent yeast work of Tobias, Mortimer, Beam, Magni, James, Wood, and Rothstein (for a review of this early research, see Tobias 1959 and the accompanying papers by the other authors just mentioned in the symposium on yeast radiobiology held during the 1958 International Congress of Radiation Research in Burlington, Vermont). I was delighted by my meeting with Latarjet and, even more, by my first glimpse of the *Quartier Latin*. I immediately asked if I could join his institute the following year. He agreed to take me on if I could obtain a suitable fellowship.

Upon my return to London, I applied to the National Cancer Institute of Canada for a fellowship and in due course I received word that my application was successful. I was elated and rushed immediately to tell my lab and pub mate, John Kirby-Smith, who was 17 years older than I, the good news. John was a physicist turned biophysicist, an apparently unreconstructed Southern aristocrat, and a senior staff member of the Oak Ridge National Laboratory then on sabbatical leave at Barts. He was a grandson of General Edmund Kirby Smith, commander of the TransMississippi Army during the War between the States, and the last Confederate General to surrender in that civil war. His reaction to my news, delivered with his dramatic Southern drawl, was quick, frank, and totally unexpected. He fixed a beady eye upon me and with a mocking laugh he exclaimed, "I know what you will do in Paris. You will waste your money eating in fancy restaurants, and waste your time sitting in sidewalk cafes drinking wine and girl-watching. You just git back to America and make something of yourself!" I was mortified. I could scarcely believe my ears. He had hit the nail on the head. I certainly wanted to work with Latarjet on cellular radiobiology, but it had not escaped my notice that life in Left Bank Paris was life that knew how to live. Thanks, but no thanks, to my Calvinistic conditioning, I was overcome with guilt. I knew I would have to do just what John suggested. I would have to find a group whose interests matched mine as well as, or better than, Latarjet's, but which was located in a serious hard-working environment where I could not be accused of, nor regularly tempted by, sybaritic indulgence. Thus it came about that I wrote to Raymond Zirkle in Chicago asking if he could accommodate me in his laboratory. It was short notice, but I already had a fellowship available to me. Some weeks later, I received a letter in which he not only agreed to take me on, but also offered me a faculty position as an Instructor in the Committee on Biophysics with my salary and research support provided initially from his U.S. Atomic Energy Commission contract! I abandoned my Canadian fellowship and, exactly a year after finishing my Ph.D., I was settled on the South Side of Chicago. In retrospect, my move there was the most significant turning point in my scientific life. I shall be forever grateful for the moral support and scientific stimulation that I received from my colleagues in biophysics (Fig. 1) and for the opportunity to savor something of the spirit of the "old college" created by Robert Maynard Hutchins. Although more by good luck than good planning, it did transpire that at Chicago I was able to

*Figure 1* Members of the Department of Biophysics, University of Chicago, 1964. (*Front row, left to right*) William Bloom, Raymond E. Zirkle, John R. Platt; (*back row, left to right*) Robert Haselkorn, Robert B. Uretz, Edwin W. Taylor, E. Peter Geiduschek, Robert H. Haynes. (*Absent from photo*) H. Fernandez-Moran.

satisfy, to some extent, Kirby-Smith's imperative to "make something of myself." Whenever we met in later years, he always recalled with a paternalistic chuckle the incident at Barts in which he "saved my soul for science."

## FROM MICROBEAMS TO MACROBEAMS

Zirkle's two main lines of research, mitotic chromosome dynamics and physical modifiers of radiobiological actions on cells (mostly fungi), matched my interests almost exactly. I was especially impressed with the potential of microbeam technology for selectively irradiating specific parts of individual chromosomes in dividing cells. Zirkle and Bloom initially developed these collimation techniques using, as a radiation source, proton beams generated by the Chicago Van de Graaff accelerator. However, at the time of my arrival, they were using only the very elegant UV microbeam invented by Uretz and Perry (1957). To have a more reliable and convenient source of ionizing radiation than that provided by the Van de Graaff, I built a polonium alpha particle microbeam device based on a recent design by C.L. Smith at Cambridge University. I had met Smith during my year in England. He showed me how he had built his microbeam and we soon became good friends. He was a delightful Cantabrigian sophisticate, a Fellow and Wine Steward of St. John's College, an Epicurean in both the classical and modern senses, and the last doctoral student to be supervised by Lord Rutherford.

Zirkle and I carried out many experiments with the UV and alpha particle microbeams on individual dividing cells, utilizing both animal (*Amblystoma* fibroblasts) and plant (*Haemanthus* endosperm) tissue cultures. We observed many interesting phenomena that amazed and fascinated me. However, there was one nagging problem that worried me very much: It was extremely laborious and time-consuming to gather *quantitative* data that could be analyzed in terms of any mathematically expressible theory known to me. Indeed, I could not intuit any useful explanations at all, physical or chemical, for what we observed: The stunning choreography of dancing chromosomes was obviously far too complicated to be understood in a simple way. Uretz had similar worries and decided to embark on a new line of experiments in radiation microbiology.

Zirkle had been trained in botany, and fungi were considered to be suitable objects of study for botanists. I believe it was largely for these reasons that he adopted *Saccharomyces* as a test organism in his radiobiological research program (Zirkle 1940; Zirkle and Tobias 1953). His student Thomas H. Wood, a physics graduate, took the trouble to develop techniques for obtaining unusually accurate and reproducible survival curves for irradiated cells (Wood 1959). For his doctoral degree with Zirkle, Uretz carried out a now classical series of experiments on the interaction between UV and X-rays in the inactivation of *S. cerevisiae* (Uretz 1955). I happened to see his paper shortly after it appeared and was greatly impressed with it for two reasons, both typical of a physicist. First, his quantitation of cell survival (measured in terms of macrocolony-forming ability) and the goodness-of-fit of the resulting dose-response data to equations derived from target theory were extremely impressive. The graphs looked like those more commonly seen in physics, rather than biology, journals. Second, I admired the tidiness of his equations as they were set out in print.

When Uretz returned to radiobiological studies on yeast (soon thereafter he also introduced *E. coli* into the laboratory), he also sought a chemically specific way to inactivate the cells. Thus, he was led to develop dye-sensitized photodynamic techniques using such supravital stains as the nucleic-acid-specific fluorochrome acridine orange. Cells treated with this dye could be inactivated by exposure to visible light. It was assumed that the light energy absorbed by the dye molecules was transferred to the nucleic acid molecules to which they were bound, thereby damaging them alone among the various macromolecular components of the cell. The dye molecules, by acting as "antennae" for energy absorption, would in effect convert a macrobeam of visible light into a molecular "microbeam." In such a system, it might be possible to relate the "sensitive sites," postulated formally to exist on the basis of target theoretical analysis of the dose-response curves, to a specific class of macromolecules. Thus, results derived from a mathematical description of the phenomena might be used to infer chemically testable ideas regarding the molecular mechanisms involved in cell killing. This approach appealed to my tastes as a physicist. It also appealed to a brilliant young physics graduate at Chicago, one of Uretz's students in biophysics, who was to become some years later the author of what I still consider to be the most cogent textbook of molecular biology so far written. This was David Freifelder, a precocious early entrant in the Hutchins program and one of the famous "Quiz Kids" of American radioland (Freifelder 1983).

Uretz and Freifelder developed the photodynamic inactivation technique (called DVL for Dye-Visible Light) using *S. cerevisiae* as their test organism. One of their first discoveries, in interaction experiments similar to those pioneered by Uretz in his doctoral work, was a new reactivation phenomenon: DVL-induced reactivation of X-ray damage in diploid yeast. In their paper describing this finding, they concluded that "Reactivation by visible light involving a dye which binds to deoxyribonucleic acid strengthens the view that part of the x-ray damage to a cell is damage to deoxyribonucleic acid molecules" (Freifelder and Uretz 1960). These findings, and our discussions of them, probably provided the spark for my subsequent interest in DNA damage and its repair in yeast.

In 1959, I set up shop in the laboratory recently vacated by Aaron Novick (where he and Leo Szilard ran chemostats), and which was adjacent to Uretz's new radiation microbiology facility. Here, Freifelder, Uretz, and I, later joined by Uretz's student Charles E. Helmstetter and my students Michael H. Patrick, Jeremy Baptist, and Richard A. Morton, followed several lines of research simultaneously and talked about science incessantly. Another close friend and colleague, Bernard S. Strauss, who also was interested in DNA repair and the effects of alkylating agents on cellular DNA, was located only two floors below us in the Research Institutes building (cf. Strauss 1968). Freifelder and I became, and were to remain until his untimely death in March 1987, particularly close friends, in no small part because of our common enchantment with Aristotle, Newton, and other denizens of the "Great Books of the Western World," which were much revered at Hutchins' Chicago. The intellectual guru for all of us was the physicist John Rader Platt, preaching the potent scientific method he called "strong inference" (Platt 1964).

During the first 2 years of this association, three quite accidental discoveries influenced my thinking about DNA repair as I believed it to be manifest in wild-type cells exposed to UV, X-rays, or nitrogen mustard (HN2). These were (1) "liquid-holding" recovery in yeast treated with any of these mutagenic agents, (2) the quantitatively parallel patterns of survival curves for radiation-resistant (B/r) and -sensitive ($B_{s-1}$) strains of *E. coli* exposed to these agents, and (3) the strong synergistic interactions for cell killing among all possible pair-wise combinations of UV, X-rays, and HN2 in both *E. coli* B/r and diploid yeast. I reviewed the implications of these results in some detail at the Wakulla Springs (Florida) Symposium on Molecular Mechanisms in Photobiology organized by Michael Kasha (Haynes 1964b). My main conclusion was that both diploid yeast and *E. coli* B/r possess enzymatic, energy-requiring mechanisms capable of repairing DNA structural defects caused by UV, X-rays, and HN2, three very diverse inactivating and mutagenic agents. In view of the existence of the interactions and the cross-sensitivities among three mutagenic agents that produce chemically different sorts of damage in DNA, I suggested that the enzymes involved in the underlying repair processes might not recognize the altered bases themselves, as in photoreactivation (Rupert 1961), but rather act on associated secondary structural distortions in the phosphodiester backbone. This conception of a "general" DNA-repair system, initiated by nuclease attack at, or near, sterically altered phosphodiester bonds, was dramatically illustrated a few years later by a *Scientific American* artist (Hanawalt and Haynes 1967). Our hypothetical excision-repair complexes were portrayed as close-fitting protein "sleeves" which moved processively along the double helix gauging its "closeness-of-fit" to the normal Watson-Crick structure as a means of recognizing damage-induced sites of strand distortion (Hanawalt and Haynes 1965). Recently, Hanawalt has reviewed these early speculations in the context of what is known today about the complex enzymology involved in the recognition of structural defects in duplex DNA (Hanawalt 1993). He concludes by pointing out the similarity between our "close-fitting sleeve" or "molecular calipers" and the "sliding clamp" feature of the β-subunit of the *E. coli* DNA polymerase III holoenzyme (Kong et al. 1992).

The discovery of liquid-holding recovery in yeast was the most important factor in stimulating my interest in studying DNA repair and in convincing me that it might be something more than an experimental artifact. This discovery came about quite by chance. Late one Friday afternoon, very early in our collaboration, Uretz and I completed a long and tedious series of X-irradiations of a suspension of stationary-phase diploid yeast. It was late in the day, and being tempted by the prospect of a visit to the Faculty Club bar, we decided to postpone plating the cells until the following Monday. After all, the cells were not growing and we felt sure they would not die, or be adversely changed, by storage over the weekend in our darkened laboratory. Accordingly, we left for the Club and plated the cells on rich growth medium when we returned Monday morning. A few days later, when we took the plates out of the incubator, I got the shock of my life: They were covered with thousands of colonies instead of the 200 we were expecting! We could not believe we had made sufficiently serious dilution errors to get such a result. We immediately repeated the experiment, once again delaying plating a few days but using a series of plating dilutions in order to get an accurate estimate of the magnitude of the "recovery" effect. It emerged that it was surprisingly large: The survival curves for immediate versus delayed plating differed by a constant dose-modifying factor of 2; i.e., the recovery process appeared to eliminate, in effect, one half of the original damage to the population. However, we were concerned that we might be witnessing an artifact of growth in our suspensions: Perhaps the dead cells in the population were lysing and their contents were being used by the undamaged cells to multiply. We carried out many control experiments to convince ourselves that the cells really were recovering from the lethal effects of X-irradiation. My student, Michael Patrick, was fascinated with this phenomenon and so he opted to explore it in detail for his doctoral research (Patrick and Haynes 1964; Patrick et al. 1964).

At the time Patrick began this research, I was for the most part unfamiliar with the large number of similar phenomena that had been reported previously in other organisms (Hollaender 1960). In particular, I had no idea that precisely this same recovery phenomenon in diploid yeast had been discovered 5 years previously by V.I. Korogodin in the (former) Soviet Union. I discovered his work when I picked up a copy of a Russian popular science magazine (*Priroda*) in the university library. As one with an addiction to such magazines, I was curious to see how the Russians designed and illustrated theirs, even though I could not read the language. As I flipped the pages, some illustrations in one of the papers caught my eye. It appeared that it might be an article on delayed plating of effects in irradiated yeast! I grabbed a Russian-English dictionary from a nearby reference shelf and puzzled out a few words. It was indeed a short article on what we thought was "our" effect (Korogodin and Malumina 1959). Patrick was nearing the end of his experiments and beginning to write his thesis and some papers for publication. Still, I was intrigued to learn more about Korogodin's work, to confirm that he really had discovered liquid-holding recovery and if so, to see how he interpreted his data. Thus, Patrick and I did our best to search the Russian literature, a nontrivial task in those days. We were modestly successful and our first paper (Patrick et al. 1964) contains references to nine papers by Korogodin, whom I was delighted to meet about 9 years later (1972) during my first visit to Moscow. It warmed my heart to learn, *sotto voce*, from one of his students that Korogodin played an important role in keeping genetics alive during the dreadful Lysenko period when the Communist authorities were suppressing genetics and killing geneticists. Evidently, Korogodin had surreptitiously taught orthodox genetics in his laboratory under the rubric of "radiobiology."

Most of the previous work on recovery and reactivation phenomena in irradiated cells could be interpreted only in terms of the formal concepts of target theory. In part, this was because no one knew specifically what the sensitive targets might be, even though Zirkle's partial cell irradiation studies (among others) indicated that in eukaryotes, the nucleus was more sensitive than the cytoplasm. It had of course been suggested that DNA was an important target, certainly for UV, and quite plausibly for X-rays. When I was in England, there was much debate over the role of DNA as a target for ionizing radiation. However, I was one of those who supported the DNA hypothesis, and during the early 1960s, more and more evidence seemed to be accumulating in favor of it (Haynes 1964a). Thus, when we observed similar radiobiological responses in yeast and *E. coli* for X-rays, as well as UV and HN2 (for which the question of a DNA target was less controversial), we simply *assumed* that DNA damage was indeed involved in all of these effects. Thus, we began to speculate about possible mechanisms of repair on the basis of our knowledge of DNA structure.

One afternoon early in 1963, during one of our many lively discussions in the laboratory, I suggested to Patrick and Morton that we try to write down all of the ways in which DNA damage *might* be repaired, without benefit of photoreactivating light, in diploid yeast. Three mechanisms, which seemed to exhaust all logical possibilities, came to mind: (1) direct enzymatic reversal of the defects in analogy with photoreactivation (Rupert 1961), (2) replacement of damaged single-strand segments by local DNA synthesis, and (3) genetic exchange between either homologous or sister DNA duplexes (Patrick et al. 1964). Because of these preliminary speculations, we were well primed for Setlow's announcement of his discovery of excision repair of UV-induced pyrimidine dimers in UV-resistant strains of *E. coli*, which came just 1 month after our yeast paper appeared in *Radiation Research* (Setlow and Carrier 1964).

The immediate confirmation of Setlow's excision data by Boyce and Howard-Flanders (1964), a mere 67 pages later in the same journal, together with the complementary discovery of repair-replication by Pettijohn and Hanawalt (1964), laid the foundations for the new and important branch of molecular biology that the study of DNA repair has become today.

The first plenary symposium on DNA repair at a major national meeting took place on February 26, 1964 at the Chicago meeting of the Biophysical Society, during the very month that excision repair was first reported in the *Proceedings of the National Academy of Sciences*. The speakers were Setlow, myself, and Howard-Flanders, with the late Henry S. Kaplan of Stanford University (Radiology Department) in the chair. The symposium attracted considerable attention, both from scientists and from the media. The *Chicago Tribune* reported on it the next day in a surprisingly long article under the headline "Medics Probe How Cancers Defeat Rays." Other headlines following a talk I

*Figure 2* Participants in the Conference on Structural Defects in DNA and Their Repair in Microorganisms, sponsored by the National Academy of Sciences/National Research Council and held at the University of Chicago, October 18–20, 1965. This was the first conference specifically devoted to the molecular mechanisms and genetic implications of DNA repair (for the published proceedings, see Haynes et al. 1966).

(*First row, left to right*) E.P. Geiduschek, D. Freifelder, R.B. Setlow, C.S. Rupert, R.F. Hill, E.M. Witkin, W. Szybalski

(*Second row*) J.K. Setlow, L. Grossman, A. Loveless, S. Wolff, W. Harm, R.F. Kimball, R.B. Uretz, R.P. Boyce

(*Third row*) H.E. Johns, R.E. Zirkle, R.K. Mortimer, H.I. Adler, L.D. Hamilton, D. Billen, P.C. Hanawalt, D.R. Krieg, H.E. Kubitschek

(*Fourth row*) J.E. Till, M. Delbrück, R.A. Deering, F.L. Haas, E.C. Pollard, A. Rörsch, E. Freese, B.S. Strauss

(*Fifth row*) H. Marcovich, K.A. Stacey, P. Howard-Flanders, H.S. Kaplan, R.H. Haynes, K.C. Smith

(*Absent from photo*) K.C. Atwood, V.P. Bond, R.S. Caldecott, J.W. Drake, A. Hollaender.

gave in England some months later proclaimed "Body Cells Are Saving Human Life," and, in a more sinister vein, "Professor Warns On Cure For Radiation."

Many of our scientific friends and colleagues were excited by these new findings and so Sheldon Wolff (then at the Oak Ridge National Laboratory) and I organized a small international conference to discuss the molecular mechanisms and biological implications of DNA repair. It was held in October 1965 at the University of Chicago (Fig. 2); the proceedings were published the following year (Haynes et al. 1966). Today, 27 years later, I am amazed to see that so many of the basic concepts, which continue to nourish the field, were put forward in nascent form at that symposium. My impression is that many of these ideas were "in the air" in the early 1960s. Thus, only a few key discoveries at the molecular level were required to coalesce several otherwise independent lines of research into a new and coherent body of knowledge. Work in this area has expanded dramatically since 1964. Only 43 people attended the Chicago meeting, not all of whom actually were engaged in research on repair. In subsequent years, meetings devoted to this subject have regularly attracted several hundred participants (cf. Hanawalt and Setlow 1975; Hanawalt et al. 1978; Friedberg and Hanawalt 1988).

Just as we began to entertain the possibility that DNA repair might be effected by the replacement of damaged single-strand segments by local DNA synthesis using the complementary strand as template, a prominent cell biologist visited Chicago to give a seminar on mitosis. After his talk, he made the customary round of informal visits to local faculty members and students to discuss their research. I sketched out for him the three possible mechanisms (see above) I thought might be responsible for various reactivation and recovery phenomena observed in cells treated with UV, X-rays, or HN2. He did not approve of the idea of nucleotide excision repair. He argued that cutting out and resynthesizing single-strand segments of DNA was tantamount to "DNA turnover." Didn't I know that DNA did not undergo any kind of "turnover" in cells, because after all, DNA was the genetic material? I had to confess, with considerable embarrassment, that I did not know this and so he referred me to several papers, including a lengthy review he had written about 10 years previously (for a skeptical assessment of these early ideas regarding DNA turnover, see Strauss 1960). I often describe this incident to students as illustrating the "importance of being ignorant" in science. It is rare to have one's curiosity aroused, and hard to think freshly on any topic, if one has read too much of the "relevant" background material. Had I been familiar with the previous literature, especially Korogodin's experiments, at the time Uretz and I stumbled upon liquid-holding recovery in yeast, I doubt if I would have been sufficiently intrigued to follow up on our fluke "discovery" of this phenomenon. And had I known that credible, senior scientists had argued that lack of "metabolic turnover" was an "expected" characteristic of genes, then I doubt if I would have been prepared to accept so readily the inferences drawn by Setlow and Howard-Flanders regarding excision repair.

## FROM BIOPHYSICS TO GENETICS

The genesis of my work with yeast occurred during the years I spent at the University of Chicago (1958–1964). I had learned to work with yeast but not to do any serious genetics with it. I regarded it initially as a convenient black-box for which the quantitative relationships between well-defined physical stimuli (mutagen doses) and elementary, all-or-none biological responses (cell death or mutation) could be measured accurately and analyzed mathematically. I think such attitudes were not uncommon among physicists who entered biology in those days. (My current views on the merits and limitations of this approach are summarized in Haynes [1989].) However, I "got religion" as a result of the biochemical revelation that DNA repair is a critically important process in the life of the cell, with implications for mutation, recombination, and evolution (Haynes et al. 1968; Strauss 1968). Suddenly, *biology* became more important to me than the *physics* of biology. This conversion occurred just before I left Chicago to join the Department of Medical Physics and the Lawrence Radiation Laboratory at the University of California, Berkeley, where I remained until 1968 when I became chairman of the newly established Biology Department at York University in Toronto.

David Freifelder had preceded me to Berkeley and our offices and laboratories were adjacent. Robert K. Mortimer's laboratory was a few steps down the hall, and Seymour Fogel came later to the Genetics Department. However, the only work I did with yeast at Berkeley, in collaboration with Cornelius A. Tobias and John T. Lyman, was to establish that the magnitude of liquid-holding recovery is independent of the LET (mean ionization density per unit track length of the ionizing particles) of the radiation employed. Freifelder soon stimulated a new interest in bacteriophage λ. The only Ph.D. student I supervised directly at Berkeley, Raymond M. (Bud) Baker, carried out several important studies with this phage, including a demonstration of UV-induced enhancement of recombination in excision-defective bacterial hosts (Baker and Haynes 1967, 1972).

Philip Hanawalt, who was by then at Stanford, was as intrigued as I was by the fact that the repair-deficient strain *E. coli* $B_{s-1}$ was sensitive not only to UV and X-rays, but also to a chemical mutagen (HN2). Furthermore, the degree of sensitization to HN2, in comparison with B/r, was remarkably similar to that observed for UV. This suggested to us that HN2 might be regarded as a "UV-like" mutagen, even though the chemical natures of the defects produced in DNA by UV and HN2 were known to be very different. It also suggested that HN2 damage in DNA was subject to excision repair, perhaps in much the same way as UV-induced pyrimidine dimers. This was consistent with the idea that nucleotide excision repair might be a "general" surveillance system necessary to maintain the informational integrity of DNA in the face of chemical, as well as UV, attack. Thus, we were led to ask if repair replication can be detected in *E. coli* treated with HN2, just as it can in UV-irradiated cells. The experiments gave positive results, and so we obtained direct molecular evidence that DNA damaged by a *chemical* mutagen also could be repaired (Hanawalt and Haynes 1965).

Upon my arrival in Toronto, I was faced with the task of building a modern research-oriented biology department in a new university. I was also faced with a large, but empty, laboratory. For a few months, I was afraid that my scientific career was coming to an end and that I would disappear down the seductive path of university administration. Fortunately, I was saved from this peculiar form of purgatory by the arrival of two postdoctoral fellows, Martin Brendel from the University of Frankfurt and Nasim A. Khan from

Brooklyn College. Brendel was trained as a bacterial geneticist and Khan as a yeast geneticist.

We decided to do some simple, straightforward experiments, sure to yield publishable results, just to get the laboratory started. Although I had virtually abandoned yeast during my years in Berkeley, my interest in it had been kept alive through contact with Robert Mortimer and his student, Michael A. Resnick. Thus, I knew that Resnick (1969), as well as Cox and Parry (1968) at Oxford, had isolated three apparently distinct phenotypic classes of radiation-sensitive yeast mutants, strains that were (1) sensitive to UV but resistant to X-rays, (2) sensitive to X-rays but resistant to UV, and (3) sensitive to both UV and X-rays. In light of my previous work on the pattern of cross-sensitivities among HN2, UV, and X-rays in *E. coli*, it seemed that the availability of these strains would provide an opportunity to determine whether cellular sensitivity to monofunctional (methyl methane sulfonate, MMS) and bifunctional (HN2) alkylating agents is correlated with sensitivity to UV or to X-rays or to both. We found that the X-ray-sensitive strain was also sensitive to MMS but not to HN2, that the UV-sensitive strains were sensitive to HN2 but not to MMS, and that the strain sensitive to both X-rays and UV was likewise sensitive to both MMS and HN2 (Brendel et al. 1970). Thus, contrary to my initial expectation, it appeared that there was no single, wholly promiscuous, dark repair system in yeast. Rather, there could be three different biochemical "pathways" for repair that might either function independently or be somehow interconnected. Even more surprising to us was that Cox and Parry (1968) had identified no fewer than 22 complementation groups (based on 96 recessive mutants) controlling radiation sensitivity. Clearly, there were many more genetic loci putatively involved in repair than seemed to be required by the number of enzymatic steps in the current model for excision.

My conditioning as a physicist predisposed me to eschew the kind of baroque complexity that was emerging from this work. It seemed that the only way out of this looming morass was to see if these many loci could be classified into a small number of biologically meaningful groups. Thus, we decided to examine the interactions among various pairwise combinations of loci in specially constructed double-mutant strains. We reasoned, quite simplistically, that if there were three multistep pathways (or enzyme complexes) for repair in yeast, then double mutants blocked in two competing pathways should be *supersensitive* to radiation, whereas double mutants blocked at two steps in the same pathway should be no more sensitive than their most sensitive single-mutant parent. We first constructed strains mutant at loci belonging to the phenotypic classes (1) and (3) described above. They proved to be supersensitive to UV and HN2 but not to MMS (Khan et al. 1970). We concluded from these and other studies that the three phenotypic classes probably represented three independent pathways for repair (Brendel and Haynes 1973). We did not know at the time we did this work that John Game and Brian Cox at Oxford were thinking along similar lines and would soon report on the other implication of this theory regarding the organization of repair pathways in yeast, namely, that strains mutant at two loci in the same class were no more UV sensitive than their single-mutant parents (Game and Cox 1972). Thus, what had been defined merely as three phenotypic classes of radiation-sensitive mutants could now be characterized genetically as three "epistasis groups" of loci presumably controlling three independent pathways for DNA repair. Martin Brendel carried out a comprehensive survey of the behavior of such double mutants with respect to inactivation by UV, HN2, X-rays, and MMS, and together, we developed the dose-response criteria necessary to distinguish epistatic (single pathway) from "synergistic" (two competing pathways) interactions between loci conferring radiation sensitivity in double mutants (Brendel and Haynes 1973).

Many additional loci affecting mutagen sensitivity, liquid-holding recovery, sporulation, meiosis, spontaneous and induced mutation, and mitotic recombination have been identified over the past 25 years. I do not know what the current count is, but our 1981 review of DNA repair in yeast contains a table listing 115 such loci and a summary of their (often pleiotropic) properties (Haynes and Kunz 1981). Although the concept of epistasis groups is today not as clear-cut genetically or biochemically as it seemed to be in the early 1970s, it still provides a useful framework for classifying loci involved in repair. Many of these genes have now been cloned, and the functions of their gene products have been identified (Friedberg et al. 1991).

At the 1970 International Conference on Yeast Genetics held in Chalk River, Ontario, Brendel and I heard a startling report from Wolfgang Laskowski's institute in Berlin: A strain of yeast had been found that was capable of taking up exogenous thymidylate (dTMP), thereby making it possible to label yeast DNA conveniently and specifically (Jannsen et al. 1970). I could hardly believe that any strain of yeast could take up nucleotides. (Such strains, of which there are now many, were later designated *tup* mutants by Wickner [1974].) However, if Laskowski's report was true, it might be possible to measure repair replication in yeast. This had not been attempted previously because yeast, in common with other fungi, was known to lack thymidine kinase. Thus, mutants lacking thymidylate synthase could not be selected because they could not utilize exogenous thymidine to form dTMP, and apparently no one had thought of supplementing them directly with dTMP. Brendel was able to repeat Laskowski's experiments in Toronto (Brendel and Haynes 1972), and this ushered in a new line of research that was soon to become a major theme in my laboratory. A few years later, J.G. Little, who had come initially as a postdoctoral fellow from Hanawalt's group, together with his student, Orna Landman, succeeded in detecting repair replication in yeast. However, much more important for later developments was the isolation of dTMP *auxotrophs* independently by Brendel in Frankfurt and by Gerry Little in Toronto (Brendel and Fäth 1974; Little and Haynes 1979). The isolation of various nucleotide permeable strains and Bernard Kunz's extensive work on mitotic recombination have stimulated new interest in the genetic consequences of deoxynucleotide pool imbalances (Kunz et al. 1980, 1986; Eckardt et al. 1983). In addition, it was this work on dNTP pool balance that led us to adopt the more broadly biochemical picture of the molecular basis of genetic stability and change outlined earlier in this essay (cf. Haynes and Kunz 1988) and furthermore, to initiate a program of study on the structure and expression of genes involved in the biosynthesis of pyrimidine nucleotides in yeast (McIntosh and Haynes 1986; Lagosky et al. 1987; Taylor et al. 1987). Thus, I seem to have become, against all youthful expectations, a physicist-*manqué* mired in biochemistry—although today, with genetic engineering techniques, it no longer seems messy.

## FROM GENETICS TO MATHEMATICS: CLOSING THE CIRCLE

When I first realized that cells possess complex enzymatic mechanisms for the dark repair of structural defects produced in DNA by various mutagenic agents, it seemed clear that the classical target theory for the interpretation of radiation dose-response curves would have to be amended if it was to be of any use at all in radiation biology. I outlined one way in which this might be done at the Chicago symposium on repair (Haynes 1966). Some years later, this approach was used to *predict* quantitatively a survival curve for UV-irradiated yeast cells on the basis of dose-response data for pyrimidine dimer excision (Wheatcroft et al. 1975). I thought that this was a rather neat exercise in radiation biophysics, but it attracted relatively little interest from others: Mathematics is a game few biologists like to play. Still, I had long toyed with the idea of extending the equations I developed for survival curves to dose-response curves for induced mutagenesis. However, until Friederike Eckardt joined me as a postdoctoral fellow, I did not have available any suitable mutation induction data that could be used for such calculations. As part of her doctoral research with Laskowski in Berlin, Friederike had carried out an extensive series of very accurate measurements of UV-induced mutation frequencies for forward and reverse mutations in UV-resistant and excision-deficient strains of yeast. These data were rather perplexing because they exhibited two qualitatively different departures from linearity at moderate to high UV doses: In some cases, the mutation frequency curves were quadratic at high doses, and in others, the mutation frequency reached a maximum at moderate doses and then declined. Friederike urged me to try to describe these curves mathematically. Thus, one afternoon a few months after her arrival, I sat down and began fiddling with some equations. I soon realized that even in the simplest case (exponential survival and linear mutation induction), mutation frequency curves (mutants per survivor) need not remain linear with increasing dose if, for some reason, the induced mutants had probabilities of clone formation that differed even slightly from those of the nonmutated cells in the population. This provided an almost trivial explanation for very substantial departures from low-dose linearity in the curves for excision-deficient yeast (Eckardt and Haynes 1977). We immediately realized that much further mathematical work might be done on the analysis of complex dose-response patterns for induced mutagenesis. Such analysis could also be useful in the increasingly important area of environmental mutagenesis. We therefore embarked on a major research effort along these lines (cf. Haynes and Eckardt 1979). This work did not merely pander to our taste for mathematical analysis, but it also provided some early clues regarding the inducibility of mutagenic responses in yeast (Haynes et al. 1985).

## CONCLUDING REFLECTIONS

Biologists choose to work with specific organisms, or groups of organisms, for reasons ranging from the immediately practical to the purely intellectual—or sometimes for no reason other than that they just happen to like birds or bees or even bacteria. It would be interesting, and perhaps constitute a revealing contribution to the psychosociology of science, to survey these various motivations among a representative sample of working biologists. The case of yeast would be of considerable interest in this regard. For many years, mycologists, geneticists, biochemists, and radiation biologists have made good use of this remarkable fungus for reasons sometimes unrelated to beer, bread, or wine. However, in the mid 1970s, the sudden popularity of *Saccharomyces* as a model system for studies on the molecular biology of eukaryotes took even me by surprise. What I had thought to be a wallflower organism in the eyes of mainstream molecular biologists suddenly became fashionable among the *avant garde*. I had begun work with yeast for no better reason than its convenience as a biological "black box." Thus, I feel fortunate to have been carried along and supported in my research by the truly remarkable advances in yeast genetics that have been made in recent years. And I feel sure that yeast still has much to teach us about the molecular basis of life, even if *Quantenbiologie* remains the will-o'-the-wisp it seems ever to have been.

## ACKNOWLEDGMENTS

I am deeply grateful to my family and personal friends for their affection and moral support over many years, especially at times when they would have been well advised to behave otherwise. Numerous students and colleagues, in addition to those named in the text or cited in the reference list, have contributed to my research activities in significant and memorable ways. I thank them all and regret that it was not feasible to mention everyone, despite the importance of their contributions to the development of yeast genetics. I feel that my work at York over the past 5–10 years does not qualify as being part of the "early days" of our field. It is for this reason that I have emphasized the events of the 1970s and earlier, even though I seem always to be most excited by what happened yesterday and the prospects for tomorrow. I have written this memoir, with its many technicalities and much jargon, primarily for my fellow scientists. However, I hope that it speaks also, in some personal way, to Joanne and Jane, to Mark, Geoffrey, and Paul, and someday to Jennifer Haynes.

## REFERENCES

Baker, R.M. and R.H. Haynes. 1967. UV-induced enhancement of recombination among lambda bacteriophages in UV-sensitive host bacteria. *Mol. Gen. Genet.* **100:** 166–177.

———. 1972. UV-induced recombination and repair of parental lambda bacteriophages labeled by means of host-controlled modification. *Virology* **50:** 11–26.

Barlow, N. 1958. *The autobiography of Charles Darwin*. Collins, London.

Bernstein C. and H. Bernstein. 1991. *Aging, sex and DNA repair*. Academic Press, New York.

Boyce, R.P. and P. Howard-Flanders. 1964. Release of ultraviolet light-induced thymine dimers from DNA in *E. coli* K12. *Proc. Natl. Acad. Sci.* **51:** 293–300.

Brendel, M. and W.W. Fäth. 1974. Isolation and characterization of mutants of *Saccharomyces cerevisiae* auxotrophic and conditionally auxotrophic for 5′-dTMP. *Z. Naturforsch.* **29c:** 773–738.

Brendel, M. and R.H. Haynes. 1972. Kinetics and genetic control of the incorporation of thymidine monophosphate in yeast DNA. *Mol. Gen. Genet.* **117:** 39–44.

———. 1973. Interactions among genes controlling sensitivity to radiation and alkylation in yeast. *Mol. Gen. Genet.* **125:** 197–216.

Brendel, M., N.A. Khan, and R.H. Haynes. 1970. Common steps in the repair of alkylation and radiation damage in yeast. *Mol. Gen. Genet.* **106:** 289–295.

Burton, A.C. 1975. Variety—The spice of science as well as of life. *Annu. Rev. Physiol.* **37:** 1–12.

Burton, A.C. and R.H. Haynes. 1955. A new "vortex" theory in the mechanism of mitosis. *J. Cell. Comp. Physiol.* **46:** 360. (Abstr.). (Our full paper on this subject, withdrawn from publication [1957] in this same journal at the page proof stage was: Haynes, R.H. and A.C. Burton, A theory of chromosome movements based on protoplasmic streaming [20 pages]. I would be glad to provide a photocopy of the proofs to anyone interested.)

Cox, B.S. and J.M. Parry. 1968. The isolation, genetics and survival characteristics of ultraviolet sensitive mutants in yeast. *Mutat. Res.* **6:** 37–55.

Dancoff, S.M. and H. Quastler. 1953. The information content and error rate of living things. In *Information theory in biology* (ed. H. Quastler), pp. 263–273. University of Illinois Press, Urbana.

Dessauer, F. 1954. *Quantenbiologie.* Springer-Verlag, Berlin.

Dirac, P.A.M. 1930. *The principles of quantum mechanics.* Clarendon Press, Oxford.

Echols, H. 1982. Mutation rate: Some biological and biochemical considerations. *Biochimie* **64:** 571–575.

Eckardt, F. and R.H. Haynes. 1977. Kinetics of mutation induction by ultraviolet light in excision deficient yeast. *Genetics* **85:** 225–247.

Eckardt, F., B.A. Kunz, and R.H. Haynes. 1983. Variation of mutation and recombination frequencies on a range of thymidylate concentrations in a diploid thymidylate auxotroph. *Curr. Genet.* **7:** 399–402.

Freifelder, D. 1983. *Molecular biology: A comprehensive introduction to prokaryotes and eukaryotes.* Science Books International, Boston and Van Nostrand Reinhold, New York.

Freifelder, D. and R.B. Uretz. 1960. Dye-sensitized photo-reactivation of X-ray damage in diploid yeast. *Nature* **187:** 953–954.

Friedberg, E.C. 1985. *DNA repair.* W.H. Freeman, New York.

———. 1988. Deoxyribonucleic acid repair in the yeast *Saccharomyces cerevisiae. Microbiol. Rev.* **52:** 70–102.

Friedberg, E.C. and P.C. Hanawalt. 1988. *Mechanisms and consequences of DNA damage processing.* A.R. Liss, New York.

Friedberg, E.C., W. Siede, and A.J. Cooper. 1991. Cellular responses to DNA damage in yeast. In *The molecular and cellular biology of the yeast* Saccharomyces: *Genome dynamics, protein synthesis, and energetics* (ed. J.R. Broach et al.), pp. 147–192. Cold Spring Harbor Laboratory Press, Cold Spring Harbor, New York.

Game, J.C. and B.S. Cox. 1972. Epistatic interactions between four *Rad* loci in yeast. *Mutat. Res.* **16:** 353–362.

Hanawalt, P.C. 1993. "Close-fitting" sleeves—Recognition of structural defects in duplex DNA. *Mutat. Res.* **289(1):** 5–13.

Hanawalt, P.C. and R.H. Haynes. 1965. Repair replication of DNA in bacteria: Irrelevance of chemical nature of base defect. *Biochem. Biophys. Res. Commun.* **19:** 462–467.

———. 1967. The repair of DNA. *Sci. Am.* **216(2):** 36–43.

Hanawalt, P.C. and R.B. Setlow, eds. 1975. *Molecular mechanisms for repair of DNA*, parts A and B. Plenum Press, New York.

Hanawalt, P.C., E.C. Friedberg, and C.F. Fox, eds. 1978. *DNA repair mechanisms.* Academic Press, New York.

Haynes, R.H. 1957. "The rheology of blood." Ph.D. thesis, University of Western Ontario, London, Ontario, Canada.

———. 1964a. Molecular localization of radiation damage relevant to bacterial inactivation. In *Physical processes in radiation biology* (ed. L. Augenstein et al.), pp. 51–72. Academic Press, New York.

———. 1964b. Role of DNA repair mechanisms in microbial inactivation and recovery phenomena. *Photochem. Photobiol.* **3:** 429–450. (As a result of a printer's error, the full summary was omitted from the paper but published later [1965] in *Photochem. Photobiol.* **4:** 839–840.)

———. 1966. The interpretation of microbial inactivation and recovery phenomena. *Radiat Res.* (suppl. 6), pp. 1–29.

———. 1985. Molecular mechanisms in genetic stability and change: The role of deoxyribonucleotide pool balance. In *Genetic consequences of nucleotide pool imbalance* (ed. F.J. deSerres), pp. 1–23. Plenum Press, New York.
———. 1987. The "purpose" of chance in light of the physical basis of evolution. In *Origin and evolution of the universe: Evidence for design?* (ed. J.M. Robson), pp. 1–31. McGill-Queen's University Press, Kingston and Montreal.
———. 1989. Mutations and mathematics: The allure of numbers. *Environ. Mol. Mutagen.* **14:** 200–205.
Haynes, R.H. and A.C. Burton. 1958. Axial accumulation of cells and the rheology of blood. In *Proceedings of the 1st National Biophysics Conference* (ed. H. Quastler and H.J. Morowitz), pp. 452–459, Yale University Press, New Haven.
———. 1959. Role of the non-Newtonian properties of blood in hemodynamics. *Am. J. Physiol.* **197:** 943–950.
Haynes, R.H. and F. Eckardt. 1979. Analysis of dose-response patterns in mutation research. *Can. J. Genet. Cytol.* **21:** 277–302.
Haynes, R.H. and G. Froese. 1957. Idealized body contours in rotation dosimetry. *Acta Radiol.* **48:** 209–226.
Haynes, R.H. and B.A. Kunz. 1981. DNA repair and mutagenesis in yeast. In *The molecular biology of the yeast* Saccharomyces: *Life cycle and inheritance* (ed. J.N. Strathern et al.), pp. 371–414. Cold Spring Harbor Laboratory, Cold Spring Harbor, New York.
———. 1988. Metaphysics of regulated deoxyribonucleotide biosynthesis. *Mutat. Res.* **200:** 5–10.
Haynes, R.H., R.M. Baker, and G.E. Jones. 1968. Genetic implications of DNA repair. In *Energetics and mechanisms in radiation biology* (ed. G.O. Phillips), pp. 425–465. Academic Press, London.
Haynes, R.H., F. Eckardt, and B.A. Kunz. 1985. Analysis of non-linearities in mutation frequency curves. *Mutat. Res.* **150:** 51–59.
Haynes, R.H., S. Wolff, and J.E. Till, eds. 1966. Structural defects in DNA and their repair in microorganisms. *Radiat. Res.* (suppl. 6), pp. 1–243.
Hollaender, A., ed. 1960. *Radiation protection and recovery.* Pergamon Press, New York.
Hurst, A. and A. Nasim, eds. 1984. *Repairable lesions in microorganisms.* Academic Press, New York.
Jannsen, S., E.-R. Lochmann, and R. Megnet. 1970. Specific incorporation of exogenous thymidine monophosphate into DNA in *Saccharomyces cerevisiae. FEBS Lett.* **8:** 113–115.
Jehle, H., J.M. Yos, and W.L. Bade. 1959. The biological significance of charge fluctuation forces. In *Proceedings of the 1st National Biophysics Conference* (ed. H. Quastler and H.J. Morowitz), pp. 86–92. Yale University Press, New Haven, Connecticut.
Khan, N.A., M. Brendel, and R.H. Haynes. 1970. Supersensitive double mutants in yeast. *Mol. Gen. Genet.* **107:** 376–378.
Kimball, R.F. 1987. The development of ideas about the effect of DNA repair in the induction of gene mutations and chromosomal aberrations by radiation and chemicals. *Mutat. Res.* **186:** 1–34.
Kirkwood, T.B.L., R.F. Rosenberger, and D.J. Galas, eds. 1986. *Accuracy in molecular processes.* Chapman and Hall, London.
Kong, X.-P., R. Onrust, M. O'Donnell, and J. Kuriyan. 1992. Three-dimensional structure of the β-subunit of *E. coli* DNA polymerase III holoenzyme: A sliding DNA clamp. *Cell* **69:** 426–437.
Korogodin, V.I. and T.S. Malumina. 1959. Recovery of viability of irradiated yeast cells. *Priroda* **48:** 82–85 (in Russian).
Kunz, B.A. 1982. Genetic effects of deoxyribonucleotide pool imbalance. *Environ. Mutagen.* **4:** 695–725.
Kunz, B.A., G.R. Taylor, and R.H. Haynes. 1986. Intrachromosomal recombination is induced in yeast by inhibition of thymidylate biosynthesis. *Genetics* **114:** 375–392.
Kunz, B.A., B.J. Barclay, J.C. Game, J.G. Little, and R.H. Haynes. 1980. Induction of mitotic recombination in yeast by starvation for thymine nucleotides. *Proc. Natl. Acad. Sci.* **77:** 6057–6081.
Lagosky, P.A., G.R. Taylor, and R.H. Haynes. 1987. Molecular characterization of the

*Saccharomyces cerevisiae* dihydrofolate reductase gene (*DFR1*). *Nucleic Acids Res.* **15:** 10355–10371.
Latarjet, R. and B. Ephrussi. 1949. Courbes de survie de levures haploides et diploides soumises aux rayons. *C.R. Acad. Sci.* **229:** 306–308.
Little, J.G. and R.H. Haynes. 1979. Isolation and characterization of yeast mutants auxotrophic for thymidine monophosphate. *Mol. Gen. Genet.* **168:** 141–151.
MacPhee, D.G., R.H. Haynes, B.A. Kunz, and D. Anderson, eds. 1988. Genetic aspects of deoxyribonucleotide metabolism. *Mutat. Res.* **200:** 1–256.
McIntosh, E.M. and R.H. Haynes. 1986. Sequence and expression of the dCMP deaminase gene in *Saccharomyces cerevisiae. Mol. Cell. Biol.* **6:** 1711–1721.
Muller, H.J. 1954. The nature of the genetic effects produced by radiation. In *Radiation biology* (ed. A. Hollaender), vol. 1, part 1, pp. 351–473. McGraw-Hill, New York.
Patrick, M.H. and R.H. Haynes. 1964. Dark recovery phenomena in yeast. II. Conditions that modify the recovery process. *Radiat. Res.* **23:** 564–579.
Patrick, M.H., R.H. Haynes, and R.B. Uretz. 1964. Dark recovery phenomena in yeast. I. Comparative effects with various inactivating agents. *Radiat. Res.* **21:** 144–163.
Pettijohn, D. and P.C. Hanawalt. 1964. Evidence for repair replication of ultraviolet damaged DNA in bacteria. *J. Mol. Biol.* **9:** 395–410.
Platt, J.R. 1964. Strong inference. *Science* **146:** 347–353.
Reanney, D.C., D.G. MacPhee, and J. Pressing. 1983. Intrinsic noise and the design of the genetic machinery. *Aust. J. Biol. Sci.* **36:** 77–91.
Resnick, M.A. 1969. Genetic control of radiation sensitivity in *Saccharomyces cerevisiae. Genetics* **62:** 519–531.
Rupert, C.S. 1961. Repair of ultraviolet damage in cellular DNA. *J. Cell. Comp. Physiol.* (suppl. 1) **58:** 57–68.
Schrödinger, E. 1944. *What is life?* Cambridge University Press, Cambridge.
Setlow, R.B. and W.L. Carrier. 1964. The disappearance of thymine dimers from DNA: An error-correcting mechanism. *Proc. Natl. Acad. Sci.* **51:** 226–231.
Strauss, B.S. 1960. *An outline of chemical genetics.* W.B. Saunders, Philadelphia.
———. 1968. DNA repair mechanisms and their relation to mutation and recombination. *Curr. Top. Microbiol. Immunol.* **44:** 1–85.
Taylor, G.R., P.A. Lagosky, R.K. Storms, and R.H. Haynes. 1987. Molecular characterization of the cell cycle regulated thymidylate synthase gene of *Saccharomyces cerevisiae. J. Biol. Chem.* **262:** 5298–5307.
Timoféeff-Ressovsky, N.W. and K.G. Zimmer. 1947. *Das Trefferprinzip in der Biologie.* Hirzel, Leipzig.
Timoféeff-Ressovsky, N.W., K.G. Zimmer, and M. Delbrück. 1935. Über die Natur der Genmutation und der Genstruktur. *Nachr. Ges. Wiss. Göttingen FG VI Biol. N.F.* **1:** 189–245.
Tobias, C.A. 1959. Attempted analysis and correlation of various radiobiological actions on the same kind of cell (yeast). *Radiat. Res.* (suppl. 1), pp. 326–331.
Uretz, R.B. 1955. Additivity of X-rays and ultraviolet light in the inactivation of haploid and diploid yeast. *Radiat. Res.* **2:** 240–252.
Uretz, R.B. and R.P. Perry. 1957. Improved ultraviolet microbeam apparatus. *Rev. Sci. Instr.* **28:** 861–866.
Wheatcroft, R., B.S. Cox, and R.H. Haynes. 1975. Repair of UV-induced DNA damage and survival in yeast. I. Dimer excision. *Mutat. Res.* **30:** 209–218.
Wickner, R.B. 1974. Mutants of *Saccharomyces cerevisiae* that incorporate deoxythymidine-5′-monophosphate into deoxyribonucleic acid *in vivo. J. Bacteriol.* **117:** 252–260.
Witkin, E.M. 1969. Ultraviolet induced mutation and DNA repair. *Annu. Rev. Genet.* **3:** 525–552.
Wood, T.H. 1959. Some aspects of cellular radiobiology. *Rev. Mod. Phys.* **31:** 282–288.
Zirkle, R.E. 1940. The radiobiological importance of the energy distribution along ionization tracts. *J. Cell. Comp. Physiol.* **16:** 221–235.
———. 1957. Partial-cell irradiation. *Adv. Biol. Med. Phys.* **5:** 103–146.
Zirkle, R.E. and C.A. Tobias. 1953. Effects of ploidy and linear energy transfer on radiobiological survival curves. *Arch. Biochem. Biophys.* **47:** 282–306.

# Some Recollections on Forty Years of Research in Yeast Genetics

ROBERT K. MORTIMER
*Department of Molecular and Cell Biology*
*Division of Genetics*
*University of California, Berkeley, California 94720*

I was raised in Didsbury, a small (<1000 population) town in the Canadian prairie province of Alberta. The town served a farming community that extended approximately 5–20 miles in all directions. Didsbury's main claim to fame was its seven grain elevators, a much larger number than most of the nearby towns could boast. I learned quite a bit about the dairy business while in Didsbury and even became a licensed cream and milk grader and tester. In addition to moving around a lot of milk cans, I had the responsibility of determining the butterfat content of the milk and cream brought in by the farmers. They were paid by butterfat content times the volume. These tests involved the addition of sulfuric acid to an aliquot of the product, addition of water, and centrifugation in a steam-turbine-driven centrifuge. When the fat rose to the neck of a graduated centrifuge tube, I read the fat level off with calipers. While attending school in Didsbury, I benefited from some very dedicated teachers who taught me the basics of science and mathematics. Biology, as far as I remember, was not taught in this school. I did very well in mathematics and physics and went on to the University of Alberta in Edmonton where I majored in Honors Physics. I suppose I chose this major more because it did not require humanities beyond scientific German than because I had a deep interest in physics. At Edmonton, I took math, physics, and chemistry for 3 years and stayed on the honor list, but I was not happy with my chosen field. So in my fourth and final year, I was given "permission" to enroll in organic chemistry, zoology, and medical physiology as long as I agreed to take the remaining math and physics courses for my major. I happily agreed. I was becoming interested in the field of biophysics, although I really did not know what the term meant. I just knew that I did not want to work in the field of physics, at least as it was presented to me.

Alberta at that time had just discovered that it was sitting on top of some very productive oil fields, and people with training in physics were being sought for oil exploration using geophysical techniques. I had spent the summer before my senior year in this activity and was offered, upon graduation, a chance to join a new oil exploration firm. At about the same time, I had applied to the University of California at Berkeley to do graduate studies in biophysics and was accepted with the offer of a teaching assistant position at $120 per month. I opted for the oil company at a considerably larger salary and wrote Berkeley to tell them about my decision. After 2 months of pouring over seismic charts, I decided that this was not the way I wanted to spend my life. Fortunately, about the same time, I received a letter from Professor


 0-87969-378-9/93 $5 + .00

Cornelius Tobias at Berkeley urging me to reconsider. I did. On June 18, 1949, Mary and I were married, and 2 weeks later, we headed south for Berkeley.

My first month in Berkeley was spent asking a lot of questions and trying to determine what was going on. Raymond Zirkle, a visiting professor from the University of Chicago, who had arrived only a short time before me, was working with Tobias on some strains of yeast that they had recently obtained from Carl Lindegren. They were trying to answer what was at that time an important question in radiation biology: Are the lethal effects of ionizing radiation due to nuclear or to cytoplasmic damage? Zirkle and Tobias reasoned that if the lethal damage was nuclear, haploid and diploid cells should respond differently, but if it was cytoplasmic, these cells should respond similarly. They responded differently. Haploid cell survival decreased exponentially with increasing dose, whereas diploid cells showed a sigmoidal type of survival curve. Being biophysicists, Zirkle and Tobias soon developed a mathematical model to explain these differences. It was proposed that haploid cells would be killed if a single event, such as a primary ionization, occurred in any one of a number (20–64) of "sites," whereas a diploid cell needed an event in both members of any one pair of these sites. This was equivalent to assuming that the damage was recessive lethal mutations. I was assigned the job of testing this model.

As a graduate student with primarily a physics background, I needed to expand my knowledge in biology so I took several courses, including a course in genetics from Curt Stern and a course in cytology from Richard Goldschmidt. I remember Stern presenting with great excitement Joshua Lederberg's demonstration of crossing-over in bacteria. We also learned a lot about basic genetics in this course. Goldschmidt of course had his own radical ideas of how chromosomes and cells were organized. This was 1949–1950, when our knowledge of biology was still in the classical phase. DNA was certainly known, but there was no agreement about its role in the cell. I even learned in one of my courses that DNA, or thymus nucleic acid, was organized as a repeating tetranucleotide. RNA was called yeast nucleic acid. In addition to biophysics, biochemistry, and cell biology courses, I took graduate courses in physics and mathematics. My thesis committee was set up during my second year with Cornelius Tobias as chairman and Curt Stern and Edward McMillan as the other two members. Tobias was an assistant professor in the Physics Department and he and a few other faculty members had already established a graduate program in biophysics. I soon learned that Stern was a highly respected geneticist and that McMillan was a well-known physicist who had invented the synchrotron. He was to receive the Nobel prize for this work a few years later.

## POLYPLOIDY AND RADIATION GENETICS

To test the Zirkle-Tobias model, I reasoned that diploid survivors should be heterozygous for recessive lethal mutations and also that triploid and tetraploid cells should be more resistant to the lethal effects of X-rays than were diploids. To carry out the experiments to test these predictions, I needed to use genetic procedures, but no one at Berkeley knew anything about yeast genetics. I was given a copy of Lindegren's book, *The Yeast Cell, Its Genetics and Cytology* (Lindegren 1949), but it confused more than clarified matters. I learned from Carl Beam, a friend of Barbara Bachman who was then in the Bacteriology Department, that Edward Tatum at Stanford University had a graduate student, Sheldon Reaume, who was working on yeast genetics. I drove our 1934 Plymouth to Stanford and met with Reaume who gave me several tips about tetrad dissection as well as two genetically marked strains, YO2022 (*a met1*) and YO2587 (*a trp1 ura1 thi1*). The mutations had been produced by Seymor Pomper at Yale in the same two strains that Lindegren had also sent to Berkeley, Seattle, and Stanford (Mortimer and Johnston 1986). Although Reaume had produced the *ade1* mutation earlier, I did not receive this mutation at that time.

With some help from Lindegren's book, I was able to isolate mating diploids (**aa** and αα) from each of these haploids and, by appropriate crosses, soon had what I thought were triploids and tetraploids. The cell sizes were consistent with this assumption, which eventually was confirmed by genetic analyses. Contrary to the predictions of the model, however, the tetraploids were more sensitive to X-rays than were the diploids or triploids (Mortimer 1958). I also dissected diploid survivors of X-ray treatment and indeed found recessive lethals but not nearly at the frequency predicted from the assumption that haploids were killed entirely by this type of damage (Mortimer and Tobias 1953). This suggested some kind of repair, but it was many years before we and other groups established that repair did occur in yeast. To explain the higher sensitivity of tetraploids, I proposed that there was also some dominant lethal damage and that the frequency of this damage increased with the ploidy of the irradiated cell. Dominant lethality was thought at that time to involve chromosomal damage and most likely dicentrics and resultant anaphase bridges. By pairing individual haploid cells of opposite mating type, one of which was irradiated, I showed that some of the resultant zygotes (I studied thousands of such zygotes) were unable to divide to form colonies. This was operationally equivalent to dominant lethal damage (Mortimer 1955). Maureen Owen and I next used irradiated diploid cells and mated these to unirradiated haploid or diploid cells and found that, at a given dose, more dominant lethality was induced in diploids than in haploids (Owen and Mortimer 1956). This confirmed my hypothesis and explained the radiation response of cells of different ploidy up to hexaploid. I reasoned that if I was going to understand this phenomenon further, I would need to study the yeast chromosomes which I already knew from other work were too small for a cytological approach. So I decided to take a year off from these other studies to develop a genetic map of yeast. That was in about 1955, and I am still working on this project 37 years later.

## DEVELOPMENT OF S288C AND ITS USE IN MUTANT HUNTS

In 1956, I constructed the cross X345 that had, as one of its parents, the strain SC7. This strain was Lindegren's 1Cα (now called EM93-1C) but had been renamed by Zirkle and Tobias after it and its α mating-type partner were sent to Berkeley in 1949. These same strains were also sent to Seattle, Stanford, and Yale in the same year and were given different names at each place (Mortimer and Johnston 1986). The other parent, S177A, came from a cross between 1198-1b (from Don Hawthorne) and S139D, which came directly from the two strains sent out by Lindegren. X345 was segregating for *met1 ura1 trp2 his1 ade1*, and these were most of the nutritional markers available at that time. In addition, this cross was segregating for mating type, clumpiness, arsenic resistance, and galactose and maltose fermentation. This diploid yielded about 10% diamond-shaped asci and had high spore viability. I was able to score the segregation of all but the two fermentation markers. At that time, I was experimenting with indicator plates to score fermentation and presumably the results were not clear. I could have used Durham tubes but was probably not set up for the use of these at that time. I dissected S280 to S297 (asci were numbered consecutively with a S prefix) from X345 and scored the corresponding markers. In particular, I scored clumpiness by microscopic examination. I wanted to obtain a haploid strain that was completely prototrophic and that dispersed into single cells when suspended in water. S288C and one other segregant of X345 met these requirements, but apparently S288C was the better of the two. Within the next 2 months, I had used this strain in six experiments to isolate nutritional mutants. Several years later, this strain diploidized spontaneously to a strain I called X2180. Haploid derivatives of this diploid, X2180-1A (**a** mating type) and X2180-1B (α mating type), have also become standard "wild-type" laboratory strains (Mortimer and Johnston 1986). S288C is now being used in the European Genome Sequencing Project.

The mutant hunts were all of the same type. I exposed cells to various doses of UV light and looked for mutants among the colonies formed by surviving cells. Replica-plating procedures were not well developed, at least in my laboratory, so each colony was sampled with a loop and streaked on minimal medium and YPD. The minimal medium I used was based on Lynferd J. Wickerham's recipe (Difco Manual) and contained only salts, trace elements, nitrogen source, carbon source, and biotin. This medium was not commercially available at that time. I was concerned that mutant colonies might be sectored so each colony was sampled from a small sector at its edge. We know now that this was unnecessary; most mutant colonies are not sectored. Colonies that failed to grow on minimal medium were tested first on minimal plus 20 amino acids, minimal plus purines and pyrimidines, and minimal plus vitamins. Each presumptive auxotroph was then identified by auxanographic procedures. The presumptive mutants were spread over the surfaces of two or three minimal plates and at defined areas around the edge of the plates were placed a few crystals of different amino acids, purines, or pyrimidines or vitamins. Positive growth at one of these areas identified the growth requirement of the mutant. By July 1956, I had identified nearly 100 nutritional mutants. Over the next year or two, I incorporated these various mutants, plus many more isolated later, into crosses. Segregants of these crosses were used as allelism testers against new mutants. In this way, I characterized most of the mutants in the arginine, histidine, isoleucine-valine, leucine, threonine-methionine, tryptophan, uracil, and other pathways. The only new factor about this work, however, was that it was being done in yeast. I relied heavily on knowledge and procedures that had been developed earlier for *Escherichia coli*, *Neurospora*, and *Aspergillus*. In fact, this work was never published in the open literature and only appeared in an internal report of the UC Radiation Laboratory (now Lawrence Berkeley Laboratory) (Mortimer et al. 1957).

The set of arginine testers that I had developed were sent to J.M. Wiame and his group in Belgium, and the strains requiring histidine and aromatic amino acid were sent to Norman Giles at Yale. The histidine testers were used by Gerry Fink in his doctoral studies and the aromatic amino acid testers were used similarly by Antoine de Leeuw. François Lacroute used our uracil testers for part of his studies on pyrimidine biosynthesis, and Huguette de Robichon-Szulmajster and I characterized the aspartate-homoserine biosynthetic pathway (de Robichon-Szulmajster and Mortimer 1966). The tryptophan genes were eventually worked on by Tom Manney in my laboratory and by Ralph and Jack DeMoss.

Herschel Roman had identified the genes involved in adenine biosynthesis and had given me a set of tester strains for each of the adenine genes (Roman 1956). The genotypes of these strains, combined with those of the testers that I had developed for the other biosynthetic pathways, were published in the Microbial Genetics Bulletin and eventually provided the core from which the Yeast Genetics Stock Center was built.

The set of nutritional tester strains that had been developed was announced at the First International Conference on Yeast Genetics held in Carbondale, Illinois, in November 1961. From that time on, they were made available to anyone who wanted them. Throughout this initial period, and for several years following, I was very ably assisted by Ruth Lerner, wife of the distinguished geneticist, I. Michael Lerner.

## DEVELOPMENT OF THE GENETIC MAP

In 1955, I visited Don Hawthorne at the California Institute of Technology where he was doing postdoctoral studies with Sterling Emerson. We agreed at that time to cooperate on developing the genetic map of yeast. I had naively planned to "take a year off" and do this work myself but had discovered, during a trip to Seattle to visit Herschel Roman, that Don had just completed a doctoral dissertation in Seattle on this topic and I was pleased to have help. Don had concentrated on mapping the fermentation genes, some metal resistance genes, and the few nutritional markers that were around at that time. Curiously, there was no serious effort at Seattle or anywhere else toward inducing nutritional mutants so my efforts complemented what had been going on there.

In the course of testing the various mutants for allelism, I also generated some mapping data. For example, several of the mutants that I had isolated were found to be centromere-linked, e.g., *leu1*, *leu2*, *his4*, *trp5*, *arg4*, *his6*, and *lys1*. Earlier work of Carl and Gertrude Lindegren had identified *ade1*, *gal1*, *MAT*, and *ura3* as centromere markers, and these defined chromosomes I, II, III, and V. Don Hawthorne (Hawthorne 1955) showed that *trp1* (isolated by Seymor Pomper at Yale) was centromere-linked and defined chromosome IV. Don had also found a new centromere-linked histidine mutant in one of his *his1* strains and this mutant, *his2*, defined chromosome VI. I then used *leu1*, *arg4*, and *his6* to define chromosomes VII, VIII, and IX. I also found that *leu2* was linked to *MAT* and to a new histidine mutant, *his4*, on the left arm of chromosome III, that *his6* was across the centromere from *lys1* on chromosome IX, and that *trp5* was linked to *leu1* on chromosome VII. In addition, I found that *leu1* was linked to *ade6* which had been sent to me by Herschel Roman. I was reproached by Herschel for not telling him immediately that *leu1* and *trp5* were on the same chromosome on which he had been carrying out his mitotic-crossing-over study. I was not trying to be secretive and probably thought that he would likely not be interested. I think I told Don Hawthorne about this, but Don would only have mentioned it to Hersch if he had been asked specifically whether there were any new genes on chromosome VII. After this occasion, I tried to keep Hersch fully informed about what I was doing.

About 4–5 years after we started this project, Don Hawthorne and I published our first version of the genetic map (Hawthorne and Mortimer 1960), where we described the locations of 26 genes on 10 chromosomes. The pace of scientific research at that time was much more relaxed. We were in no hurry to publish and we let anyone who was interested know what we were doing. The importance of this publication went beyond the mapping information that it presented. There was a general conception in the biological sciences community at that time that yeast was somewhat anomalous in its behavior and this was reinforced by the controversy involving Lindegren and Winge (see corresponding chapters in this volume). Don and I showed that yeast conformed to all of the rules of genetics and in addition offered some advantages over other microbial systems such as *Neurospora* and *Aspergillus*. We published several other versions of the genetic map during the next 15 years (Hawthorne and Mortimer 1968; Mortimer and Hawthorne 1966, 1973, 1975), and I have been continuing this effort, with other collaborators, to the present.

Herschel Roman had shown that mitotic crossing-over occurs in yeast and that the frequency of this event could be increased by exposure of cells to UV light (Roman 1956). He had shown, by these techniques, that *ade6*, *ade3*, and *MAL1* were linked even though meiotic linkage could not be detected between *ade6* and the other two genes. I found out that X-rays also induced mitotic crossing-over and used this to find new groups of linked genes. For example, John Johnston during his doctoral studies showed that *trp4*, *hom2*, and *aro1* were mitotically linked to *ade8*, and Sayaka Nakai later linked up *met13*, *ade5*, *lys5*, and *cyh2* by mitotic analysis (Nakai and Mortimer 1969). This group of genes on the left arm of chromosome VII was then linked to *trp5* and *leu1* by meiotic analyses. In a later study with Don Hawthorne, several other genes were added to the map by mitotic crossing-over procedures (Mortimer and Hawthorne 1973).

## SNAILS, PETITES, AND TETRAD ANALYSIS

Very early in my experiments on the genetic analysis of yeast, I discovered that certain yeast strains when crossed were incapable of sporulation. I was quite excited to then show that this sporulation defect was inherited as a cytoplasmic factor. One of my graduate students, Ralph Lee Gunther, showed me a section in Ephrussi's book on cytoplasmic inheritance (Ephrussi 1953) that in effect said that *petite* diploids do not sporulate. This disclosure brought that project to a very abrupt end. Gunther's thesis project, which was carried out in the mid 1950s, involved determining the "oxygen effect" for X-ray-induced mutation. Ionizing radiations were known to be about two to three times more effective for killing cells if the cells were exposed in the presence of oxygen rather than in its absence. Although we had some reason to believe that mutation induction would be different in this respect, this was not the case. Induction of most mutations by X-rays showed a comparable enhancement by oxygen. In the course of his studies, Gunther surveyed a large number of mutant loci for X-ray-induced revertibility and found that *trp1-1* and *his5-2* were the most revertible. These were, respectively, amber and ocher alleles of these genes and we were inducing mostly suppressor mutations, but of course we did not know this at that time. Gunther also reasoned that reversion could be studied better in diploids than in haploids because killing in diploids was minor at doses that caused considerable reversion. Throughout these studies, we were frustrated and hampered by the difficulty of making synthetic media. Wickerham had devised a set of synthetic media, which was marketed by Difco, that contained all the essential salts, trace elements, vitamins, nitrogen source, and carbon sources. He called this Yeast Morphology Agar, which was also available in versions that lacked the carbon source, nitrogen source, or vitamins. Yeast Nitrogen Base was closest to our needs because it contained everything except the carbon source. Unfortunately, this and the other media also included tryptophan, histidine, and methionine. Gunther (spelled Guenther in the Difco manual) finally wrote to Difco and asked them to modify their Yeast Nitrogen Base by leaving out these amino acids. They agreed, and this became Yeast Nitrogen Base Special and later Yeast Nitrogen Base w/o amino acids; this medium is used in nearly all yeast laboratories today.

Fred Sherman, another one of my early graduate students, worked on *petite* induction both by heat shock and by growth at elevated temperatures. He showed that growth of yeast cells at 38°C led to the production of *petites* and that only the bud was *petite*; the mother cell remained *grande* (Sherman 1959a). Heat shock caused by exposure of cells to elevated temperatures (60°C) for short periods of time also induced *petites* at high frequencies (Sherman 1959b). At some time during his tenure as a graduate student, Fred decided that he would try to transform a *petite* strain using mitochondria from a *grande* strain. From the work of Ephrussi, it was pretty well established that the mitochondria were the site of the *petite* mutation. In addition, procedures for making yeast protoplasts and for regenerating vegetative cells from these protoplasts had been worked out by Něcas (1956) and by Eddy and Williamson (1957). This procedure involved the use of digestive juices from snails, so Fred collected several dozen snails (*Helix aspersa*) from various gardens in Berkeley and kept them in some kind of a snail cage in our laboratory. He made a very good enzyme preparation from these snails, but he never got around to trying this preparation on mitochondrial transformation. Had he done so, and succeeded, he would have been even more famous than he now is. However, the snails escaped from their cage one day and caused a lot of consternation in the laboratory and probably distracted Fred from his important goal. In the meantime, another graduate student John Johnston and I were struggling with ascus dissection trying to develop strains for a mitotic crossing-over study. We had a rectangular dissection chamber covered with two square plastic cover glasses. A thin agar slab was placed on one cover glass, and the asci on an agar wedge were placed on the other coverglass. The asci were dragged with the needle (long and pointed and nearly horizontal; the needle was sharpened with hydrofluoric acid) onto the cover glass surface, the ascus was cut with the needle, and the released spores were picked up and moved to the agar slab. Although this was very tedious work, we could still do a fairly large number of tetrads in 1 day. One day, we asked Fred for some of his snail juice and placed some under a coverslip where there was a suspension of asci. Within a few minutes, we could see the asci pop open, but the spore tetrads remained together. Within a few hours, we were dissecting tetrads by this procedure, a method that is still in use today in nearly all yeast laboratories (Johnston and Mortimer 1959).

## GENE CONVERSION

I was not directly involved in the controversy surrounding gene conversion among Øjvind Winge, Herschel Roman, and Carl Lindegren. However, during my genetic characterization of the various nutritional genes, I had seen many irregular segregations that could not be explained by any of the mechanisms proposed by Winge or Roman. For example, in crosses with several markers segregating, it was often seen that one marker would occasionally segregate 3:1 or 1:3 while all of the other markers segregated 2:2. Many times, this involved markers on the same chromosome, only one of which segregated irregularly. I became convinced quite early that gene conversion was a real, but as yet unexplained, phenomenon. I prepared a table of these irregular segregations, which occurred in the 1–10% range, and gave it to Herschel Roman who included it in his 1956 Cold Spring Harbor Symposium presentation (Roman 1957). By then, gene conversion had been accepted as a real genetic phenomenon even though it was some time before that term was fully accepted.

During the 1960s, Tom Manney, another of my graduate students, and I had developed a procedure for determining genetic fine-structure maps using X-ray-induced interallelic recombination. These events were induced linearly, and the slope of induction of revertants turned out to be a reliable parameter for the distance between alleles (Manney and Mortimer 1964). We made detailed maps of both the *trp5* and *arg4* genes, and Tom used data from his *trp5* map to characterize nonsense alleles and deduce that they were chain terminating (Manney 1964). Several other groups subsequently used this procedure to produce fine-structure maps of their genes (e.g., *his4, cyc1, ade3, ade6, ade8*).

In 1959, I made a cross, X901, involving two alleles in the *arg4* gene, *arg4-1* and *arg4-2*, dissected 116 tetrads, and scored each allele by X-ray-induced allelic recombination. The gene conversion events I observed involved one or both of these two alleles as well as crossing-over between the alleles at an overall frequency of around 10%. About half of the events were coconversions. A few years later (1967–1968), Sy Fogel came to spend a sabbatical leave with me and we continued this study. About the same time, Sergei Inge Vechtomov, Öestein Strömnaes, and Jack von Borstel were also on sabbatical leave in my laboratory. Sy and I first showed that conversion did not generate new information; ocher alleles and ocher suppressors were converted into the same type of allele or suppressor (Fogel and Mortimer 1970). This study involved two ocher alleles some distance apart in *arg4*, and we noticed that coconversions for this pair of alleles were relatively much less frequent than in the cross I had studied earlier. This suggested an obvious experiment in which we tested two alleles that were known to be near each other. As expected, coconversions were observed most of the time, which led us to propose that gene conversion involved a transfer of a stretch of information from one homolog to the opposite homolog and that the transfer occurred with fidelity

(Fogel and Mortimer 1969). Sy Fogel and I continued these studies for several years hoping to establish which model of recombination best fit our results. We thought we had eliminated the original Holliday model, but a slight variation on this model fit most of our results. Although the Meselson-Radding model gave the best fit to our data, it could not account for some observations, such as the failure of gene conversion without crossing-over to interfere with other gene conversion events (Mortimer and Fogel 1974), and the position of the associated crossover not corresponding to that predicted from the model. Our results, which were based on the analysis of more than 25,000 meiotic tetrads, were presented at the Cold Spring Harbor Symposium in 1978 (Fogel et al. 1979).

## SUPPRESSORS

Suppressor mutations had been defined genetically quite early and such mutations were known in *Drosophila* and *E. coli.* One of my students had isolated a suppressor of a *trp5* allele, *trp5-2*. The initial mapping results indicated that this suppressor was on the mating-type chromosome and I made crosses to incorporate it into a cross with other markers on chromosome III. I was trying to build a strain with several genes on chromosome III for a study of interference. To my surprise, I found that the suppressor also acted on *his4-2* and *leu2-1* but failed to act on *thr4-1*, which was also in the cross, and I duly recorded this in my notebook. Our knowledge of protein synthesis in 1960 was in its infancy so this observation was merely a curiosity. I went on to what I thought were more important things and did not return to this observation for more than 1 year. We then screened several other mutant alleles and found that the suppressor acted on some alleles but not on others. These suppressors were allele-specific and locus-nonspecific. Quite independently, Don Hawthorne had isolated similar suppressors and had worked on them for a couple of years. During one of my trips to Seattle, we compared notes and realized that we were working on similar genes. We finally published this work a few years later (Hawthorne and Mortimer 1963). I thought that the suppressors had to be acting at some step in protein synthesis to affect so many different pathways, but Don was sure that they were modifying the osmotic pressure within the cell; he had shown that certain alleles could be phenotypically reversed by high osmotic pressure. We had some good debates on this topic. Tom Manney showed convincingly in his studies on suppression that suppressible mutations were chain-terminating. Some suppressible alleles were also complementing, but these always mapped at one end of the gene. Tom correctly reasoned that translation proceeded toward this end and that complementation could occur if most of the protein chain had been synthesized (Manney 1964). Dick Gilmore, as part of doctoral studies, isolated a large number of suppressors that acted on a set of five ocher alleles and showed genetically that they fell into eight genes (Gilmore 1967). He then did a postdoctoral stint with Fred Sherman and showed that these suppressors all inserted tyrosine into ocher sites in the cytochrome *c* gene (Gilmore et al. 1971). Jeremy Bruenn decided to establish that nonsense suppressors were due to mutations in genes encoding tRNA molecules (Bruenn 1972). He developed an in vitro translation system that worked very well for bacteria but failed when yeast tRNAs were added. He and Ray Gesteland, using another in vitro system, later showed that indeed suppression did involve tRNA genes.

## DNA REPAIR

Mike Resnick isolated some DNA-repair mutants as part of his doctoral research while in my laboratory. These mutants included *rad52, rad18, rad2,* and *phr1* (Resnick 1969). About the same time, several other mutants had been isolated by Cox and Perry (1968) and by Nakai and Matsumoto (1967). Nakai had isolated some radiation-sensitive mutants incidental to his studies on mitotic crossing-over while in my laboratory. John Game carried out the necessary genetics on these mutants and placed them all in three epistatic groups based on the response of various double-mutant combinations (Game and Cox 1971, 1972, 1973). Following up on Mike's work, Karen Ho showed that *rad52* strains were blocked in the repair of DNA double-strand breaks (Ho 1975) while Mike was also doing the same thing at the University of Rochester (Resnick and Martin 1976). At least Karen thought that I should have more control over my students and former students to prevent such an overlap of effort. Jeff Lemmont isolated mutants blocked in induced mutation (*rev1, rev2, rev3*) during his doctoral studies (Lemontt 1976), and these mutations are still being worked on today by different groups. I have continued the studies on DNA-repair genes, and three of my current students are still working on this subject. Since this history is to cover only up to the 1970s, I will not go into further detail about their work.

John Game joined my group in the the 1970s and we worked out the genetics of the *rad50* to *rad57* group of mutants (Game and Mortimer 1974). We are still working on several of these genes at the present time. Rebecca Contopoulou joined my group in 1974 and continues to the present as Curator of the Yeast Genetics Stock Center. David Schild came as a postdoctoral fellow in 1978 and worked with me on compilations of the genetic map as well as on cloning of various DNA-repair genes. We published a mapping compilation in 1980 (Mortimer and Schild 1980) that included all the mapping information available to that date and this was updated in 1985 (Mortimer and Schild 1985) and 1989 (Mortimer *et al.* 1989). David is currently a Staff Scientist at Lawrence Berkeley Laboratory.

## EPILOGUE

I had some difficulty writing parts of this article because I did not have easy access to all the publications of my students. It is not that they published in obscure journals; I insisted that their work be of a quality that would allow publication in solid refereed journals. However, it was the tradition, at that time, to encourage graduate students to publish their thesis work by themselves and I followed this tradition for many years. Thus, the thesis works of Lee Gunther, Fred Sherman, John Johnston, Tom Manney, Gary Jones, Mike Resnick, Jeff Lemmont, Richard Gilmore, Jeremy Bruenn, and some to follow were published as sole authors. Although I kept track of my own publications, I was not as careful about those of my students and finding a publication 20–30 years old can be a problem. I have not followed this publication procedure for the past several years and I am not entirely sure why. The rigors of academic review procedures in most American universities discourage this practice, and this was undoubtedly an important factor in my case. Certainly it is not the general tradition now to encourage graduate students to publish their theses as sole authors, but maybe it would be a good time to start this practice again.

I joined the Berkeley faculty in 1953 in what was to become the Biophysics Department. I served as chairman of this department for 9 years and also saw service on several major campus committees. I had the opportunity of working with many outstanding graduate students and postdoctoral fellows during my 40 years on the faculty. I also enjoyed four sabbatical leaves. In 1962–1963, my wife, four children, and I traveled to Pavia, Italy, where I spent a sabbatical year at the Istituto di Genetica working with Giovanni Magni. In 1972–1973, we went to Seattle where Don Hawthorne and I made a major effort on the genetic map (Edition 4). During 1978–1979, I was in John Carbon's laboratory at Santa Barbara to learn some molecular procedures. I also injected a little yeast genetics into their program. They were very actively engaged in cloning the centromere of chromosome III and I helped them some with strain construction and the mysteries of *cdc10*. Finally, in 1986, I spent part of a sabbatical year at Genentech working with Ron Hitzeman. I had intended to be on still another sabbatical leave during the time I was in Basel in 1991, but the University of California made a retirement offer that I felt I could not refuse. So, on July 1, 1991, I retired as an active faculty member, and a few days later my wife and I left for Basel where I worked on this chapter as well as on chapters about Lindegren and Winge. Being retired relieves me from teaching and committee duties, but I still remain actively involved in yeast genetics research and hope to continue to do so for several years.

## REFERENCES


Bruenn, J. 1972. Characterization of a recessive lethal amber suppressor strain of *Salmonella typhimurium* by in vitro synthesis of T4 lysozyme. *Biochim. Biophys. Acta* **269:** 162–169.

Cox, B.S. and J.M. Parry. 1968. The isolation, genetics, and survival characteristics of ultraviolet light-sensitive mutants in yeast. *Mutat. Res.* **6:** 37–55.

de Robichon-Szulmajster, H. and R.K. Mortimer. 1966. Genetic and biochemical studies of genes controlling the synthesis of threonine and methionine in *Saccharomyces cerevisiae. Genetics* **53:** 609–619.

Eddy, A.A. and D.H. Williamson. 1957. A method of isolating protoplasts from yeast. *Nature* **179:** 1252–1253.

Ephrussi, B. 1953. *Nucleo-cytoplasmic relations in microorganisms, their bearing on cell heredity and differentiation.* Clarendon Press, Oxford.

Fogel, S. and R.K. Mortimer. 1969. Informational transfer in meiotic gene conversion. *Proc. Natl. Acad. Sci.* **62:** 96–103.

———. 1970). Fidelity of meiotic gene conversion in yeast. *Mol. Gen. Genet.* **109:** 177–185.

Fogel, S., R.K. Mortimer, K. Lusnak, and F. Tavares. 1979. Meiotic gene conversion: A signal of the basic recombination event in yeast. *Cold Spring Harbor Symp. Quant. Biol.* **43:** 1325–1341.

Game, J.C. and B.S. Cox. 1971. Allelism tests of mutants affecting sensitivity to radiation in yeast and a proposed nomenclature system. *Mutat. Res.* **12:** 329–331.

———. 1972. Epitstatic interactions between four *rad* loci in yeast. *Mutat.Res.* **16:** 353–362.

———. 1973. Synergistic interactions between *rad* mutations in yeast. *Mutat. Res.* **20:** 35–44.

Game, J.C. and R.K. Mortimer. 1974. A genetic study of X-ray sensitivie mutants in yeast. *Mutat. Res.* **24:** 281–292.

Gilmore, R.A. 1967. Super-suppressors in *Saccharomyces cerevisiae. Genetics* **56:** 641–658.

Gilmore, R.A., J.W. Stewart, and F. Sherman. 1971. Amino acid replacements resulting from super-suppression of nonsense mutants of iso-1-cytsochrome *c* from yeast. *J. Mol. Biol.* **61:** 157–173.

Hawthorne, D.C. 1955. The use of linear asci for chromosome marking in *Saccharomyces. Genetics* **40:** 511–518.

Hawthorne, D.C. and R.K. Mortimer. 1960. Chromosome mapping in *Saccharomyces*: Centromere-linked genes. *Genetics* **45:** 1085–1110.

———. 1963. Super suppressors in yeast. *Genetics* **48:** 617–620.

———. 1968. Genetic mapping of nonsense suppressors in yeast. *Genetics* **60:** 735–742.

Ho, K.S.Y. 1975. Induction of DNA double-strand breaks by X-rays in a radio-sensitive strain of yeast. *Mutat. Res.* **20:** 45–51.

Johnston, J.R. and R.K. Mortimer. 1959. Use of snail digestive juice in isolation of yeast spore tetrads. *J. Bacteriol.* **78:** 292.

Lemontt, J.F. 1976. Mutants of yeast defective in mutation induced by utraviolet light. *Genetics* **68:** 21–33.

Lindegren, C.C. 1949. *The yeast cell, its genetics and cytology*. Educational Publishers, St. Louis, Missouri.

Manney, T.R. 1964. Action of a supersuppressor in yeast in relation to allelic mapping and complementation. *Genetics* **50:** 109–121.

Manney, T.R. and R.K. Mortimer. 1964. Allelic mapping by X-ray induced mitotic reversion. *Science* **143:** 581–583.

Mortimer, R.K. 1955. Evidence for two types of X-ray induced lethal damage in *Saccharomyces cerevisiae. Radiat. Res.* **2:** 361–368.

———. 1958. Radiobiological and genetic studies on a polyploid series (haploid to hexaploid) of *Saccharomyces cerevisiae. Radiat. Res.* **9:** 312–326.

Mortimer, R.K. and S. Fogel. 1974. Genetical interference and gene conversion. In *Mechanisms in recombination* (ed. R. Grell), p. 236. Plenum Press, New York.

Mortimer, R.K. and D.C. Hawthorne. 1966. Genetic mapping in *Saccharomyces. Genetics* **53:** 165–173.

———. 1973. Genetic mapping in *Saccharomyces*. IV. Mapping of temperature-sensitive genes and use of disomic strains in localizing genes. *Genetics* **74:** 33–54.

———. 1975. Genetic mapping in yeast. *Methods Cell Biol.* **11:** 221–233.

Mortimer R.K. and J.R. Johnston. 1986. Geneology of principal strains from the Yeast Genetics Stock Center. *Genetics.* **113:** 35–43.

Mortimer, R.K. and D. Schild. 1980. Genetic map of *Saccharomyces cerevisiae. Microbiol. Rev.* **44:** 519–571.

———. 1985. Genetic map of *Saccharomyces cerevisiae,* edition 9. *Microbiol. Rev.* **49:** 181–213.

Mortimer, R.K. and C.A. Tobias. 1953. Evidence for X-ray induced recessive lethal mutations in yeast. *Science* **118:** 517.

Mortimer, R.K., R.S. Lerner, and J.K. Barr. 1957. Ultlraviolet-induced biochemical mutants of *Saccharomyces cerevisiae*. U.S.A.E.C. *Doc.* UCRL **3746:** 1–10.

Mortimer, R.K., D. Schild, C.R. Contopoulou, and J. Kans. 1989. Genetic map of *Saccharomyces cerevisiae,* edition 10. *Yeast* **5:** 321–404.

Nakai, S. and S. Matsumoto. 1967. Two types of radiation-sensitive mutants in yeast. *Mutat.Res.* **4:** 129–136.

Nakai, S. and R.K. Mortimer. 1969. Studies on the genetic mechanism of radiation-induced mitotic segregation in yeast. *Mol. Gen. Genet.* **103:** 329–338.

Něcas, O. 1956. Regeneration of yeast cells from naked protoplast. *Nature* **177:** 898–899.

Owen, M.E. and R.K. Mortimer. 1956. Dominant lethality induced by X-rays in haploid and diploid *Saccharomyces cerevisiae. Nature* **177:** 625–626.

Resnick, M.A. 1969. Genetic control of radiation sensitivity in *Saccharomyces cerevisiae. Genetics.* **62:** 519–531.

Resnick, M.A. and P. Martin. 1976. The repair of double-stranded breaks in the nuclear DNA of *Saccharomyces cerevisiae* and its genetic control. *Mol. Gen. Genet.* **143:** 119–129.

Roman, H. 1956. A system selective for mutations affecting the synthesis of adenine in

yeast. *C.R. Trav. Lab. Carlsberg Ser. Physiol.* **26:** 299–314.
———. 1957. Studies of gene mutation in *Saccharomyces*. *Cold Spring Harbor Symp. Quant. Biol.* **21:** 175–185.
Sherman, F. 1959a. The effects of elevated temperatures on yeast. I. Nutrient requirements for the growth at elevated temperatures. *J. Cell. Comp. Physiol.* **54:** 29–35.
———. 1959b. The efects of elevated temperatures on yeast. II. Induction-respiratory deficient mutants. *J. Cell. Comp. Physiol.* **54:** 37–52.

# Taming the Oldest Domesticated Organism

ROBERT C. "JACK" VON BORSTEL
*Department of Genetics*
*University of Alberta, Edmonton*
*Alberta, Canada T6G 2E9 and*
*Basel Institute for Immunology*
*CH-4058 Basel, Switzerland*

*It is highly creditable to the ingenuity of our ancestors that the peculiar property of fermented liquids, in virtue of which they "make glad the heart of man," seems to have been known in the remotest periods of which we have any record.*

Thomas Henry Huxley, 1871

## INTRODUCTION TO YEAST GENETICS

In 1950, John Preer taught a course on the genetics of microorganisms at the University of Pennsylvania. It was there that I learned of the remarkable happenings in the genetics of phage and bacteria, where Max Delbrück and Josh Lederberg were the respective kings, and about the cytoplasmic inheritance circus in paramecium in a world led by Tracy Sonneborn. I also was fascinated by the weird wonders of yeast genetics, then the fiefdom of Øjvind Winge and Carl and Gertrude "Gerry" Lindegren. I read papers on adaptive enzyme regulation by Sol Spiegelman, who then was just emerging from Carl Lindegren's stimulating influence. Boris Ephrussi was studying cytoplasmic inheritance in yeast, and since his papers were in French, this was one way to practice for the French examination, which every graduate student was required to pass in the halcyon days of yore.

I was being trained in *Habrobracon* genetics by Anna Rachel Whiting and in *Sciara* cytogenetics by Charles W. Metz. Since *Habrobracon* was more tractable, I concentrated my efforts there to find out how cells were killed by mutagens. Dominant lethals and cell killing by radiation and chemical mutagens are what I thought about as a graduate student, and what I continued to work on at Cold Spring Harbor (1952–1953) and Oak Ridge (1953–1971).

At the annual meetings of the Radiation Research Society during the decade of the 1950s, I met Bob Mortimer, who was interested in differences between yeast haploids and diploids in the expression of radiation damage. As a way into radiation killing, haploidy and diploidy were among my arsenals as well. Yeast and *Habrobracon* had something in common. Sex was determined by an azygous haploid and a diploid, with a heterozygous mating-type locus in both organisms. In addition, both Bob and I could make triploids; diploids of homozygous mating type could also be made. Moreover, as Carroll Williams pointed out after a lecture I presented at Harvard University many years later, "Yeast is a kind of an insect. After all, it has a chitinous exterior."

At the 1st International Congress of Radiation Research in 1958, Bob

Mortimer challenged us to devise a way to determine whether the haploid or diploid sperm of *Habrobracon* was the more sensitive to the induction of dominant lethals. Both haploid and diploid males could be obtained in *Habrobracon* if triploid females were bred unmated. In yeast, Bob had shown that at a given dose, more dominant lethality was induced in diploid cells than in haploid cells. In those days, DNA repair was barely understood. I told him that the wasp *Mormoniella* would be a better organism because we could get diploid sperm from fully fertile diploid males. Diploid *Habrobracon* males tend to be somewhat sterile.

So we began our collaboration using a wasp that neither of us had reared before. I flew out to Berkeley carrying strains of *Mormoniella* and the host blowfly pupae on which the wasps were reared. Phineas Whiting contributed the wasps and the host. Naturally, nothing worked the first time through, but we did enough to learn how to use the wasps. We then carried out separate experiments in Berkeley and Oak Ridge and joined forces in Pavia to write the paper (Mortimer and von Borstel 1963). We found that, like yeast, diploid *Mormoniella* sperm was twice as sensitive as the haploid sperm for the induction of dominant lethality. This was my second paper in collaboration with a yeast geneticist; the first was with Giovanni Magni (Magni and von Borstel 1962).

## IL ISTITUTO DI GENETICA, UNIVERSITÀ DI PAVIA

I made another bargain at the 1958 International Congress of Radiation Research with Giovanni Magni to work with him in Pavia to learn about yeast. I was into chromosomal translocations in *Habrobracon* at that time, using inherited partial sterility as the detector. Mary-Lou Pardue and I were busily demonstrating that the wasp embryos containing translocations as products of adjacent I and adjacent II meiotic segregation were dying in a manner that mimicked one type of dominant lethality. So I proposed to Giovanni that we study translocations in yeast. He said, "OK, come on, but with that experiment there will be problems."

The National Science Foundation was feeling its way toward building scientific capability in the United States as an answer to Sputnik, and in 1959, I was awarded an NSF senior postdoctoral fellowship, a good tradition that died after a few years but not before I was able to benefit from it. The Institute of Genetics was lodged in an old monastery within the city walls, and Adriano Buzzati-Traverso was Il Professore. It was there that I learned my first Italian scientific word, "*grummi*," which means clumps. Giovanni had obtained his strains while working in Winge's laboratory, and clumpy they were. I also learned "*piove*" and "*nebbia*"; rain and fog are Po Valley fall and winter weather. Franco Guerrini took me in hand and taught me an obscene word each day without translating it. He made me pronounce it correctly and then would laugh wickedly and go away. I was purposely given a desk in an office shared with a graduate student, Clara Ghini, who was stunning to look at, but spoke no English. I was told that this would give me an incentive to learn Italian. At the end of the year, Clara spoke English perfectly.

Giovanni was right; chromosomal translocations were impossible to study because of sterility problems probably caused, ironically enough, by translocations. So in my clumsy attempt to learn something about yeast genetics, we started measuring spontaneous mutation rates during mitosis and meiosis, because Giovanni had the notion that we might find something interesting. The trick was to measure the spontaneous mutation rate in diploid mitotic cells and then to send the cells directly into meiosis without any intervening vegetative growth. High sporulation frequency was therefore highly desirable. One of the very good things about Giovanni's strains was their high sporulation frequency. Since asci were dissected only as a last resort in Pavia, there was an unconscious selection for strains with high sporulation frequencies, but the frequency of four-spored asci tended to be low. Even so, it was necessary to concoct a new sporulation medium that contained no YEPD in order to obtain high frequencies of sporulation for some of the less tractable strains, and it was essential for sporulation to take place in all strains without intervening mitoses. That was my job. It was just the sort of witchcraft for me. It only required patience and a loose imagination. Silvio Sora and I worked closely together in Giovanni's laboratory, and it was essential for us to understand each other perfectly. So I spoke slow Italian to him, and he answered in slow English. Silvio told me not long ago that VB medium, which he named after my child-like efforts, is still used in his laboratory as a sporulation medium.

After a number of experiments with lots of controls, we observed that some of the mutants had much higher spontaneous mutation rates during meiosis than during vegetative growth. This started us off on a new project, the origin of spontaneous mutations, an area of study that has kept me occupied ever since.

## YEAST GENETICS AT OAK RIDGE

Seymour Pomper was the first yeast geneticist at Oak Ridge, but he had taken a position at Standard Brands before I came to Oak Ridge in 1953. Pomper's interesting contribution to biology was being the first scientist to bring evolution to a stop deliberately. At Standard Brands, he froze a culture of yeast that was used to reseed all of the Standard Brands fermenters when they cleaned out the vats two or three times a year. Don Hawthorne once told me that he had analyzed the Standard Brands yeast from the grocery shelf and found what he thought were nearly all of Pomper's mutants. Don hypothesized that Seymour had mixed all of his stocks together, which then became Standard Brands Yeast forevermore. Only Seymour Pomper knows for sure.

I returned to Oak Ridge from Pavia in 1960 and started to work on yeast. Giovanni and I had decided we would work on slightly different aspects of the same problem. He worked more rapidly than I, but when he visited me in Oak Ridge in 1961, we had enough data to write a paper on the meiotic effect, with all of the controls in place to demonstrate that the spontaneous reversion rate was higher during meiosis than during mitosis with many of the mutants (Magni and von Borstel 1962). This paper was among the first to give credence to the notion that spontaneous mutation rates were not fixed by the physical principle of keto-enol shifts but depended on the metabolic state of the organism itself. The realization was just beginning that keto-enol shifts could produce only transitions, if anything, and the wonderful garden of mutational genetics was beginning to yield all sorts of other variants (von Borstel 1969a; Drake 1970). The paper was also the first example of what is now known as hypermutation, where Mother Nature used increased mutation rates for her own diabolical purposes, such as increasing antibody diversity.

Giovanni and I talked about a number of ideas during his visit. He told me about testing hydroxylamine mutagenesis in yeast. He had had trouble getting enough hydroxylamine into solution to be mutagenic, so he increased the salt content in an effort to get the hydroxylamine into solution. Giovanni and Silvio found that this increased salt content caused plenty of mutagenesis. They also used salt alone as a control and, to their astonishment, found the same amount of mutagenicity with salt alone as with salt plus hydroxylamine. Giovanni published the salt results in a report to a granting agency in 1964, not in a full publication. At the Cold Spring Harbor Symposium in 1966, Giovanni met Ernst Freese, who had first reported hydroxylamine mutagenesis in 1961 and had reasoned that the specific mutagen receptor was cytosine. Giovanni asked Ernst why he had used so much pyrophosphate (1 M) when he carried out hydroxylamine mutagenesis. Giovanni told me that Ernst just smiled. In 1986, Kenneth Parker picked up where Giovanni and Silvio had left off (Parker et al. 1986; Parker and von Borstel 1987). Salt is probably the safest mutagen to work with in the laboratory, and Ken has shown that 1 M or 2 M solutions increase mutation frequency up to 100-fold in yeast cells in logarithmic phase, and at high concentrations, the mutagenic effect is in good correlation with the molarity (or osmolality).

I plugged along, working on too many different projects, as usual. When Charley Steinberg joined the Biology Division at Oak Ridge in 1962, he straightened out my thinking on some of the problems, and we published an abstract in 1964 that gave the data for a known base substitution, a supersuppressible mutant (von Borstel et al. 1964). Charley also taught me economy; he said "if it all can be said in an abstract, don't bother to write a paper." Anyway, the supersuppressible mutant did not demonstrate the "meiotic effect," which correlated well with Giovanni's demonstration that the meiotic-effect mutants exhibited outside marker exchange whenever a reversion took place. In addition, Magni (1964) found that temperature-sensitive mutants do not exhibit a meiotic effect. For many years, we assumed the genetic evidence to be true that meiotic-effect mutants were frameshifts, but we always were careful to refer to them as putative frameshifts. Even as recently as 5 or 6 years ago, I received an outraged review of a manuscript of mine saying that we had no evidence that mutants exhibiting the meiotic effect were frameshifts, even putative ones. We now have DNA sequence data showing that at least one of our favorites, *hom3-10*, is a +1 frameshift (Wang et al. 1990), and of course we had known since 1963 that base substitutions do not show the meiotic effect.

I had once offered to collaborate with Fred Sherman, who had the only system at the time that was any good for back-translating base sequences in yeast, to look at spontaneous mutation rates in mitosis and meiosis for some of his well-defined frameshifts, but he said he wanted to do it himself to prove that we were wrong. I always like it when people say they will do the experiment I offer to do, so that I can go ahead and do something else. (It was the biophysicist Bill Arnold who taught me that the best experiment is the one you can look up in the library, the second best is the one you can talk someone else into doing, and the third, and worst, is the one you have to do yourself.) But Fred never got around to it, or merely found evidence that we were right. Anyway, he never published his findings, and we were left in molecular limbo until we entered the new era of molecular genetics with special reference to DNA sequencing.

Tom Manney came to the Biology Division from Berkeley in 1964 to add to the yeast endeavors. His careful work combining the genetics and biochemistry of the tryptophan pathway was a welcome addition at Oak Ridge. In those days, it was the type of research needed to make the leap from the prokaryotes to the eukaryotes. Tom always selected fat and juicy strains that grow perfectly, and his strains are useful to me even now.

One of my favorite publications from the Oak Ridge years of the 1960s was a note that Charley Steinberg, Giovanni, and I published in the *Journal of Molecular Biology* in 1966 using tight genetic evidence to argue that supersuppressors had to encode mutants of tRNA, or sRNA as it was known then (Magni et al. 1966). Mario Capecchi had published a paper stating that it had to be either a tRNA or a tRNA-modifying enzyme but argued that the suppressors encoded tRNA. Norton Zinder published similar data but argued that the suppressors encoded tRNA-modifying enzymes. Giovanni and I knew we had the data that could demonstrate genetically that supersuppressors really encoded tRNA and were wrestling with how to say it when Charley stuck his head in the door of my tiny office and asked us what we were doing. We explained our problem. He said "Oh, that's easy," and dictated the paper off the top of his head. As I recall, we made one small change in the text before mailing it a few days later (Magni et al. 1966). Sal Luria sent us a postcard disputing our reasoning, but being a phage geneticist, he had forgotten the power of dominance. Gunther Stent told Giovanni that he knew that any one of us could have written the paper but he said that he believed that Giovanni had provided the data. Actually, I had provided one third of the data with the prodding of Charley's sharp goad. Francis Crick remarked to Giovanni that the paper was well written, but good-naturedly added "We don't believe it!" In those days, the hypotheses capable of being generalized from genetic reasoning were permitted to originate only in Cambridge, Paris, or Pasadena (cf. Judson 1980).

About 1964, I became fed up with making residual growth measurements, so Charley suggested that we devise a compartmentalization method for measuring spontaneous mutation rates, and it worked just right. The cells were not plated from growth medium with the confounding residual growth happening after the cells hit the plate of omission medium; the cells simply ran out of the limiting amino acid or base in liquid medium, and the revertants continued to grow. This was a substantial step forward. We used 100-compartment boxes that Fred de Serres had had designed for studying complementation in *Neurospora*, which we taped to keep evaporation low. We now use 24-compartment tissue-culture trays and keep the samples from evaporating by keeping them in petri dish bags that we tape shut or we seal them in resealable sandwich bags from our local grocer.

In 1967, Mike Resnick sent me some repair-deficient mutants. Two summer students, Kathy La Brot and Dale Graham, helped me to demonstrate that these mutants were mutators (von Borstel et al. 1968). In 1968, Bob Mortimer invited me to Berkeley to assist in his Radiation Biology course. It was there that I completed the paper on the three Rs of the origin of spontaneous mutations: recombination, replication, and repair. The paper was presented in 1968 at the International Congress of Genetics in Tokyo (von Borstel 1969a).

Seymour Fogel was on sabbatical leave in Berkeley working with Bob on their first paper in that wonderful series of studies on recombination while I was there, and the place buzzed with models of gene conversion and recombination. I had been invited as a visiting lecturer to the University of California, and I was told that it was essential to sign a loyalty oath to the State of California. I found this a bit distasteful, since only people who carry guns, or would like to, consider loyalty oaths a proper method of ensuring uniform subservience. Charley Steinberg suggested I prepare a codicil to submit with the loyalty oath. So we wrote one that stated two exceptions: First, I must support the State of Tennessee in case of border disputes despite any position taken by the State of California; second, I would not carry out any loyalist activities for the State of California which might jeopardize my becoming a Tennessee Colonel. Sy Fogel and Bob Mortimer went with me as witnesses to the office for the signing of the loyalty oath, and we presented a sober and solemn demeanor throughout.

In early 1969, while I was sitting in my office busy with busy work, Charley came in and asked me to come into his laboratory for a minute. He said, "Jack, you've been talking for too long a time about selecting for mutator mutants, and you're not going back into that office until we've done it." So we did it, mutagenizing and selecting mutants, coming in at all hours to lift up the lids of plates and cheering the little monsters along. We had read somewhere that chlorophenylalanine was an aphrodisiac; Parkinson's disease patients treated with the drug chased their nurses around the room. So we thought it fitting to look for chlorophenylalanine resistance in yeast as an assay for mutator ability. That did not work very well, so we used fluorophenylalanine instead, and it worked wonderfully, but we didn't test its aphrodisiac properties. Winnie Palmer, one of the more energetic research associates in the Biology Division, insisted that we put some chlorophenylalanine in the communal coffee urn, but we didn't follow her advice.

During our mutator screening process, I noted one day that revertants of *lys1-1* grew up through the lawn on the Mortimer Complete medium we were using. When I added more lysine to the medium, the lawn became thicker, and no more revertants poked their heads through the lawn. For some unremembered reason I had marked "LAS" on the plates that showed the revertants. So Charley immediately invented the term "lassie test" as a qualitative way to show whether a strain was a mutator or not. We used the lassie test to screen for mutator mutants. Since it was slow and tedious to use the conventional glass spreaders (which in Canada are called "hockey sticks") used by microbiologists everywhere, Charley and I pioneered the use of glass pipets and a turntable to spread cells on plates. All that was needed for this technique was lots of 1-ml pipets, but it certainly sped up the spreading of cells.

When Alex Hollaender retired as Director of the Biology Division in 1968, the glorious scientific atmosphere that he had built began to wither. One summer day in 1970, I realized that it was not just sick, it was dead, nevermore to be revived. An example of the paranoia that had been creeping into the place was characterized by the plastic and tape sculpture that was hung in one of the stairwells by one of the inventive staff members: Certain members of the administrative staff felt so unloved that each believed independently that he had been hung in effigy. It seemed like a good idea, but that was not the intent. Lucien Caro and Grete Kellenberger were the first to depart, and Charley was next. I followed shortly thereafter, and 20 or 30 more geneticists left within the next 2 or 3 years. The 200–250 scientists in the Biology Division in 1969 have now shrunk to about 30–50 scientists to inhabit the ruins. *Sic transit gloria mundi.*

## THE FIRST GATHERINGS OF YEAST GENETICISTS

At the beginning of the Genetics Society of America in 1961, Seymour "Sy" Fogel suggested to me that we should hold a yeast meeting. I liked the notion, and we decided that the best place to hold the meeting would be in Carbondale and that the objective of the meeting would be to straighten out the nomenclature. Magni was finding mutant genes and naming them, as were Bob Mortimer and Don Hawthorne. Of course, the Lindegrens and Winge had been naming mutants for a number of years before. Bob and Don had pooled their information, so Sy and I decided we should call everyone together in Carbondale, as a bow to Carl and Gerry Lindegren. Also, Maurice Ogur was on the staff at Carbondale; he had been a postdoc at the University of Pennsylvania when I was a graduate student there. I had known him to be a very sound biochemist, and I was happy that he was working on yeast. He showed his clout by working out some of the principles of *petite* detection.

In September 1961, Giovanni came to Oak Ridge for our yearly collaboration, and we began our trek to Carbondale. We flew first from Knoxville to Lexington and then changed planes. It was raining exceedingly hard. As we sat in the DC-3 waiting to take off, Giovanni noticed a man putting gasoline into the wing tank in this terrible rainstorm. Finally he stopped and tried to put the cap on the tank. He hit the cap with his fist and it bounced right off. He grabbed the cap again, jammed it in place without banging on it, and then ran for cover. We were on our way to Paducah. The DC-3 flew low, and it bounced a lot. The plane's atmosphere was that of depression, heat, and tropical humidity. When we arrived in Paducah, we had to get off even though the flight was supposed to go on to Carbondale. It was announced that the plane was mysteriously out of fuel and could not be refueled in Paducah. Apparently, the cap had come off and the gasoline had siphoned out of the tank. We were provided with a taxi to travel on to Carbondale in the heavy rain. We were very late in arriving, so it was necessary to share a room with Giovanni at the motel. Giovanni announced to me "I hope that you are a sound sleeper, because I snore!" He kept his promise.

The treetop trip with its possible frightening outcome plus the dismal noisy night in the motel did not augur well for the meeting. The first session began the next morning. It was more of a haggling session than a meeting. There were 11 people there: Carl and Gerry Lindegren, Maurice Ogur, David Pittman, and Ernie Shult were from Carbondale; Bob Mortimer, Don Hawthorne, Sy Fogel, and I represented the United States, Giovanni Magni was from Italy, and Robert Doyle was from Canada. This group represented as much of the world of yeast genetics as we could muster. Also present at the meeting from time to time were the enthusiastic students who worked with Carl and Gerry, e.g., J.K. Bhattacharjee and six or seven others.

We listed all of the genes, but then nothing happened. After 2 days, Giovanni got up, went to the blackboard, and said "We shall call this gene *adenine*

*1*." Pointing his finger at each person present, he went around the room saying accusatively, "Right! right! right! right! right!... . Okay! We shall call this gene *adenine* 2; right! right! right! right!... . Okay!" In about 30 minutes, all of the genes were named, with everyone in concordance, some cheering silently, others with dismay or relief written on their faces.

In 1986, at the 13th International Yeast Genetics Conference in Banff, Giovanni told me that the biennial international conferences had been misnumbered. I asked him why, and he told me that the 1st Conference had been in reality Conference Number Zero.

The second gathering was arranged in 1963 at Gif-sur-Yvette by Piotr Slonimski just before the International Congress of Genetics in The Hague. This was a real meeting, where we each discussed our current work. It was clear then that yeast was headed for a great future as the transfer organism of prokaryotic discoveries into eukaryotic systems. It was really good to meet the people whose work I had read with deep interest.

The third gathering was called by Herschel Roman in Seattle at the inauguration of the new building for the Department of Genetics in 1966. The meetings had now begun to take on an aspect of a series of conferences. It was at this meeting that Giovanni declared, during an evening round of libation, that the yeast community was somewhat like the true Church: Herschel Roman was the Pope, and anyone who had reached the age of 40 in the yeast genetics business was a Cardinal, each taking some small responsibility as keepers of the churchly duties. In addition, the conferring of Cardinalship at the age of 40 was a small compensation for the leftover memory from everyone's childhood that 40 was *really old*. Presumably, although the originators of yeast genetics, Winge and Lindegren, were still alive and well, they already had passed on to some mysterious sainthood where their lives would be wrapped in the blissful kisses of the heavenly seraphim and cherubim.

Although the international yeast genetics community is indeed managed in a manner somewhat less than what one could call participatory democracy, there are no dues to be paid, and it gives ample room for everyone to do what they want; in addition, when yeast is no longer needed as the most representative eukaryotic cell, the community can quietly disappear. Fragility from a no-dues society also provides strength. The last time an intervention was needed was when Hersch retired as Pope, and an election as mysterious as a Vatican election was held with Urs Leupold officiating by mail. Piotr Slonimski ascended to the papal role. As was announced to the congregation at the 12th International Conference in Edinburgh in 1984 after the vote count where Piotr was elected the Chairman of the Steering Committee, "There is more than one organization with a Polish Pope!"

After conferring with Bob, I went ahead and attempted to push through a variation of the three-letter nomenclature that Milislav Demerec had proposed for mutants of *Escherichia coli* and *Salmonella typhimurium*. I knew there would be resistance in the yeast community, but I was astonished to find that I was encountering resistance from Demerec as well. After a series of letters sent with copies to longer and longer lists of recipients, Demerec called a meeting in conjunction with the 1966 Cold Spring Harbor meeting where lots of folks made lots of suggestions. The result was many of the genetic nomenclature rules we use today.

The fourth International Conference was held in Osaka in 1968 as a satellite meeting for the International Conference of Genetics in Tokyo. The Japanese scientists had not yet shifted into well-spoken English. One speaker, H. Kasahara, haltingly said he could not speak English well, but he had studied in Germany in pre-war days, so he would speak in German about Sake yeast. It was difficult for many of us to understand his discussion of segregation of respiration deficiency in natural populations of Sake yeast. After a short wait and no questions, Gordin Kaplan asked him, in excellent German, which species of yeast he had studied. The reply was "*Saccharomyces cerevisiae*!" We all understood that.

A young speaker from Japan told of his studies of yeast. He seemed to be quite intelligent, and spoke very well, but the experiments seemed to have started from nowhere and were going to the same place. In the question period, I asked him why he had done this experiment. He replied, "Because I was ordered to do so by my Professor." It was plain that he wasn't happy about it either.

In Japan, Gordin Kaplan announced his intentions to hold the next Yeast Genetics Conference in Canada in 1970, with he and Allen James as organizers. Gordin and I counted up how many meetings had already been held, so the Chalk River meeting was known as the 5th International Conference, the first one to have a number. The meeting was informal and the wine and whiskey were free and plentiful. In the evenings, Gordin Kaplan, Pierre Meuris, and other members of the French contingent always sat at a table drinking wine and harmoniously and quietly singing French songs. The continuous melodious harmonization in the background made me liken the singing to that of the elves at their final departure in Tolkien's *Lord of the Rings*, a book which everyone had been reading at that time. One evening I was sitting at a table with Denise Duphil, and she made a snarky remark about the singers. I listened for a moment and suddenly realized they were, and had been, pummeling the welkin with rich ribaldry that was strong even by today's standards.

The final comment at the meeting by Sy Fogel, while he was thanking the Canadian organizers, summed up everyone's feelings when he said that he had never before really appreciated the signs he had seen on billboards around Canada: "Drink Canada Dry!"

My duty at those first conferences was to write a little meeting report, usually for *Science* (von Borstel 1963a, 1966a, 1969b; Esposito et al. 1972), and to list the genes, the gene products by Enzyme Commission number and description, and the gene nomenclature (von Borstel 1963b, 1964, 1966b, 1968, 1969c; von Borstel et al. 1971b; Plischke et al. 1976). Bob and Don were to make new editions of the genetic map. Bob has kept this tradition alive and is now up to the 11th edition. Jim Broach took over the listing of genes and gene products for the 1981 yeast books, and, hopefully, now that the oars are in his hands, he will keep rowing the boat.

## EDMONTON

When I was asked for an interview for the position of Chairman of the Department of Genetics at the University of Alberta in December 1970, I had no intention of taking the job. So on the plane to Edmonton, I decided that I had to ask for something important but impossible. I settled on the notion that I would take the position only if the University would provide a technician not only for me, but also for everyone in the Department of Genetics. This is what we had at Oak Ridge, so that would be the minimum for Edmonton. The Acting Chairman, John Kuspira, invited me to dinner at his home upon my arrival. As subtly as I could, I pushed my outrageous notion forward. He stated, "We already have that; is there anything else you would like?" So in my amazed state, I asked, "What other things do you have?"

I was astonished to find that the Dean of Science was a man who had just received a prize for his current research, that the Vice President Academic was helping a student convert one of the Vice President's short stories into a play, and that the President's manuscript had just been accepted by a mathematics journal. What kind of a place was this? The traditional unusable and unsuccessful people were not in the administrative jobs. The good things that every department should have were already there. Moreover, the Department of Genetics harbored such geniuses as Phil Hastings, Rustem Aksel, and Mike Russell, with other quite respectable geneticists present and working as well. Something could be done. So I took the position. The main office was on the second floor and my office and laboratory were on the fifth floor; I installed Dennis Wighton, the Administrative Professional Officer, in the Chairman's office. He was very good at talking with the random visitors who appeared too often. The Departmental Secretary came to my office for 2 hours each day so that administrative duties could be fulfilled (I learned that trick from Demerec), and laboratory business went on almost as usual. Nevertheless, administrators have about a 3-minute attention span, and I eventually discovered that all day Saturday and every Tuesday evening were mine to do uninterrupted research.

I gradually found out that the University was just like every other institution on earth in that improvement of fringe benefits was the first order of business. Nevertheless, there was something to build with, and build we did. Genetics was always popular in Edmonton due to the enthusiastic teaching done by John Kuspira, and genetics was being taught to about one quarter of the students who attended the University, a percentage unchanged even now that the University has grown to about 25,000 undergraduate students. During the two 5-year terms of my Chairmanship, we were able to make research dollars per person spiral upward at a rate six times faster than the University as a whole, until we had more dollars per person for research than any department in the Faculty of Science. It was time to quit being Chairman.

Right away, after arriving in Edmonton, I had a tiny group. Fred Flury arrived in Edmonton from Urs Leupold's laboratory to demonstrate that *petite* strains had higher spontaneous mutation rates for nuclear genes than *grande* strains (Flury et al. 1976); Siew-Keen Quah had been a graduate student of Charlotte Auerbach and was visiting Mike Russell, her long-time friend from Edinburgh days. She decided to stay in Edmonton to carry out research on yeast. Rita Schuller was a remarkable undergraduate whose first experiments on a color test, using reduced adenine in the medium in conjunction with *ade2-1* to show red locus mutants of *arg4-17* versus white suppressors, worked well (Schuller and von Borstel 1974), and Gail Meadows came into my laboratory as a graduate student. Rita and Gail were the first to cross the bridge from our Department to the Basel Institute of Immunology, where Rita was the technician with the strong intellectual and technical capabilities to help Susumu Tonegawa begin his long journey to Stockholm. It was in Edmonton that we worked on the mutator genes that Charley and I had selected in Oak Ridge, and, in 1972, with the help of Siew-Keen Quah, we developed strain XV185-14C that had the useful markers *hom3-10*, *his1-7*, *arg4-17*, *lys1-1*, *trp5-48*, and *ade2-1*. The mutant *hom3-10* is a frameshift mutation, as defined by the meiotic effect (Magni 1969), by intercalation mutagenesis (Magni et al. 1964; Meadows et al. 1973; Hennig et al. 1987), and now by DNA sequence analysis (Wang et al. 1990). The mutant *his1-7* is a base substitution revertible by internal missense suppression (Fogel et al. 1978; Lax and Fogel 1978). The other four markers are ocher mutants. By genetic reasoning, I knew that *hom3-10* could revert by both +1 and –1 frameshift mutations, *his1-7* could be reverted by base substitutions of any variety, and *arg4-17* could be reverted by three kinds of transversions, AT→TA, AT→CG, and GC→TA (Gottlieb and von Borstel 1976). The mutant *trp5-48* is a useful mutant for detection of the cytoplasmically inherited ψ factor during the crossing of strains. Since *trp5-48* is suppressed by the ψ factor, which floats through the strains of many laboratories, we needed to know when it had found its way into our strains. The ψ factor screws up mutation rate measurements when suppression is one of the endpoints.

In 1978, I used the lassie test to screen for mutators and antimutators under conditions where all surviving colonies were tested. The work was quantitated by screening for auxotrophs at the same time (von Borstel and Lynch 1978). Assuming Poisson relationships, rather than assuming that there was only one great big gene, I was able to estimate that the yeast genome had approximately 200 different possible mutator genes and about 20 different possible antimutator genes by making a reasonable guess about the total number of possible auxotrophs. The work was never published because at that time I was being glued together with medication in preparation for cardiac surgery, and my able technician, Audrey Lynch, had followed her husband to the States, leaving me in the lurch.

## THE USE OF YEAST FOR ENVIRONMENTAL MUTAGENESIS STUDIES

An international crisis of a sort propelled many of us at Oak Ridge into environmental mutagenesis: Hycanthone was found to be an excellent antischistosomal agent, but after about 2 million people in tropical nations had been administered the drug, an international fear arose that hycanthone might be a powerful carcinogen, teratogen, and mutagen. As a radiation research center, we were positioned to test the chemical in many ways, and Alexander Hollaender was there pushing us into it. Like everyone else at Oak Ridge with one or more mutagen testing systems, we demonstrated that hycanthone is indeed a powerful mutagen. Charley Steinberg, Roger Smith, and I did not stop there. Another national problem was the mace that was being used widely to quell riots in the late 1960s. We thought we could have some fun with the purchasing department with their self-proclaimed role of company censorship by ordering some for scientific use, as Alan Conger had fun earlier by ordering gross lots of condoms for mixing gases for radiation studies on *Tradescantia*. The mace arrived after a few administrative delays, and we found it was an excellent insecticide for *Habrobracon*, and a rather poor carbon source for yeast. It made me remember my postdoctoral days at Cold Spring Harbor Laboratory 15 years earlier, where, for the fun of it, Berwind Kaufmann was demonstrating daily that anything from hamburgers to tomato juice to distilled water would break *Allium* root-tip chromosomes. The charred hamburgers turned into the charred tryptophan which made the Cancer Research Center in Tokyo world famous, and only Mother Nature knows what's in tomato juice and the distilled water at Cold Spring Harbor.

Milislav Demerec, Alexander Hollaender, and Adriano Buzzati-Traverso had one feature in common as successful administrators: They carried out science that was useful at the local, national, and international levels. The national and international levels took care of themselves if proper publicity for the good science was done, but the local level taxed the imagination because it had to have practical as well as local implications. As far as the local acceptance of science was concerned, for Milislav, it was high levels of production of penicillin, for Alex, it was radiation protection and provision of lecturers to visit Southern colleges and universities, and for Adriano, it was the use of mutational techniques for rapid evolution (mutagen treatment and selection in alternate generations) to be used by Renzo Scossiroli as methods for obtaining useful traits in the domesticated plants and animals of Italy.

In light of the examples illustrated by the great administrators of genetics in the 20th Century, my choice at Edmonton was to continue studies on environmental mutagens that might be of significance at local and national levels. The local newspaper always was asking for advice, and mutagen testing was useful for abating the furor over the latest local environmental outrage; i.e., it gave the impression that the Department of Genetics was stepping in as a white knight to slay, allay, or tame the dragons.

Our favorite strain for mutagen testing, XV185-14C, developed in 1972, became a particularly useful adjunct to the magnificent D5 and D7 strains developed by Fritz Zimmermann for testing potential mutagens and carcinogens. His strains are useful for assaying presumptive chemical mutagens for gene conversion and somatic crossing-over, and ours were good for detecting different types of mutations. In this way, the exact mutagenic activity could be assigned for any chemical that was mutagenic or carcinogenic. Gail Meadows was the first to show the power of this combination of mutants (Meadows et al. 1973).

Sándor Igali came from Budapest for 1 year (1973–1974) and helped us to set up a Mutagenesis and Putative Carcinogenesis Research Unit in the Department of Genetics, to be funded only from external sources. The environmental mutagenesis studies were continued, first by Majdi Shahin and then by Ram Mehta for about 10 years. After I had taken two 5-year turns as Chairman, and Ram Mehta had departed, I stopped applying for funds for mutagen testing. We still did testing from time to time, most notably on the water supplies for the city of Edmonton, but I could now concentrate fully on studying the mutational responses of yeast using only a few well-defined mutagens. In other words, let each Chairman justify the Department's existence to the autochthons in his/her own way.

## MORIRE IN PACE

Since 1959, my research has centered most closely around the multiple origins of spontaneous mutations, with analysis of potentially mutagenic chemicals being done to show the power of our mutational methodology. Reasonably important findings about every 3–5 years have shown that we have been plowing a fertile field. Some examples are the meiotic effect (Magni and von Borstel 1962), the lack of it for base substitutions (Magni et al. 1964; von Borstel et al. 1964), the mutator activity (von Borstel et al. 1968, 1971a, 1973) and antimutator activity (Quah et al. 1980) of repair-deficient mutants, and the channeling of lesions into nonmutagenic and mutagenic DNA-repair pathways (Hastings et al. 1976; Brychcy and von Borstel 1977; von Borstel et al. 1977; von Borstel and Hastings 1977).

The stunning series of experiments begun by Anwar Nasim and Charlotte Auerbach (1967), followed by Hannan et al. (1976), on the origin of sectored mutant colonies of yeast was brought to fruition brilliantly by Allen James, Brian Kilbey, and Anwar Nasim (James and Kilbey 1977; James et al. 1978; Kilbey and James 1978; Kilbey et al. 1978) using James's method for pedigree analysis. This work led to our theoretical paper on situation-dependent DNA repair (von Borstel and Hastings 1985) and also has helped us to determine a likely origin of UV-induced hot spots (von Borstel and Lee 1991).

The molecular nature of the spontaneous mutations in the *URA3* locus (Lee et al. 1988), the role of spontaneous mutations as an apparent threshold for mutagenesis (von Borstel 1976; von Borstel et al. 1978), and the possible facts and artifacts we have found that underlie the Cairns deviation in yeast (von Borstel 1978; von Borstel et al. 1992) have given us much food for thought. I am grateful for invitations to work at Mill Hill with Don Williams, John Game, and Lee Johnston; at Orsay with Ethel Moustacchi, Francis Fabre, and Roland Chanet; and at Basel with Georges Koehler, Klaus Karjaleinen, and Charley Steinberg, as well as a special invitation from Iliana Ferrero and Pier-Paolo Puglisi to present a lecture series on spontaneous mutations at Parma. Moreover, invitations to Darmstadt to sip dozens of different wines with Fritz Zimmermann have been gratefully received. Fritz once said to me as we were driving in the Moselle Valley, "Jack, look at all those grapes growing on those hills! It's your and my responsibility to drink up last year's wine so that we can be ready for what's coming!"

During the past 33 years, Bob gave me encouragement and lots of strains, Giovanni propped me up during the formative years, Charley set me up with reliable and accurate methods, Phil Hastings volunteered ideas at a rapid rate, Brian Cox, Peter Dickie, Jürgen Heinisch, Ursula Hennig, Grace Lee, Gary Ritzel, and Rosaura Rodicio developed our storehouse of biochemical and molecular techniques; Martha Bond, Katherine Cain, Dorma Jean Gottlieb, Siew-Keen Quah, Elizabeth Savage, and Ursula Hennig kept the mutational ship afloat, and a multitude of undergraduate and graduate students, technicians, postdoctoral fellows, visiting scientists, and research associates have contributed to all of the above and to much more as well.

Moreover, besides my colleagues in Edmonton, Bob Haynes at York University, Dianne Cox at the University of Toronto, Gordin Kaplan at the Universities of Ottawa and Alberta, Dave Suzuki at the University of British Columbia, and Carol Miller, Earl Nestmann, George Douglas, and Anwar Nasim in governmental institutions in Ottawa, were constant friends and supporters at the national level, and we combined forces in different ways at numerous times to support the best genetic and mutational research that Canada could produce.

I cannot forget that Fred Flury and Qi Wang were absolutely tops in quantitating spontaneous mutation rates by their extraordinarily careful control of factors indigenous to their experiments and that Jennie Chui became the world champion at spreading cells on plates; using the pipet-turntable technique, she could spread 0.2 ml on each of ten plates every minute.

My role seems to be that of fund-raiser and cheerleader.

## REFERENCES


Brychcy, T. and R.C. von Borstel. 1977. Spontaneous mutability in UV-sensitive excision defective strains of *Saccharomyces*. *Mutat. Res.* **45:** 185–194.

Drake, J.W. 1970. *Molecular mutagenesis*. Holden-Day, San Francisco.

Esposito, M.S., R.E. Esposito, and R.C. von Borstel. 1972. Report on the Fifth Yeast Genetics Conference. *Yeast News Letter* **20:** 67–73.

Flury, F., R.C. von Borstel, and D.H. Williamson. 1976. Mutator activity of petite strains of *Saccharomyces cerevisiae*. *Genetics* **83:** 645–653.

Fogel, S., C. Lax, and D.D. Hearst. 1978. Reversion at the *his1* locus of yeast. *Genetics* **90:** 489–500.

Gottlieb, D.J.C. and R.C. von Borstel. 1976. Mutators in *Saccharomyces cerevisiae: mut1-1, mut1-2,* and *mut2-1*. *Genetics* **83:** 655–666.

Hannan, M.A., P. Duck, and A. Nasim. 1976. UV-induced lethal sectoring and pure mutant clones in yeast. *Mutat. Res.* **36:** 171–176.

Hastings, P.J., S.-K. Quah, and R.C. von Borstel. 1976. Spontaneous mutation by mutagenic repair of spontaneous lesions in DNA. *Nature* **264:** 719–722.

Hennig, U.G.G., L.G. Chatten, R.A. Pon, and R.C. von Borstel. 1987. The detection of chemical impurities by high pressure liquid chromatography and the genetic activity of medical grades of pyrvinium pamoate in *Saccharomyces cerevisiae* and *Salmonella typhimurium*. *Arch Toxicol.* **60:** 278–286.

Huxley, T.H. 1871. Yeast. In *Critiques and addresses*, pp. 71–91. MacMillan, London. (Taken from *The Contemporary Review*.)

James, A.P. and B.J. Kilbey. 1977. The timing of UV mutagenesis in yeast: A pedigree analysis of induced recessive mutation. *Genetics* **87:** 237–248.

James, A.P., B.J. Kilbey, and G. Prefontaine. 1978. The timing of UV mutagenesis in yeast. Continuing mutation in an excision defective (*rad1-1*) strain. *Mol. Gen. Genet.* **165:** 207–212.

Judson, H.F. 1980. *The eighth day of creation*. Touchstone, Simon and Shuster, New York.

Kilbey, B.J. and A.P. James. 1978. The mutagenic potential of unexcised pyrimidine dimers in *Saccharomyces cerevisiae rad1-1*. Evidence from photoreactivation and pedigree analysis. *Mutat. Res.* **60:** 163–171.

Kilbey, B.J., T. Brychcy, and A. Nasim. 1978. Initiation of UV mutagenesis in *Saccharomyces cerevisiae*. *Nature* **274:** 889–891.

Lax, C. and S. Fogel. 1978. Novel interallelic complementation at the *his1* locus of yeast. *Genetics* **90:** 501–516.

Lee, G.S-F., E.A. Savage, R.G. Ritzel, and R.C. von Borstel. 1988. The base-alteration spectrum of spontaneous and ultraviolet radiation-induced forward mutations in the *URA3* locus of *Saccharomyces cerevisiae*. *Mol. Gen. Genet.* **214:** 396–404.

Magni, G.E. 1964. Origin and nature of spontaneous mutations in meiotic organisms. *J. Cell. Comp. Physiol.* (suppl. 1) **64:** 165–172.

———. 1969. Spontaneous mutations. *Proc. Int. Congr. Genet.* **3:** 247–259.

Magni, G.E. and R.C. von Borstel. 1962. Different rates of spontaneous mutation during mitosis and meiosis in yeast. *Genetics* **47:** 1097–1108.

Magni, G.E., R.C. von Borstel, and S. Sora. 1964. Mutagenic action during meiosis and antimutagenic action during mitosis by 5-amino-acridine in yeast. *Mutat. Res.* **1:** 227–230.

Magni, G.E., R.C. von Borstel, and C.M. Steinberg. 1966. Super-suppressors as addition-deletion mutations. *J. Mol. Biol.* **16:** 568–570.

Meadows, M.G., S.-K. Quah, and R.C. von Borstel. 1973. Mutagenic action of hycanthone and IA-4 on yeast. *J. Pharmacol. Exp. Ther.* **187:** 444–450.

Mortimer, R.K. and R.C. von Borstel. 1963. Radiation-induced dominant lethality in haploid and diploid sperm of the wasp *Mormoniella*. *Genetics* **48:** 1545–1549.

Nasim, A. and C. Auerbach. 1967. The origin of complete and mosaic mutants from mutagenic treatment of single cells. *Mutat. Res.* **4:** 1–4.

Parker, K.R. and R.C. von Borstel. 1987. Base-substitution and frameshift mutagenesis by sodium chloride and potassium chloride in *Saccharomyces cerevisiae*. *Mutat. Res.* **189:** 11–14.

Parker, K.R., E. Fuog, S. Sora, G.E. Magni, and R.C. von Borstel. 1986. The mutagenic action of sodium chloride and potassium chloride on *Saccharomyces cerevisiae*. *Yeast* (suppl.) **2:** s287.

Plischke, M.E., R.C. von Borstel, R.K. Mortimer, and W.E. Cohn. 1976. Genetic markers and associated gene products in *Saccharomyces cerevisiae*. In *Handbook of biochemistry and molecular biology: Nucleic acids,* 3rd edition (ed. G. Fasman), vol. 2, pp. 767–832. The Chemical Rubber Company Press, Cleveland.

Quah, S.-K., R.C. von Borstel, and P.J. Hastings. 1980. Spontaneous mutations in yeast. *Genetics* **96:** 819–839.

Schuller, R.C. and R.C. von Borstel. 1974. Spontaneous mutability in yeast. I. Stability of lysine reversion rates to variation of adenine concentration. *Mutat. Res.* **24:** 17–23.

von Borstel, R.C. 1963a. Yeast Genetics Conference. *Science* **142:** 1594.

———, ed. 1963b. Carbondale Yeast Genetics Conference. *Microb. Genet. Bull.* (suppl.) **19:** 1–21.

———, ed. 1964. Yeast genetics supplement. *Microb. Genet. Bull.* (suppl.) **20:** 1–20.

———. 1966a. Yeast genetics. *Science* **152:** 1287–1288.

———, ed. 1966b. Yeast genetics supplement. *Microb. Genet. Bull.* **25:** 1–23.

———. 1968. Genetic markers and associated enzymes in *Saccharomyces*. In *Handbook of biochemistry: Selected data for molecular biology* (ed. H.A. Sober), pp. 159–163. The Chemical Rubber Company Press, Cleveland.

———. 1969a. On the origin of spontaneous mutations. *Jpn. J. Genet.* (suppl.) **44:** 102–105.

———. 1969b. Yeast Genetics Conference. *Science* **163:** 962–964.

———. 1969c. Yeast genetics supplement. *Microb. Genet. Bull.* **31:** 1–28.

———. 1976. Partial-target mutagenesis and carcinogenesis. In *Biological and environmental effects of low-level radiation,* vol. 1, pp. 361–368. International Atomic Energy Agency, Vienna.

———. 1978. Measuring spontaneous mutation rates in yeast. *Methods Cell Biol.* **20:** 1–24.

von Borstel, R.C. and P.J. Hastings. 1977. Mutagenic repair pathways in yeast. In *Research in photobiology: Proceedings of the VII International Congress of Photobiology,* Rome Italy, August 29–September 3, 1976 (ed. A. Castellani), pp. 683–687. Plenum Press, New York.

———. 1985. Situation-dependent repair of DNA damage in yeast. In *Basic and applied mutagenesis* (ed. A. Muhammed and R.C. von Borstel), pp. 121–145. Plenum Press, New York.

von Borstel, R.C. and G.S.-F. Lee. 1991. Origin of hotspots for mutations induced by ultraviolet radiation in yeast. In *Abstracts from Cellular Responses to Environmental Damage Meeting,* Banff, December 1–6, 1991. Abstr. B-25. American Association for Cancer Research, Philadelphila.

von Borstel, R.C. and A. Lynch. 1978. Mutators and antimutators in *Saccharomyces*. In *Proceedings of the 14th International Congress of Genetics,* Moscow, August 21–30 1978 (ed. D.K. Belyaev), vol. II, p. 210. Nauka (Soviet Publication).

von Borstel, R.C., M.J. Bond, and C.M. Steinberg. 1964. Spontaneous reversion rates of super-suppressible mutant during mitosis and meiosis. *Genetics* **50:** 293. (Abstr.)

von Borstel, R.C., K.T. Cain, and C.M. Steinberg. 1971a. Inheritance of spontaneous mutability in yeast. *Genetics* **69:** 17–27.

von Borstel, R.C., P.J. Hastings, and C. Schroeder. 1977. Comparison of replication errors and mutagenic repair as a source of spontaneous mutations. In *Environmental mutagens: Proceedings of the 6th Annual Meeting of the European Environmental Mutagen Society,* Gernrode, GDR, September 27–October 1, 1976 (ed. H. Böhme and J. Schöneich), vol. 9, pp. 89–93. Akademie-Verlag, Berlin.

von Borstel, R.C., R.K. Mortimer, and W.E. Cohn. 1971b. Genetic markers and associated gene products in *Saccharomyces*. In *Handbook of biochemistry: Selected data for molecular biology*, second edition (ed. H.A. Sober), pp. 182–187. The Chemical Rubber Company Press, Cleveland.

von Borstel, R.C., E. Moustacchi, and R. Latarjet. 1978. Spontaneous mutations rates and the rate-doubling dose. In *Late biological effects of ionizing radiation*, vol. II, pp. 277–290. Atomic Energy Agency, Vienna.

von Borstel, R.C., D.E. Graham, K.J. La Brot, and M.A. Resnick. 1968. Mutator activity of an X-radiation-sensitive yeast. *Genetics* **60:** 233. (Abstr.)

von Borstel, R.C., U.G.G. Hennig, Q. Wang, K.R. Parker, and C.M. Steinberg. 1992. The Cairns deviation *Saccharomyces cerevisiae. Environ. Mol. Mutagen.* **19**(s20)**:** 68. (Abstr.)

von Borstel, R.C., S.-K. Quah, C.M. Steinberg, F. Flury, and D.J.C. Gottlieb. 1973. Mutants of yeast with enhanced spontaneous mutation rates. *Genetics* (suppl.) **73:** 141–151.

Wang, Q., U.G.G. Hennig, R.G. Ritzel, E.A. Savage, and R.C. von Borstel. 1990. Double-stranded base-sequencing confirms the genetic evidence that the *hom3-10* allele of *Saccharomyces cerevisiae* is a frameshift mutant. *Yeast* (suppl.) **6:** S76. (Abstr. 02–10A.)

# Chemically Induced Genetic Change in Yeast Cells

FRIEDRICH K. ZIMMERMANN
*Institut für Mikrobiologie*
*Technische Hochschule Darmstadt*
*D-6100 Darmstadt, Germany*

## MY APPRENTICESHIP IN YEAST GENETICS

My doctoral thesis at the University of Freiburg under the supervision of Professor Hans Marquardt at the Institute of Forest Botany was on the relative biological and genetic effectiveness of different types of ionizing radiations in *Neurospora crassa*. Professor Marquardt was a former student of Friedrich Oehlkers who pioneered chemical mutagenesis when he discovered that urethane induced chromosome aberrations in plants (Oehlkers 1943). Marquardt (1938a,b, 1940, 1942) himself had pioneered the study of effects of X-rays on plant mitosis.

Every type of research on the biological effects of ionizing radiation was well funded in those days. However, the threat of chemical mutagens to human health was obvious only to a few informed scientists who had read the isolated publications reporting on mutation induction in *Drosophila* by bis(2-chloroethyl) sulfide, mustard gas (Auerbach and Robson 1947), dimethyl and diethyl sulfate (Rapoport 1947), diazomethane, *N*-methyl-*N*-nitrosourethane, and ethyleneimine (Rapoport 1948). In addition, Westergaard (1957) published an impressive list of chemicals that had been shown to induce reverse mutation in *N. crassa*. Marquardt and his co-workers had demonstrated the induction of chromosome aberrations in plant cells by a number of chemicals and wanted to include microbial mutation tests in a more comprehensive testing program.

When I was completing my thesis, Marquardt proposed that I visit a number of different laboratories in the United States to acquaint myself with actual state-of-the-art mutation research and genetics. The choice of laboratories was left to me, and so my obvious choice was the laboratory of Ernst Freese and his wife Elisabeth Bautz-Freese who had obtained her doctoral degree as a student of Marquardt.

In addition to my interest in science, I also had a strong passion for mountain climbing. I therefore started with a map of the United States to identify regions with interesting mountains. To my absolute delight, I discovered that the University of Washington was located in a most desirable area and was scientifically most suited for my purposes. David Stadler, Department of Genetics, worked there on intragenic recombination in *N. crassa*. Moreover, in addition to *Neurospora* genetics, there was also genetics of the yeast *Saccharomyces cerevisiae* represented by Herschel Roman. Consequently, at Seattle, I could expect to get a thorough training in the genetics of two microorganisms and at the same time indulge in mountain climbing. Both expectations were more than fulfilled.

Starting in December 1960, I worked for 6 months with Ernst Freese at the University of Wisconsin in Madison on mutation induction in phage T4. During this period, I learned about phage genetics, but, more importantly, I was exposed to a molecular geneticist's mode of thinking, which differed considerably from that of a "classical" biologist. A classical biologist deals with a system of black boxes, whereas a molecular biologist wants to analyze and ultimately explain all of the biological phenomena at the level of chemical reactions and laws of physics. I must say, it took me years to reach this level of thinking, but it had begun in Madison. The fascinating aspect of Freese's mutation work on phage T4 was that it first made possible the identification of molecular changes induced by a given mutagen. Mutations were induced by mutagens that preferentially caused defined types of chemical changes in the bases of DNA (Freese et al. 1961). Such mutations could then be caused to revert to wild type only by specific types of mutagens. Base analogs that cause mutation by incorporation into DNA in place of specific normal bases were also very important in this experimental design. Taking these factors into account, it was possible to identify the types of genetic changes induced at the DNA level by an "unknown" mutagen. With this on my mind, I went to Seattle in June 1961 to try the same approach in *N. crassa*.

The flight from Chicago to Seattle provided a breath-taking view of the Cascade Mountains. At the Department of Genetics, I was greeted by Satya Kakar and then met the other members of the Department. I immediately felt very happy there.

I started to hunt for nitrous-acid-induced forward mutants in *Neurospora*, hoping that they would all be caused by base-specific transitions. But I was not very successful with the filtration enrichment technique used for this experiment, and Herschel Roman suggested that I use the elegant yeast mutation system that he had already exploited in 1956. Yeast mutants with defects in the genes *ADE1* or *ADE2* not only require adenine for growth, but also form red colonies due to the accumulation of a red pigment derived from an intermediate in purine biosynthesis accumulating in front of the blocked reactions. Mutations in genes coding for enzymes involved in synthetic reactions preceding the steps catalyzed by the *ADE1* and *ADE2* enzymes block formation of this precursor and convert the colonies to white. This approach allowed the identification of mutations induced in a number of adenine genes, some of which were distinctly more mutable than others (Roman 1957).

I began working with Elizabeth Jones who, as part of her Ph.D. thesis, was using Roman's red-adenine mutant system to study the differential mutability of *ADE3* and *ADE6* (Jones 1964). We were excited to find that nitrous acid worked very well in inducing *ade* mutations, until a publication by Gutz (1961) appeared in which he reported on mutation induction by nitrous acid in *Schizosaccharomyces pombe*. Until that time, nitrous acid had been used to induce mutations only in viruses and bacteria and not in eukaryotic cells.

In 1961, the Department of Genetics had acquired an analytical centrifuge, which made it possible to band DNA in a CsCl density gradient. Norman Eaton and Jørgen Friis used this machine to band yeast DNA, and with their help, I tried the same with *Neurospora* DNA. I had hoped to demonstrate the incorporation of 5-bromouracil into *Neurospora* DNA. I could not. But neither yeast nor *Neurospora* DNA gave a single band as had been found with bacterial DNA. Despite trying all sorts of tricks, we always obtained one major band and at least one or even two minor bands. We finally gave up since several experts assured us that we were looking at artifacts. Shortly after we had given up, Kit (1961) reported that mammalian DNA always gives more than one band in a CsCl density gradient. As was later shown, mitochondrial DNA has a buoyant density different from that of nuclear DNA (Bernardi et al. 1968), and even nuclear DNA consists of different density classes. However, many years later, Norman Eaton and I resumed a very successful collaboration at Brooklyn College in New York on the genetics of maltose utilization in yeast.

On the way back to Freiburg in March 1962, I spent a few days with Satya Kakar, who by that time was a postdoc with Robert P. Wagner at the University of Texas at Austin. With one exception, the same enzymes are required for the synthesis of valine and isoleucine in yeast (Kakar and Wagner 1964). I had hopes that reverse mutations in different *ILV* genes would generate not only full revertants not requiring isoleucine and valine, but also half revertants requiring only valine or only isoleucine. Different mutagens might show differences in the induction of full revertants and the two types of half revertants. The idea was that if there are base-specific mutagens, some would induce, for example, only full revertants, whereas others would also induce half revertants. This would have provided a simple test to identify the base specificities of newly discovered mutagens. We indeed obtained half revertants but equally well with UV light and nitrous acid (Kakar et al. 1964). I did not pursue this any further.

## CARCINOGENS INDUCE POINT MUTATIONS IN YEAST

Back at Freiburg with Professor Marquardt at the Institute of Forest Botany, I worked together with Roland Schwaier, a doctoral student. We also had the support of a highly efficient technician, Uta von Laer. Dr. Eberhardt Gläss, a long-time co-worker of Marquardt's, with ample experience in chemical-induced chromosome aberrations (Gläss 1955, 1956) studied the induction of chromosome breaks in the lily *Bellevalia romana*, which had four easily distinguishable chromosomes and asymmetrically located kinetochores so that the distribution of induced breaks could easily be studied even within a given chromosome. In addition, Marquardt had been collaborating with Hermann Druckrey at the Institute of Preventive Medicine studying the induction of rat tumors by alkylating chemicals. Marquardt as a geneticist had always advocated that somatic mutation could be a major mechanism in tumorigenesis; his co-worker Gläss (1960a,b) had studied chromosomal changes in the rat liver during butter-yellow-induced carcinogenesis. Their goal was to show that the primary and inducing event in tumorigenesis was a mutation in somatic cells. However, this was not a very popular concept at a time when most biologists thought that genetics played a role only in meiosis and did not even think about somatic genetics. Consequently, their advocacy of the somatic mutation theory of cancer met deaf ears. Marquardt used to say, "If you talk to them about genetics, they always get a glassy stare in their eyes."

Beth Jones had given me an *ade2-1* mutant, A1327A, that could be used in the red-to-white forward mutation system introduced by Herschel Roman

(1957) and a set of *ade3* and *ade6* mutants that we tested for the induction of reverse mutation. One of them, with allele *ade6-45*, turned out to be most useful. Thus, the Institute of Forest Botany had a program of testing chemicals for the induction of genetic change in the eukaryotic model *S. cerevisiae* and cytogenetic effects in *B. romana*. The chemicals were selected on the basis of their carcinogenic activity, established mostly in Druckrey's laboratory, and also according to their expected carcinogenic potential.

The reverse mutation system using mutant allele *ade6-45* revealed a number of interesting results. We used a set of *N*-methyl-*N*-nitrosamides, all of which released diazomethane in aqueous solution but at widely different rates. Some of these had already been shown to be carcinogenic in rats by Druckrey and his co-workers. The mutagenic efficiencies (strength of the mutagenic effects) and the mutagenic effectivenesses (concentrations required for a mutagenic effect) did not correlate with the rates of diazomethane production in vitro. Apparently, a transport form determined mutagenic efficiency and effectiveness. However, these two parameters need not be correlated. We also used dimethylnitrosamine and diethylnitrosamine, which had been shown to induce tumors in rats. These two agents were stable in aqueous solution and were not even toxic to yeast. This was considered to be due to the absence in yeast of a metabolic activation required to generate a reactive metabolite (Marquardt et al. 1964).

The reverse mutation assay using *ade6-45* would signal induction of mutation most likely at a suppressor locus as later shown by Roland Schwaier (1965). In contrast, Roman's red-to-white system signals mutation induction in different loci. Therefore, we could test different mutagens/carcinogens also for mutagen specificity at the level of different genes. We found that forward mutations of spontaneous origin and those induced by seven methyl nitrosamides and nitrous acid showed the same distribution patterns with respect to the different genes (Marquardt et al. 1966).

Even though there was no indication for mutagen specificity at the level of different gene loci, we observed a striking case of mutagen specificity with respect to reversion of the *ade6-45* defect and two mutant alleles in strain K3/17 of *N. crassa* (Jensen et al. 1949). We had two methylating mutagens, methylmethanesulfonate and *N*-methyl-*N*-nitrosourethane, and two ethylating agents, diethyl sulfate and *N*-ethyl-*N*-nitrosourethane. The two nitrosamides induced very high mutation frequencies in yeast, whereas methylmethanesulfonate and diethyl sulfate only killed cells without inducing reverse mutation. In *N. crassa*, diethyl sulfate induced high frequencies of reversion of an adenine requirement but not of an inositol requirement, whereas *N*-ethyl-*N*-nitrosourethane did not induce any reverse mutations. This clearly indicated mutagen specificity at the level of reversion of a specific mutation and suggested that there are different mechanisms of alkylation (Marquardt et al. 1967).

Druckrey had taught us that a chemical enters an organism in a transport form but that the final effects may be exerted only by an active form. We intended to identify the chemical parameters that determine the influence of the transport form on the mutagenicity. For this, we needed a variety of different mutagens that all generated the same active form. Rudolf Preussmann, a chemist and co-worker of Druckrey, synthesized all of the chemicals we considered to be interesting, with the goal being to establish structural-functional relationships of mutagens/carcinogens. He synthesized *N*-methyl-*N*-nitrosamide derviatives of fatty acids with a chain length of 3–8, 10, 12, and 14 carbon atoms. Fortunately, their rates of hydrolysis, release of mutagenic diazomethane, were almost the same and so this parameter was constant. They were tested using the *ade6-45* reverse mutation assay. We could study two effects. One was the mutational effectiveness, the threshold concentration required for mutation induction—the lower the threshold concentration, the higher the effectiveness. The effectiveness increased with chain length and reached a maximum at a length of ten carbon atoms with a possible decline beyond that. The difference amounted to a factor of $10^4$. The other type of effect was the extent of mutation induction, mutants per survivors or lethal hits, which was called mutagenic efficiency. This could not be determined with the same degree of accuracy, but the highest and unparalleled mutation frequencies were induced at a chain length of five and six carbon atoms (Schwaier et al. 1966).

At first glance, it appeared to be obvious that the active form released by *N*-methyl-*N*-nitrosamides was diazomethane. Nevertheless, we also tested the influence of external pH on their mutagenic effects in the *ade6-45* reverse mutation assay and compared it to that of sodium nitrite. A solution of sodium nitrite at low pH contains undissociated $HNO_2$, considered to be the active form. As expected, sodium nitrite was neither toxic nor mutagenic at pH 6; also, low pH by itself was not mutagenic. Nitrosamides can decompose to nitrous acid at lower pH, a reversal of their synthesis from *N*-acylamides and sodium nitrite. All of seven nitrosamides were indeed mutagenic even at pH 2 and formed nitrous acid at pH 2.2–2.3. Deamination of adenine to hypoxanthine was demonstrated for three of the nitrosamides. This suggested that, depending on the pH, *N*-acyl-*N*-nitrosamides release two active forms, diazomethane and nitrous acid (Zimmermann et al. 1965).

In the early days of mutation research, it was not always clear whether the observed genetic change after mutagenic treatment was the immediate effect of a primary chemical change induced in DNA or the result of a mutagenic pathway called mutation fixation. We had observed that the frequencies of mutation induction increased with increasing treatment temperature (Schwaier et al. 1965). On the other hand, posttreatment incubation temperature was also shown to affect the yields from mutagenic treatments (Witkin 1957). We had observed that mutation yield was reduced with increasing posttreatment temperature and to different extents with different mutagens (Zimmermann et al. 1966a). We exploited this effect to determine the temperature-sensitive phase. Even though nitrous acid did not show the strongest posttreatment temperature effect, we used this mutagen because induced mutation frequencies were very reproducible. The standard posttreatment temperature was 25°C; a high-temperature treatment of 35°C reduced the yield of nitrous-acid-induced mutation frequencies sevenfold. Temperature-shift experiments demonstrated that a critical phase determining the mutation yield started at 8 hours after plating, and mutation fixation was completed by about 20 hours. Further experiments showed that the first cell divisions were completed between 18 and 20 hours. From this, we concluded that mutation fixation began after starved cells had resumed their metabolic activities at about 8 hours after plating and was completed by the time DNA had been replicated and cells started to divide. We could not identify what was going on at the molecular level during this period, but it was obvious that the new genotype resulting from mutagenic treatment required processing of premutational lesions to a new genotype (Zimmermann et al. 1966a). We now know that the induction of point mutations in yeast by most mutagens requires the functions of repair genes such as *RAD6* as shown by Louise Prakash (1974). Most mutagenic chemicals do not induce a new genotype directly; they cause damage in the DNA structure, which can then lead to a new genotype through error-prone repair.

Experiments on posttreatment modification of mutation yields gave impressive results, but they did not help to identify any molecular mechanisms involved in mutation fixation. A different approach was provided when Cox and Parry (1968) identified 22 complementation groups of mutations that caused an enhanced sensitivity to UV light. At that time, considerable progress had been made toward the elucidation of the molecular events involved in UV-light mutagenesis. As a first step, I tested the set of these UV-sensitive mutants and found that most of them were cross-sensitive to nitrous acid and methylmethanesulfonate (Zimmermann 1968b; for comments, see Cox 1968). In retrospect, the data showed that excision and error-prone repair functions are also involved in the repair of genetic lesions induced with nitrous acid and methylmethanesulfonate.

## POINT MUTATIONS ARE NOT NECESSARILY RECESSIVE

One of the major conceptual obstacles with the mutation theory of cancer was the commonly accepted notion that most mutations are recessive and have no phenotypic effect in a heterozygous diploid. However, this view is symptomatic of "classical"—or better—premolecular genetics. A gene codes for a polypeptide chain that functions as an enzyme or as a regulatory or a structural protein. Many of these proteins have an oligomeric structure and consist of several identical subunits. A heterozygous cell with a wild-type and a defective mutant allele coding for a tetrameric protein, for example, could then form, in addition to the pure mutant and the pure wild-type tetramer—three classes of hybrid tetramers. In the case of strictly random aggregation, each of the pure tetramers would account for only 1/16 of all tetrameric forms. Consequently, an allele coding for an inactive protein could affect the phenotype in a heterozygous condition. However, many enzymes are formed in saturating amounts so that reduced activities do not cause deficiencies or cause only weak effects that require quantitative determinations as shown by Reichert (1967). Simple replica plating of mutants with a heterozygous nutritional requirement does not always reveal reduced growth rates.

Interactions between different alleles are most impressively highlighted by allelic or intragenic complementation, a commonly observed phenomenon in fungal genetics summarized in a most stimulating book by Fincham (1966). Allelic complementation is based on a mutual correction of mutational defects in hybrid oligomers.

Mutants of the *ADE2* gene show allelic complementation (Woods and Bevan 1966). Two doctoral students, Naguib Nashed from Egypt and Gibrail Jabbur from Lebanon, our Arab League, isolated a large number of *ade2* mutants and tested them for a conditionally functional gene product using the osmotic remedial test of Hawthorne and Friis (1964) and restoration of function at 35°C (Nashed and Jabbur 1966). There were combinations of osmotic or temperature osmotic remedial mutants that showed negative allelic complementation; i.e., even though each partner by itself could grow on an adenine-free medium with 1 M KCl, the heteroallelic combination could not (Nashed et al. 1967). In subsequent tests with two large sets of *ade2* mutants, Nashed (1968a) showed that negative complementation of a mutant allele is not an intrinsic property but depends on specific allelic combinations. Another almost unbelievable result was that there was not a single mutant allele obtained by induction with nitrous acid or alkylating *N*-nitroso-imidazolidinone-2 that would not show positive allelic complementation with at least one other allele (Nashed 1968b).

A more thorough analysis of allelic complementation was possible with mutants of the *ILV1* structural gene of threonine dehydratase (Kakar and Wagner 1964; de Robichon-Szulmajster and Magee 1968) for which a convenient spectrophotometric enzyme assay had been described by Holzer et al. (1963). We isolated *ilv1* mutants that were recessive by simple growth tests on isoleucine-free media. Several allelic combinations showed complementation and some had detectable threonine dehydratase activity, which in two cases was resistant to feedback inhibition by isoleucine (Zimmermann and Gundelach 1969). Threonine dehydratase was assayed in all mutant/wild-type heterozygotes. On the basis of simple gene-dosage relationships, enzyme activities should be 50% that of homozygous wild type. However, this was rather the exception than the rule and several heterozygous wild-type/mutant diploids showed specific activities of less than 10% of the wild-type level (Zimmermann et al. 1969). On the other hand, in one case, heterozygous activity was about one-third that of wild type, but 40% of this residual enzyme activity determined under standard conditions (Holzer et al. 1963) was resistant to feedback inhibition by isoleucine.

The above results showed that a mutant allele coding for an inactive protein not only can reduce enzyme activities in a heterozygous diploid, but can do so by more than 50%. Moreover, such a mutant allele not only can have a quantitative effect at the level of enzyme activity, but can also create, in combination with wild-type monomers, hybrid proteins with new properties.

## INDUCTION OF MITOCHONDRIAL MUTATION

Even though the nucleus contains the vast bulk of genetic information, the mitochondrial genome carries a few genes required for mitochondrial function. Ephrussi et al. (1949) described a type of mutation characterized by the trait "*petite colonie*" (or *petite* for short) on glucose media and an inability to grow on a nonfermentable carbon source. These mutants did not show a Mendelian segregation, which indicated that a cytoplasmic genetic factor was affected. They were deficient in respiration (Slonimski and Ephrussi 1949) and called $\rho^-$ (respiratory factor) or RD (respiratory deficient) for short. Some $\rho^-$ mutants were "neutrals," i.e., they were completely recessive when crossed to a normal RC (respiratory competent) strain. Other $\rho^-$ mutants rapidly converted the diploid into an RD type, and these were called suppressive (Ephrussi et al. 1955). In addition, there were also segregational or nuclear *petites* caused by mutations in nuclear genes. Some of the segregational *petites* can cause the loss of the mitochondrial genetic determinants (Sherman 1963).

We isolated large numbers of RD mutants on a color indicator medium developed by Nagai (1963) after treatment with two nitrosamide carcinogens and nitrous acid. Mitochondrial RD mutants should not complement a mitochondrial RD tester strain and only suppressives would yield an RD diploid when crossed to a normal RC tester strain. A nuclear *petite* mutant will complement a mitochondrial RD tester strain. Lacroute (1966) had already reported that nitrous acid does not induce mitochondrial *petites*. However, our complementation tests seemed to indicate that there were quite a number of nitrous-acid-induced mitochondrial *petites*. Tetrad analysis then showed that they were all of a segregational type that induces loss of the mitochondrial genetic factors. On the other hand, the nitrosamide-induced mitochondrial *petites* were real mitochondrial mutants. This showed that nitrous acid could not induce mitochondrial mutation—a case of nuclear versus organelle mutagen specificity (Schwaier et al. 1968).

## CARCINOGENS INDUCE MITOTIC CROSSING-OVER AND GENE CONVERSION

Mitotic crossing-over was first detected in *Drosophila* by Stern (1936) and in *Aspergillus nidulans* by Pontecorvo et al. (1953). James and Lee-Whiting (1955) showed that UV light induces mitotic crossing-over in *S. cerevisiae*, and Roman and Jacob (1959) demonstrated this for mitotic gene conversion. Becker (1957) induced mitotic crossing-over with X-rays in *Drosophila*. Most importantly, Morpurgo (1963) reported on the induction of mitotic crossing-over in *Aspergillus* by bifunctional alkylating agents. Holliday (1964a) induced mitotic crossing-over in *Saccharomyces* and *Ustilago madis* with mitomycin C, and Esposito and Holliday (1964) induced mitotic crossing-over in *Ustilago* with 5-fluorodeoxyuridine. This demonstrated that mutation is not the only mechanism of induced genetic change but that mitotic crossing-over and mitotic gene conversion are additional mechanisms of genetic change in heterozygous diploid cells. Therefore, we started to test for the induction of mitotic crossing-over and mitotic gene conversion.

Don Hawthorne, Department of Genetics, University of Washington, kindly sent us a haploid strain with appropriate markers, a dominant allele conferring resistance to cycloheximide proximal to convenient color marker *ade2-1* and *his8* (*his8* is now *his3*) as the most distal marker. This strain was used to make heterozygous diploid D1. Mitotic crossing-over could then be demonstrated in colonies where the red sector expressed *his8* and the white sector was sensitive to cycloheximide. It turned out that nitrous acid, diethyl sulfate, and five nitrosamides, most of them established carcinogens, induced mitotic crossing-over. The homozygous condition of the red and white sectors were confirmed by tetrad analysis in representative samples. We also performed tests to show that what we thought was mitotic crossing-over was not stimulated by induction of meiosis. We could induce mitotic crossing-over in *petite* derivatives of strain D1, which could not sporulate and perform meiosis. In addition, the incidence of expression of additional heterozygous recessive markers on other chromosomes was almost the same among colonies, with crossing-over leading to the expression of *ade2–his8* and normal white colonies

(Zimmermann et al. 1966b). In this publication, we also explicitly discussed mitotic recombination "as a possible process causing cancer because it leads to homozygosis of recessive genes which, when phenotypically expressed, might lead to malignant growth."

Assays for mitotic crossing-over with the recovery of the two reciprocal products are quite laborious. In contrast to this, mitotic intragenic recombination, which is mostly caused by nonreciprocal gene conversion, can be studied using heteroallelic diploids with two different defective mutant alleles and selection for prototrophic recombinants (Roman and Jacob 1959: Sherman and Roman 1963).

We obtained the diplold D45 (Sherman and Roman 1963) which is heteroallelic for *ilv1-a/ilv1-b* from Fred Sherman (University of Rochester) and a pair of haploids from Tom Manney (at that time at Western Reserve University), one *ade2-2* and *trp5-12* and the other *ade2-1* and *trp5-27*, which we crossed together to obtain diploid D4. Diploid D45 had additional markers, *thr3* proximal and *trp2* distal to *ilv1-a*, whereas *his1* was located proximal to *ilv1-b*. Induction of mitotic gene conversion by several mutagens, some established carcinogens, was very strong even at doses that were barely causing lethality. The yield of convertants did not depend on residual growth, and no enhanced coincident induction of gene conversion at the two loci *ADE2* and *TRP5* was observed in the stationary-phase cultures we used. Tetrad analysis showed that all of 22 mutagen-induced convertants obtained in strain D45 were caused by mitotic gene conversion, not crossing-over. Moreover, in 16 of 22 cases, recombinational events were restricted to the *ILV1* gene, only six convertants showing rearrangements for flanking markers. The important aspect of mitotic gene conversion was considered to be the fact that it increases mitotic instability by orders of magnitude beyond that caused by mutation (Roman 1957). Moreover, in the case of two sites of heterozygosity within a gene, gene conversion generates novel alleles in a predetermined way (Zimmermann and Schwaier 1967).

At that time, Robin Holliday's hybrid DNA model of recombination as originally formulated (Holliday 1964b) had begun having an impact on our ways of thinking. According to this model, single strands of DNA exchange partners between homologous regions of nonsister chromatids. Mismatches in the resulting hybrid DNA would be repaired, and if this were the case, gene conversion between two differently defective mutant alleles would generate either a double mutant or an intact and perfectly wild-type allele. The precision of gene conversion between the mutant alleles *ilv1-a* and *ilv1-b* could be studied at the gene product level by testing the properties of threonine dehydratase in "wild-type" convertants. Threonine dehydratase in haploid "wild-type" segregants from 23 mitotic gene convertants was tested for a series of enzyme parameters such as feedback inhibition by isoleucine or activation by valine. Even though there was considerable spread between strains of different recombinational origins, none of the convertant alleles showed a consistent deviation from the properties of a genuine wild-type allele (Zimmermann 1968a).

Marquardt had tried to convince the public of the value of mutation assays for the detection of carcinogens. He also pointed out that mutagen "precursors" often must be activated metabolically. However, since metabolic activation is not observed in yeast, yeast could allow identification of the ultimate and active form of mutagens. Hans Dannenberg at the Max Planck Institute for Biochemistry at Munich had been studying the carcinogenic activity of aromatic amines in rats and identified the ultimately active metabolic derivatives. He offered to supply us under code a large number of chemical samples among which the mitotic gene conversion test identified what was considered to be the ultimate carcinogen (Marquardt et al. 1970).

At long last, we did something that could be considered as appropriate for The Institute of Forest Botany. Dr. Edmund Lemperle, a chemist at the State Institute of Viticulture at Freiburg, and I supervised the thesis project of Dietrich Siebert, a student of biology, on testing 14 chemicals present in commonly used fungicides and found that two, folpet (ortho-phaltan) and metiram (polyram combi), strongly induced mitotic gene conversion in strain D4, whereas cignolin (1,8-dihydroxyanthranole) was a powerful *petite* mutagen (Siebert et al. 1970).

As it had turned out, mitotic gene conversion was a convenient test for the identification of mutagens. Comparison of induction of mitotic gene conversion in strain D4 heteroallelic at two loci, *ADE2* and *TRP5*, indicated that there was no mutagen specificity, in contrast to many reverse mutation assays including our own with mutant allele *ade6-45* (Zimmermann 1971a).

In 1969, Peter N. Magee, one of the major researchers in the field of chemical carcinogenesis, at that time at Middlesex Hospital in London, invited me to a small conference at Magdalen College in Oxford, where I presented a list of 24 chemicals that had been shown to be both mutagenic and carcinogenic and 13 chemicals (partly overlapping the 24 mutagenic ones) that induced mitotic recombination (crossing-over and/or gene conversion) in yeast (Zimmermann 1971b).

In September 1970, I moved to Brooklyn College of the City University of New York where I held the position of an associate professor and worked together with Norman Eaton and Nasim Khan on the genetics of maltose utilization. From that time on, I worked in the field of mutagenesis only as a sideline.

## LOOKING BACK 20 YEARS LATER

I still remember my excitement when I counted the first *N. crassa* reverse mutants after X-irradiation. During my thesis work, I scored more than 14,000 mutants. Another highlight was when Beth Jones and I scored the plates of the first nitrous acid treatment for white sectors and white colonies among the usually red *ade2-1* colonies. We never established the elegant forward-reverse mutation system in yeast that Freese and his associates had so successfully exploited in phage T4. However, I feel a deep satisfaction when I see all the thousands of publications decribing attempts to establish and validate short-term tests for the identification of carcinogens. Our speculations in the 1960s on the role of mitotic recombinational events, both reciprocal crossing-over and gene conversion, in tumorigenesis and our efforts to establish a correlation between the recombinogenic and carcinogenic activities have been substantiated in recent years when restriction-fragment-length polymorphism directly confirmed their role in cases of inherited predisposition to cancer like retinoblastoma (Cavenee et al. 1986).

This essay was not intended as a comprehensive review of chemical mutagenesis in the 1960s. Therefore, I did not mention and quote the work of many dear colleagues and friends who influenced, encouraged, and stimulated my research: Helmut Böhme, R.C. (Jack) von Borstel, Colin H. Clarke, Rudolf Fahrig, Frederick J. de Serres, Tomáš Gichner, Brian J. Kilbey, Nicola Loprieno, Giovanni E. Magni, Heinrich V. Malling, Vernon W. Mayer, Georges R. Mohn, Jörg Schöneich, Jiří Velemlnsky, and Gösta Zetterberg.

## REFERENCES


Auerbach, C. and J.M. Robson. 1947. Chemical production of mutations by chemical substances. *Proc. R. Soc. Edinb. B* **62:** 271–283.

Becker H.J. .1957. Über Röntgenmosaikflecken und Defektmutationen am Auge von *Drosophila* und die Entwicklungsphysiologie des Auges. *Z. Indukt. Abstammungs Vererbungsl.* **88:** 333–373.

Bernardi, G., F. Carnevali, A. Nicolaieff, G. Piperno, and G. Tecce. 1968. Separation and characterization of a satellite DNA from a cytoplasmic "petite" mutant. *J. Mol. Biol.* **37:** 493–499.

Cavenee, W.K., A. Koufos, and M.F. Hansen. 1986. Recessive mutants predisposing to cancer. *Mutat. Res.* **168:** 3–14.

Cox B.S. 1968. Comment on the difference in sensitivity to various mutagens shown by certain mutants. *Mol. Gen. Genet.* **102:** 256.

Cox B.S. and J.M. Parry. 1968. The isolation genetics and survival characteristics of ultraviolet light sensitive mutants in yeast. *Mutat. Res.* **6:** 37–55.

de Robichon-Szulmajster, H. and P.T. Magee. 1968. The regulation of isoleucine-valine biosynthesis in *Saccharomyces cerevisiae*. I. Threonine deaminase. *Eur. J. Biochem.* **3:** 492–501.

Ephrussi, B., H. Hottinger, and A.M. Chimenes. 1949. Action de l'acriflavine sur les levure. I. La mutation "petite colonie." *Ann. Inst. Pasteur* **76:** 351–368.

Ephrussi, B., H. de Margerie-Hottinger, and H. Roman. 1955. Suppressives: A new factor in the genetic determinism of the synthesis of respiratory enzymes in yeast. *Proc. Natl. Acad. Sci.* **41:** 1065–1071.

Esposito. R.E. and R. Holliday. 1964. The effect of 5-fluoro-deoxyuridine on genetic replication and mitotic crossing over in synchronized cultures of *Ustilago maydis*. *Genetics* **50:** 1009–1017.

Fincham, J.R.S. 1966. *Genetic complementation*. W.A. Benjamin, New York.

Freese, E., E. Bautz, and E. Bautz-Freese. 1961. The chemical and mutagenic specificity of hydroxylamine. *Proc. Natl. Acad. Sci.* **47:** 845–855.

Gläss, E. 1955. Untersuchungen über die Einwirkung von Schwermetallsalzen auf die Wurzelspitzenmitose von *Vicia faba* I. *Z. Bot.* **43:** 359–403.

———. 1956. Untersuchungen über die Einwirkung von Schwermetallsalzen auf die Wurzelspitzenmitose von *Vicia faba* II. *Z. Bot.* **44:** 1–58.

———. 1960a. Die chromosomalen Veränderungen in der Rattenleber während der krebsigen Entartung nach Verfütterung von Buttergelb. *Z. Krebsforsch.* **63:** 294–310.

———. 1960b. Die Chromosomenzahlen in der durch Buttergelbverfütterung krebsig entarteten Rattenleber. *Z. Krebsforsch.* **63:** 362–371.

Gutz, H. 1961. Distribution of X-ray- and nitrous acid-induced mutations in the genetic fine structure of the $ad^7$ locus of *Schizosaccharomyces pombe*. *Nature* **191:** 1125–1126.

Hawthorne. D.C. and J. Friis. 1964. Osmotic remedial mutants. A new classification for nutritional mutants of yeast. *Genetics* **49:** 212–222.

Holliday, R. 1964a. The induction of mitotic recombination by mitomycin C in *Ustilago* and *Saccharomyces*. *Genetics* **50:** 323–335.

———. 1964b. A mechanism for gene conversion in fungi. *Genet. Res.* **5:** 282–304.

Holzer, H., M. Boll, and C. Cennamo. 1963. Zur Biochemie der Threonine-Desaminase aus Hefe. *Angew. Chem.* **75:** 894–900.

James, A.P. and B. Lee-Whiting. 1955. Radiation induced genetic segregations in vegetative cells of yeast. *Genetics* **40:** 826–831.

Jensen, K.A., G. Kolmark, and M. Westergaard. 1949. Back mutations in *Neurospora crassa* induced by diazomethane. *Hereditas* **35:** 521–525.

Jones. E.W. 1964. "A comparative study of two adenine loci in *Saccharomyces cerevisiae.*" Ph.D. thesis, University of Washington, Seattle.

Kakar, S.N. and R.P. Wagner. 1964. Genetic and biochemical analysis of isoleucine-valine mutants of yeast. *Genetics* **49:** 213–222.

Kakar, S.N., F. Zimmermann, and R.P. Wagner. 1964. Reversion behavior of isoleucine-valine mutants of yeast. *Mutat. Res.* **1:** 381–386.

Kit, S. 1961. Equilibrium sedimentations in density gradients of DNA preparations from animal tissues. *J. Mol. Biol.* **3:** 711–716.

Lacroute, F. 1966. "Régulation de la chaine de biosynthese de l'uracile chez *Saccharomyces cerevisiae.*" Ph.D. thesis, Université de Paris.

Marquardt, H. 1938a. Die Röntgenpathologie der Mitose I. Der Primäreffekt weicher Röntgenstrahlen auf die Mitose von *Scilla campanulata. Z. Bot.* **32:** 401–429.

———. 1938b. Die Röntgenpathologie der Mitose II. Der Primär- und Sekundäreffekt der Röntgenstrahlen auf die haploide Mitose von *Bellevalia romana.* Die Chromatidpathologien. *Z. Bot.* **32:** 429–482.

———. 1940. Die Röntgenpathologie der Mitose III. Weitere Untersuchungen des Sekundäreffekts der Röntgenstrahlen auf die haploide Mitose von *Bellevalia romana. Z. Bot.* **36:** 273–386.

———. 1942. Die Verteilung röntgeninduzierter Veränderungen auf den Chromosomen von *Bellvalia romana. Ber. Dtsch. Bot. Ges.* **60:** 98–124.

Marquardt, H., R. Schwaier, and F.K. Zimmermann. 1967. Die Wirkung von Methyl- und Äthylnitrosourethan, Diäthylsulfat und Methylmethansulfonat auf zwei Genorte von *Neurospora* sowie einen Genort von *Saccharomyces. Mol. Gen. Genet.* **99:** 1–4.

Marquardt, H., U. von Laer, and F.K. Zimmermann. 1966. Das spontane, Nitrosamid- und Nitrit-induzierte Mutationsmuster von 6 Adenin-Genloci der Hefe. *Z. Vererbungsl.* **98:** 1–9.

Marquardt, H., F.K. Zimmermann, and R. Schwaier. 1964. Die Wirkung krebsauslösender Nitrosamine und Nitrosamide auf das Adenin-6-45-Rückmutationssystem von *Saccharomyces cerevisiae. Z. Vererbungsl.* **95:** 82–96.

Marquardt, H., F.K. Zimmermann, H. Dannenberg, H.-G. Neumann, A. Bodenberger, and M. Metzler. 1970. Die genetische Wirkung von aromatischen Aminen und ihren Derivaten: Induktion mitotischer Konversionen bei der Hefe *Saccharomyces cerevisiae. Z. Krebsforsch.* **74:** 412–433.

Morpurgo, G. 1963. Induction of mitotic crossing-over in *Aspergillus nidulans* by bifunctional alkylating agents. *Genetics* **48:** 1159–1163.

Nagai, S. 1963. Diagnostic color differentiation plates for hereditary respiration deficiency in yeast. *J. Bacteriol.* **86:** 299–302.

Nashed, H. 1968a. On the nature of dominance in the $ad_2$ locus of *Saccharomyces cerevisiae. Mol. Gen. Genet.* **102:** 285–289.

———. 1968b. A lack of noncomplementation among chemically induced $ad_2$ mutants of *Saccharomyces cerevisiae. Mol. Gen. Genet.* **102:** 348–352.

Nashed, N. and G. Jabbur. 1966. A genetic and functional chracterization of adenine mutants induced in yeast by 1-nitroso-imidazolidone-2 and nitrous acid. *Z. Vererbungsl.* **98:** 106–110.

Nashed, N., G. Jabbur, and F.K. Zimmermann. 1967. Negative complementation among $ad_2$ mutants of yeast. *Mol. Gen. Genet.* **99:** 69–75.

Oehlkers, F. 1943. Die Auslösung von Chromosomenmutationen in der Meiosis durch Einwirkung von Chemikalien. *Z. Indukt. Abstammungs. Vererbungsl.* **83:** 313–341.

Pontecorvo, G., L.M. Hemmons, K.D. MacDonald, and A.W.J. Bufton. 1953. The genetics of *Aspergillus nidulans. Adv. Genet.* **5:** 141–238.

Prakash, L. 1974. Lack of chemically induced mutation in repair deficient mutants of yeast. *Genetics* **78:** 1101–1118.

Rapoport, J.A. 1947. Inheritance changes taking place under the influence of diethyl sulfate and dimethyl sulfate. *Dokl. Vses. Akad. Skh. Nauk.* **12:** 12–14 (in Russian).

———. 1948. Alkylation of the gene molecule. *Dokl. Akad. Nauk. SSSR* **59:** 1183–1186 (in Russian).

Reichert, U. 1967. Gendosiswirkungen in einem $ad_2$-Mutantensystem bei *Saccharomyces cerevisiae. Zentralbl. Bakteriol. Parasitenkd. Infektionskr. Hyg. Abt. I Orig.* **205:** 63–68.

Roman, H. 1957. Studies on gene mutation in *Saccharomyces*. *Cold Spring Harbor. Symp. Quant. Biol.* **21:** 175–185.
Roman, H. and F. Jacob. 1959. A comparison of spontaneous and ultraviolet-induced allelic recombination with reference to the recombination of outside markers. *Cold Spring Harbor Symp. Quant. Biol.* **23:** 155–160.
Schwaier, R. 1965. Vergleichende Mutationsversuche mit sieben Nitrosamiden im Rückmutationstest an Hefen. *Z. Vererbungsl.* **97:** 55–67.
Schwaier, R., N. Nashed, and F. K. Zimmermann. 1968. Mutagen specificity in the induction of karyotic versus cytoplasmic respiratory deficient mutants in yeast by nitrous acid and alkylating nitrosamides. *Mol. Gen. Genet.* **102:** 290–300.
Schwaier, R., F.K. Zimmermann, and R. Preussmann. 1966. Chemical constitution and mutagenic efficiency: Mutation induction in *Saccharomyces cerevisiae* by a homologous series of N-nitroso-N-methylcarbonamides. *Z. Vererbungsl.* **98:** 309–319.
Schwaier R., F.K. Zimmermann, and U. von Laer. 1965. The effect of treatment temperature on the mutation induction in yeast by N-alkylnitrosamides and nitrous acid. *Z. Vererbungsl.* **97:** 72–74.
Sherman, F. 1963. Respiratory deficient mutants of yeast. I. Genetics. *Genetics* **48:** 375–385.
Sherman, F. and H. Roman. 1963. Evidence for two types of allelic recombination in yeast. *Genetics* **48:** 255–262.
Siebert, D., F.K. Zimmermann, and E. Lemperle. 1970. Genetic effects of fungicides. *Mutat. Res.* **10:** 533–543.
Slonimski, P.P. and B. Ephrussi. 1949. Action de l'acriflavine sure les levures. V. Le système des cytochromes des mutants "petite colonie." *Ann. Inst. Pasteur* **77:** 47–57.
Stern, C. 1936. Somatic crossing over and segregation in *Drosophila melanogaster*. *Genetics* **21:** 625–730.
Westergaard, M. 1957. Chemical mutagenesis in relation to the concept of the gene. *Experientia* **13:** 31–34.
Witkin, E.M. 1957. Time, temperature and protein synthesis: A study of ultraviolet-induced mutation in bacteria. *Cold Spring Harbor Symp. Quant. Biol.* **21:** 123–140.
Woods, R.A. and E.A. Bevan. 1966. Interallelic complementation at the $ad_2$ locus of *Saccharomyces cerevisiae*. *Heredity* **21:** 121–130.
Zimmermann, F.K. 1968a. Enzyme studies on the products of mitotic gene conversion in *Saccharomyces cerevisiae*. *Mol. Gen. Genet.* **101:** 171–184.
———. 1968b. Sensitivity to methylmethanesulfonate and nitrous acid of ultraviolet light-sensitive mutants in *Saccharomyces cerevisiae*. *Mol. Gen. Genet.* **102:** 247–256.
———. 1971a. Induction of mitotic gene conversion by mutagens. *Mutat. Res.* **11:** 327–337.
———. 1971b. Genetic aspects of carcinogenesis. *Biochem. Pharmacol.* **20:** 985–995.
Zimmermann, F.K. and E. Gundelach. 1969. Intragenic complementation, hybrid enzyme formation and dominance in diploid cells of *Saccharomyces cerevisiae*. *Mol. Gen. Genet.* **103:** 348–362.
Zimmermann, F.K. and R. Schwaier. 1967. Induction of mitotic gene conversion with nitrous acid, 1-methyl-3-nitro-1-nitrosoguanidine and other alkylating agents in *Saccharomyces cerevisiae*. *Mol. Gen. Genet.* **100:** 63–76.
Zimmermann, F.K., I. Schmiedt, and A.M.A. ten Berge. 1969. Dominance and recessiveness at the protein level in mutant x wild-type crosses in *Saccharomyces cerevisiae*. *Mol. Gen. Genet.* **104:** 321–330.
Zimmermann, F.K., R. Schwaier, and U. von Laer. 1965. The influence of pH on the mutagenicity in yeast of N-methylnitrosamides and nitrous acid. *Z. Vererbungsl.* **97:** 68–71.
———. 1966a. Nitrous acid and alkylating nitrosamides: Mutation fixation in *Saccharomyces cerevisiae*. *Z. Vererbungsl.* **98:** 152–166.
———. 1966b. Mitotic recombination induced in *Saccharomyces cerevisiae* with nitrous acid, dimethylsulfate and carcinogenic, alkylating nitrosamides. *Z. Vererbungsl.* **98:** 230–246.

# MITOCHONDRIA AND CYTOPLASMIC INHERITANCE

P.P. Slonimski (Louvain-la-Neuve, Belgium 1980)

P.P. Slonimski and F. Kaudewitz (Louvain-la-Neuve, Belgium 1980)

G. Schatz (1990)

B.S. Cox (Berlin 1989)

P.P. Slonimski (Louvain-la-Neuve, Belgium 1980)

# Psi Phenomena in Yeast

BRIAN COX
*Department of Plant Sciences*
*Oxford University, United Kingdom*

*Felix qui potuit rerum cognoscere causas*
Virgil

Yeast geneticists belong to a club that was probably founded by Herschel Roman but acquired its atmosphere from its first real meeting in Gif-sur-Yvette. I was still a student when I received a casual invitation from someone I had never heard of called Slonimski, suggesting that it was perhaps opportune for a gathering of "yeast people," and would we care to try the hospitality of the CNRS after the XIth International Congress of Genetics in the Hague? When we arrived there, on the morning after a champagne reception in the Chateau, Piotr Slonimski opened the proceedings slumped in a chair to one side of a blackboard on an easel and asked: "Well, what shall we talk about?"

The French were wonderful. They dined and wined us in restaurants and in their homes, and I made, in a few days, friends for life—*santé*, Piotr. The informality, hospitality, and friendship set a pattern that was reinforced at the next meeting in Seattle. The success of our enterprise in the exploration of the yeast cell owes so much to the sense of cooperation, interest, and mutual encouragement that Herschel Roman engendered at this meeting and whenever young people just beginning their careers visited his department. Getting willing help from friends when doing research is worth hours of work in the laboratory, not to mention the happiness it brings.

The science was good, too. This was the first time ψ was mentioned in public. I remember Don Hawthorne voicing a doubt that is still with us. "But that doesn't make it cytoplasmic," he said, anxiously protecting his territory of supersuppressors. "It's just some regulatory phenomenon." There is no answer for this, but Hersch was very kind and came to my defense.

What follows is mostly a tribute to my students and postdocs who have tilled the ψ field industriously, but I want to use this paper to acknowledge the help, interest, and advice of many colleagues first met and befriended at meetings of yeast people.

I first observed the genetic segregations that define the ψ factor of yeast in 1964 (Cox 1965). Since then, six Ph.D. students and four postdoctoral research assistants have devoted their energies to the problem it presents, and I must say that I feel no confidence that I am closer to an understanding than I was at the beginning. Of course, we have accumulated a lot more information about the control of translational fidelity as a result, but we also have many more questions.

The ψ mutation in yeast first appeared as sectors on colonies formed from ascospores. These spores were isolated during tetrad analyses of strains in

*The Early Days of Yeast Genetics*

which a suppressor mutation was segregating. There were altogether five such sectors on a total of about 100 colonies. The sectors were conspicuous, being red on colonies that were expected to be completely white because, as the tetrads were intended to show, a "supersuppressor" mutation was segregating, suppressing an *ade2* mutation. Normally, *ade2*$^-$ strains are red, and these red sectors suggested that a mutation had occurred that abolished the suppression.

Mutations, however, are not that common. When these tetrads were dissected, I had recently completed a Ph.D. project that involved many hours of scanning white yeast colonies with a low-power microscope for signs of red sectors. I thus had a very good idea of how many colonies one might have to scan to find red sectors due to mutation or to mitotic recombination or to chromosome nondisjunction. The tetrad sectors were in the latter range of frequency, and I thought it best to pick them to find out what was going on. They were the result of none of these events. The "mutation" failed to segregate in further crosses, although it could be shown that the supersuppressor did; it was just that in the red revertants, it was, it seemed, inactive.

A mutation that fails to segregate in crosses is likely to present difficulties for genetic analysis; if the mutation cannot be moved into other genetic backgrounds, nothing much can be found out about it. Fortunately, however, I found that other strains of yeast had a similar property, namely, that they did not support the activity of this suppressor. These strains were sent to me by Don Hawthorne, because I wanted to see if the *ade2* suppressor was one of the supersuppressors he and Bob Mortimer had described recently (Hawthorne and Mortimer 1963). The crosses I did with these strains showed first that the suppressor *SUQ5* fitted the definition of supersuppressor, and second that yeast strains all seemed to fall into two clear categories: those in which the supersuppressor was active ($\psi^+$) and those in which it was not ($\psi^-$). When crossed, the hybrid diploids were $\psi^+$ and segregated 100% $\psi^+$ progeny. In 1965, the model for this kind of result was that of the *petite* mutation and so it was easy to conclude that the difference between $\psi^+$ and $\psi^-$ strains was due to a cytoplasmic determinant. Of course, these words overinterpret the data, and they still do.

At that time it was uncertain what supersuppressors were. Hawthorne (1969) and Manney (1969) had drawn the analogy with nonsense suppressors in phage and bacteria. This made $\psi$ a particularly interesting discovery because it impinged on two areas in which yeast genetics was becoming exciting. On the one hand, supersuppressor mutations and their modifiers were promising to become the basis of a genetic analysis of the genetic code and translation in eukaryotes that would extend these wonderful discoveries in phage and bacteria to higher organisms. On the other hand, $\psi$ was only the second example of a non-Mendelian determinant to be found in yeast, encouraging the belief in the significance of extranuclear genomes in the life of cells. At about the same time, back in Oxford, Alan Bevan and Mallory Makower (cited in Somers and Bevan 1969) were describing the killer factor. In due course, the conspicuous double-stranded RNA (dsRNA) that determines the killer factor was found (Bevan et al. 1973), which finally made cytoplasmic determinants respectable as a category of genes. As for translation and the genetic code, the single-minded and brilliant work of Stewart and Sherman (1972) on the iso-1-cytochrome *c* gene provided the first direct access to the molecular biology of eukaryotic genes. Among other features, these authors were able to show that supersuppressible mutations of *cyc1* were indeed "ocher" (UAA), or UAG or UGA (Stewart et al. 1972; Stewart and Sherman 1972). It was some time before the suppressing tRNAs or their genes were sequenced (Olson et al. 1979; Broach et al. 1981; Waldron et al. 1981).

From the beginning therefore, the ψ factor became two problems. One was the problem of genetics and heredity: What determined the $\psi^+$ and $\psi^-$ phenotypes and how were they so stably inherited? The other was the problem of how the ψ factor worked.

> The next yeast conference I attended was a sort of turning point, since it marks the time when other researchers started helping ψ research along by taking the strains into their laboratories and applying their own particular expertise to the problem. This yeast conference held at Chalk River was the last fully informal meeting in the sense that no program was announced in advance, nor were any abstracts presented. It was also pretty informal in other ways. Sy Fogel made the conference dinner speech and proclaimed the conference motto, taken from a hoarding (billboard) we had passed on the drive up from Ottawa: Drink Canada Dry. I also remember Bob Haynes engaging some of the local color in conversation at the conference pork and beans supper in a remote Ontario (or possibly Quebec) hunting lodge. Three rough and dangerous looking Quebecois trappers had come in and the intrepid RHH tried them out in his unique french. As it turned out, he was never in danger—they were from Cleveland, Ohio.
>
> Although Bob never engaged in ψ work, he was, as they say, an instrument. Yeast meetings then were able largely to subsidize overseas participants, and Bob had provided the money for me to attend. It was the first time we met and we enjoyed it. A consequence was that I spent a sabbatical 6 months in Toronto a year later, doing DNA-repair research. Toronto is not far from Rochester, New York, at least not as far as Oxford is, and I paid several visits to Fred Sherman's laboratory. Once he had the appropriate strains, Fred and his student Sue Liebman were able to make the first molecular connections to ψ. I hope that elsewhere in this volume appropriate recognition is given to the iso-1-cytochrome *c* gene and the numerous genetic questions Fred was able to put on a molecular basis using his mutations and John Stewart's protein chemistry. What they were able to deduce in the days before cloning and DNA sequencing is astonishing; at the time, it was significantly more subtle than that being done with *Escherichia coli* and phage.

Perhaps at this point it is worth explaining the basic properties of the system: (1) A number of mutant tRNAs that suppress ocher mutations do so far more efficiently in $\psi^+$ strains than they do in $\psi^-$ strains (Liebman and Sherman 1979). ψ therefore affects the fidelity of translation. (2) The $\psi^+/\psi^-$ difference is not inherited in a Mendelian fashion, as if there were two alleles of a gene on a chromosome. Instead, it behaves exactly as if there were an extrachromosomal determinant that is present in a high enough copy number to be transmitted to 100% of the haploid progeny segregating in tetrads. The behavior of ψ in heredity is illustrated in Figure 1.

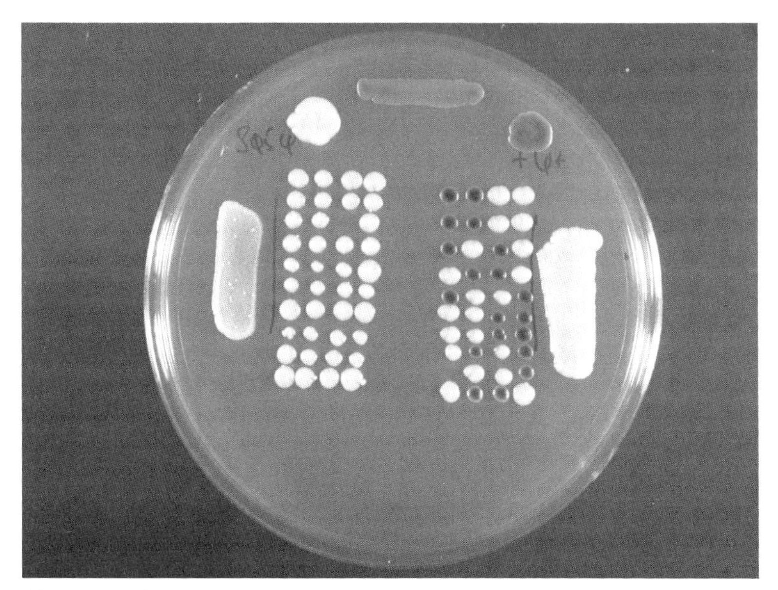

*Figure 1* Inheritance of $\psi$ in the sexual cycle of yeast. (*Right*) This cross, *suq5*$^+$ $\psi^+$ x *SUQ5* $\psi^-$, illustrates the fact that suppression is dependent on two determinants that complement each other in the diploid, namely, the suppressor gene *SUQ5* and $\psi^+$. The 2:2 segregation traces the inheritance of *SUQ5*; $\psi^+$segregates 4:0. (*Left*) This cross, in which *SUQ5* is homozygous (*SUQ5* $\psi^+$ x *SUQ5* $\psi^-$) illustrates the 4:0 segregation of $\psi$. The parent cultures are growing at the top of the plate, the very top one being common to both diploids, i.e., *SUQ5* $\psi^-$ (but see Fig. 5).

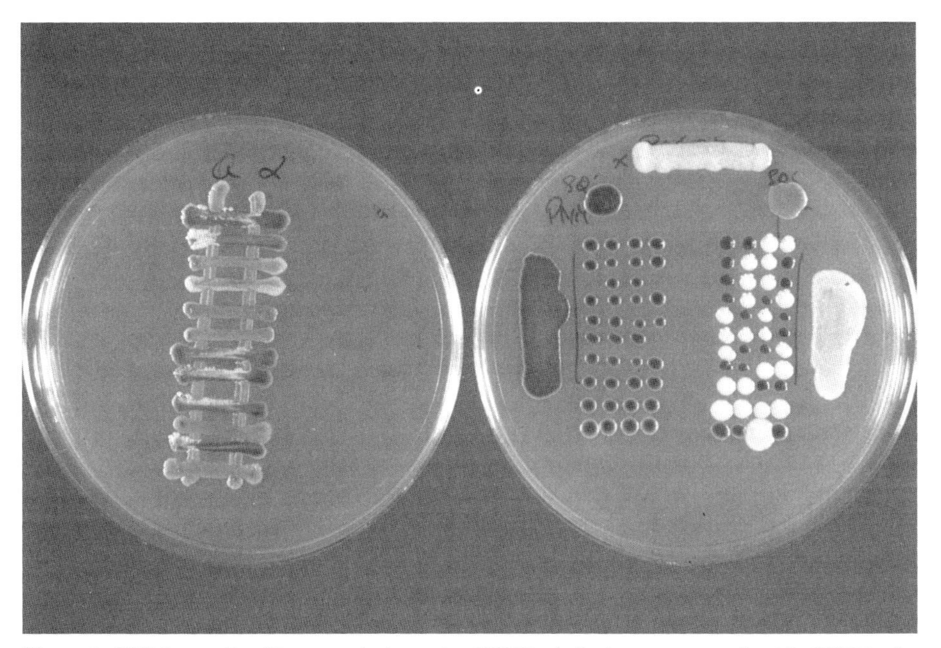

*Figure 2* *PNM* genetics illustrated. A strain *SUQ5* $\psi^+$ (*top*) was crossed with *SUQ5* $\psi^+$ *asu* (*left*) and with *SUQ5 PNM2* (*right*). The segregants from the latter cross are all $\psi^-$, and further crosses show that the *PNM2* gene segregates 2:2. This is illustrated in the plate on the left, where segregants from three tetrads have been cross-streaked with "+, $\psi^+$ *MAT*a" or "+, $\psi^+$ *MAT*$\alpha$" cultures. Two cultures from each tetrad complement these tester strains.

Both of the problems posed by these properties have been pursued in tandem, and it is worth exemplifying this by the work embodied in the first ψ thesis, written by Hamish Young (1969). Three papers all laying the foundations of future work emerged from this slim volume. Two concerned heredity, and the third, translational fidelity. Hamish's project brief was to address the question of the nature of ψ by two approaches. One was to see if ψ could be induced to mutate by conventional mutagens. This would be expected of it if it were a nucleic-acid-based determinant. The second approach was to examine its relationship with the other extrachromosomal determinant then known in yeast—mitochondrial DNA.

The first paper described a chromosomal mutation that eliminated ψ. We called it *R*. The pursuit of the genetics of ψ has been greatly facilitated by the ocher allele of the *ADE2* gene, *ade2-1*. This mutation causes strains to accumulate a red pigment, and when it is suppressed, they are white. $\psi^+$ and $\psi^-$ are thus distinguishable by their colors in our strains, and no doubt *R* means red. When crossed with a (white) $\psi^+$ strain, the diploid was, unexpectedly, red. Normally, of course, $\psi^+ \times \psi^-$ diploids are $\psi^+$ and white. Tetrads from the diploid with the dominant red mutations segregated 4 red:0 white (see Fig. 2) (Young and Cox 1971).

It turned out on analysis of the red haploid segregants by a second generation of crosses to $\psi^+$ (white) strains that *R* was not a dominant version of $\psi^-$, but chromosomal. Two members of every tetrad were ordinary $\psi^-$ strains: They gave white diploids that segregated 4 white:0 red tetrads. The other two were exactly like the original *R* parent: They gave red diploids that segregated 4 red:0 white.

## PARTICULATE INHERITANCE OF $\psi^+$ AND THE GENETIC CONTROL OF ITS EXPRESSION

We thus saw the classic ingredients of a phenotype determined by both nuclear and cytoplasmic determinants. The discovery of the *R* nuclear gene proved both informative and useful in all sorts of ways. For example, before the discovery of *R*, it was possible only to turn $\psi^-$ strains to $\psi^+$ by genetic crosses. With *R*, $\psi^+$ strains could be made $\psi^-$.

*R* also showed us that the expression and inheritance of ψ were independently controlled and that its inheritance was particulate. This was because *R* affected both the ψ phenotype and the ψ genotype, but independently of one another. Thus, it became clear in the course of crosses with the *R* mutant that repression of the $\psi^+$ phenotype was immediate, since all $R \times \psi^+$ diploids were phenotypically $\psi^-$, unsuppressed. However, the $\psi^+$ determinant was not immediately altered or eliminated. As a rare event, $\psi^+$ unsuppressed segregants appeared from $R \times \psi^+$ unsuppressed diploids after meiosis. The $\psi^+$ factor clearly survived, unexpressed, in these strains. Furthermore, the survival of the determinant in the diploid was time-dependent, i.e., the longer the period of growth of the diploid before being arrested on sporulation medium, the fewer $\psi^+$ segregants among the progeny (Fig. 3). This is consistent with a model in which there are a number of self-replicating particles in the $\psi^+$ cell whose expression *and* replication are simultaneously blocked by the *R* mutation in the cell with which it is mated. Consequently, the diploid would be phenotypically $\psi^-$ and, as it multiplied, would segregate cells with fewer and

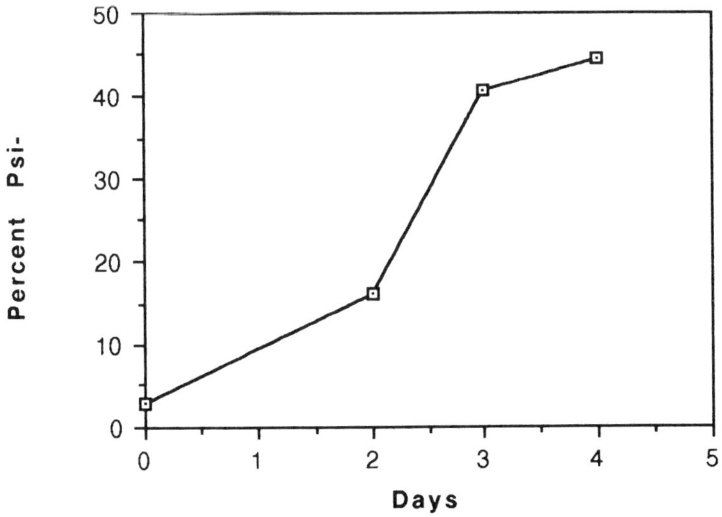


*Figure 3* Loss of ψ during the growth of diploids heterozygous for *R* (*PNM*). *MAT*α $\psi^+$ cells mixed with *MAT***a** *PNM* cells were transferred to sporulation medium at various times after mixing. When asci formed, several were dissected, and the segregants were scored for suppression ($\psi^+$) or nonsuppression ($\psi^-$).

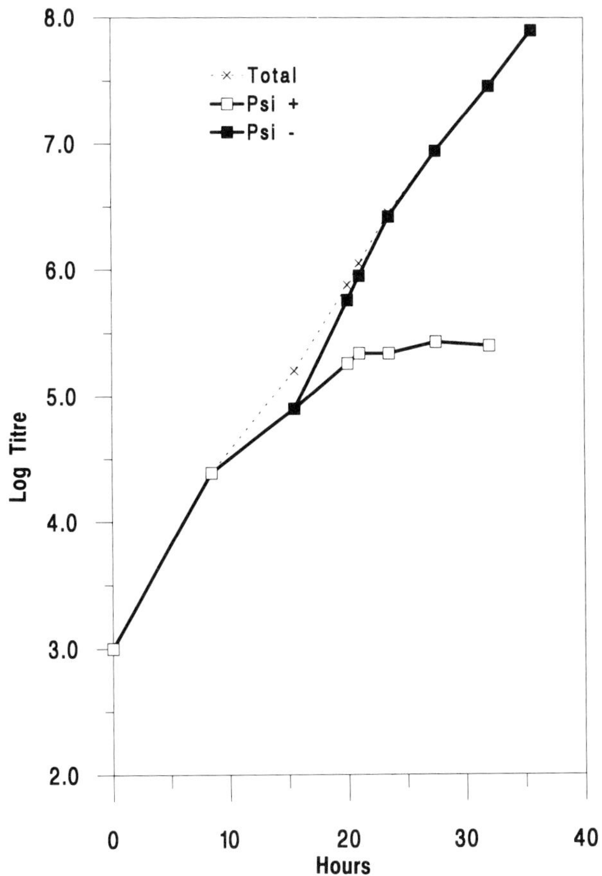


*Figure 4* Loss of ψ during the growth of $\psi^+$ haploid cells in 4 mM guanidine HCl. The plots are of $\psi^+$, $\psi^-$, and total colonies as a function of time.

fewer $\psi^+$ particles, so that in due course, when meiosis occurred, some or all of the spores would contain no $\psi^+$ particles. However, at all stages of growth, meiosis would give some spores that still contained $\psi^+$, but in which *R* had segregated away. These spores would grow into white, suppressed colonies of cells.

Whether or not the model is to be believed, a particularly telling indication that $\psi$ is a segregating determinant occurred in these analyses. A few of these non-*R* segregants segregated into $\psi^+$ and $\psi^-$ in a subsequent mitotic division, giving red/white-sectored colonies. We were clearly observing a phenomenon that had no intermediate states; there is no blending inheritance of the $\psi$ phenotype. This model was reduced to numerical proportions in a study some years later by another of the $\psi$ students, Shirley McCready. While she was a postdoc in Cal McLaughlin's laboratory, she measured the number of generations it took an *R* gene to eliminate $\psi$ on the simple assumption of this model and arrived at a number of 58 (McCready et al. 1977). I would be more inclined to be skeptical of this number if it were not that I recently got a similar number by another method. In this case, "elimination" of $\psi^+$ is achieved by growing cells in 4 mM guanidine HCl. The kinetics of induction of $\psi^-$ cells in this experiment strongly suggest a similar model and, with it taking six to seven generations to segregate $\psi^-$ cells, a starting number of 64–128 particles is obtained (Fig. 4).

The particulate nature of $\psi$ was also suggested by another observation by Hamish Young in the second of his papers (Young and Cox 1972). This paper described experiments in which the independent segregation of $\psi$ and the mitochondrial genome markers $\rho$ and $ery^r$ was demonstrated. In one of the crosses where a suppressive *petite* parent was involved, it was necessary to sporulate newly formed zygotes, since diploid cultures (had they been allowed to grow) would have consisted mostly of *petite* cells unable to sporulate, and the test of independent segregation would have been worthless. In the tetrads formed by these zygotes, both $\psi^+$ and $\psi^-$ spores segregated in a range of ratios from 4:0 through 3:1, 2:2, and 1:3 to 0:4 suppressed:unsuppressed. As with *R* crosses, sectored colonies were observed, demonstrating mitotic segregation (Fig. 5). These observations were repeated in crosses where both parents were $\rho^+$ and also where one was a neutral *petite*. The high frequency of $\psi^-$ segregants suggests either a low copy number of $\psi^+$ particles or a degree of compartmentalization.

The independent segregation of $\psi$ and mitochondrial markers is illustrated in the 2 x 2 contingency tables derived from two of these crosses (Table 1). In due course, we returned to this problem when more extrachromosomal elements had been identified in yeast. A biochemical test of the independence of $\psi$ from other elements was applied. The question was asked whether the $\psi$ phenotype was changed when the nucleic acid (mitochondrial DNA, dsRNA, or 2-µm DNA) of the element was eliminated. $\psi$ survived the loss of all or any of these nucleic acids from cells (Tuite et al. 1982). The rigor of this test was limited by the sensitivity of methods for detecting specific nucleic acid molecules, which in 1982 was several orders of magnitude worse than it is today.

At each of the yeast meetings, I made new friends who soon became colleagues. Among others, Chalk River gave me Jack von Borstel and

Cal McLaughlin. Cal and Lee Hartwell had just published those wonderful papers describing the isolation and preliminary characterization of hundreds of temperature-sensitive mutants of yeast with the declared aim of identifying all of the genes involved in the essential macromolecular metabolism of the cell. I came across the two of them having coffee after dinner and Cal expressed an interest in ψ. They had been parcelling up the temperature-sensitive mutant territory, with Cal taking on the macromolecular metabolism and Lee, the control of the cell cycle. I remember looking at the first published pictures of these mutants in their terminal arrest morphologies and thinking that it would require a prodigious intelligence to make any analytical sense of them. We take it all for granted now.

Cal was interested in ψ because he was looking for a parasitic cellular mRNA that could provide an experimental system for yeast that could be used much as phage were used in bacteria. ψ never met the case, but I am glad to say he became interested in it for its own sake. He took on two of my students as postdocs—Shirley McCready and Mick Tuite—and was a generous host to me on several sabbatical leaves. In collaboration with Kivie Moldave, Cal perfected the yeast cell-free translation system that allowed Mick Tuite to assay tRNA-mediated readthrough in vitro. Shirley McCready learned how to use the electron microscope in Cal's laboratory and, using it and cesium gradients to look for plasmids, they discovered both 2-μm and 3-μm circles (in both $\psi^+$ and $\psi^-$ cells).

## ψ AND THE NATURE OF TRANSLATIONAL FIDELITY

One may question the quality of the information about the nature of the ψ factor implied by the properties of the *R* gene. Be that as it may, it raised questions that prompted future projects, and it is these questions that make it such an important mutation.

One such question is that of how *R* (now renamed *PNM* to bring it into line with the conventional triliterate nomenclature of yeast genes) controls the suppression phenotype. Is it through controlling the expression of the ψ factor or of suppression itself? Indeed, one may ask whether the two can be distinguished.

The approach to this was to see if *PNM* affected other ocher suppressors. It does not. *SUP11*, *SUP2*, and *SUP3*, all tyrosine-inserting "strong" ocher suppressors, are unaffected by *PNM*. Out of the controls for this experiment (which was itself never published, because the controls left the original question unanswered) came three other projects, however.

One control was to establish whether these strong ocher suppressors were affected by ψ. It turned out that they were, but in a manner dramatically different from that of *SUQ5*, the serine-inserting "weak" suppressor involved in all of our genetic assays of ψ. The strong *SUP11*, *SUP2*, and *SUP3* suppressors are lethal in combination with $\psi^+$; *SUP11* is a recessive lethal, and *SUP2* and *SUP3* are dominant lethals. This was interpreted as being due to a universal effect of ψ in modulating the activity of tRNA suppressors. Experiments followed to establish this information directly. Liebman and Sherman (1979) measured readthrough in $\psi^+$ and $\psi^-$ strains using ocher mutations of the iso-

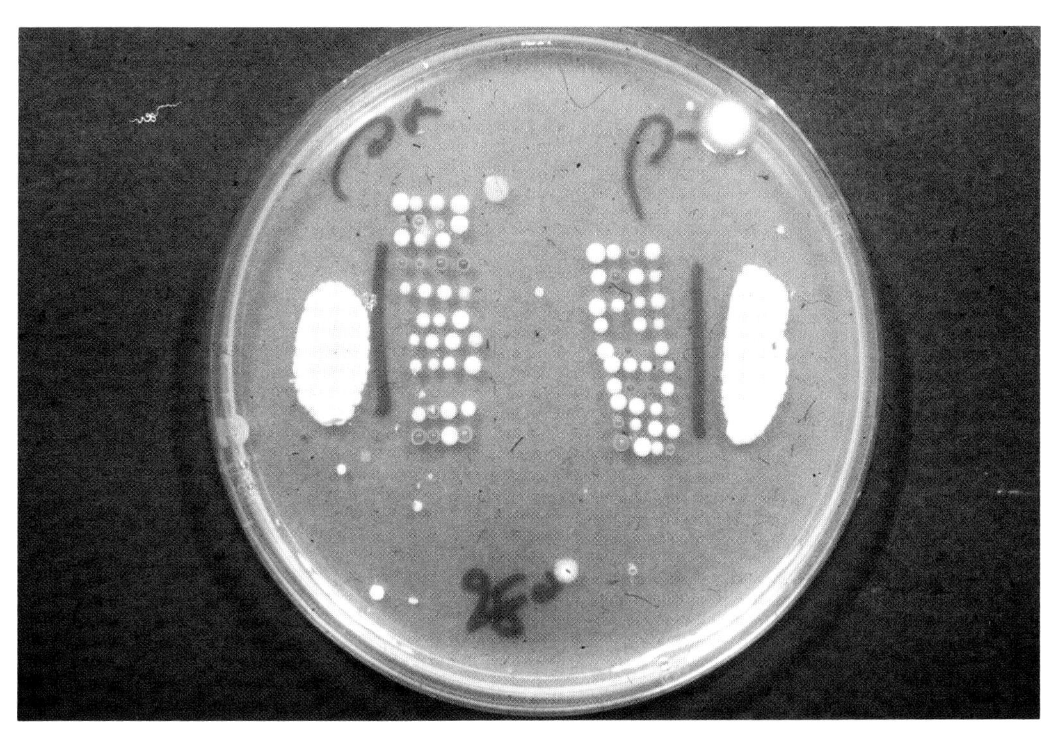


*Figure 5* Segregation of $\psi^+$ and $\psi^-$ in tetrads formed by the sporulation of newly formed zygotes. The parents were similar in genotype to those described in Fig. 2, but in this case, segregations of red, $\psi^-$ spore cultures and cultures with red sectors are observed.

1-cytochrome *c* gene and measuring the amount of protein produced. Mick Tuite and Cal McLaughlin devised an in vitro assay for tRNA-mediated readthrough of natural chain-termination codons and showed that all tRNA-mediated readthrough, whether of ocher, amber, or UGA chain-termination codons, depended on the cell-free translation extracts being made from $\psi^+$ cells (Tuite et al. 1983, 1987). In addition, Mandy Firoozan and her colleagues have now developed another in vivo assay of readthrough based on a *lacZ*

*Table 1* Independent Segregation of $\psi$ and Mitochondrial Markers

| | $\psi^+$ | $\psi^-$ | Totals |
|---|---|---|---|
| (a) $\psi$ *and respiratory competence* | | | |
| $\rho^+$ | 47 | 13 | *60* |
| $\rho^-$ | 10 | 6 | *16* |
| *Totals* | *57* | *19* | *76* |
| (b) $\psi$ *and erythromycin resistance* | | | |
| $E^r$ | 40 | 9 | *49* |
| $E^S$ | 7 | 4 | *11* |
| *Totals* | *47* | *13* | *60* |

The independent segregation of (a) $\psi$ and respiratory competence and (b) of $\psi$ and erythromycin resistance in spores isolated from tetrads from a diploid ($\psi^+$, $E^r$, $\rho^+$ x $\psi^-$, $E^s$, $\rho^-$).

gene fusion, with *lacZ* expression being protected by stop codons (Firoozan et al. 1991). All three assays confirm that readthrough is enhanced in $\psi^+$ strains—about ten times enhanced in vivo and absolutely dependent on $\psi^+$ in vitro.

## GENES CONTROLLING SUPPRESSION

The second control is a corollary: One needs to show that *SUP11*, *SUP2*, and *SUP3* can indeed be modulated and if so, to ask whether these modulators also affect *SUQ5* without affecting $\psi$. Shirley McCready, our second $\psi$ student, undertook a formal search for such modulators in her Ph.D. project. She isolated nearly 100 mutants in which suppression of *ade2-1* and *can1-100* by *SUQ5* $\psi^+$ had been eliminated. Complementation tests (with *SUQ5* $\psi^-$) disclosed 25 of these mutants that were recessive and still $\psi^+$. These she assigned to eight loci by complementation and recombination analysis and then showed that mutations at all eight loci also abolished the ability of *SUP2* to suppress *ade2-1* and *can1-100* and also prevented the *SUP11* gene from being lethal in a $\psi^+$ background (McCready and Cox 1973).

This kind of mutation, which we called *antisuppressors* (*asu*), clearly establish that there are two categories of genes affecting tRNA-mediated suppression: those that affect the suppression itself through some interaction with the tRNAs and those that affect $\psi$. Antisuppressors are the former and *PNM* the latter, and they make it possible to study either phenomenon independently of the other (Fig. 6).

$\psi$ students three and four, Mick Tuite and Chris Mundy, shared the laboratory with $\psi$ postdoc Clegg Waldron. All had completely different personalities and communication was not among their gifts, which did not matter too much as far as the phenomena were concerned. Although their paths through their assorted projects crossed, diverged and recrossed, ran parallel, or ran together for various lengths of time, their results never conflicted. However, their scientific methods and the details of their methodologies were consistently different and idiosyncratic so that writing papers based on their joint discoveries became an exercise in diplomacy as much as science. Trying to write sentences and describe experiments that were simultaneously acceptable to Chris, Mick, and the editors of *Genetics* often required a delicate turn of phrase. I am afraid I cheated by not showing the drafts to either Chris or Mick. Clegg was fortunately not involved in all this—while the others were doing only genetics, he was looking for biochemical differences between $\psi^+$ and $\psi^-$ cells. His endearing demeanor of consistently cheerful pessimism exactly suited the succession of negative results that he enthusiastically generated (see, e.g., Waldron and Cox 1978) and did a lot to keep the laboratory together and productive. Mick's thesis was as fat as Hamish's was slender. His external examiner, Alan Bevan, phoned me to complain, and it is true that he could have, in the Oxford phrase, satisfied the examiners with any one third of the material he included. However, as I explained to Bev, if I did not insist on all of the material going into the thesis, a great deal of information about $\psi$ would disappear into forgotten notebooks. I was quite right. To this day, Mick and I nag each other about writing up more of his thesis for publication. The essentials of Mick's thesis were, however, published.

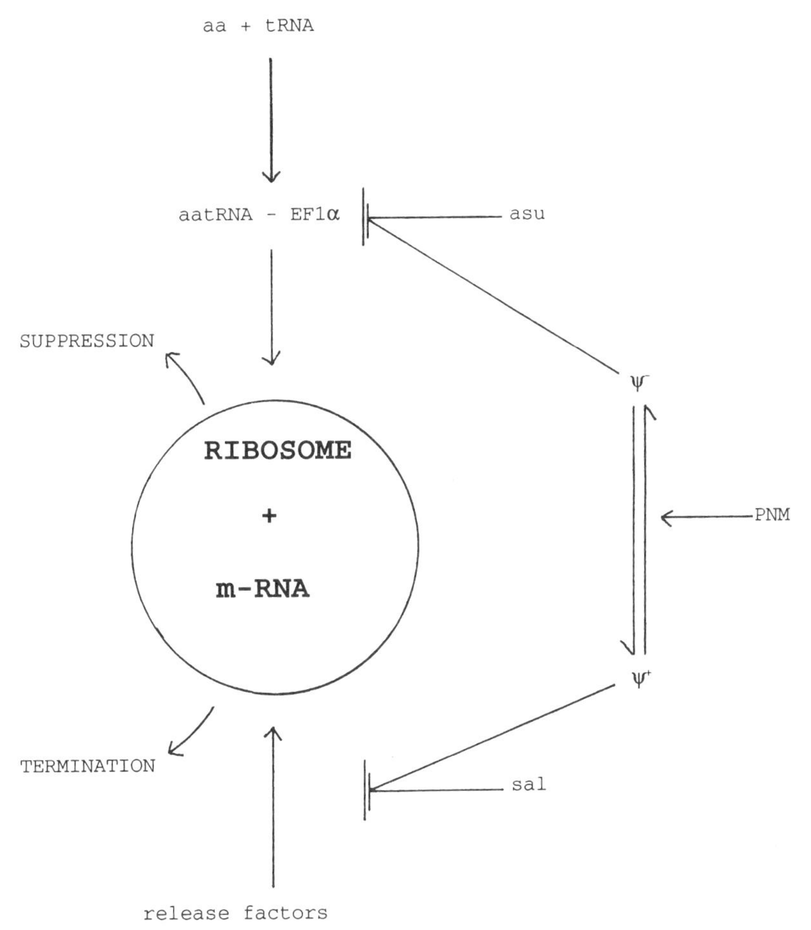


*Figure 6* Diagram illustrating the independence of components of systems modifying suppression in yeast from those modifying ψ.

Chris Mundy, in addition to opening a Pandora's box of agents that mutate ψ, did an extensive and detailed analysis of modifiers of ψ, showing that yeast strains are full of genes that affect suppression. He found it was possible to derive strains by mutation and meiotic segregation in which suppression of even the most easily suppressed ocher mutation by *SUQ5* was absolutely dependent on ψ and that in these genetic backgrounds, ψ showed no trace of any ability to act as a suppressor on its own. I think this was our first indication that in most yeast strains, there is a low level of endogenous suppression. Some of the causes of this were later put on a molecular footing by other workers, who have found native tRNA species that may act as very weak suppressors. Chris was also the first student to suggest in his thesis that ψ might be some sort of regulation phenomenon (Mundy 1979).

As Mick's and Chris' publications make clear, Hamish's *R* mutation (*PNM*) and Shirley's *asu* mutations were only part of the spectrum of nonsuppressed revertants that arose spontaneously or after induction by mutagens in *SUQ5* $\psi^+$ strains. Mick and Chris defined those spectra by exhaustive genetic analysis and devised clever complementation tests (using me as some kind of go-between) to make it easy to sort them out. The result was the demonstration of both dominant and semidominant antisuppressors in addition to the recessive category on which Shirley had focused, and recessive *pnm* mutations (Cox et al. 1980).

Recessive *pnm* mutations posed a real problem for ψ genetics. They were defined as mutations that when crossed with a $\psi^+$ strain gave rise to a suppressed $\psi^+$ diploid and when crossed with a $\psi^-$ strain gave unsuppressed diploids, as do ordinary $\psi^-$ mutations. Tetrads from the latter, however, segregated suppressed spore cultures, usually 2:2, but with some 1:3 and 0:4 suppressed:nonsuppressed. The problem (quite apart from how such mutations work) is that these "recessive *pnm* mutations" behave in complementation tests exactly like ordinary $\psi^-$ mutations: Only tetrads show that they are not. Since Mick wanted to carry out a mutagenesis study of ψ, a simple method of discrimination was essential. This was provided by the *kar1* mutation, which prevents nuclear fusion in zygotes, allowing heterokaryons and haploid cells to segregate from them (Conde and Fink 1976). Nuclear mutations and ψ behave differently in *kar* crosses. ψ, as a cytoplasmic determinant, segregates to all the cell progeny of a *kar* zygote, so they are all $\psi^+$. Nonsuppressing nuclear mutations segregate only to some progeny. If one parent in the cross is *kar1 suq5*$^+$ $\psi^+$, i.e., nonsuppressed, then the only nonsuppressed type of mutant that will, after crossing with this *kar* strain, segregate suppressed cells from the zygotes will be the ψ type (by acquiring $\psi^+$ particles from the *kar* parent): All mutants that are not suppressed because of a nuclear mutation will segregate only nonsuppressed cell progeny (Cox et al. 1980).

The *kar1* mutation provided another demonstration, if one was needed, that the determinant of ψ is cytoplasmic. Mick showed that *all* of the cells segregating from *kar* crosses between $\psi^+$ and $\psi^-$ cells were $\psi^+$, regardless of which parental nucleus they inherited from the heterokaryotic zygote. These experiments with *kar1* strains contributed significantly to one of the ψ paradoxes (Tuite 1978).

Shortly after this laboratory with Mick, Chris, and Clegg in it went into its diaspora, Jack von Borstel invited me for a sabbatical. In return for running some mutagenicity assays on assorted compounds, he allowed me time and facilities to try to purify the serine-accepting tRNA coded by *SUQ5*. This included a month in Ray Gesteland's laboratory at Cold Spring Harbor Laboratory to learn the method. Despite this lofty tuition, I never managed it, but after Clegg Waldron left my laboratory, he joined Ray at the Howard Hughes Medical Research Institute in Salt Lake City, where he completed a difficult job by sequencing the normal and mutant *SUQ5* tRNAs (Waldron et al. 1981).

## ψ PARADOXES

The main object of Mick Tuite's work was to determine whether ψ was DNA-based, which he did by showing that it is mutable by coventional mutagens such as UV and nitrosoguanidine (NTG) and that its mutagenesis is dependent on the DNA-repair genes, including photoreactivating enzyme, in exactly the same way as are nuclear genes. The paradox is that the induction kinetics of $\psi^-$ mutations by UV are single-hit and exactly parallel the induction of the nuclear gene mutations scored at the same time (Tuite and Cox 1980). It follows that, because the mutation is recessive, $\psi$ is a single target, not the multiparticle determinant suggested by its segregation in *PNM* strains; 100% segregation of $\psi^+$ to spores in meiosis is also incompatible with a single-copy determinant. The cytoduction experiments reinforced the discrepancy. To achieve 100% cytoduction of the $\psi^+$ phenotype with a single determinant would not seem to be feasible. Neither 2-μm circles (copy number of 70) nor mitochondrial DNA (copy number of 20–100) shows 100% cytoduction (Livingston 1977; Cox et al. 1988).

It soon became clear that another property of $\psi$ was paradoxical. Chris Mundy discovered by accident that ethanol caused $\psi^+$ to $\psi^-$ mutations at a very high rate. The conventional mutagens induced $\psi^-$ mutations at rates comparable to those of other genes. In dramatic contrast, ethanol and, it became clear in due course, other treatments induced $\psi^-$ in as many as 10–20% of the cells during growth. Singh et al. (1979) discovered the same to be true of substances (like molar KCl) causing osmotic stress. The most remarkable of these substances is guanidine HCl, which is toxic to growth above about 7 mM, but at 4–5 mM, it completely converts a $\psi^+$ culture to $\psi^-$ without affecting the overall growth rate. The conversion requires growth and the kinetics are as if, like *PNM*, guanidine HCl stops a $\psi^+$ determinant from replicating (or being regenerated). $\psi^-$ cells start to segregate six to eight generations after adding the compound (see Fig. 4). On the other hand, treatment with acridine dyes or with ethidium bromide, both of which bind to DNA and induce cytoplasmic *petite* mutations, has no effect on $\psi$ (Tuite et al. 1981).

We are left with two paradoxes. UV induction of $\psi^-$ mutations suggests a single-copy DNA determinant, particularly as this depends on the normal DNA-repair processes. Genetic segregation in meiosis, in *PNM* and in *kar* crosses, and from guanidine HCl treatment suggests a multicopy cytoplasmic determinant. The susceptibility of $\psi$ to compounds like organic salts, alcohols, guanidine HCl, and dimethylsulfoxide (DMSO), which do not cause gene mutations, suggests that $\psi$ may not be DNA.

The question of whether or not $\psi$ was DNA, whether or not the mutations were real, led to the projects undertaken by Pat Lund, the next $\psi$ student, and Hwa Dai who visited from Taiwan a little later, with some dabbling in $\psi$ on the side by Bruce Futcher. Pat undertook a classic Seymour Benzer approach, there being two questions: Are $\psi^-$ mutations revertible to $\psi^+$ and do they recombine to allow a genetic map of $\psi$ to be made?

As far as mutability goes, $\psi$ behaves like any other gene in that it appears to mutate to alleles that have characteristic rates of reversion. $\psi^-$ mutations induced with UV, NTG, ethylmethanesulfonate (EMS), methanol, molar salt DMSO, *PNM*, or spontaneously acquired nearly all reverted to $\psi^+$ at rates comparable to reversions of mutant genes. An exception was the set of $\psi^-$ mutations induced by 5 mM guanidine HCl, which did not revert at all. Treating revertible $\psi^-$ mutants with guanidine HCl rendered them nonrevertible (Lund and Cox 1981).

Recombination, however, could not be demonstrated. The requirement is for $\psi^+$ revertants to appear from a cross between two nonrevertible $\psi^-$ mutants. Indeed, in two such crosses, revertants were found both in the diploids and among their haploid progeny. If this were the rII gene of T4 phage, one would have defined these $\psi^-$ mutations as nonoverlapping deletions. Unfortunately, an attempt to develop a deletion map foundered on the fact that no more "recombining" combinations were ever found.

It was clear that the time was ripe for the simple, direct approach to the problem, namely, transformation. By this time, the early 1980s, transformation of yeast was commonplace, so both positive and negative controls could be included in the experiments. If the recipient strain is a nonrevertible $\psi^-$, induced by guanidine HCl, and transformation with DNA from the $\psi^+$ strain gives $\psi^+$ revertants, this is *prima facie* evidence for the DNA basis of the phenotype.

Pat Lund succeeded in doing this experiment with relatively crude preparations of DNA. At this time, we were joined for a year by Hwa Dai from the Botany Department of Academica Sinica, Taipei, Taiwan. She became convinced that $\psi$ must be determined by the only known plasmid of yeast we had not managed to eliminate from $\psi^+$ strains, namely, the 3-μm circle (Tuite et al. 1982). She set about purifying 3-μm circles from a $\psi^+$ strain and effectively showed that a $\psi^+$ transforming activity copurified with the DNA, down to the level of a band cut from a gel (Dai et al. 1986).

The 3-μm circles are circularized rDNA repeats. There are higher multimers present in cells and probably 10% or more of the rDNA is in this episomal form. One can define the characteristics expected of rDNA 3-μm circles if it is the case that some occur that determine $\psi$. First, there must be a mutation in such a DNA that directly or indirectly affects translational fidelity. It may do so directly by affecting the interaction of 18S or 28S rRNA with tRNA or with termination factors, like the mutations in *E. coli* 16S rRNA described by Dahlberg (1989) and Prescott and Goringer (1990). Indirectly, the mutation may affect the binding of an editing protein such as the product of the *SUP35* gene (q.v.) is likely to be.

Second, either this mutation itself makes the difference between $\psi^+$ and $\psi^-$ strains or the sequence carrying the mutation is differentially transcribed in $\psi^+$ and $\psi^-$ strains. Differential transcription would have to be associated with some form of DNA imprinting, loosely defined, to account for transformation. The $\psi^+$ "imprinting" need be merely the existence of the sequence in the episomal form (Cox et al. 1988).

The behavior of $\psi^+$ and $\psi^-$ in heredity would then require that the sequence be confined to the chromosome in $\psi^-$ strains and not transcribed. In $\psi^+$ strains, transcription and the episomal form would be mutually interdependent and self-sustaining. The effect of crossing $\psi^+$ and $\psi^-$, or of cytoducing or transforming $\psi^+$ episomes into $\psi^-$, would be to instigate the transcription and episomal state of the sequence in the $\psi^-$ genome and to substitute for it, so that all subsequent progeny would be $\psi^+$.

This model accommodates both the heredity and the mutagenesis data. $\psi^-$ mutations may be point mutations or deletions in a chromosomal sequence or the consequence of treatments (e.g., DMSO) or gene mutations (e.g., *PNM*) that repress expression. It makes both soft and hard predictions.

One genetic prediction is that a Mendelian segregation ought to be observed if the correct situation is analyzed. Such a situation might occur if, for example, a nonrevertible $\psi^-$ (guanidine-HCl-induced) were crossed by a revertible $\psi^-$: Revertibility might segregate 2:2. A hard prediction is that it should be possible to clone a 3-μm DNA sequence that gives the $\psi^+$ phenotype.

A serious attempt to do this was made by Sheila Doel, who painstakingly built libraries of rDNA sequences in a yeast shuttle vector. It was painstaking because it turned out that complete rDNA sequences from yeast were difficult to transform into *E. coli*. Transformation frequencies of cloned rDNA, either from libraries of rDNA or from 3-μm preparations, or of recovered rDNA clones, were three or four orders of magnitude lower than those from random DNA libraries. It was necessary to pick the few individual transformants for pooling and amplification. Collections of about 100 failed to reveal $\psi^+$ activity.

## THE FUTURE OF ψ

This is where the ψ problem stands at present: The phenomenon is phenotypically clear-cut, is repeatably measurable in vivo using either native proteins such as iso-1-cytochrome *c* or artificial constructs, and has a very clear biochemical effect demonstrable in vitro.

The determinant, however, is elusive. There exist two physically accessible handles. The most definite is the "fidelity factor," which washes off ribosomes prepared from $\psi^-$ cells, that inhibits in vitro readthrough in $\psi^+$ cell-free extracts. If this is a reasonably abundant, reasonably stable macromolecule, it should be possible to purify it and from there clone the gene that produces it. The second, less definite entity is the 3-μm circle. If a species exists that codes for ψ, it too should be clonable.

Meanwhile, it may be that Hamish's third paper contains the seeds of another solution to the ψ problem (Young and Cox 1975). He had observed that colonies of *ade*⁻ strains with the genotype *ade2-1 SUQ5* $\psi^-$ threw an unusually large number of white papillae compared with *ade2-1* strains. He showed that most of these were due to secondary weak suppressors that in $\psi^-$ were only expressed in the presence of the *SUQ5* mutation. This led to a genetic analysis of these "suppressors" which demonstrated that they had no suppressor activity in their own right but merely enhanced the activity of *SUQ5* and other ocher suppressors. They were called "allosuppressors" and 100 or more independent isolates fell into no more than five loci (*sal1–sal5*) (Cox 1977). These mutations therefore also distinguish between effects on ψ and effects on suppression: They are the converse of antisuppressor mutations, and no doubt some of the loci in the two categories are allelic, although this has not been directly tested. Nevertheless, allosuppressors are not altogether unequivocal in affecting suppressor activity rather than ψ. Allosuppressor mutations at two of the loci, *sal3* and *sal4*, have a number of interactions with ψ, with each other, and with *PNM* mutations, which suggest that these loci are the fulcrum through which ψ affects translational fidelity and is itself determined.

First, all alleles of *sal3* and *sal4* have lethal interactions with $\psi^+$ regardless of the presence or absence of a suppressor gene. *sal3* and *sal4* are also lethal to each other. *sal3* was linked to *PNM2* but did not otherwise appear to have anything to do with it or with the other locus, *PNM1*, in the sense that the *PNM* mutations did not affect the *sal* phenotype. *sal4* was linked to neither. Nevertheless, *sal3* and *sal4* do interact with *PNM* in a more subtle way: They render *PNM* recessive. Diploids formed by crossing, for example, a *PNM2 sal3* strain with a $\psi^+$ strain often remain white and the white colonies often sector red, and when tetrads are dissected, $\psi^+$ spores segregate, at least among the $SAL^+$ segregants. Another way of looking at the same results is to say that

*Table 2* Some Properties of the ψ Phenomenon

(A) *Physiology*

| | $\psi^+$ | $\psi^-$ |
|---|---|---|
| No suppressor tRNA | weak suppression | no suppression |
| Ser-$tRNA_{UUA}$ | strong suppression | weak suppression |
| Leu-$tRNA_{UUA}$ | strong suppression | weak suppression |
| Tyr-$tRNA_{UUA}$ | lethal | strong suppression |
| Gly-$tRNA_{CCCC}$ | strong frameshift suppression | weak frameshift suppression |

(B) *Interactions*

| | $\psi^+$ | $\psi^-$ | *sal3* | *sal4* |
|---|---|---|---|---|
| *sup35* | omni | omni | — | — |
| *sal3* | lethal | suppression | — | lethal |
| *PNM* | becomes $\psi^-$ | none | becomes recessive | becomes recessive |

they make $\psi^+$ resistant to the *PNM* mutations. Table 2 summarizes the *PNM*-*sal*-ψ complex of behavior.

When Sheila Doel concluded that enough was enough of cloning rDNA, she decided that a *PNM* gene would be at least easier, and in view of the ambiguity implicit in *sal* and *PNM* interactions, quite possibly informative. Mick Tuite had cloned the wild-type allele of *sal3* by complementation of its cold sensitivity and discovered that its restriction map was identical to that of *SUP35*, already cloned by Kushnirov et al. (1988), complementing an omnipotent suppressor (*sup2* = *sup35*), and by Wilson and Culbertson (1988), complementing a frameshift suppressor *suf12*. Since we knew that *sal3* and *PNM2* were linked, Sheila decided on a chromosome walk to find *PNM2*, using Mick's *SAL3* clone as a probe. What she found was that *PNM2* was on the same fragment that hybridized to the probe, and it now turns out that *sal3*, *PNM2*, and *sup35* are all alleles of the same gene. Yet another clone of *SUP35* has been isolated by complementing a cell-cycle-arrest mutation (Kikuchi et al. 1988).

The gene is essential for cell viability. Its product seems to function at the heart of the translation machinery, but it is not one of the factors essential for polypeptide synthesis identified hitherto by biochemical methods. However, it is clearly concerned, most probably directly so, with translational fidelity. A detailed molecular and mutational analysis has been carried out by Ter-Avanesyan and his colleagues (Dagkesamanskaya et al. 1990). Sheila Doel's analysis was less detailed, sufficient only to establish that what she had was indeed both *SAL3* and *PNM2*. Both sets of analyses demonstrate that there are two functional domains that are to some extent antagonistic. The amino-terminal end of the protein promotes translational infidelity, and the carboxy-terminal end reduces it, increasing fidelity. Some of the known properties of the gene, its mutations, and its protein product are summarized in Figure 7.

Although the mutational analysis clearly establishes the existence of two domains in the protein, none of the mutations created by Dagkesamanskaya et al. (1990) exactly mimic the mutant alleles of *SUP35* isolated in yeast. The only

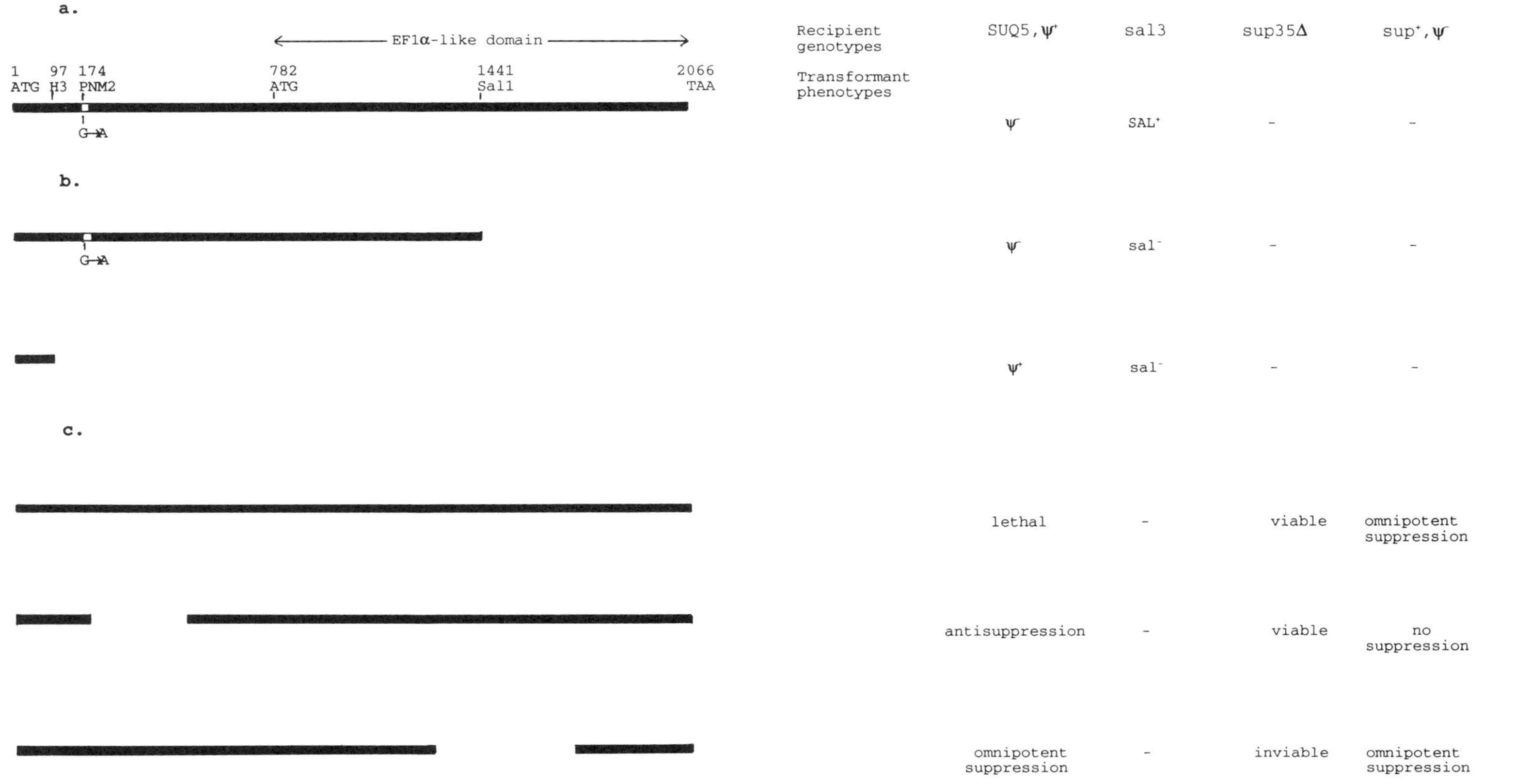


*Figure 7* (*a*) Diagram of the *SUP35* gene showing the location of the *PNM2* mutation; (*b*) consequences of deletions in the cloned *PNM2* gene on the phenotype of transformed strains of two genotypes: $\psi^+$ and *sal3*; (*c*) consequences of deletions in the two domains of the wild-type *SUP35* gene on the phenotypes of transformed strains. (Drastically abbreviated and summarized from Kushnirov et al. [1987], Dagkesamanskaya [1990], and V. Kushnirov [pers. comm.].)

one of these so far sequenced is *PNM2*, which is a point mutation in the amino-terminal domain of the gene, where deletions cause a dominant antisuppressor phenotype. There seems little doubt that this complex and pleiotropic essential gene holds the key to the ψ factor. Let us explore how it may do so. The basic observation is that the two domains of the *SUP35* gene have antipathetic effects, i.e., mutations disrupting the amino-terminal domain cause antisuppression phenotypes and mutations disrupting the carboxy-terminal domain result in infidelity. One may thus describe the carboxy-terminal end of the protein as a fidelity factor. This end is also clearly the bit that is essential for cell viability. What does the amino-terminal end do? By analogy, one would say it causes infidelity. However, it seems irrational to suppose that an infidelity function might be adaptive, so I would retreat to the position that it has some useful function; but incidentally, it interferes with the fidelity functions performed by its carboxy-terminal portion. How may it do that?

Inspection of Northern blots shows that two transcripts exist, one of which probably represents transcription of the whole coding sequence and the other representing a transcription of only the carboxy-terminal end of the sequence. Suppose therefore that there are, from these transcripts, two protein products, one being translated from the complete open reading frame (ORF) and the other from a shorter transcript, starting at a downstream AUG, probably that at position 782. This translation product, which is, remember, EF1α-like, is the key protein that ensures fidelity at the point of elongation. The effect of the longer protein containing the amino-terminal "infidelity" function is to compete for ribosome sites with the short protein. However, when in place, it fails to promote fidelity. The balance between suppression and fidelity is thus struck by the ratio of these two proteins in the cell.

Most of the observations so far published or rumored about are compatible with this model. For example, all overexpression experiments, which lead to omnipotent suppression, have been done by putting the whole ORF either under the control of a high-expression promoter or on a multi-copy plasmid. The model predicts that if, in contrast, the carboxy-terminal domain only is overexpressed, this phenotype would not be observed. The model also predicts that *sup35* (omnipotent suppressor) and *sal3* (allosuppressor) mutations will map in this domain. The distinction between them will be interesting. Omnipotent suppression implies *mis*reading of codons by tRNAs; allosuppression implies instead a higher affinity of anticodons for codons, thus competing out release factors that must read the same codons. A protein that regulates codon-anticodon interaction would be expected to acquire either kind of property by mutation.

It is the relationship of this protein with the ψ factor that is most interesting. At this point, we have two observations and two others that need confirmation. The observations are (1) that mutations which affect the amino-terminal end of the ORF, i.e., mutations in the pro-suppressor end, cause $\psi^+$ to become $\psi^-$ and (2) that *sal3* mutations which cause this protein to allow ocher readthrough and overexpression of the whole ORF are incompatible with the cells being $\psi^+$. One intriguing possibility is that the maintenance of $\psi^+$ is promoted by infidelity specifically mediated by this protein. Since ψ promotes infidelity of this kind, the system could consist of a positive feedback loop in which $\psi^+$ is indefinitely maintained. The combination of ψ-promoted readthrough and

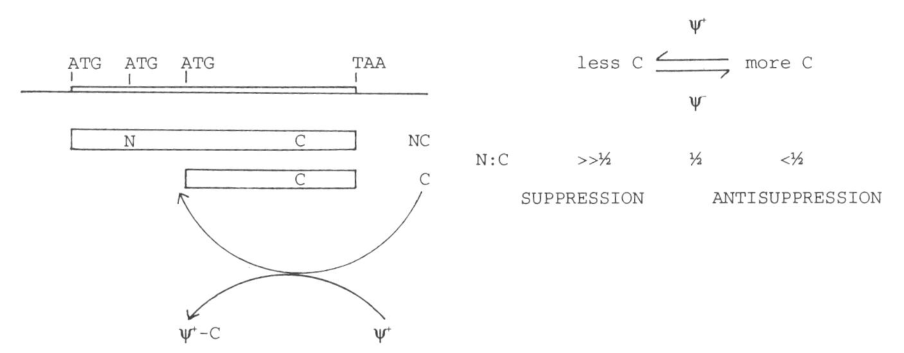


*Figure 8* Marie Nierras' model for ψ. The model describes the interactions of ψ and the *SUP35* gene products which might account for the effect of ψ on suppression. In Marie's opinion, ψ in the model is a special form of rRNA.

C promotes its own transcription: This creates a positive feedback loop, decreasing N:C and promoting antisuppression.

$\psi^+$ binds C, acting as a sink, increasing N:C and modulating the positive feedback, promoting suppression.

$\psi^+$ is maintained by high N:C ratios, so a counter-C positive feedback is established in $\psi^+$ cells.

The *PNM2* mutation makes NC bind to $\psi^+$ so that $\psi^+$ becomes a sink for NC: The N:C ratio falls. The released C starts to promote its own transcription: This drives the N:C ratio lower still, the positive feedback is reestablished and the cell becomes $\psi^-$ in phenotype and genotype.

*sal*-promoted readthrough is lethal, and so the model is immune to direct test; however, the two other observations suggest there may be some truth in this notion. The first is that in a *sal3 PNM* strain, i.e., one in which the amino-terminal end of the *SUP35* ORF is mutated, but suppression is mediated by the *sal3* mutation, presumably in the carboxy-terminal end, $\psi^+$ is maintained. The second is that S. Inge-Vectomov (unpubl.) has observed that $\psi^-$ strains transformed with the wild-type *SUP35* gene on a multicopy plasmid (which promotes omnipotent suppression) occasionally revert to $\psi^+$ when the plasmid is lost. It would be a further refinement of the model if it were supposed that the $\psi^+$ phenotype were itself a consequence only of the presence and activity of the amino-terminal portion of this gene.

No ψ article is complete without a model. Figure 8 is the model suggested now by the last of the ψ postdocs, Marie Nierras, who, with Shirley McCready, sequenced the *PNM2* mutant allele of *SUP35*. It cleverly incorporates rRNA and therefore implicates 3-μm circles. It does not explain ψ inheritance, of course. That is still another problem.

The club of yeast geneticists is still in existence but has suffered a degree of dispersion and lost its exclusivity. However, a number of clubs within the club have formed informally, like eddies in a tide pool, and one of the most esoteric is the little coterie interested in cytoplasmic or extrachromosomal inheritance, and the more obscure the particle, the greater the bond. We do not have a handshake, but sometimes an eyebrow is raised. One of the pleasures that endures at the huge modern conventions on yeast genetics is meeting each other, especially those who did and still do genetics with their phenotypes, to exchange details of the latest rococco experiments. Here is to the killer people, especially Reed Wickner and Alan Bevan, to the mitochondrial people and their doyen, Piotr, to Jim Haber for giving us 20S and congratulations to them all on their graduation into molecular respectability. ψ and *URE3*, I am sure Francois Lacroute and Michel Aigle would agree, *te salute cumque gratias*.

## REFERENCES


Bevan, E.A., A.J. Herring, and D.J. Mitchell. 1973. Preliminary characterisation of two species of dsRNA in yeast and their relationship to the killer character. *Nature* **245:** 81–87.

Broach, J.R., L.R. Friedman, and F. Sherman. 1981. Correspondence of yeast UAA suppressors to cloned $tRNA^{ser}/_{UCA}$ genes. *J. Mol. Biol.* **150:** 375–387.

Conde, J. and G.R. Fink. 1976 A mutant of *Saccharomyces cerevisiae* defective for nuclear fusion. *Proc. Natl. Acad. Sci.* **73:** 3651–3655.

Cox, B.S. 1965. ψ, a cytoplasmic suppressor of super-suppressor in yeast. *Heredity* **20:** 505–521.

———. 1977 Allosuppressors in yeast. *Genet. Res.* **30:** 187–205.

Cox, B.S., M.F. Tuite, and C.S. McLaughlin. 1988. The ψ factor of yeast: A problem in inheritance. *Yeast* **4:** 159–178.

Cox, B.S., M.F. Tuite, and C.R. Mundy. 1980. Reversion from suppression to nonsuppression in SUQ5 [psi$^+$] strains of yeast: The classification of mutations. *Genetics* **95:** 589–609.

Dagkesamanskaya, A.R., V.V. Kushnirov, S.A. Didichenko, and M.D. Ter-Avenesyan. 1990. Deletion analysis of SUP2 (SUP35) suppressor gene. *Yeast* (special issue) **6:** S398.

Dahlberg, A.F. 1989. The functional role of ribosomal RNA in protein synthesis. *Cell* **57:** 525–529.

Dai, H., B.S. Cox, S.-H. Tsay, and P.M. Lund. 1988. Transformation of *ψ—Saccharomyces cerevisiae* $ψ^+$ with DNA co-purified with 3-μm circles. *Curr. Genet.* **11:** 79–82.

Firoozan, M., C.M. Grant, J.A. Duarte, and M.F. Tuite. 1991. An *in vivo* assay for readthrough of chain termination codons in *Saccharomyces cerevisiae. Yeast* **7:** 173–183.

Hawthorne, D.C. 1969. Identification of nonsense codons in yeast. *J. Mol. Biol.* **43:** 71–75.

Hawthorne, D.C. and R.K. Mortimer. 1963. Super-suppressors in yeast. *Genetics* **48:** 617–620.

Kikuchi, Y., H. Shimatake, and A. Kikuchi. 1988. A yeast gene required for the G1 to S transition encodes a a protein containing an A-kinase target site and a GTPase domain. *EMBO J.* **7:** 1175–1182.

Kushnirov, V.V., M.D. Ter-Avanesyan, A.P. Surguchov, V.N. Smirnov, and S.G. Inge-Vechtomov. 1987. Localization of possible functional domains in SUP2 gene product of the yeast *Saccharomyces cerevisiae. FEBS Lett.* **215:** 257–260.

Kushnirov, V.V., M.D. Ter-Avenesyan, M.V. Telekov, A.P. Surguchov, V.N. Smirnov, and S.G. Inge-Vectomov. 1988. Nucleotide sequence of the SUP2 (SUP35) gene of *Saccharomyces cerevisiae. Gene* **66:** 45–54.

Liebman, S.W. and F. Sherman. 1979. The extrachromosomal $ψ^+$ determinant suppresses nonsense mutations in yeast. *J. Bacteriol.* **139:** 1068–1071.

Livingston, D.M. 1977. Inheritance of the 2 μm DNA plasmid from *Saccharomyces cerevisiae. Genetics* **86:** 73–84.

Lund, P.M. and B.S. Cox. 1981. Reversion analysis of [psi$^-$] mutations in *Saccharomyces cerevisiae. Genet. Res.* **37:** 173–182.

Manney, T.R. 1969. Evidence for chain termination by super-suppressible mutants in yeast. *Genetics* **60:** 719–753.

McCready, S.J. and B.S. Cox. 1973. Antisuppressors in yeast. *Mol. Gen. Genet.* **124:** 305–320.

McCready, S.J., B.S. Cox, and C.S. McLaughlin. 1977. The extrachromosomal control of nonsense suppression in yeast: An analysis of the elimination of [psi$^+$] in the presence of a nuclear gene, PNM. *Mol. Gen. Genet.* **150:** 265–270.

Mundy, C.R. 1979. "The genetics of translation in yeast." Ph.D. thesis, Oxford University, England.

Olson, M.V., B. Hall, J. Cameron, and R. Davis. 1979. Cloning of the yeast tyrosine transfer RNA genes in bacteriophage lambda. *J. Mol. Biol.* **127:** 285–295.

Prescott, C.D. and H.U. Goringer. 1990. A single mutation in 16S RNA that affects mRNA binding and translation termination. *Nucleic Acids Res.* **18:** 5381–5386.

Singh, A.C., C. Helms, and F. Sherman. 1979. Mutation of the non-Mendelian suppressor $\psi^+$ in yeast by hypertonic media. *Proc. Natl. Acad. Sci.* **76:** 1952–1956.

Somers, J.M. and E.A. Bevan. 1969. The inheritance of the killer character in yeast. *Genet. Res.* **13:** 71–83.

Stewart, J.W. and F. Sherman. 1972. Demonstration of UAG as a nonsense codon in bakers yeast by amino-acid replacements in iso-1-cytochrome *c*. *J. Mol. Biol.* **68:** 429–443.

Stewart, J.W., F. Sherman, M. Jackson, F.L.X. Thomas, and N. Shipman. 1972. Demonstration of the UAA codon in baker's yeast by amino-acid replacements in iso-1-cytochrome *c*. *J. Mol. Biol.* **68:** 83–96.

Tuite, M.F. 1978. "Genetics of nonsense suppressors in yeast." Ph.D. thesis, Oxford University, England.

Tuite, M.F. and B.S. Cox. 1980. Ultraviolet mutagenesis studies of [psi], a cytoplasmic determinant of *Saccharomyces cerevisiae. Genetics* **95:** 611–630.

Tuite, M.F., B.S. Cox, and C.S. McLaughlin. 1983. In vitro nonsense suppression in [psi$^+$] and [psi$^-$] cell-free lysates of *Saccharomyces cerevisiae. Proc. Natl. Acad. Sci.* **80:** 2824–2828.

———. 1987. A ribosome-associated inhibitor of in vitro nonsense suppression in [psi$^-$] strains of yeast. *FEBS Lett.* **225:** 205–208.

Tuite, M.F., C.R. Mundy, and B.S. Cox. 1981. Agents that cause a high frequency of genetic change from [psi$^+$] to [psi$^-$] in *Saccharomyces cerevisiae. Genetics* **98:** 691–711.

Tuite, M.F., P.M. Lund, A.B. Futcher, M.J. Dobson, B.S. Cox, and C.S. McLaughlin. 1982. Relationship of the [psi] factor with other plasmids of *Saccharomyces cerevisiae. Plasmid* **8:** 103–111.

Waldron, C.J. and B.S. Cox. 1978. Ribosomal proteins of yeast strains carrying mutations which affect the efficiency of nonsense suppression. *Mol. Gen. Genet.* **159:** 223–225.

Waldron, C., B.S. Cox, N. Wills, R.F. Gesteland, P.W. Piper, D. Colby, and C. Guthrie. 1981. Yeast ochre suppressor *SUP-o1* is an altered $tRNA^{ser}/_{UCA}$. *Nucleic Acids Res.* **9:** 3077–3088.

Wilson, P.G. and M.R. Culbertson. 1988. SUF12 suppressor protein of yeast: A fusion protein related to the EF-1α family of elongation factors. *J. Mol. Biol.* **199:** 559–573.

Young, C.S.H. 1969. "A study of genetic suppression in *Saccharomyces*." Ph.D. thesis, Oxford University, England.

Young, C.S.H. and B.S. Cox. 1971. Extrachromosomal elements in a super-suppression system of yeast. I. A nuclear gene controlling the inheritance of the extrachromosomal elements. *Heredity* **26:** 413–422.

———. 1972. Extrachromosomal elements in super-suppression system of yeast. II. Relations with other extrachomosomal elements. *Heredity* **28:** 189–199.

———. 1975. Extrachromosomal elements in a super-suppression system of yeast. III. Enhanced detection of weak suppressors in certain non-suppressed strains. *Heredity* **34:** 83–92.

# From "Granules" to Organelles: How Yeast Mitochondria Became Respectable

GOTTFRIED SCHATZ
*Biocenter, University of Basel*
*CH-4056, Basel, Switzerland*

I could not believe my ears. I had just finished my first research seminar and the initial question from the audience was "What makes you think these yeast granules are real mitochondria?" Not only was the question deflating, but it was to rear its ugly head after nearly all of my seminars for years to come.

This was in 1962—history for me, Stone Age for today's students. The heroic age of mitochondrial research had just come to an end (Ernster and Schatz 1981). During the past 15 years, G.H. Hogeboom, W.C. Schneider, and G.E. Palade had worked out methods for isolating liver mitochondria by differential centrifugation in isotonic sucrose solutions; E.P. Kennedy and A.L. Lehninger had discovered that mitochondria contained all of the enzymes for oxidative phosphorylation, the tricarboxylic acid cycle, and fatty acid oxidation; G.E. Palade, F. Sjostrand, and K.R. Porter had obtained the first high-resolution electron micrographs of mitochondria; B. Chance had analyzed electron flow in the mitochondrial respiratory chain by his elegant, ultrasensitive spectroscopic methods; D.E. Green had isolated several respiratory enzymes from huge amounts of beef heart mitochondria, thanks to the generosity of the many slaughterhouses around Wisconsin; and E. Racker had purified $F_1$-ATPase, the first defined catalyst of mitochondrial ATP synthesis. Mitochondrial researchers ("mitochondriacs") considered themselves, and were regarded as, an elite among biochemists. The function of the respiratory chain was reasonably well understood, and everybody was convinced that, a few years down the road, the same would be true for oxidative phosphorylation.

Occasionally, at the end of one of the numerous meetings on mitochondria that took place in the 1960s, somebody would ask "How are mitochondria made?" At that time, however, an answer to this question seemed beyond reach. In 1958, M.V. Simpson and his colleagues had made the startling discovery that isolated rat liver mitochondria could incorporate labeled amino acids into proteins (McLean et al. 1958), but it seemed hopeless to try to identify these proteins. Protein synthesis could only be assayed with expensive $^{14}C$-labeled amino acids and insensitive gas-flow counters; SDS-polyacrylamide gels had not yet burst upon the scientific scene; biochemists still believed that the proteins of biological membranes were bound to the polar headgroups of phospholipids, and were blissfully ignorant of the horrors of working with membrane proteins; and researchers trying to sequence their favorite proteins had to feed 10–100-mg samples of the pure protein to some overworked protein chemist living among bulky and temperamental machines. Simpson's discovery *was* a foot in the door to the problem of mitochondrial biogenesis, but this door firmly resisted all further efforts; biochemists just lacked the necessary technology.

Enter yeast genetics. In France, a long steamer's journey removed from the bustle of American "mitochondriology," B. Ephrussi and P. Slonimski were studying strange mutations that abolished the respiration of yeast cells. The history surrounding these mutants is recounted more expertly elsewhere in this book; here, it is relevant only that these mutations were irreversible and that they were not inherited according to Mendel's laws. Because the respiration-deficient yeast mutants utilized glucose less efficiently than respiring cells, they formed smaller colonies on plates containing low glucose levels. Ephrussi, an aristocratic Russian émigré with an excellent command of French, referred to them as *petite* mutants. We must be grateful for Ephrussi's linguistic skills, for had he stuck to his native tongue, we would now know these mutants by the epithet М А Л Е Н К И Е, which would be pronounced something like *mahlenkeeyee* by Russians, and like God knows what by our colleagues from France or Texas. Ephrussi and Slonimski were convinced that these mutations reflected the inactivation or the loss of some extrachromosomal factor that controlled the formation of the respiratory system; they took it for granted that the respiratory system of yeast was housed in typical mitochondria and suspected that this might also be true for the mysterious genetic factor.

To those of us who studied mitochondrial formation in yeast at that time, the slim monographs by Ephrussi and Slonimski on this topic (Ephrussi 1953; Slonimski 1953) were read like the Holy Scriptures. But few biochemists knew about them, and their general impact was relatively small. There are several reasons for this neglect, which tell us much about how biology has changed during the past few decades.

In the early 1960s, biology had not yet fully emerged from its feudal era. Scientific fiefdoms were still clearly demarcated and often well defended. In particular, yeast genetics was still considered an arcane calling reserved for the chosen few. This suited yeast geneticists just fine; they loved to intimidate outsiders with their obfuscating terms, many of them quite unnecessary. On the other hand, mitochondriacs were so busy chasing after nonexistent intermediates of oxidative phosphorylation that they had neither the time nor the inclination to read journals like *Genetics* or the *Journal of Molecular Biology*.

In addition, most biochemists considered yeast to be just another "microbe" which was not much different from *Escherichia coli*, then the unchallenged pet of molecular biologists. There was no general awareness of the distinction between prokaryotes and eukaryotes. This awareness spread only slowly in the late 1960s, as more cells were examined with the electron microscope or with the powerful new tools of molecular biology. Today, we know that respiring membrane vesicles isolated from bacteria are vesicular fragments of the plasma membrane, but at that time, they were often considered to be preexisting intracellular organelles resembling mitochondria.

In addition to these conceptual limitations, there were also technical hurdles that excluded the "respiring yeast granules" from the club of well-bred mitochondria. First, electron micrographs of fixed and thin-sectioned yeast cells often failed to reveal mitochondrial profiles. To this day, yeast is difficult to study by conventional electron microscopy: The thick cell wall interferes with fixation and sectioning, the high RNA content of the cytosol results in high background staining, and growth of the cells on fermentable sugars represses the formation of well-developed mitochondria. Second, yeast mitochondria could only be isolated by disrupting yeast cells with glass beads. As a revenge for this unfriendly procedure, the mitochondria usually lost many of their soluble proteins, appeared as a jumble of nondescript membrane fragments in the electron microscope, and usually refused to couple respiration to ATP synthesis. Indeed, they appeared to be quite similar to bacterial membrane vesicles!

Finally, the study of mitochondria in the early 1960s was still very much an American enterprise. Europe was still suffering from the scientific hemorrhage caused by the preceding war, and Europeans trying to do sophisticated measurements on mitochondria had to embark on a pilgrimage to New York, Madison, Philadelphia, or Baltimore. The mitochondrial research centers at Amsterdam and Munich had already been set up, but it took a while until their impact was fully felt.

In sum, when I started to work on yeast mitochondria as a young Assistant Professor in Vienna, I had no idea that I would be separated from the mainstream of mitochondriology by two formidable barriers, the Atlantic ocean and the continental divide between yeast and mammals. Toward the mid 1960s, however, the general awareness about yeast mitochondria changed, and the questions after my seminars about "respiratory granules" gradually subsided. I cannot identify a single turning point, but several findings all happened within a few years.

One of these findings was the discovery of mitochondrial DNA. Around 1962, it became clear that chloroplasts, the green brethren of mitochondria, contained their own DNA. H. Tuppy (who was then my boss) and I reasoned that if mitochondria also contained DNA, this DNA might well be the long-sought extrachromosomal factor that controlled mitochondrial formation in yeast. Together with E. Haslbrunner, my first graduate student, we decided to look for DNA in yeast mitochondria. Just at that time, S. Brenner and his colleagues at Cambridge University had discovered mRNA, using sucrose density gradient centrifugation as one of their major tools. In the course of this work, they had devised a simple gadget for generating linear sucrose gradients. By the time-honored Viennese method of brandishing a bottle of wine, I convinced our local mechanic to copy this device from a hand-drawn sketch. I then purified yeast mitochondria by isopycnic centrifugation in a 20–80% sucrose gradient, collected 15 fractions by puncturing first my finger and then the centrifuge tube, and analyzed each fraction for DNA by the color reaction according to the method of Z. Diesche. Did I find DNA! Every fraction gave a deep blue color, without any visible peak where the mitochondria had equilibrated as a discrete turbid band. Clearly, DNA had leaked out of the nucleus and had not been separated from the mitochondria by centrifugation in a sucrose gradient. I wondered whether I could do better if I replaced sucrose with something else.

After much trial and error, I begged one of the young interns in the university hospital across the street to give me a vial of "Urografin," a concentrated, dense solution used as an X-ray contrasting agent in renal examinations. I reasoned that anything that could be injected into a patient without killing him might also be innocuous for yeast mitochondria. The result was startling. After centrifuging yeast mitochondria in a Urografin density gradient, the mitochondria formed an extremely sharp band, and DNA was present in only two fractions. Most of it was in the bottom fraction, and a very small amount was in the mitochondrial fraction. The DNA in the bottom fraction was easily digested by DNase and probably represented nuclear DNA. The DNA in the mitochondrial fraction was not readily digested by DNase, unless the fraction was first mistreated with trichloroacetic acid, and presumably represented mitochondrial DNA enclosed by mitochondrial membranes. Its amount per milligram of mitochondrial protein was very constant between different experiments and a simple calculation showed that it was not enough to code for all mitochondrial proteins (Schatz et al. 1964). A few months before, Nass and Nass from Philadelphia had published beautiful electron micrographs that showed DNase-sensitive threads in the matrix of chicken mitochondria (Nass and Nass 1963). These parallel findings may have helped to get mitochondriacs interested in DNA and to gain yeast mitochondria an entry ticket to the main arena of mitochondrial research.

Help also came from another front. In 1966, Ohnishi finally worked out a reliable method for isolating intact yeast mitochondria by gentle lysis of yeast spheroplasts (Ohnishi et al. 1966). The respiration of these mitochondrial preparations was not only efficiently coupled to ATP synthesis, but also controlled by the availability of ADP. Since ADP control of respiration is a particularly sensitive indicator of mitochondrial integrity, this preparative procedure opened the way for functional studies on normal and genetically altered yeast mitochondria. A few years later, the freeze-etch technique developed by H. Mohr and K. Mühlethaler reproducibly visualized mitochondria in yeast cells regardless of how the cells had been grown. In 1969, this method helped us to refute the claim that yeast cells grown in the absence of oxygen reversibly lose their mitochondria. We showed that anaerobically grown yeast cells contain respiration-deficient "promitochondria" and that respiratory adaptation involves the differentiation of these structures into respiring mitochondria (Plattner and Schatz 1969). It is now generally accepted that yeast cells, like all eukaryotic cells, cannot reversibly lose their mitochondria and that mitochondria are indispensable organelles for all but a few eukaryotes (Baker and Schatz 1991).

In the late 1960s, many scientists who had been trained as classical "mitochondriacs" decided to work on yeast mitochondria. They had been trained by the masters and had acquired an intuitive feeling for how to work with mitochondria. Almost a decade before, Linnane had been the first to make this switch after leaving Green's laboratory. Back in Australia, he reported the first detailed study of mitochondrial energy coupling in yeast (Vitols and Linnane 1961) and founded an active school devoted to the study of mitochondrial biogenesis. Tzagoloff, also a Green disciple, moved to New York City and started his innovative experiments on the biogenesis of the yeast mitochondrial ATP synthase complex (Tzagoloff 1969). Ohnishi used her training with Chance to explore some unique features of the yeast mitochondrial respiratory chain (Capeillere Blandin and Ohnishi 1982). And I, as one of Racker's postdoctoral fellows at the cockroach-infested Public Health Research Institute in New York City, first worked on mammalian oxidative phosphorylation and then purified the yeast mitochondrial $F_1$-ATPase (Schatz et al. 1967). Later on, in Vienna and then at Cornell, I showed that $F_1$ is made in the cytoplasm but its association with the mitochondrial inner membrane is controlled by the mitochondrial genetic system (Schatz 1968). At Cornell, we used classical techniques of mitochondriology to isolate cytochrome oxidase and cytochrome $c_1$ from yeast. Exploiting SDS-polyacrylamide gels and other recently developed methods, we then defined the subunit composition of these heme proteins and showed that cytochrome $c_1$ was a single polypeptide chain made in the cytoplasm, whereas cytochrome oxidase contained three subunits made in mitochondria and at least four subunits made in the cytoplasm. A few years later, these findings prompted us to study how cytoplasmically made proteins are imported into mitochondria (Pon and Schatz 1991; Schatz 1991). Others, such as H. Mahler, P. Borst, and R.A. Butow, were trained in classical mitochondriology and then went on to make fundamental discoveries about the biochemical and genetic properties of mitochondrial DNA.

Today's young scientists would rather work on mitochondria from yeast than on mitochondria from rat liver. Yeast mitochondria are definitely "in," and a vast body of knowledge about mammalian mitochondria that had accumulated in the 1950s and 1960s is rapidly being lost. But fashion today is obsolescence tomorrow. A few decades from now, reviewers may tell students about another Stone Age and its arrogant yeast mitochondriacs.

## REFERENCES


Baker, K.P. and G. Schatz. 1991. Mitochondrial proteins essential for viability mediate protein import into yeast mitochondria. *Nature* **349:** 205–208.

Capeillere Blandin, C. and T. Ohnishi. 1982. Investigation of the iron-sulfur clusters in some mitochondrial mutants of *Saccharomyces cerevisiae.* A possible correlation between Rieske's iron-sulfur cluster and cytochrome *b. Eur. J. Biochem.* **122:** 403–413.

Ephrussi, B. 1953. *Nucleo-cytoplasmic relations in microorganisms: Their bearing on cell heredity and differentiation.* Clarendon Press, Oxford.

Ernster, L. and G. Schatz. 1981. Mitochondria: A historical review. *J. Cell Biol.* **91:** 227s–255s.

McLean, J.R., G.L. Cohn, I.K. Brandt, and M.V. Simpson. 1958. Incorporation of labeled amino acids into the protein of muscle and liver mitochondria. *J. Biol. Chem.* **233:** 657–663.

Nass, S. and M.M.K. Nass. 1963. Intramitochondrial fibers with DNA characteristics. II. Enzymatic and other hydrolytic treatments. *J. Cell Biol.* **19:** 613–629.

Ohnishi, T., K. Kawaguchi, and B. Hagihara. 1966. Preparation and some properties of yeast mitochondria. *J. Biol. Chem.* **241:** 1797–1806.

Plattner, H. and G. Schatz. 1969. Promitochondria of anaerobically-grown yeast. II. Morphology. *Biochemistry* **8:** 339–343.

Pon, L. and G. Schatz. 1991. Biogenesis of yeast mitochondria. In *The molecular and cellular biology of the yeast* Saccharomyces: *Genome dynamics, protein synthesis, and energetics* (ed. J.R. Broach et al.), vol. 1, pp. 333–406. Cold Spring Harbor Laboratory Press, Cold Spring Harbor, New York.

Schatz, G. 1968. Impaired binding of mitochondrial adenosine triphosphatase in the cytoplasmic "petite" mutant of *Saccharomyces cerevisiae. J. Biol. Chem.* **243:** 2192–2199.

———. 1991. Transport of proteins into mitochondria. *Harvey Lect.* **85:** 105–121.

Schatz, G., E. Haslbrunner, and H. Tuppy. 1964. Desoxyribonucleic acid associated with yeast mitochondria. *Biochem. Biophys. Res. Commun.* **15:** 127–132.

Schatz, G., H. Penefsky, and E. Racker. 1967. Partial resolution of the enzymes catalyzing oxidative phosphorylation. XIV. Interaction of purified mitochondrial adenosine triphosphatase from baker's yeast with submitochondrial particles from beef heart. *J. Biol. Chem.* **242:** 2552–2560.

Slonimski, P.P. 1953. *La formation des enzymes respiratoires chez la levure.* Masson, Paris.
Tzagoloff, A. 1969. Assembly of the mitochondrial membrane system. II. Synthesis of the mitochondrial adenosine triphosphatase. *J. Biol. Chem.* **244:** 5027–5033.
Vitols, E. and A.W. Linnane. 1961. Studies on the oxidative metabolism of *Saccharomyces cerevisiae.* II. Morphology and oxidative phosphorylation capacity of mitochondria and derived particles from baker's yeast. *J. Biophys. Biochem. Cytol.* **9:** 701–710.

# From *MIT* to *PET* Genes

ALEXANDER TZAGOLOFF
*Department of Biological Sciences*
*Columbia University*
*New York, New York 10027*

In 1962, Charlie Morlang asked me if I would teach biochemistry to a group of New York high school students. This was shortly after the Sputnik success when the United States government decided to infuse money into science education. Charlie, a fellow graduate student in the Department of Botany at Columbia University, was in charge of organizing a series of biology courses and in dispensing some of the funds made available through the National Science Foundation for this crash program. The payment for the course was handsome even by today's standards and I was delighted to accept. The only problem was my almost complete ignorance of the subject. I had taken a course in plant biochemistry in college, but the material was taught from a highly unorthodox perspective with an esoteric treatise on birefringence as a text. It was obvious that some serious homework would be needed. I spent a good month going over the few biochemistry texts available then and for obscure reasons became interested in mitochondria. Several excellent review articles had been published that year on the respiratory system of this organelle. The articles by David Green that I particularly enjoyed reading summarized in a lucid way the complexities of a still burgeoning field of studies. David Green was one of the directors of the Institute for Enzyme Research at the University of Wisconsin. In the spring of that year, I wrote to him about a postdoctoral position. He promptly replied that indeed there were openings in the laboratory and that I was welcome to come as soon as the details of my thesis were taken care of.

The time spent in Madison was full of excitement and occasional moments of high drama. Green was a charismatic individual with an unusual ability for dramatizing ideas and stirring up his listeners. Beneath the actor's exterior, however, he was highly analytical and rigorous in his approach to science. Earlier in his career, Green had found that the entire TCA cycle could be isolated in a particulate fraction even though under other conditions the same enzymes behaved as soluble proteins. This observation led him to propose that the TCA cycle enzymes and several other related pathways are housed in an organized subcellular particle which he termed cyclophorase (Green et al. 1948). Several years later, Lehninger and Kennedy (1949) demonstrated that much of the oxidative metabolism of cells takes place in mitochondria, a subcellular structure that up to that time had been more familiar to electron microscopists than to biochemists.

Green's section of the institute was an active place full of young people eager to leave their mark on the scientific landscape. Slow pipettes were disliked equally by all. Incoming postdoctoral fellows, generally fresh out of graduate school, were assigned to work with more senior members of the laboratory for the first year. I was delegated to David Wharton, who was in the midst of testing the role of copper as an obligatory electron carrier of

*The Early Days of Yeast Genetics*

cytochrome oxidase. The project was interesting but not without serious hazards. Dave wanted to remove copper without affecting the quaternary structure of the enzyme so that the metal could be reintroduced into an essentially native protein, hopefully with restoration of catalytic activity. The first part worked out quite well but required enormous amounts of enzyme to be dialyzed against staggering volumes of concentrated cyanide solutions. The reconstitution never materialized. Fortunately, I developed a timely sensitivity to the reagent that armed me with the perfect excuse to try another project. This was a turbulent period due to a series of papers authored by a member of the laboratory in which claims were made for the identity of several intermediates of oxidative phosphorylation. Eventually, the papers were retracted but the entire process palled the spirits of many people in the laboratory. I later viewed this episode as illustrative of the precariousness of large laboratory groups where by necessity much of the work is done with minimal supervision and example. It is instructive that in all the recent discussion of how to best police fraud in research, little mention is made of the time-honored solution not to stray too far or for too long from the laboratory bench.

Green did not object to having me try something new and suggested some experiments that might help to establish a functional correlate for a morphological entity of mitochondria discovered by Fernandez-Moran, who had pioneered the use of negative staining as a means of visualizing ultrastructural features of membranes. The particle in question was a 90-Å sphere that lined the entire matrix side of the inner membrane (Fernandez-Moran et al. 1964). Green was convinced that the spheres contained all of the respiratory chain complexes and had gone through some fairly arcane calculations to support this contention. His interpretation was soon shown to be incorrect by Kagawa and Racker (1966), who conclusively demonstrated the identity of the inner membrane spheres with the $F_1$ ATPase. My venture into this arena did lead to an unexpected finding. In the course of trying to isolate the 90-Å unit described by Fernandez-Moran, I noticed that it was possible by purely physical means to shear the inner membrane into small globular lipoprotein particles. Under some conditions, the dispersed particles would aggregate into vesicles not too different in their physical and enzymatic properties from the starting membranes. This serendipitous finding suggested that individual enzymes of the inner membrane might hold within them the potential to form membranes.

Together with David McConnell, who was spending a sabbatical leave in the laboratory, and David MacLennan, we started a project to test this idea. Only a few years earlier, Youssef Hatefi, a masterful enzymologist, had devised methods for isolating the four complexes of the respiratory chain (Hatefi 1963). We chose cytochrome oxidase for the reconstitutions because it could be obtained in high purity, and we soon showed that this respiratory complex free of other proteins was capable of forming membranes in the presence of phospholipids when proper care was taken to remove the dispersing detergent (McConnell et al. 1966). This result convinced us that cytochrome oxidase was a unit not only of function, but also of membrane structure capable of organizing a lipid bilayer into a unilamellar membrane vesicle. Conceptually, the finding was significant since it dispelled the notion of special structural proteins as determinants of membrane structure. The membrane-forming property of cytochrome oxidase turned out to be an attribute of other respiratory complexes and of the ATPase as well. This experimental circumstance was taken advantage of to its fullest by Racker and his colleagues, who not long afterward succeeded in reconstituting oxidative phosphorylation from the purified enzyme complexes of the mitochondrial inner membrane (Racker and Kandrach 1971).

At the end of my fifth year at the Enzyme Institute, many of the more seasoned members of the laboratory had left, and David Green, perhaps inspired by Peter Mitchell, steered his somewhat smaller laboratory toward more theoretical pursuits. The time seemed right to leave, but I was not sure what geographic or scientific direction to follow. Green thought that there was not much of a future left in mitochondria and suggested that my best course would be to find another membrane system, preferably one not contested by other laboratories. Since I had no intention of leaving the mitochondrial field, it was advice I chose to ignore.

A few years earlier, DNA had been detected in mammalian mitochondria (Nass and Nass 1963) and in yeast mitochondria (Schatz et al. 1964), and by 1967, several laboratories had obtained convincing evidence for the identity of Ephrussi's ρ factor with this newly discovered DNA (Mounolou et al. 1966). The resolution of the controversy surrounding earlier reports of mitochondrial protein synthesis and the demonstration of the existence of mitochondrial ribosomes (Borst and Grivell 1971) made the role of this genetic system all the more intriguing. My mind was pretty much made up to join the ranks of the few people who had created this new and exciting field. The opportunity to set up an independent laboratory came when Maynard Pullman offered me a position at the Public Health Research Institute in New York, where he had succeeded Efraim Racker as head of the Department of Biochemistry. Maynard's offer was especially attractive because he told me that Jeff Schatz would also be joining the department. I was familiar with Jeff's work on yeast mitochondria and thought that being in the same department with him would be immensely helpful in breaking into this new area. Unfortunately, this did not come to pass as Jeff had in the meantime chosen to settle in the quieter surroundings of Ithaca.

By the time I started in this area, almost 10 years had elapsed since the first report of protein synthetic activity in mitochondria (McLean et al. 1958), and still there were hardly any clues about the identity of the products. It was a challenging and interesting problem and I decided to give it a try. Thanks to the efforts of Schatz and Linnane, baker's yeast was obviously the organism of choice for looking at problems related to mitochondrial morphogenesis. With their successful use of inhibitors to dissect the two translational systems, Clark-Walker and Linnane (1966) had paved the way for the analysis of the mitochondrial products in yeast. Since there was evidence pointing to the respiratory and ATPase complexes as beneficiaries of this translation system, I chose to first concentrate on these enzymes. All of my experience up to that time had been with beef heart mitochondria. The realization that the respiratory capacity is some ten times lower in yeast mitochondria came as a real shock. While attempting to purify cytochrome oxidase and some of the other complexes, I often wished that yeast had beef heart mitochondria. Nevertheless, work progressed at a reasonable pace, and it soon became apparent that some of the subunits of the $F_0$ factor of the ATPase were synthesized on mitochondrial ribosomes (Tzagoloff and Meagher 1972). Parallel studies on coenzyme $QH_2$-cytochrome *c* reductase by Weiss (1972) and on cytochrome oxidase by Mason and Schatz (1973) provided an almost complete list of the mitochondrially translated polypeptides of yeast by 1973. Most of this information was summarized in a Bari meeting held in Italy in the summer of that year (see Kroon and Saccone 1974).

At the same meeting, several papers were presented on recombination of drug resistance markers in yeast mitochondrial DNA. The mutants used in the studies were resistant either to ribosomal inhibitors such as erythromycin (Linnane et al. 1968) or to oligomycin (Avner and Griffiths 1973), a well-established inhibitor of the $F_1$-$F_0$ ATPase. The drug-resistant mutants were the first available markers for studying the behavior of mitochondrial DNA and as such contributed in an important way to the development of mitochondrial genetics (Coen et al. 1970). They were also instructive from the standpoint of providing the first glimpses into the genetic makeup of mitochondrial DNA. After hearing these papers, I became intrigued with the possibility of obtaining mutants functionally defective in respiration or oxidative phosphorylation as a result of point mutations in mitochondrial DNA. A discussion I had with Piotr Slonimski, however, dampened my enthusiasm. He pointed out that earlier attempts to isolate such strains had failed and only yielded cytoplasmic *petite* mutants similar to those first described by Ephrussi et al. (1949). Piotr and others had shown such cytoplasmic *petites* to have extensive deletions in mitochondrial DNA, thereby accounting for the pleiotropic absence of respiratory enzymes and of mitochondrial protein synthesis. Although the cytoplasmic *petite* mutation remains of intrinsic interest, in the period preceding methods for sequencing DNA, such mutants were of limited usefulness in studying the coding properties of this genome.

The news about mitochondrial point mutants was certainly disappointing, but I was not totally discouraged. The following summer, while spending a few relaxed days at Sarah Ratner's vacation home, I gave the problem some more thought. It occurred to me that since one of the hallmarks of cytoplasmic *petites* is the absence of mitochondrial protein synthesis, this could be taken advantage of to enrich for point mutants. The idea was to select for respiratory-defective mutants on the basis of growth phenotype, subtract the preponderant class of cytoplasmic *petites* with an in vivo assay for mitochondrial protein synthesis, and further eliminate mutations in nuclear genes by a simple complementation test with a strain lacking mitochondrial DNA. Mutants surviving the screen should be enriched for the desired genotype. Using some makeshift gadgets for processing multiple samples of yeast, we screened that same summer upward of several thousand respiratory-deficient strains. As anticipated, most turned out to be cytoplasmic and nuclear *petites*, but a handful of strains tested positive for mitochondrial protein synthesis and had genetic properties consistent with mutations in mitochondrial DNA. One of these turned out to be defective in cytochrome oxidase and, as we found out later, had a lesion in the mitochondrial gene for subunit 2 of the enzyme. These first results were encouraging, but it was clear that the frequency of mutations needed to be improved. Richard Needleman, who was in the laboratory at that time, had read a paper in which manganese chloride was reported to be an efficient mutagen for inducing mitochondrial mutations conferring resistance to antibiotics (Putrament et al. 1973). The first mutagenesis trial with manganese resulted in a dramatic increase in the number of mutants, and before long, we were swimming in a sea of cytochrome oxidase, coenzyme $QH_2$-cytochrome *c* reductase, and even a few ATPase mutants (Tzagoloff et al. 1975a,b). Significantly, the new mutants were defective precisely in those enzymes known to have mitochondrially translated subunits. This fact supported earlier suspicions that the products of mitochondrial protein synthesis were also encoded in mitochondrial DNA.

Shortly after obtaining the first mutants, I contacted Piotr Slonimski to see if he might be interested in mapping the mutations relative to some of the antibiotic resistance markers that he and others had already located on the circular genome of yeast. Piotr was enthusiastic about the idea, and the following spring I spent several months in his laboratory in Gif-sur-Yvette, where I first learned about the difference between complementation and recombination. Piotr had outlined several mapping strategies, and in a short space of time, we were able to map the new *mit*$^-$ mutations relative to the drug resistance markers (Slonimski and Tzagoloff 1976). The short stay in Piotr's laboratory filled a serious gap in my knowledge of genetics, and I benefited greatly from the example of his intellectual vigor and strong devotion to science. During the next few years, the map was refined through the genetic and physical mapping studies in the laboratories of Rabinowitz, Slonimski, Schweyen, and Linnane (see Saccone and Kroon 1976). Some of the gene products were also identified (Cabral et al. 1978), and the first harbingers of the peculiarities of the mitochondrial genetic system started to surface (Foury and Tzagoloff 1978; Slonimski et al. 1978).

In 1977, I received a call from Cy Levinthal with an offer to join the Department of Biology at Columbia University which he together with Jim Darnell and Sherman Beychok had shaped from the separate Departments of Zoology and Botany. A move to a more academic setting was appealing, and the following year, I was back on the familiar campus of my student days. This was a very exciting period when as the result of new methods for manipulating nucleic acids, it became possible for the first time to examine genes in great detail. There were many talented students and associates in the laboratory, and we began to apply the new DNA sequencing technology to look at the genes of yeast mitochondria. Our first surprise came when Giuseppe Macino obtained the sequence of the *oli1* gene, whose product Walter Sebald had shown a few years earlier to be a subunit of ATPase (Wachter et al. 1977). Since the primary structure of the protein was known, it was possible to compare it with the sequence derived from the gene. The two matched perfectly except at one position where the DNA sequence indicated a CUA codon for leucine and the reported protein sequence had a threonine (Hensgens et al. 1979; Macino and Tzagoloff 1979). After writing to Walter about the discrepancy, I learned that he had already been informed of this fact by Piet Borst and was in the process of rechecking the residue which from his data appeared to be the most ambiguous in the sequence. The anomaly was resolved when May Li, a graduate student in the laboratory, characterized a threonine tRNA with an unusual anticodon loop and a 3′-GAU-5′anticodon (Li and Tzagoloff 1979). My faith in the uniqueness of mitochondria was confirmed.

At the same time, Francisco Nobrega and Gloria Coruzzi were analyzing the genes for cytochrome *b* and subunit 2 of cytochrome oxidase. The sequences, which we knew from genetic data to be in the coding regions of the genes, only made sense if the UGA opal terminator that consistently occurred in one of the reading frames was assumed to be an amino acid codon. Fortunately, the previous year, Buse et al. (1978) had published a partial sequence of the beef cytochrome oxidase subunit 2. Gloria Coruzzi was able to align the deduced sequence with the partial protein sequence, which happened to have several tryptophan residues at positions of the yeast gene with UGA codons (Macino et al. 1979). This fact did not escape the attention of Bart Barrell and Fred Sanger, who had sequenced the homologous gene of human mitochondrial DNA (Barrell et al. 1979), and of Tom Fox, who had also obtained the sequence of the yeast subunit 2 (Fox 1979). The advantage of having a distinct genetic code in mitochondria remains unexplained. For one thing, it acts as barrier for the productive exchange of nuclear and mitochondrial genetic information. Perhaps the present-day distribution of genes between the two compartments is particularly favorable and has been fixed, at least for the time being, by the differences in the two codes.

Giuseppe Macino and I visited Fred Sanger's laboratory in the summer of 1979. He was away at the time, but we spent a day with Bart Barrell who first made us aware of their extensive work on the human and bovine genomes. Later that year, Fred Sanger was in New York to accept a prize awarded by Columbia University and we agreed to compare the yeast and mammalian sequences. This proved to be very useful in recognizing the intron regions of the cytochrome *b* and cytochrome oxidase subunit 1 genes.

The outline of the mitochondrial genetic system of yeast and the differences between yeast and other organisms were worked out in only a few years. In 1981, Slonimski, Borst, and Attardi organized a Cold Spring Harbor symposium dedicated to the memory of Boris Ephrussi (Slonimski et al. 1982). The new information reported at this *fin de siècle* meeting not only has influenced our understanding how mitochondria propagate and maintain their functional status during cell growth and development, but has also opened the way for probing wide-ranging questions related to evolution and the molecular basis of inherited diseases. Mitochondria, because of the many different mechanisms they employ to express a fairly constant and small number of genes, have also brought to light new aspects of the versatility of nucleic acids as conveyers of genetic information. The efficient use of single tRNAs to read entire codon families (Barrell et al. 1980; Bonitz et al. 1980; Heckman et al. 1980), the novel mechanisms for the coordinate processing of different RNAs through intron-encoded maturases (Lazowska et al. 1980), and the newly discovered phenomenon of RNA editing (Blum et al. 1990) are a few examples illustrative of the newly discovered capabilities of DNA and RNA.

After the 1981 meeting, it became evident to many people in the field that questions about the function of mitochondrial DNA which had prevailed since Ephrussi's description of the *petite* mutation were now answered, and the time was opportune to switch attention to the much vaster pool of mitochondrion-specific genes in the nucleus. For many years, it was generally recognized that mitochondria arise not by a de novo process but rather increase in mass through addition of new lipids and proteins to a population of preexisting organelles (Luck 1965; Plattner et al. 1970). The advances made during the past decade in our understanding of protein transport (Hartl et al. 1989; Pon and Schatz 1991) and the pivotal importance of compartment-targeted import for enzyme assembly and membrane topology have emphasized the notion of mitochondria serving as an organizing principle for their own propagation.

The ability of cytoplasmic *petite* mutants to maintain morphologically and functionally recognizable mitochondria attests to the fact that the basic mitochondrial template is determined entirely by genetic information originating in the nucleus. The importance of nuclear genes for aerobic growth of yeast dates back to a 1950 article from Ephrussi's laboratory in which the genetic lesion responsible for the respiratory-defective phenotype of a mutant termed as a "segregation *petite*" was found to segregate in classical Mendelian fashion (Chen et al. 1950). A more complete genetic analysis of such nuclear "*pet*" mutants was published by Sherman (1963). Sherman and Slonimski (1964) reported the nuclear respiratory-defective *pet* mutants to have a variety of phenotypes that included deficiencies in cytochrome oxidase, cytochrome *b*, and cytochrome *c*. In the intervening years, many laboratories have made use of such strains to look at a broad spectrum of mitochondrially related phenomena (Ebner et al. 1973; Kovac 1974; Tzagoloff et al. 1975a; Schweitzer et al. 1977; Pillar et al. 1983).

As a side product of our searches for mitochondrial mutants, we had earlier isolated a sizable number of nuclear *pet* strains with phenotypes similar to those first observed by Sherman and Slonimski. Some years later after having spoken to a group of students attending the yearly Cold Spring Harbor course on yeast genetics, I met Gerald Fink who with great excitement informed me of his success in developing a yeast transformation system that made it possible to clone genes routinely as long as one had appropriate mutants and could select for the transformed phenotype. When I asked him if it would work with respiratory-defective mutants, he seemed surprised at my ignorance and replied "why not?". This chance encounter made me realize that the *pet* mutants languishing in our freezer could now be used to learn something about the nuclear genes impinging on the synthesis of cytochrome oxidase and the other complexes of the respiratory chain. At the time, however, we were still too preoccupied with mitochondrial DNA, and the mutants remained frozen until 1981 when Carol Dieckmann cloned the gene for a protein she had earlier found to be essential for processing of the cytochrome *b* pre-mRNA (Dieckmann et al. 1982). The previous year, I had also started to screen for additional *pet* mutants. My idea at the time was to obtain a strain collection that would come reasonably close to describing most of the nuclear genes affecting the fate of mitochondria in yeast. For technical reasons, I did not wish to deal with conditional mutants even though this meant that some important functions were going to be missed. As this work progressed, it became apparent that the usual calculations for gauging total number of genes controlling a complex biological phenomenon were not applicable to this system. With upward of several thousand strains on hand, instead of seeing a Gaussian distribution, the majority of complementation groups had considerably fewer than the expected number of mutants (Tzagoloff and Dieckmann 1991). The main explanation for the skewed distribution lies in the inherent instability of mitochondrial DNA in a large proportion of *pet* mutants, particularly those defective in mitochondrial protein synthesis (Myers et al. 1985). As a consequence, many *pet* mutants were scored as cytoplasmic *petites* or *mit*$^-$ during the genetic screen. Despite this circumstance, the investment of time in the genetic and biochemical screens did reward us with a set of strains that is probably representative of a large subset of the events governing the maintenance of respiratory competence in yeast.

There were a number of immediate reasons for the interest in having a comprehensive collection of *pet* mutants. The importance of RNA processing for mitochondrial respiration became apparent with the discovery that many genes are cotranscribed from a relatively small number of promoters (Christianson and Rabinowitz 1983). Despite the great interest in mitochondrially encoded maturases, I thought that *pet* mutants could tell us something about RNA maturation, much of which was likely to be determined by proteins of nucleocytoplasmic origin. I was also interested in getting back to the problem of how some of the heteroligomeric enzymes of the mitochondrial respiratory chain are assembled from subunits derived from genes in the two separate subcellular compartments. Mutants have traditionally served as important tools for deciphering complex assembly pathways, and there was every reason to think that this would also be true in the case of mitochondria. This expectation has been fully borne out. In the past 10 years, many laboratories have exploited *pet* mutants to study the diverse aspects of mitochondrial function and biogenesis.

In trying to understand the biochemical defects in just a sample of the mutants, it has become only too painfully clear that we are still grossly unaware of the most fundamental processes carried out by mitochondria. Why is it that one half of the genes affecting the protein synthetic activity of mitochondria code for totally unknown proteins? By the same token, how can we rationalize the necessity of 20 odd genes to ensure the assembly of a single enzyme like cytochrome oxidase? It is clear that many exciting chapters about mitochondria remain to be written and much of it will be told from further genetic and biochemical studies of *pet* mutants. In a sense, the strains collected over the years in different laboratories have created a mini-genome project, out of which hopefully will emerge a blueprint for the future course of investigations of the biogenesis process. Sometimes the work seems to be open-ended and is remindful of the story about a Messiah and a small village in Palestine. One day, the villagers of this remote town learned that the Messiah was expected to appear soon. Because of their isolation, they became fearful of missing out on this event and decided to build a tall tower from which they could watch out for the Messiah. A local man was hired to keep watch. After several weeks, a friend of his who was passing by inquired how he liked his job. The watchman thought for a few minutes and replied that it didn't pay much but it was going to be steady employment.

## ACKNOWLEDGMENTS

Over the years, it has been my fortune to have worked with many talented students and associates, only a few of whose names were mentioned in this article. To them, I owe a great debt for their enthusiasm, intelligence, and dedication to some of the goals we have striven to achieve together. Alphabetically, they are Sharon Ackerman, Anna Akai, Roberta Berlani, Susan Bonitz, Josephine Collins, Jacky Cosson, Gloria Coruzzi, Mary Crivellone, Carol Dieckmann, Francoise Foury, Domenico Gatti, John Hill, Gregory Homison, T.J. Koerner, Alexandra Gampel, May Li, Giuseppe Macino, Patricia McGaw, Pauline Meagher, Ivor Muroff, Alan Myers, Richard Needleman, Marina Nobrega, Francisco Nobrega, Louise Pape, Clara Pentalla, Barbara Repetto, Meryl Rubin, Marcelino Sierra, Barbara Thalenfeld, Kaye Trembath, and Mian Wu. I also wish to express my gratitude to the National Heart and Lung Institute of the National Institutes of Health for their loyal and generous support during the past 23 years.

## REFERENCES

Avner, P.R. and D.E. Griffiths. 1973. Studies of energy-linked reactions: Isolation and characterization of oligomycin-resistant mutants of *Saccharomyces cerevisiae*. *Eur. J. Biochem.* **32:** 301–311.

Barrell, B.G., A.T. Bankier, and J. Drouin. 1979. A different genetic code in human mitochondria. *Nature* **282:** 189–194.

Barrell, B.G., S. Anderson, A.T. Bankier, M.H.L. deBruijn, E. Chen, A.R. Coulson, J. Drouin, I.C. Eperon, D.P. Nierlich, B.A. Roe, F. Sanger, P.H. Schreier, A.J.H. Smith, R. Staden, and I.G. Young. 1980. Different pattern of codon recognition by mammalian tRNAs. *Proc. Natl. Acad. Sci.* **77:** 3164–3166.

Blum, B., N. Bakalara, and L. Simpson. 1990. A model for RNA editing in kinetoplastid mitochondria: "Guide" RNA molecules transcribed from maxicircle DNA provide the edited information. *Cell* **60:** 189–198.

Bonitz, S.G., R. Berlani, G. Coruzzi, M. Li, G. Macino, F.G. Mobrega, M.P. Nobrega, B.E. Thalenfeld, and A. Tzagoloff. 1980. Codon recognition rules in yeast mitochondria. *Proc. Natl. Acad. Sci.* **77:** 3167–3170.

Borst, P. and L.A. Grivell. 1971. Mitochondrial ribosomes. *FEBS Lett.* **13:** 73–88.

Buse, G., G.J. Steffens, and G.C.M. Steffens. 1978. Studies on cytochrome *c* oxidase. III. Relationship of cytochrome oxidase subunits to electron carriers of photophosphorylation. *Hoppe-Seyler's Z. Physiol. Chem.* **360:** 1011–1013.

Cabral, F., M. Solioz, Y. Rudin, G. Schatz, L. Clavilier, and P.P. Slonimski. 1978. Identification of the structural gene for yeast cytochrome *c* oxidase subunit II on mitochondrial DNA. *J. Biol. Chem.* **253:** 297–304.

Chen, S.Y., B. Ephrussi, and H. Hottinguer. 1950. Nature genetique des mutants a déficience respiratoire de la souche B-11 de la levure de boulangerie. *Heredity* **4:** 337–351.

Christianson, T. and M. Rabinowitz. 1983. Identification of multiple transcriptional initiation sites on the yeast mitochondrial genome by *in vitro* capping with guanylyltransferase. *J. Biol. Chem.* **258:** 14025–14033.

Clark-Walker, G.D. and A.W. Linnane. 1966. *In vivo* differentiation of yeast cytoplasmic and mitochondrial protein synthesis with antibiotics. *Biochem. Biophys. Res. Commun.* **25:** 8–13.

Coen, D., J. Deutsch, P. Netter, E. Petrochilo, and P.P. Slominski. 1970. Mitochondrial genetics. I. Methodology and phenomenology. *Symp. Soc. Exp. Biol.* **24:** 449–496.

Dieckmann, C.L., L.K. Pape, and A. Tzagoloff. 1982. Identification and cloning of a yeast nuclear gene (*CBP1*) involved in expression of mitochondrial cytochrome *b*. *Proc. Natl. Acad. Sci.* **79:** 1805–1809.

Ebner, E., T.L. Mason, and G. Schatz. 1973. Mitochondrial assembly in respiration-deficient mutants of *Saccharomyces cerevisiae*. II. Effect of nuclear and extrachromosomal mutations on the formation of cytochrome *c* oxidase. *J. Biol. Chem.* **248:** 5369–5378.

Ephrussi, B., H. Hottinguer, and J. Tavlitzki. 1949. Action de l'acriflavine sur les levures. II. Etude genetique du mutant "petite colonies." *Ann. Inst. Pasteur* **79:** 419–450.

Fernandez-Moran, H., T. Oda, P.V. Blair, and D.E. Green. 1964. A macromolecular repeating unit of mitochondrial structure and function. *J. Cell Biol.* **22:** 63–100.

Foury, F. and A. Tzagoloff. 1978. Assembly of the mitochondrial membrane system. Genetic complementation of *mit*$^-$ mutations in mitochondrial DNA of *Saccharomyces cerevisiae*. *J. Biol. Chem.* **253:** 3792–3797.

Fox, T.D. 1979. Five TGA "stop" codons occur within the translated sequence of the yeast mitochondrial gene for cytochrome *c* oxidase subunit II. *Proc. Natl. Acad. Sci.* **76:** 6534–6538.

Green, D.E., W.F. Loomis, and V.H. Auerbach. 1948. Studies on the cyclophorase sys-

tem. I. The complete oxidation of pyruvic acid to carbon dioxide and water. *J. Biol. Chem.* **172:** 389–403.
Hatefi, Y. 1963. The pyridine nucleotide-cytochrome *c* reductases. *The Enzymes* (2nd edition) **7:** 495–515.
Hartl, F.-U., N. Pfanner, D.W. Nicholson, and W. Neupert. 1989. Mitochondrial protein import. *Biochim. Biophys. Acta* **988:** 1–45.
Heckman, J.E., J. Sarnoff, B. Alzner-DeWeerd, S. Yin, and U.L. RajBhandary. 1980. Novel features in the genetic code and codon reading patterns in *Neurospora crassa* mitochondria based on sequences of six mitochondrial tRNAs. *Proc. Natl. Acad. Sci.* **77:** 3159–3163.
Hensgens, L.A.M., L.A. Grivell, P. Borst, and J.L. Bos. 1979. Nucleotide sequence of the mitochondrial structure gene subunit 9 of yeast APTase complex. *Proc. Natl. Acad. Sci.* **76:** 1663–1667.
Kagawa, Y. and E. Racker. 1966. Partial resolution of the enzymes catalyzing oxidative phosphorylation. X. Correlation of morphology and function in submitochondrial particles. *J. Biol. Chem.* **241:** 2475–2482.
Kovac, L. 1974. Biochemical mutants: An approach to mitochondrial energy coupling. *Biochim. Biophys. Acta* **346:** 101–135.
Kroon, A.M. and C. Saccone, eds. 1974. *The biogenesis of mitochondria.* Academic Press, New York.
Lazowska, J., C. Jacq, and P.P. Slonimski. 1980. Sequence of introns and flanking exons in wild type and *box3* mutants of cytochrome *b* reveals an interlaced splicing protein coded by an intron. *Cell* **22:** 333–348.
Lehninger, A.L. and E.P. Kennedy. 1949. Oxidation of fatty acids and tricarboxylic acid cycle intermediates by isolated rat liver mitochondria. *J. Biol. Chem.* **179:** 957–972.
Li, M. and A. Tzagoloff. 1979. Assembly of the mitochondrial membrane system. Sequences of yeast mitochondrial valine and an unusual threonine tRNA gene. *Cell* **18:** 47–53.
Linnane, A.W., G.W. Saunders, E.B. Gingold, and H.B. Lukins. 1968. The biogenesis of mitochondria. V. Cytoplasmic inheritance of erythromycin resistance in *S. cerevisiae. Proc. Natl. Acad. Sci.* **59:** 903–910.
Luck, D.J.L. 1965. Formation of mitochondria in *Neurospora crassa.* A study based on mitochondrial density changes. *J. Cell Biol.* **24:** 461–470.
Macino, G. and A. Tzagoloff. 1979. Assembly of the mitochondrial membrane system. The DNA sequence of a mitochondrial ATPase gene in *Saccharomyces cerevisiae. J. Biol. Chem.* **254:** 4617–4623.
Macino, G., G. Coruzzi, F.G. Nobrega, M. Li, and A. Tzagoloff. 1979. Use of the UGA terminator as a tryptophan codon in yeast mitochondria. *Proc. Natl. Acad. Sci.* **76:** 3784–3785.
Mason, T.L. and G. Schatz. 1973. Cytochrome oxidase from baker's yeast. II. Site of translation of protein components. *J. Biol. Chem.* **248:** 1355–1360.
McConnell, D.G., A. Tzagoloff, D.H. MacLennan, and D.E. Green. 1966. Studies of the electron transfer system. LXV. Formation of membranes by purified cytochrome oxidase. *J. Biol. Chem.* **241:** 2373–2382.
McLean, J.R., G.L. Cohn, I.K. Brandt, and M.V. Simpson. 1958. Incorporation of labeled amino acids into the protein of muscle and liver mitochondria. *J. Biol. Chem.* **233:** 657–663.
Mounolou, J.C., H. Jacob, and P.P. Slonimski. 1966. Mitochondrial DNA from yeast "petite" mutants: Specific changes of buoyant density corresponding to different cytoplasmic mutations. *Biochem. Biophys. Res. Commun.* **24:** 218–224.
Myers, A.M., L.K. Pape, and A. Tzagoloff. 1985. Mitochondrial protein synthesis is required for maintenance of intact mitochondrial genomes in *Saccharomyces cerevisiae. EMBO J.* **4:** 2087–2092.
Nass, S. and M.M.K. Nass. 1963. Intramitochondrial fibers with DNA characteristics. I. Fixation and electron staining reactions. *J. Cell Biol.* **19:** 543–611.
Pillar, T., B.F. Lang, I. Steinberger, B. Vogt, and F. Kaudewitz. 1983. Expression of the "split gene" *cob* in yeast mtDNA: Nuclear mutations specifically block the excision of different introns from its primary transcript. *J. Biol. Chem.* **258:** 7954–7959.
Plattner, H., M.M. Salpeter, J. Saltzgaber, and G. Schatz. 1970. Promitochondria of

anaerobically grown yeast. IV. Conversion into respiring mitochondria. *Proc. Natl. Acad. Sci.* **66:** 1252–1259.

Pon, L. and G. Schatz. 1991. Biogenesis of yeast mitochondria. In *The molecular and cellular biology of the yeast* Saccharomyces: *Genome dynamics, protein synthesis, and energetics* (ed. J. Broach et al.), pp. 333–406. Cold Spring Harbor Laboratory Press, Cold Spring Harbor, New York.

Putrament, A., H. Baranowska, and W. Prazmo. 1973. Induction by manganese of mitochondrial antibiotic resistant mutations in yeast. *Mol. Gen. Genet.* **126:** 357–366.

Racker, E. and A. Kandrach. 1971. Reconstitution of the third site of oxidative phosphorylation. *J. Biol. Chem.* **246:** 7069–7071.

Saccone, C. and A.M. Kroon, eds. 1976. *The genetic function of mitochondrial DNA*. North-Holland, Amsterdam.

Schatz, G., E. Haslbrunner, and H. Tuppy. 1964. Deoxyribonucleic acid associated with yeast mitochondria. *Biochem. Biophys. Res. Commun.* **15:** 127–132.

Schweizer, E., W. Demmer, W. Holzner, and H.W. Tahedl. 1977. Controlled mitochondrial inactivation of temperature-sensitive *Saccharomyces cerevisiae* nuclear *petite* mutants. In *Mitochondria 1977: Genetics and biogenesis of mitochondria* (ed. W. Bandlow et al.), pp. 91–105. de Gruyter, Berlin.

Sherman, F. 1963. Respiration-deficient mutants of yeast. I. Genetics. *Genetics* **48:** 375–385.

Sherman, F. and P.P. Slonimski. 1964. Respiration-deficient mutants of yeast. II. Biochemistry. *Biochim. Biophys. Acta* **90:** 1–15.

Slonimski, P.P. and A. Tzagoloff. 1976. Localization in yeast mitochondrial DNA of mutations expressed in a deficiency of cytochrome oxidase and/or coenzyme $QH_2$-cytochrome *c* reductase. *Eur. J. Biochem.* **61:** 27–41.

Slonimski, P.P., P. Borst, and G. Attardi, eds. 1982. *Mitochondrial genes.* Cold Spring Harbor Laboratory, Cold Spring Harbor, New York.

Slonimski, P.P., P. Pajot, C. Jacq, M. Foucher, G. Perrodin, A. Kochko, and A. Lamouroux. 1978. Mosaic organization and expression of the mitochondrial DNA region controlling cytochrome *c* reductase and oxidase. I. Genetic, physical, and complementation maps of the *box* region. In *Biochemistry and genetics of yeast. Pure and applied aspects* (ed. M. Bacila et al.), pp. 339–368. Academic Press, New York.

Tzagoloff, A. and C.L. Dieckmann. 1991. *PET* genes of *Saccharomyces cerevisiae. Microbiol. Rev.* **54:** 211–225.

Tzagoloff, A. and P. Meagher. 1972. Assembly of the mitochondrial membrane system. VI. Mitochondrial synthesis of subunit proteins of the rutamycin-sensitive adenosine triphosphatase. *J. Biol. Chem.* **247:** 594–603.

Tzagoloff, A., A. Akai, and R.B. Needleman. 1975a. Assembly of the mitochondrial membrane system. Characterization of nuclear mutants of *Saccharomyces cerevisiae* with defects in mitochondrial ATPase and respiratory enzymes. *J. Biol. Chem.* **250:** 8228–8235.

Tzagoloff, A., A. Akai, R.B. Needleman, and G. Zulch. 1975b. Assembly of the mitochondrial membrane system. Cytoplasmic mutants of *Saccharomyces cerevisiae* with lesions in enzymes of the respiratory chain and in the mitochondrial ATPase. *J. Biol. Chem.* **250:** 8236–8242.

Wachter, E., W. Sebald, and A. Tzagoloff. 1977. Altered amino acid sequence of the DCCD-binding protein in the oil1 resistant mutant D273-10B/A21 of *Saccharomyces cerevisiae.* In *Mitochondria 1977: Genetics and biogenesis of mitochondria* (ed. W. Bandlow et al.), pp. 441–449. de Gruyter, Berlin.

Weiss, H. 1972. Cytochrome *b* in *Neurospora crassa* mitochondria. A membrane protein containing subunits of cytoplasmic and mitochondrial origins. *Eur. J. Biochem.* **30:** 469–478.

# Early Recollections of Fungal Genetics and the Cytoplasmic Inheritance Controversy

DAVID WILKIE
*Department of Biology*
*University College London*
*London WC1E 6BT, United Kingdom*

## SOME BASIC PRINCIPLES IN GENETIC ANALYSIS

The Genetics Department of the University of Glasgow was inaugurated in 1948 under the direction of Guido Pontecorvo, who was then elevated to the title of Reader. Research was concentrated on the genetics of the ascomycete *Aspergillus nidulans*, with the identification of mutant sites within genes being the main line of investigation. This involved the demonstration of intragenic recombination that in turn necessitated a reappraisal of the definition of the gene in terms of a unit of recombination. Later when it was established that the genetic material was DNA, Pontecorvo was able to claim that the fine genetic analysis of mutant sites had a greater resolving power than the biochemical analysis of the time, since it was capable of getting down to the level of individual base pairs, at least theoretically. Another line of investigation concerned the phenomenon of mitotic recombination in heterozygous diploids, a process originally proposed by Curt Stern in the 1930s to explain the spontaneous appearance of mutant areas on the body of the fruit fly *Drosophila*. The findings at Glasgow not only supported Stern's theory, but also gave a detailed description of the phenomenon underlining its significance in linkage studies and predicting its use in generating recombinant types in otherwise asexual species. Extrapolating the findings to diploid organisms generally, Pontecorvo felt justified in attempting to use mitotic recombination to map genes in human cells, the successful culture of which was a recent achievement. Unfortunately, known genes were not expressed in culture, and cells remained undifferentiated for the most part.

When I moved to University College in 1954 to take up a lectureship, I continued to work with *Aspergillus*, concentrating on a mutant aperithecial form that showed cytoplasmic transmission. It can be appreciated that my transition to yeast was relatively straightforward. This took place during the years 1959–1960 in the Genetics Department of the University of Washington, Seattle, under the inspiring guidance of Herschel Roman and Don Hawthorne. This visit was made possible by the award of a Rockefeller Foundation Visiting Research Fellowship. The achievements of this department in yeast genetics were already receiving wide acclaim and were epitomized in the elegant studies elucidating the genetic control of adenine biosynthesis in *Saccharomyces cerevisiae*. This work exploited the red-pigmented adenine-requiring mutant designated $ad_2$ (now *ade2*) in which the step controlled by this gene was blocked, leading to the accumulation of precursor substance. This was channeled into the production of the red pigment whose chemical identity was unknown, as was the series of biochemical steps leading to its synthesis. The important feature was not so much the nature of the pigment but rather its use in the identification of spontaneous mutants that blocked adenine biosynthesis at stages prior to the $ad_2$ lesion, thus precluding the formation of the precursor necessary for red pigment synthesis. On plating, these mutants therefore gave rise to white colonies that were still adenine-requiring, being of genotype $ad_2$ $ad_x$. In crosses to wild type, the recombinant $AD_x$ $ad_2$ with the red phenotype would reemerge among the ascospore progeny. It was a straightforward genetic test to establish which mutants were nonallelic to $ad_x$ by setting up a series of crosses of the type $ad_x$ $ad_2$ x $ad_y$ $ad_2$ and scoring the phenotype of the diploids. If $ad_x$ and $ad_y$ were different genes, the diploid would produce the red pigment, mutant genes being recessive. Experiments with precursor compounds could establish the order of the genes in the biosynthetic pathway along the lines of the classical studies of George Beadle and Edward Tatum with *Neurospora*.

"Leaky" mutants at the $ad_2$ locus were of interest in further exploiting the visual marker. These produced limited amounts of red pigment so that colonies were pale pink, developing slowly on adenineless medium. It was thought that these mutations were small aberrations of the gene giving partially defective products with some functional capability. The possibility was also considered that the leaky phenotype was due to mutation of other genetic elements that affected transcription rates of the $ad_2$ gene. However, crosses between several leaky mutants that I was given to analyze produced only leaky progeny both in diploids and among the ascospores on tetrad analysis, indicating that the same gene was involved in each of the mutants. This technique of microdissection of meiotic tetrads was by then widely used in genetic analysis and not more so than by the Seattle group. Indeed, competitions were held from time to time to see who could dissect the most tetrads in a given time. The record holder was Satya Kakar, a graduate student from India, now Professor of Genetics, Hissar, Punjab (in his CV, I doubt whether he ever listed this among his achievements). I cannot remember what his score was, but it was impressive. In tetrad analysis, as everyone knows, to establish single gene control of a mutant character, it is necessary only to demonstrate 2:2 segregation of mutant to wild type in the cross mutant x wild type, as described by Øjvind Winge more than 50 years ago, with the full support of the cytological evidence of chromosome segregation at meiosis. Any departure from the 2:2 ratio such as 3:1, which appeared with some regularity, was ascribed to errors of technique and listed as false tetrads. It was to the credit of investigators such as Carl Lindegren with yeast and Mary Mitchell with *Neurospora*, whose insistence that the aberrant tetrad was a genuine departure from the expected, that it became recognized as such, and so required an explanation. Lindegren described the phenomenon as "gene conversion" in which the conformation of one allele could be superimposed on the other during meiotic pairing and recombination. Thus, in the heterozygote *a*/*A*, a tetrad of 3*a*:1*A* would result from the transformation of *A* into *a* under the influence of *a* itself by an unknown mechanism probably involving a recombinational event, giving a new dimension to the concept of crossing-over. Lindegren was so intrigued by aberrant tetrads that on his visit to Glasgow in 1952, he maintained that they were more important than so-called normal tetrads and that we should concentrate on them and record any factors influencing their formation as they held the key to the nature of the gene (it should be remembered that this was before Watson and Crick). It was not until much later when the evidence of the clustering of exchanges within the gene in studies with *Aspergillus* and the citing of instances of nonreciprocal recombination were brought to bear on the problem that it was realized that Lindegren had a point. It was then that a model of recombination embodying all its features, proposed by Robin Holliday on the basis of exchanges between single strands of DNA, provided a satisfactory explanation of the origin of the aberrant tetrad.

During this time in Seattle, mitotic recombination was under scrutiny and being used mostly in mapping genes. Yeast was more amenable than *Aspergillus* in studies of this phenomenon, since in the filamentous fungus, the diploid condition had to be induced and was unstable, whereas the diploid state is the natural one for yeast. In considering the mechanism of the event, it appeared that in heterozygous diploids undergoing mitosis, individual pairs of homologous chromosomes accidentally synapsed, exchanged segments as in meiosis, separated, and continued to divide mitotically. It can be appreciated that this procedure would generate diploid daughter nuclei homozygous with respect to marker genes and that all genes distal to the crossover would be homozygous, or, in other words, there would be a positive correlation between the frequency of homozygosity and map distance from the centromere.

## STUDIES WITH UV LIGHT

It was known that irradiation of diploid cells with UV light increased the frequency of mitotic recombination, but whether this was due to an increase in the aberration of pairing or induction of the crossover event after pairing was an open question. In view of the fact that UV light was a known mutagen, i.e., had a direct effect on the structure of DNA, the latter mechanism was generally accepted. However, in a series of experiments carried out by myself and Don Hawthorne, a negative correlation between the frequency of homozygosity of linked genes and map distance from the centromere was found. To explain this unexpected result, we proposed a mechanism of UV induction in which a change from a mitotic to a meiotic (chromosome pairing) condition occurred in cells that responded to the irradiation (Wilkie and Hawthorne 1961; Wilkie and Lewis 1963).

I had previously used UV light in my early studies of mutagenesis with *Aspergillus* when it was still a matter of conjecture whether the protein component of nucleoprotein had a role in the phenomenon. Using monochromatic light, particularly at the wavelengths of 260 nm and 280 nm (absorption peaks of nucleic acid and protein, respectively), action spectra were constructed and compared with absorption spectra of the chromosomal components. In the procedure, uranium glass that fluoresces in the UV was used. Spores were spread on the slide within the confines of the beam produced by the monochromator (about 2 cm high and 2 mm wide) and focused in the beam, and measured doses of UV were delivered at the individual wavelengths. This procedure overcame the problems and inaccuracies of irradiating spores in suspensions. Reversion to prototrophy and lethality were recorded, and invariably the most efficient wavelength in the induction of prototrophy was 260 nm. In the case of cell death, the results were not so clear-cut, and 280 nm could be at least as potent as 260 nm in this respect in some experiments. This system was applied to yeast cells as described below, but in general, it was concluded that mutation resulted from a photochemical change in nucleic acid.

## CYTOPLASMIC INHERITANCE

The remarkable correlation between the behavior of Mendel's factors and that of nuclear chromosomes, as revealed by cytological findings, established the authenticity of Mendel's laws and led to the chromosome theory of heredity. This was embellished by the finding of linkage in which genetic units on the same chromosome segregated together. To complete the picture of neo-Mendelian genetics, crossing-over recombined linked genes, recombination frequency being based on the distance apart of the linked markers. This phenomenon was convincingly demonstrated by cytological pictures showing chiasmata at the four-strand stage of meiosis, i.e., after chromosome replication. In the 1950s, the period under discussion, it was at best unreasonable to suggest that there could be a mechanism for the transmission of genetic elements outside this well-defined, foolproof system. However, already several authentic cases existed that departed from the Mendelian pattern of inheritance, including variegated leaf in green plants, *petite colonie* in yeast, and *poky* in *Neurospora*. I became interested in so-called cytoplasmic inheritance while investigating a mutant form of *Aspergillus* I had picked up that was unable to form perithecia, the ascospore-containing bodies. Following hyphal anastomoses in a cross with wild type, perithecia were seen to appear within the aperithecial colony. Many of these had resulted from sexual fusion of nuclei of the aperithecial parent (the organism is homothallic so that both selfed and crossed perithecia can arise from heterokaryons). Plating of ascospores from these selfed perithecia gave wild-type, perithecia-producing colonies. In other words, transformation of the asexual mutant had taken place without involving the nucleus of the wild-type parent. Transfer of a cytoplasmic genetic element seemed to be a reasonable conclusion (Mahony and Wilkie 1958, 1959).

When I started working with *S. cerevisiae*, developing an interest in the *petite* mutation was a natural progression. Instinctively, I turned to the UV induction of the mutation, which was a most efficient inducing agent. It must be remembered that the controversy of whether or not the mitochondrion was "a genetic unit" was still highly debatable. A further complication was the pronouncement by Tony Linnane's group that anaerobically grown yeast cells had no mitochondria as far as they could determine from their electron microscope studies. It was intriguing to consider what might happen on UV irradiation of anaerobic cultures: presumably you would not get any *petites*. On the contrary, *petite* mutants were readily obtained with induction curves showing single-hit kinetics, i.e., they tended to be straight lines. Induction curves of aerobic cultures, on the other hand, showed multiple-hit kinetics. Action spectra indicated nucleic acid targets. I concluded that under anaerobiosis either only one copy of the hypothetical genetic unit survived in cells or only one copy was available for transmission to daughter cells, what might be called a master copy (Wilkie 1963). Quantitative and mechanistic aspects of mitochondrial genome transmission are still a problem and are discussed further below.

Having by now become completely enmeshed in the cytoplasmic inheritance controversy, I attempted to bring all the evidence together and entered into a contract with Methuen to publish a monograph entitled *The Cytoplasm in Heredity*. This happened early in 1963 when it was considered foolhardy to be promoting the idea of genes in the cytoplasm. At a meeting of eminent geneticists of the time in Paris, the concept was ostensibly laid to rest with alternative explanations of the non-Mendelian patterns in plentiful supply. Max Delbrück, for example, favored the idea of an alteration in the physiological state of the cytoplasm affecting the functional state of certain genes in the nucleus, whereas a change in cell membranes could also be a factor in the control of nuclear gene expression. Somehow, cellular states and membrane configurations could be invoked leaving inviolate the concept of the nucleus as the sole repository of genetic information. By the time the book appeared in 1964 (Wilkie 1964), Margit Nass had uncovered in her electron microscope studies the presence of DNA strands in the mitochondria of chick cells (Nass and Nass 1963), allowing the allocation of at least some genetic autonomy to the organelle. This was even more plausible on a reconsideration of the earlier work of Don Roodyn who had shown that isolated rat liver mitochondria were capable of synthesizing protein. His work was severely criticized, unjustly as it turned out, on the grounds that his mitochondrial preparations were contaminated with bacteria that were held to be responsible for the protein synthesis. Roodyn's claim that the numbers of contaminating bacteria were insufficient to account for the amount of amino acid taken up in the system was not accepted and he gave up in despair, leaving it to others to take up the challenge. Foremost was Ab Kroon who developed a system for rearing rats under sterile conditions and making an aseptic preparation of the liver mitochondria. He then showed that they were indeed capable of incorporating amino acids into their proteins, fully vindicating Roodyn's claim. The picture was emerging of an organelle with intrinsic genetic information with a capability of expressing that information. The early analyses of the mitochondrial genome indicated that this information extended to about 20 or 30 genes contained in a single, double-stranded, supercoiled DNA molecule. Perhaps the most far-reaching discovery of the time (1966) was made by Linnane's group when they showed that the yeast mitochondrial protein-synthesizing system was selectively inhibited in cells growing in the presence of the antibacterial antibiotic chloramphenicol. It was then that Linnane invited me to visit his laboratory in Australia to collaborate on developing the genetical aspects of the system. The main drawback in their studies was the inordinately high concentration of antibiotic required to show the inhibitory effects (about 4 mg/ml, which was near the solubility limit) in the strain they were using. This more or less precluded the isolation of drug-resistant mutants, a prerequisite for genetic analysis. However, on testing the strains I had brought, it became apparent that the degree of sensitivity to antibiotics was strain-dependent so that we were able to concentrate on strains that were relatively sensitive. A number of spontaneous antibiotic-resistant mutants were obtained from these strains, and these proved invaluable in subsequent

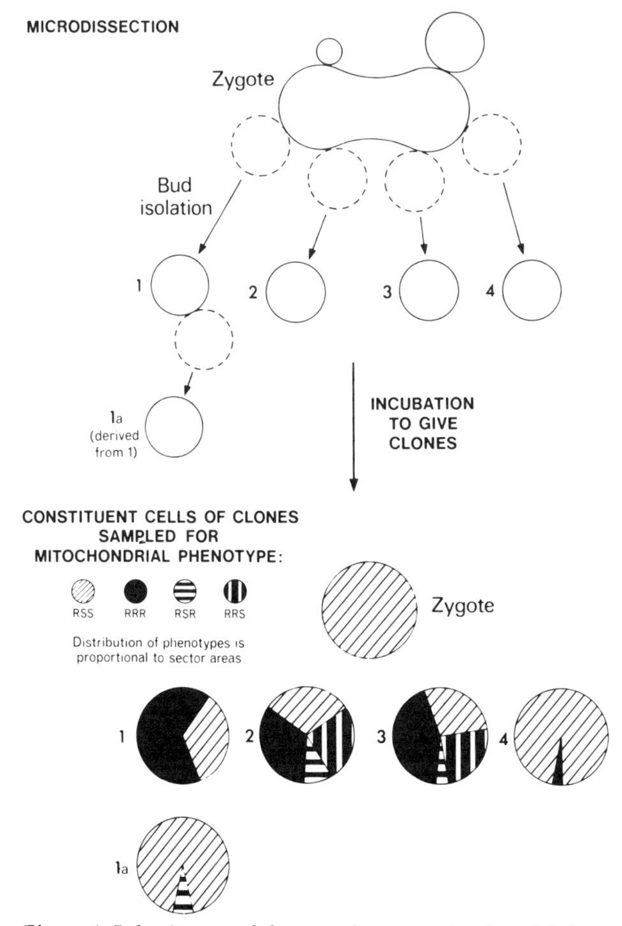


*Figure 1* Inheritance of drug resistance mitochondrial markers among diploid progeny from a zygote in the cross: R S S (oligomycin-resistant, chloramphenicol-sensitive, erythromycin-sensitive) x S R R (oligomycin-sensitive, chloramphenicol-resistant, erythromycin-resistant). (Data from Wilkie and Thomas 1973.)

studies in mitochondrial genetics both in Linnane's and in my own laboratory. Concentrating on erythromycin-resistant mutants to begin with, both groups were able to provide evidence locating the resistance genes in the mitochondrial genome, first by the non-Mendelian inheritance (Fig. 1) and second by the elimination of the genes in a *petite* mutant. These highly significant findings were presented in 1967 at the Bari symposium on mitochondrial biogenesis. This meeting was entitled "A Round Table Discussion" and indeed the participants, a total of 25, were actually seated at a single table (Fig. 2). This is very different from more recent symposia where participants can number in the hundreds. The exclusivity of the get-together of the mitochondriacs of the day was emphasized by the fact that at the end of the proceedings, the Contessa di Miani, in whose palazzo the meeting was held, presented each of us with a gold medal suitably inscribed to commemorate the occasion.

In a continuation of the genetic analysis of mitochondrially located resistance factors, my graduate student David Thomas and I set up a number

*Figure 2* Participants at the "Round Table Discussion on the Biogenesis of Mitochondria," Bari, 1967. Clockwise from the left: Tony Linnane, Sergio Papa, David Wilkie, Margit Naas, Piotr Slonimski, Eric Slater (Chairman), Igor Dawid, Bernard Kadenbach, Ab Kroon. Others not in the picture: P. Borst, H. Fukuhara, D. Neubert, W. Neupert, M. Rabinowitz, C. Saccone, M. Simpson, H. Swift, E. Wintersberger, and T.S. Work. (Photograph courtesy of D. Wilkie.)

of crosses between strains carrying markers for resistance to several antibiotics. The results clearly indicated that a mechanism operated in the zygote for recombining the markers. In three-point crosses, for example, parental types and all six recombinant classes were found segregating out among the diploid progeny. The segregation occurred early in the development of the clones from individual zygotes. The mechanism of transmission and distribution of mitochondrial genomes was complicated by the frequent finding that the distribution of reciprocal classes, both parental and recombinant, was asymmetrical. Whatever the system operating in the transmission of mitochondria, the fact remained that genetic recombination between them had occurred. Thomas and I published the data in 1968, and I think it is fair to say that this was the beginning of formal mitochondrial genetics (Thomas and Wilkie 1968a,b).

Piotr Slonimski and his co-workers (B. Dujon, M. Bolotin, D. Coen, J. Deutsch, P. Netter, E. Petrochilo) followed up these findings and made an extensive analysis of segregation patterns of antibiotic resistance genes in diploid clones. In certain crosses, asymmetry was very pronounced for a particular pair of alleles but progressively less so for other pairs. In other words, the asymmetry was polarized and reminiscent of the situation in bacterial conjugation in which the frequency of marker recombination is proportional to the proximity of the marker to the point of entry of the bacterial chromosome. Indeed, the bacterial analogy greatly appealed to the Slonimski group, who went so far as to propose that mitochondria could be classified as male and female with control vested in a mitochondrial DNA-associated sex factor that they called *omega* (ω), more or less equivalent to the F-factor in *Escherichia coli*. Crosses of the type $\omega^+ \times \omega^-$ showing the polarized asymmetry were termed "heterosexual," and the crosses $\omega^+ \times \omega^+$ and $\omega^- \times \omega^-$ that did not show the asymmetry were termed "homosexual." The idea of mitochondrial conjugation effecting a sex-mediated transfer of genetic material was intriguing, and in a series of investigations, we looked at zygotes in the electron microscope for evidence of this phenomenon. Conjugation was not apparent, but rather the mitochondria tended to lose their structure and become disaggregated in newly formed zygotes. In older but still budding zygotes, well-structured differentiated mitochondria appeared once more. These features are illustrated in Figure 3.

In further attempts to elucidate events in the zygote, we isolated daughter cells (buds) from zygotes as they were formed by microdissection. Isolated cells were cloned into small colonies whose constituent cells were sampled for mitochondrial phenotype. A typical distribution of mitochondrial types from one zygote and some of its diploid buds in a three-point cross is shown in Figure 1. In a three-point cross, there are eight possible arrangements of the three genes, namely, six recombinant classes and two parental types. In the example in the diagram, only four classes are seen, and this restriction in classes was typical for any one zygote, although the types transmitted varied with the zygote; i.e., recombinant types were seen in all zygote lineages but not all together in one lineage. Again in a four-point cross, all recombinants (14) appeared and at high frequency as in the three-point cross (see Fig. 1). In addition, members of individual reciprocal pairs tended to appear in asymmetric proportion. Overall, the results suggested that recombinational events are frequent between markers whatever their location, indicating multiple exchanges among mitochondrial DNA molecules, and that a random selection of the products is available for transmission to zygote buds. This genome pool is short-lived and only one mitochondrial type is transmitted to later buds. On reconsidering the results of the electron microscope studies (Fig. 3), it could be argued that recombinational events occur during the time mitochondria are disaggregated and that random selection of genome types takes place with their capture by reaggregating organelles. A further conclusion may be drawn from these investigations, namely, that the cell exerts a considerable degree of control over the transmission of mitochondria. Although there are apparently numerous copies of the mitochondrial chromosome, few seem to be inherited by daughter cells, since any initial heterogeneity is rapidly eliminated.

The foregoing provided the genetic evidence of recombination of mitochondrial marker genes, but it was not long before physical evidence of the process was available. Georgio Bernardi and co-workers in Paris found that two of our yeast strains differed in the fragments of mitochondrial DNA generated by certain restriction endonucleases, i.e., they differed in the location of cleavage sites. New fragment patterns were found among diploid clones from the cross between the two strains, demonstrating that recombination had taken place between restriction sites.

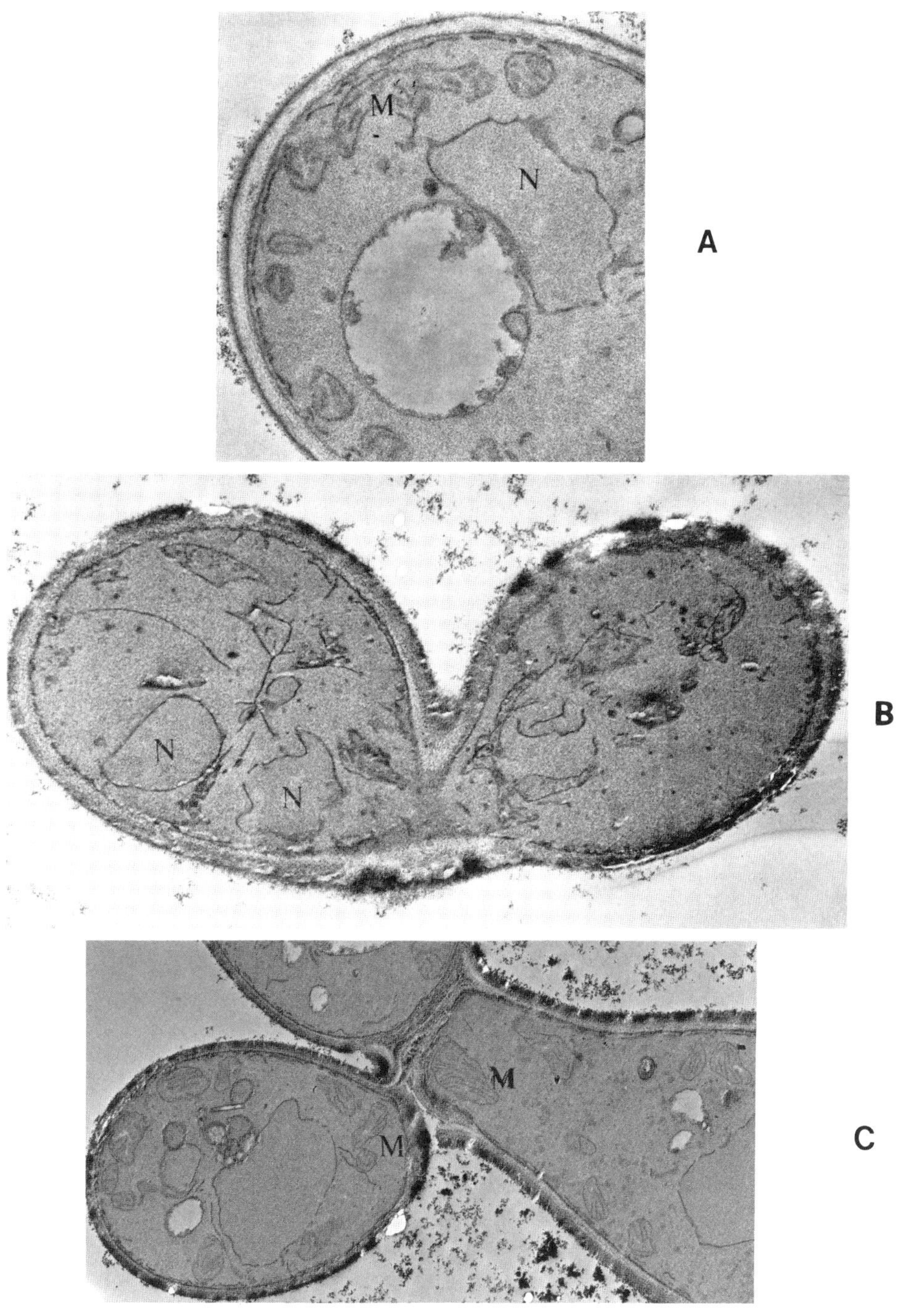


*Figure 3* Mitochondrial structure in the zygote of *S. cerevisiae*. (*A*) Parental cell prior to mating showing typical peripheral distribution of cristate mitochondria. (*B*) Young zygote just prior to nuclear fusion showing disrupted mitochondria. (*C*) Older zygote with daughter cell showing reformed mitochondria fully differentiated. (M) Mitochondrion; (N) nucleus. Magnification, 17,500x. (Reprinted, with permission, from Smith et al. 1972.)

To return to the mitochondrial *petite* ($\rho^-$) mutation, the irreversible nature of the change predicted a deletion mutation. Not only had this been seen, but the deletion was usually extensive with considerable loss of genetic complexity. Although the biochemical characteristics of the mutation had been worked out by Slonimski some 40 years ago, the events leading up to its inception are still not clearly understood. Another problem still unresolved is the phenomenon of suppressivity first described in 1956, in which a proportion of the diploid progeny from the cross $\rho^+ \times \rho^-$ are $\rho^-$. This fraction can be in excess of 90% depending on the *petite* mutant. It has been suggested that the mutant molecule has a selective advantage by virtue of a faster replication rate being less complex or having more replication sites than the wild-type molecule. These arguments are not conclusive and bring us back to the problem of the mechanism of reproduction and transmission of mitochondria and their genomes. In this context, the special case of the so-called *rho*$^o$ ($\rho^o$) *petite* mutant should be considered. This is an extreme case in which there is total loss of mitochondrial DNA, first described by Don Williamson in the late 1960s at a meeting in Strasbourg. Since it was recently demonstrated that *petite* mitochondria *had* DNA, he was accused of technical incompetence and advised to repeat his experiments by an authority who shall be nameless. Although the DNA-less mitochondria are transmitted to daughter cells during vegetative growth, they are not transmitted to daughter cells from zygotes in crosses to wild type, i.e., $\rho^o$ *petites* do not show the suppressive character. Assuming that disaggregation of mitochondria is a general feature of zygotes, reassembly may be centered on a DNA molecule that would preclude the $\rho^o$ condition. On this basis, zygotes from a cross between two $\rho^o$s would be unable to reconstitute mitochondria and so would not have viable progeny. As far as I am aware, no one has carried out this test, but it should be worth doing.

## MITOCHONDRIA AND CANCER

All of the foregoing recollections come from what were for me the early days of yeast genetics covering mainly the 1950s and 1960s. These were most productive years, even exciting at times, and a good number of very competent scientists made significant contributions to the field particularly regarding the advances in mitochondrial biogenesis. From 1970, my interests have centered more on the role of mitochondria in cancer. Reports that the mitochondria of cancer cells are defective were made by Otto Warburg as long ago as 1926, and the number of publications in support are now legion and continuing. A. Gause in the 1950s saw the *petite* cell of yeast as a model of the cancer cell in view of its respiratory deficiency. There are other features of the *petite* cell that seem to us to emphasize the cancer connection: (1) Mitochondrial DNA is much more sensitive than the nucleus to the mutagenic action of nearly all the mutagens/carcinogens we have tested; (2) the *petite* condition affects the characteristics of the plasma membrane; and (3) the defective mitochondria bring about the changes at the cell surface by modulating the activity of certain nuclear genes or the products of these genes. Presumably, these genes are specifying plasma membrane components or enzymes such as protein kinases that activate membrane proteins. It is not clear how mitochondria function in this respect, possibly by loss of control of calcium flux or some other feature of the organellar membranes. It is not unreasonable to suggest from these findings that mitochondrial mutagenesis could be yet another route to oncogenesis, particularly since changes at the cell surface are widely considered to be a prerequisite for neoplasia (for details, see Wilkie et al. 1983).

Mitochondria may also play a role as a target for chemotherapy in cancer cells in which defects in the organelle are apparently widespread, making it a weak spot. The aim would be to use nonmutagenic, membranolytic drugs of low toxicity (toxicity is a major problem with current anticancer agents) whose primary site of action is the mitochondrion. Attacking the cancer cell with these drugs could render the defective organelle more or less nonfunctional with consequent loss of viability. A good example from among a list of drugs we have placed in this category is the tricyclic antidepressant chlorimipramine (Anafranil; CIBA-GEIGY). The anticancer potential of this drug is apparent in its selective killing of human cancer cells in vitro (Wilkie and Delhanty 1970) and in the inhibition of growth of primary (inducible) tumors in mice and suppression of metastatic growth in the lungs (Wilkie 1979). The main objective now is to persuade the medical profession that Anafranil and other lipophylic agents whatever their current clinical role can be of use in cancer chemotherapy. An added attraction is that some of them have known tissue specificity, so that they are already targeted in the body. Over the years, the lack of interest has been disappointing.

It must be acknowledged that the categorizing of drugs in terms of their antimitochondrial activity has depended largely on the special attributes of the yeast cell. As a facultative anaerobe, the organism has the ability to grow and multiply in the absence of a functional respiratory system. This attribute has been invaluable to the mitochondriac and without it the *petite* mutant would never have been seen. For us, this is where it all started.

## REFERENCES

Mahony, M. and D. Wilkie. 1958. Cytoplasmic inheritance in *Aspergillus nidulans*. *Proc. R. Soc. Lond. B Biol. Sci.* **148:** 359–361.

———. 1959. Nucleo-cytoplasmic control of perithecial formation in *Aspergillus nidulans*. *Proc. R. Soc. Lond. B Biol. Sci.* **156:** 524–532.

Nass, S. and M.M.K. Nass. 1963. Intramitochondrial fibers with DNA characteristics. II. Enzymatic and other hydrolytic treatments. *J. Cell Biol.* **19:** 613–629.

Smith, D.G., K.C. Srivastava, and D. Wilkie. 1972. Ultrastructural changes in mitochondria of zygotes in *Saccharomyces cerevisiae*. *Microbios* **6:** 231–238.

Thomas, D.Y. and D. Wilkie. 1968a. Inhibition of mitochondrial synthesis in yeast by erythromycin: Cytoplasmic and nuclear factors controlling resistance. *Genet. Res.* **11:** 33–41.

———. 1968b. Recombination of mitochondrial drug-resistance factors in yeast. *Biochem. Biophys. Res. Commun.* **30:** 368–376.

Wilkie, D. 1963. Induction by monochromatic UV light of respiratory deficiency in aerobic and anaerobic yeast. *J. Mol. Biol.* **7:** 527–533.

———. 1964. *The cytoplasm in heredity.* Methuen, London.

———. 1979 Antimitochondrial drugs in cancer chemotherapy. *J. Roy. Soc. Med.* **72:** 599–601.

Wilkie, D. and J. Delhanty. 1970 Effects of chlorimipramine on human cells in tissue culture. *Br. J. Exp. Pathol.* **51:** 507–511.

Wilkie, D. and D.C. Hawthorne. 1961. Nonrandomness of mitotic recombination in yeast. *Heredity* **16:** 524.

Wilkie, D. and D. Lewis. 1963. The effect of UV light on recombination in yeast. *Genetics* **48:** 1701–1716.

Wilkie, D. and D.Y. Thomas. 1973. Mitochondrial genetic analysis by zygote cell lineages in *Saccharomyces*. *Genetics* **73:** 367–377.
Wilkie, D., I. Evans, D.C. Collier, V. Egilson, and E. Dialla. 1983. Mitochondria, cell surface and carcinogenesis. *Int. Rev. Cytol.* (suppl.) **15:** 157–189.

# MATING

I. Herskowitz (CSHL 1980)

V.L. MacKay (Schliersee, Bavaria 1976)

# **a**'s, α's, and Shmoos: Mating Pheromones and Genetics

VIVIAN L. MACKAY
*Protein Chemistry Department*
*ZymoGenetics, Inc.*
*Seattle, Washington 98105*

*Tomorrow and tomorrow and tomorrow...*
To Tom Manney and Wolfgang Duntze

My involvement in the field of yeast mating pheromones and mating-type genetics began in June 1968 as a first-year graduate student in Tom Manney's laboratory in the Department of Microbiology at Case Western Reserve University, when Tom and Wolfgang Duntze had just started work on α-factor, the diffusible factor made by mating-type α cells. To recreate the intellectual atmosphere that existed at that time and the "context of influencing events and personalities" as requested by the editors has required searching through old publications and my old laboratory notebooks, tasks that have been at times both enlightening and humbling (and frequently quite amusing). Moreover, I am reminded again how rapidly biological research has changed in two decades; in 1970, we could not just clone and sequence the gene or purify a peptide by high-performance liquid chromatography (HPLC).

The 1960s was the era of Biochemical Genetics. Bacterial operons and their regulation were accepted truth (Jacob and Monod 1961; Ames and Martin 1964), and biochemical geneticists were unraveling catabolic and biosynthetic pathways by matching genes with enzymes; leftover genes became candidates for regulatory functions. Most of the classical work had been done in bacteria (i.e., *Escherichia coli* and *Salmonella typhimurium*) and bacteriophages. When investigators turned to other genetic organisms such as *Neurospora crassa* or *Saccharomyces cerevisiae* to study amino acid biosythesis (see, e.g., Ahmed et al. 1964; Fink 1964), it rapidly became clear that the operon theory did not hold up for simple eukaryotes, so the term "regulon" was coined (and fortunately later mostly forgotten) to describe unlinked genes that were coordinately regulated.

As part of his Ph.D. dissertation (1964) with Bob Mortimer in biophysics at Berkeley, Tom Manney had investigated the *S. cerevisiae TRP5* locus, which encodes tryptophan synthetase, and allele-specific supersuppressible mutations (Manney 1964, 1968) almost simultaneously with Charles Yanofsky at Stanford, who was studying the tryptophan synthetase gene in *E. coli* (Brody and Yanofsky 1964). During his tenure at Oak Ridge and the beginning of his faculty appointment at Case Western Reserve, Tom continued his research on fine-structure mapping at the *TRP5* locus and characterization of the gene structure and enzyme. In 1967, Wolfgang Duntze, a postdoctoral fellow from Helmut Holzer's laboratory in Freiburg, Germany, joined the laboratory to study intragenic complementation between mutant alleles at *TRP5*. This re-

*The Early Days of Yeast Genetics*

search was rapidly completed and submitted for publication (Duntze and Manney 1968; Manney et al. 1969).

As Wolfgang needed a new project, Tom suggested that he follow up on observations of Winge (cited in Levi 1956), Ahmad (1953), and Levi (1956) that mating between **a** and α cells might be facilitated by hormones; perhaps Wolfgang could purify such a hormone (later renamed as a pheromone). Tom proposed that a suitable beginning project for me as a first-year graduate student would be to try to determine genetically if the hormone actually functioned to promote conjugation.

The state of the art of yeast (i.e., *S. cerevisiae*) mating physiology and genetics in 1968 can be rather briefly summarized. Ahmad (1953) and Levi (1956) had reported that when haploid cells of opposite mating type are paired close together but not touching, they elongate and eventually fuse at the narrow end of the "copulatory processes." According to folklore, Herschel Roman was responsible for naming such elongated cells "shmoos" in reference to comic strip characters in "Li'l Abner" created by Al Capp. (In our first papers [MacKay and Manney 1974a,b], we mistakenly spelled this term as "schmoo," a misspelling that persisted in other publications for a number of years.) Levi supported Winge's hypothesis that the formation of shmoos was due to hormones by observing shmoo formation in **a** cells placed on agar that had been previously covered with a mating mixture of **a** and α cells. Levi also stated that he could obtain shmoos of both mating types in liquid medium, but there were no reports of attempts to characterize these diffusible factors.

The genetics of conjugation was even less understood. As described in more detail in the chapter on Carl Lindegren by Robert Mortimer (this volume), the contributions of Lindegren and Lindegren (1943a,b), Lindegren (1949), Pomper and Burkholder (1949), Roman and Sands (1953), and Hawthorne and Mortimer (1960) established that the mating type of a haploid (i.e., heterothallic) strain was determined by which of two genetically exclusive alleles was present at the mating-type locus (*MAT*) on chromosome III. Several other *S. cerevisiae* genes that affected mating ability had been identified, including *nul3* (D.C. Hawthorne, pers. comm.), which blocked mating and mapped on chromosome IV; the "Hawthorne deletion" (Hawthorne 1963a), which joined *MAT* and *MAL2* (a distal gene on the same arm of chromosome III) and converted an α strain to an **a** mater; and various genes involved in homothallic conversion, i.e., the *HM1*, *HM2*, *HM3* genes (Takahashi 1958, 1961; Takahashi and Ikeda 1959), the *D* gene (Winge and Roberts 1949; Hawthorne 1963a,b), and the $HO_{\alpha}$ and *HM* genes (Takano and Oshima 1967, 1970). The elucidation of the homothallism genes and their effects on mating-type conversion are described in more detail and clarity in the chapter by Yasuji Oshima (this volume).

Supported with this background, Wolfgang and I pursued our projects, happily quoting Shakespeare to each other, particularly the soliloquy "*Tomorrow and tomorrow and tomorrow...*" from Macbeth. I doubt that any of us at that time understood Tom's foresight or would have predicted the extent to which these projects would grow into a major field of investigation that is still continuing. Indeed, it was generally necessary then, and for a number of years later, to justify why one worked with yeast at all, instead of *E. coli* where it was "easy" to make progress or mammalian cells and viruses where research was "relevant." Moreover, would it be possible to dissect a pathway as presumably complicated as conjugation? Twenty-three years later, we are still trying to answer that question.

## MATING PHEROMONES

### *α-Factor*

Levi's paper indicated that the hormone that acts on **a** cells would probably be the easier to isolate and purify, so Wolfgang began work on what would publicly be designated α-factor, but which carried the laboratory nicknames of "Schmoo-ogenic Hormone" or "Kickapoo Joy Juice" (again, from Al Capp). He quickly extended Levi's report that mating mixtures contain one or more diffusible substances that cause **a** cells to shmoo by showing, in a simple plate assay, that α cells alone secrete the activity. Purification of α-factor relied on this biological assay, which had the pitfalls of being somewhat inconsistent and only semiquantitative but the advantage of leading to the formation of bizarre, entertaining misshapen cells. Wolfgang achieved sufficient purification to allow characterization of α-factor as a peptide with a molecular weight between 1000 and 2000 and therefore not the same as the steroid factors reported by Yanagishima (1969) to cause expansion of cells. Nor was α-factor chemically similar to other fungal hormones that had been described previously (Chang 1968; Barksdale 1969; Konijn et al. 1969).

These results were published in a short report (Duntze et al. 1970) that also included our first genetic data. Ideally, to establish a physiological role for α-factor in conjugation, one would isolate α mutants that did not secrete active α-factor and then determine if they had simultaneously lost the ability to mate. As we were unable to devise a selection for α-factor-negative mutants and were concerned that a nonselective screen would be an unrealistic undertaking, we opted instead to obtain circumstantial evidence by selecting nonmating α mutants and asking if they could still produce α-factor (see below section on The Genetics of Mating). Of 93 mutants with defective mating ability, 60 failed to secrete detectable α-factor. (In a subsequent larger sample, 196 of 383 α mutants isolated were characterized as α-factor-negative [MacKay and Manney 1974a].) Thus, the hormone probably had a significant role in promoting conjugation. This paper also contained photomicrographs showing α-factor-induced shmoos, which one reviewer complained were not clearly distinguishable in the figures. (The reviewer was established as Gerry Fink by the Cornell watermark on the paper he used to write the anonymous review.) This criticism led to the semi-facetious but quite useful Gerry Fink test for biological assays that we still use: If you're not willing to show it to Gerry Fink as a positive, then it's not a positive.

Wolfgang and I also presented our data at the Fifth International Conference on Yeast Genetics held in 1970 in Chalk River, Ontario, Canada, where we had the opportunity to meet the 100 or so participants that included most of the world's yeast geneticists at the time. We were both encouraged by the interest shown in our work and the friendly, generally cooperative atmosphere at the Conference. It is noteworthy how many of these people are still active in the field.

By this time, Wolfgang had returned to Germany, where he and Elizabeth Throm (later Bücking-Throm) showed that DNA synthesis and cell division are reversibly inhibited in **a** cells incubated with α-factor, whereas synthesis of RNA and protein is unaffected (Throm and Duntze 1970). Moreover, cell division is specifically arrested just prior to bud emergence so that the cells accumulate in $G_1$ as single unbudded cells (Bücking-Throm et al. 1973); this observation provided a key reagent for yeast cell cycle research that was now under way in Lee Hartwell's laboratory.

Wolfgang and his colleagues continued to work on purification of α-factor and finally succeeded in obtaining enough highly purified material for manual amino acid sequencing (Duntze et al. 1973; Stötzler and Duntze 1976; Stötzler et al. 1976). (Remember, this effort was before the development of HPLC and sensitive automated peptide sequencers.) The sequence they reported (Trp-His-Trp-Leu-Gln-Leu-Lys-Pro-Gly-Gln-Pro-Met-Tyr) was later verified by synthesis of the peptide (Ciejek et al. 1977) and by cloning of the structural genes (Kurjan and Herskowitz 1982; Singh et al. 1983; see below), and, for some time, α-factor synthetic peptide has been commercially available (Sigma). The major current interests in α-factor are as a ligand for its receptor in stimulation of the pheromone response-signal transduction pathway and as a convenient marker protein for secretion studies.

In hindsight, we were fortunate that we could not think of a direct selection for α mutants defective in α-factor production, since such an effort would have been frustrated by the presence of two structural genes for α-factor in laboratory strains (Singh et al. 1983) and of several genes encoding proteases required for processing the α-factor precursor to its mature, fully active form (Julius et al. 1983, 1984). Kurjan (1985) eventually answered the question we posed in 1968 by showing that α-factor production is essential for mating of α cells. Deletion of both structural genes abolishes mating ability, and this defect cannot be overcome by the addition of exogenous α-factor, even though there is no known intracellular role for α-factor or its secretion leader sequence (Caplan and Kurjan 1991); some mysteries remain.

### **a**-*Factor*

The experiments of Ahmad (1953), Levi (1956), and T.R. Manney (unpubl.) showed that, in mixtures of **a** and α cells, shmoos of both mating types were observed and could be isolated by micromanipulation, yet attempts to detect a diffusible hormone secreted by **a** cells had been unsuccessful. However, further characterization of the nonmating α mutants and later of analogous **a** mutants (see below) led to renewed efforts in mid 1971 to develop a simple plate assay for **a**-factor production and response. After a light application of α cells very close to a heavy streak of **a** cells grown overnight on rich medium, elongated cells (similar to, but less pointed and pear-shaped than, **a** shmoos) and large unbudded cells could be seen among the α cells closest to the **a** cell streak, although frequently not all of the α cells showed a response (MacKay and Manney 1974a). After proving to be reliable and consistent in double-blind experiments (and passing the Gerry Fink test), this qualitative assay was used to characterize the mutant strains. Of 107 nonmating α mutants tested, only 1 appeared to respond to **a**-factor and only 7 of 66 **a** mutants tested could produce detectable **a**-factor. Thus, the production of and response to mating hormones (pheromones) seemed to be closely associated with mating proficiency in cells of both mating types (MacKay and Manney 1974a).

Purification and characterization of **a**-factor proved to be difficult and frustrating. Although these efforts were begun in 1974 in Wolfgang Duntze's laboratory in Bochum and in my laboratory at Rutgers, it would take more than a decade to obtain the peptide sequence and an understanding of its posttranslational modifications (Betz et al. 1987; Anderegg et al. 1988). In hindsight, the difficulties are obvious. We were apparently the first to attempt to purify a prenylated (specifically, farnesylated), carboxymethylated peptide (Anderegg et al. 1988), one of a class of eukaryotic proteins subsequently shown to include Ras proteins, nuclear lamins, and some subunits of heterotrimeric G proteins (Glomset et al. 1990; Schafer et al. 1990). The low aqueous solubility conferred by **a**-factor's amino acid sequence (Tyr-Ile-Ile-Lys-Gly-Val/Leu-Phe-Trp-Asp-Pro-Ala-Cys) and hydrophobic posttranslational modifications made its isolation from **a** cell supernatants and its biological assay in aqueous media quite inconsistent.

In 1974, however, our first purification attempts were guided by the earlier isolation of the more soluble α-factor and encouraged by the report of an extracellular substance constitutively produced by **a** cells that could transiently arrest α cells in $G_1$ (Wilkinson and Pringle 1974). We obtained a fraction partially purified by ion-exchange chromatography that could arrest α cells in $G_1$ and, at higher concentrations, induce α shmoo formation, but this activity eluted from gel-filtration columns as a void volume peak containing large amounts of carbohydrate and with an apparent molecular weight greater than 600,000 (Betz et al. 1977), hardly the characteristics expected of a pheromone. Duntze's group continued to pursue its fractionation and reported an amino acid composition of an active peptide (Betz and Duntze 1979), but it remained difficult to assay and to isolate reproducibly in high yields (leading to the occasional mournful letter from Wolfgang). In recognition of its hydrophobic nature, an alternative isolation strategy of growing **a** cells in the presence of hydrophobic polystyrene beads and eluting **a**-factor from the beads with organic reagents (Strazdis and MacKay 1982; Anderegg et al. 1988) permitted the development of a more quantitative assay and the confirmation of earlier observations that the production of extracellular **a**-factor is increased by incubation of **a** cells with α-factor (MacKay 1978; Strazdis and MacKay 1983).

On the basis of a partial amino acid sequence, gene cloning revealed that, as for α-factor, there are two structural genes, *MFa1* and *MFa2*, encoding precursors of **a**-factor (Gething 1985). As shown earlier for α-factor, synthesis of **a**-factor is essential for mating of **a** cells with wild-type α cells, and the mating deficiency of *mfa1 mfa2* double deletion strains cannot be cured with exogenous **a**-factor (Michaelis and Herskowitz 1988). Processing of the precursors to mature **a**-factor is of current major interest, as this pathway seems to have several steps in common with the posttranslational modifications that are necessary for the localization or function of other eukaryotic polypeptides, such as G proteins and mammalian Ras proteins (Hrycyna and Clarke 1990; Schafer et al. 1990), and therefore represent potential targets for drug therapy (Finegold et al. 1990).

### *Barrier Factor*

While evidence was accumulating in the mid 1970s that α-factor and **a**-factor were essential for mating, Hicks and Herskowitz (1976) reported that there was a third diffusible mating-type-specific substance, an activity constitutively produced by **a** cells that acted as a barrier to the diffusion of α-factor through agar medium, either by inhibiting the action of α-factor or by inactivating the peptide. Somehow, this factor was also associated with the conjugation pathway, since the nonmating **a** mutants in our collection that were defective in **a**-factor production also failed to secrete detectable barrier activity. As understanding of the mating-type regulatory system evolved (see below), Sprague and Herskowitz (1981) devised a method to isolate mutants defective in barrier activity and showed that these mapped in a single gene, designated *BAR1*, that was established to be allelic with the *SST1* gene (Chan and Otte 1982), mutations in which confer enhanced sensitivity to α-factor and delayed recovery from $G_1$ arrest. As **a** *bar1* (*sst1*) mutants also mate less efficiently with α cells in mass mating mixtures, barrier activity apparently functions to establish optimal pheromone concentrations for conjugation (see below). Although the barrier phenotype provided another easily scored **a**-specific characteristic, the α-factor supersensitivity conferred by *bar1* (*sst1*) mutations perhaps was (and still is) even more valuable to investigators, who can conserve α-factor by using these mutants in cell cycle experiments or studies of pheromone response and signal transduction.

In 1981, Tom Manney and I re-established our old collaboration to clone the *BAR1* gene, which we demonstrated is actually the structural gene for the extracellular activity. The usual efforts of DNA sequencing and computer genetics indicated that the polypeptide is homologous to a variety of aspartyl proteases (MacKay et al. 1988), whereas characterization of the purified enzyme showed that it is a somewhat unconventional member of the family, with a very strict substrate specificity and unusual mode of secretion (MacKay et al. 1992). (The properties of this enzyme are quite different from those of the α-factor endopeptidases described previously [Ciejek and Thorner 1979; Okada et al. 1987].)

### *Shmoo Formation*

With the availability of (semi)purified mating hormones, it was now possible to ask what physiological events led to the formation of the amusing shmoo shape. Both **a** and α factors arrested sensitive cells in $G_1$ as unbudded cells but had little or no effect on synthesis of RNA and protein, mitochondrial DNA replication, or bulk cell wall synthesis (Throm and Duntze 1970; Petes and Fangman 1973; Bücking-Throm et al. 1973; Wilkinson and Pringle 1974; Lipke et al. 1976; Betz et al. 1977). In addition, Yanagishima and colleagues had shown, in their studies of heterosexual agglutination, that many laboratory **a** strains agglutinated poorly unless they had been pretreated with α culture supernatant or with a peptide (designated α-substance I) isolated from α cultures (Sakai and Yanagishima 1972; Sakurai et al. 1974, 1975; Yanagishima et al. 1974). Radin (1976) provided evidence that α-factor and α-substance I are the same molecule. (I remember meeting David Radin in the aisle of a supermarket in Berkeley, where our enthusiastic discussion of sex hormones and mating nonplussed other shoppers.)

Macromolecule synthesis in the absence of cell division would be expected to lead to larger cells (Strazdis and MacKay 1982) but not necessarily shmoo-shaped ones. However, the wall at the tip of **a** shmoos appeared thinner, and small vesicles accumulated in the tip region (Lipke et al. 1976; W. Duntze, cited in Crandall et al. 1977), suggesting specific, localized changes in the cell wall similar to those observed in mating pairs (Osumi et al. 1974; Byers and Goetsch 1974, 1975). Lipke et al. (1976) showed that the total wall isolated from **a** shmoos had a higher glucan content and a lower mannan content; this mannan polysaccharide appeared to have fewer long side chains and a higher proportion of unsubstituted mannosyl residues in the backbone. Staining shmoos and zygotes with fluorescein-labeled concanavalin A indicated that the changes in mannan content and structure were localized to the shmoo tip (Tkacz and MacKay 1979), as were the deposition of chitin and the accumulation of newly synthesized acid phosphatase in shmoos (Schekman and Brawley 1979; Field and Schekman 1980). (Looking at vividly stained, odd-shaped cells in the microscope is perhaps why biologists tend to have a better sense of humor than chemists.) Although several proteins implicated in fusion have recently been localized to the shmoo tip (Hasek et al. 1987; Trueheart et al. 1987; Watzele et al. 1988; Baba et al. 1989; Gehrung and Snyder 1990), the pathway of directed cell-surface changes leading to shmoo formation and/or cell fusion remains to be sorted out.

The work of Sena et al. (1973), however, raised valid *caveats* about the relevance of some of these early (and current) physiological studies in which cells were bombarded with nonphysiological concentrations of mating pheromones. If high concentrations of α-factor are added to mating mixtures of unbudded ($G_1$) **a** and α cells, shmoos develop but cell fusion is nearly abolished. Thus, the shmoo may represent an aberrant form that has progressed beyond the optimal mating stage. Moreover, are changes in gene expression or cell physiology induced by pheromone concentrations 100–1000-fold higher than levels present in mating mixtures accurate reflections of the normal conjugation state?

## THE GENETICS OF MATING

### *Isolation and Physiological Characterization of Mutants*

As mentioned earlier, in 1968, we could not think of a direct selection for α mutants defective in α-factor production, so we (mostly Tom with little intellectual contribution from me at this stage) developed a method (originally suggested by Don Hawthorne) to isolate nonmating (*ste* or sterile) mutants (Duntze et al. 1970; MacKay and Manney 1974a). This initial project was to develop into my Ph.D. research, which was conducted at three universities between 1968 and 1972, as Tom moved from Case Western Reserve to Kansas State University via Berkeley. While in Berkeley, I had the opportunity to discuss my work frequently with Bob Mortimer and Sy Fogel, who lended enthusiastic encouragement to our progress.

For the isolation of nonmating mutants, mutagenized haploid α cells carrying a recessive canavanine resistance allele were mixed with a 1000-fold excess of **a** cells bearing the *CAN1* gene conferring sensitivity to canavanine (an arginine analog). The mixture was allowed to mate for 24 hours on rich medium and then resuspended and replated on medium containing canavanine to select against **a** cells and **a**/α diploids. The strategy was straightforward and direct, but the need to optimize mating conditions, to reduce the background of spontaneous canavanine-resistant **a** cells and **a**/α mitotic recombinants, and to satisfy Tom's quantitative standards required numerous preliminary experiments that were excellent training for a first-year graduate student. After obtaining many α mutants, in 1971, we turned to the isolation of analogous nonmating **a** mutants using the same selection procedure, but we incorporated into these mutant hunts selection at 36°C in attempts to obtain temperature-sensitive mutants that would be more amenable to genetic analysis (see below).

Our selection scheme limited the types of mutants we expected to isolate to those that retained viability during the mass mating and selection and precluded those that could not mate but were still arrested by α-factor or **a**-factor or those that might form inviable zygotes. As part of his research on the cell division cycle, Lee Hartwell and co-workers isolated a large sample of temperature-sensitive **a** mutants that were not arrested in $G_1$ by α-factor; all of these were nonmaters (Hartwell 1980). Other groups using less restrictive methods have isolated additional mutants that were probably not represented in our collection (Blair 1979; Rine 1979; Oshima and Takano 1980; Sprague and Herskowitz 1981; Chan and Otte 1982; Fields and Herskowitz 1985).

### *Genetic Analysis*

The association of mating deficiency and loss of α-factor production in many of the α mutants encouraged us to pursue their genetic characterization. However, the project appeared to have a fatal flaw; we had to be able to mate them in order to analyze them, and their mating defects were not cured by the addition of exogenous α-factor or helper α cells. Complementation and recombination studies appeared to be impossible. (Reliable methods for protoplast fusion had not yet been reported.) Fortunately, control experiments indicated that many of the sterile mutants would mate at a low frequency (~$10^{-5}$) with **a** cells to form diploids that could sporulate, and subsequent tetrad analysis showed that most of these fusions were not the result of reversion or suppression of the sterile mutation. Complementation tests were still not feasible, but we could do recombination and linkage studies. Using a little more foresight, we later isolated temperature-sensitive nonmating **a** mutants that greatly expanded our genetic possibilities and even permitted crosses between mutants.

For each of the 76 sterile mutants that mated at low frequency, two or three diploid clones were dissected for tetrad analysis (MacKay and Manney 1974b); during this phase of the research, Bill Whelan (another student in the laboratory) and I would have contests to see who could dissect tetrads the fastest. The 11 mutations (all derived from α mutants) that failed to recombine with the *MAT*α locus were obvious, giving 2 **a**:2 nonmater segregations and defining either the *MAT* locus itself or one or more genes within 0.9–5 cM. Those crosses that showed nonlinkage between *MAT* and the *ste* mutation were more challenging and fun to interpret, reflecting the elegance of classical genetics and predictable Mendelian segregation. After eliminating crosses that arose from reversion or extragenic suppression, we observed segregations predicted for unlinked **a**- and α-specific defects (confirmed by back-crosses), as well as mutations blocking mating (and pheromone production and response) in either cell type. However, at least in theory, all of these phenotypes could result from different mutations within a single unlinked gene. Since complementation tests would be quite difficult, the number of genes involved was determined by additional crosses between different mutants (when possible), leading to more complex segregation predictions. We also considered what patterns might be obtained if the poorly understood *HO* (*D*) homothallism gene was somehow involved. (Such crosses were the basis of Beth Jones' line of questioning in my Ph.D. qualifying exam at Case Western Reserve.)

This combination of phenotypic and genetic analyses led to the identification of several distinct genes (MacKay and Manney 1974b): (1) one or two genes closely linked to or at *MAT*α, which we designated *STE1* and *STE1-5* and suggested could be regulatory genes, but no equivalent *MAT***a** genes; (2) one **a**-specific gene (*STE2*) and one α-specific gene (*STE3*) with the phenotype expected of genes encoding surface or intracellular receptors for the pheromone from the opposite cell type, i.e., the mutants could produce but could not respond to mating pheromones; and (3) at least two nonspecific genes (*STE4* and *STE5*) which were unlinked to each other or any of the other genes; these were required for all aspects of conjugation in both mating types, including mating, pheromone production, and pheromone response. There were many other nonspecific mutations that were nonconditional, and therefore the mutants could not be crossed with the temperature-sensitive *ste4* and *ste5* strains; however, other crosses showed that there were at least two other nonspecific *STE* genes represented in a sample of temperature-sensitive **a** strains isolated by Al Teriba (in Lee Hartwell's laboratory). The existence of so many nonspecific genes that were required for pheromone production and response in both mating types indicated that conjugation, particularly as it interacted with cell cycle, would be a more difficult pathway to unravel than amino acid biosynthesis (True!).

We proposed a general model in which *MAT*α and *MAT***a** (although the latter was not defined by any mutations at this time) each regulates the expression of unlinked α-specific and **a**-specific genes, respectively, and perhaps some of the nonspecific genes. In addition, we speculated that the *MAT*α and *MAT***a** products were required together to repress at least some of the nonspecific *STE* genes and to turn on **a**/α-specific functions, such as sporulation, enhanced X-ray survival, induced mitotic recombination, and polar budding pattern.

There was one other class of nonmating mutants in our collection, i.e., those that mated at low frequency with cells of the opposite mating type to form diploids that could not sporulate. Our tetraploid analysis suggested that the recessive defects mapped at *MAT*, but it was the subsequent genetic experiments with one of these mutants, as described below, that established the existence of the *MAT*α*2* gene and the key regulatory role of its gene product.

It was at this time (1974) that Ira Herskowitz (and, shortly after, his highly productive graduate students) entered the field and moved mating-type research to center stage. (The next year was the first of the Cold Spring Harbor meetings on Yeast Molecular Biology, in which formal presentations on mating-type and animated informal discussions were major features for several years; these meetings also set a tradition for the bands and dancing after the banquet.) Ira believed that the classical fine-structure genetic logic used to dissect complex regulatory circuitry in bacteriophages could be applied to mating-type regulation in yeast. For the next few years, Ira and his coworkers conducted simple, elegant genetic experiments that led to hypotheses for regulation of and by the *MAT* genes and for homothallic conversion. These experiments were possible in part because of the timely identification by Hopper and Hall (1975) of the *CSP1* gene, which permits sporulation of diploids that do not carry the obligatory *MAT***a** and *MAT*α alleles. Later work established that *CSP1* was allelic to *RME1*, encoding a negative regulator of meiosis (Kassir and Simchen 1976; Rine et al. 1981), and that the diploids must be homozygous for the recessive allele of *CSP1* (*RME1*) to sporulate in the absence of either *MAT***a** or *MAT*α. The primary experiments and observations that led to the α1-α2 model for mating-type regulation are summarized below (for review, see Herskowitz 1982).

1. One of our α mutants that could mate with **a** cells at low frequency to form nonsporulating diploids was shown to have a mutation (initially designated *ste73* and later *mat*α*2*) that mapped at *MAT* (Strathern et al. 1981).
2. Diploids selected as fusions between *mat*α*1* and *mat*α*2* mutants mate as α's, indicating complementation between the two genes at *MAT* (Strathern et al. 1981).
3. The *mat*α*2* mutant exhibits some **a**-specific properties, e.g., an elevated mating efficiency with α cells and production of the barrier activity that degrades α-factor, observations which suggest that the Matα2 protein is a negative regulator of **a**-specific functions (Hicks and Herskowitz 1976; Tkacz and MacKay 1979; Sprague et al. 1981; Strathern et al. 1981).
4. Double mutants bearing defects in *MAT*α*2* and *BAR1* (encoding barrier activity that degrades α-factor) secrete α-factor, showing that at least some α-specific functions are expressed in *mat*α*2* mutants (Sprague et al. 1981).
5. Meiotic recombination between *mat*α*1* and *mat*α*2* occurs in about 1% of the tetrads from crosses between *mat*α*1* and *mat*α*2* mutants, yielding segregants that can mate normally as α's and, more interestingly, others that mate at normal efficiencies as **a**'s. In the latter case, the resulting "**a**"/α diploids, however, could not sporulate but could mate as α's. Backcrosses confirmed the suspected genotype: These **a** maters were *mat*α*1 mat*α*2* double mutants. Thus, the mating phenotype in **a** cells does not require any products encoded by the *MAT***a** locus (Strathern et al. 1981).

While this story was evolving, Kassir and Simchen (1976) described another key *MAT* mutation, the **a*** or *mat***a***1* defect, which has little effect on mating efficiency in **a** cells. The *mat***a***1* mutants mate with α cells to form sporulation-incompetent diploids that mate as α's, thus showing that the *MAT***a** locus is functional and required to establish the **a**/α diploid state.

These studies were the basis for the now classic model of mating-type regulation (Fig. 1), in which the Matα1 protein is a positive regulator of α-specific functions and the Matα2 protein represses expression of **a**-specific genes, whereas in **a** cells, the **a**-specific genes are constitutively expressed because of the absence of the Matα2 repressor. In **a**/α diploids, Matα2 represses **a**-specific functions as before, and the combination of the Matα2 and Mat**a**1 products have three possible activities: (1) to repress the *MAT*α*1* gene thereby repressing α-specific genes; (2) to repress nonspecific conjugation genes; and (3) to activate **a**/α-specific functions (either directly or indirectly). Almost as an anticlimax, essentially all elements of the model were later confirmed by molecular cloning and DNA sequencing of the *MAT* loci and the *STE* genes

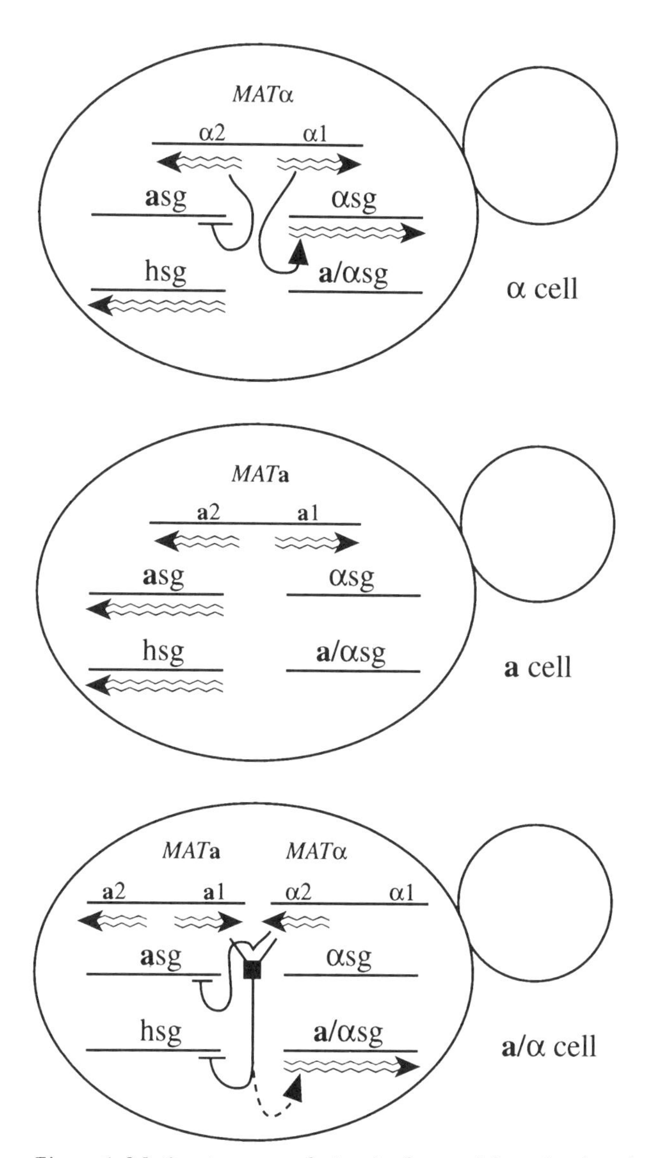


*Figure 1* Mating-type regulation in *S. cerevisiae.* α2-α1 and **a**2-**a**1 represent the alleles at the *MAT*α and *MAT***a** loci, respectively. **a**sg, αsg, hsg, and **a**/αsg are the specific genes expressed only in **a**, α, haploid, and **a**/α diploid cells, respectively. Wavy lines indicate transcription, solid lines with arrowheads indicate positive regulation of transcription, and blunted lines indicate repression of transcription. In **a**/α diploids, transcription of **a**/α-specific genes could occur via positive regulation by the **a**1-α2 heterodimer or by the absence of a repressor expressed in haploid cells (e.g., the *RME1* gene product that represses meiosis and sporulation functions).

(Hicks et al. 1979; Nasmyth and Tatchell 1980; Strathern et al. 1980; Astell et al. 1981; for review, see also Cross et al. 1988), which were essential steps for understanding the biochemical mechanisms involved.

As the cutoff date for this chapter is supposed to be approximately 1975 and since a discussion of all of the genes that are now implicated in conjugation or pheromone response/signal transduction is far beyond the page limits, I will limit this summary to those mutants that existed around 1975 or slightly later. Additional work in Hartwell's laboratory on **a** mutants resistant to α-factor ultimately led to the isolation of more mutations in the *STE2, STE4,* and *STE5* genes and the identification of the *STE7, STE8, STE9, STE11,* and *STE12* genes (Hartwell 1980). New **a**-specific (*STE6, STE14*) and α-specific (*STE13*) genes had been isolated by less restrictive schemes that did not require resistance to mating pheromones for survival (Blair 1979; Rine 1979; Sprague et al. 1981). These mutations blocked pheromone production, but not response, and were potential candidates for the pheromone structural genes. As the interest in the mating system was quite high at this time, I frequently received telephone calls from investigators in other areas of yeast research who had discovered a pleiotropic mutant with mating defects. These included *kex2* mutants that were defective in expression of the killer phenotype but also behaved as α-specific nonmaters with defects in α-factor production, but not in **a**-factor response (Leibowitz and Wickner 1976). The *tup1* mutations (and the allelic mutations *cyc9, flk1,* and *umr7*) were isolated for thymidine uptake, insensitivity to catabolite repression, or resistance to UV-induced mutagenesis, but they exhibited the same mating phenotype as *mat*α*2* mutants, as well as the other seemingly unrelated characteristics (Wickner 1974; Lemontt et al. 1980; Stark et al. 1980; Anagnost 1984).

Where are they now? With the advent of gene cloning, computer searches of databases, and insights derived from mammalian cell research, the functions of many of these genes have now been identified, although in some cases only recently. The *MAT*α*1* and *MAT*α*2* gene products do indeed regulate α- and **a**-specific gene transcription by binding to 5′ regulatory sequences in combination with the *MCM1* (PRTF) protein (Bender and Sprague 1987; Tan et al. 1988; Jarvis et al. 1989; Keleher et al. 1989; Ammerer 1990). The *STE2* and *STE3* genes encode the cell-surface receptors for α-factor and **a**-factor, respectively (Jenness et al. 1983; Burkholder and Hartwell 1985; Hagen et al. 1986), whereas *STE4* codes for the β subunit of a proposed heterotrimeric G protein complex that is thought to interact with the receptors (Whiteway et al. 1989). The α and γ subunits are products of the *SCG1* (= *GPA1*) and *STE18* genes (Dietzel and Kurjan 1987; Miyajima et al. 1987; Whiteway et al. 1989). The Ste5 protein appears to be the next step in the pheromone signal transduction pathway (Blinder et al. 1989), but its function as effector or in interacting with the effector is still unknown. *STE7* and *STE11* encode protein kinases that are required for signal transduction (Teague et al. 1986; Rhodes et al. 1990; B. Errede, pers. comm.), whereas the Ste12 protein binds to the pheromone response elements upstream of genes whose transcription is induced by a mating pheromone (Fields and Herskowitz 1987; Dolan et al. 1989; Errede and Ammerer 1989). *STE8* and *STE9* were shown to be allelic with *SIR3* and *SIR2*, respectively (Hartwell 1980), two of the four genes required for repression of the silent mating-type loci *HMR* and *HML*. *STE6* codes for a membrane protein required for secretion of **a**-factor (Kuchler et al. 1989; McGrath and Varshavsky 1989), and *STE14* is reported to be the structural gene for the methyltransferase that modifies **a**-factor (Hrycyna et al. 1991), whereas *KEX2* and *STE13* encode an endoprotease and diaminopeptidase, respectively, that (with the *KEX1* carboxypeptidase) process the precursor of α-factor to the mature active peptide (Julius et al. 1983, 1984; Dmochowska et al. 1987). Cloning and sequencing of the *TUP1* gene (MacKay 1983; Williams and Trumbly 1990) revealed interesting structural features but no clues to its function, although it appears to form a complex with the product of the *CYC8* (*SSN6*) gene that is implicated in glucose repression (Williams et al. 1991).

## EPILOGUE

In two decades, the study of yeast mating type and pheromones has been transformed from a narrow esoteric research problem of great intellectual enjoyment but little broad interest to an area recognized for its contributions to better understandings of eukaryotic cell biology with respect to transcriptional regulation, receptor binding/signal transduction, cell cycle control, secretion, posttranslational modifications, etc. Biotechnology has also borrowed from mating-type research in the use of regulated promoters and secretion leaders derived from the *MFα1* and *BAR1* genes and for the development of production strains. Since there is so much that we do not understand yet about how yeast cells mate, such contributions should continue. *Tomorrow and tomorrow and tomorrow....*

## ACKNOWLEDGMENTS

I thank Betsy Moore, Donna Prunkard, and Gary McKnight for their useful criticism of the manuscript, Margo Rogers for producing Figure 1, and Molly Bernard for literature searches.

## REFERENCES

Ahmad, M. 1953. The mating system in *Saccharomyces*. *Ann. Bot.* **17:** 329–342.

Ahmed, A., M.E. Case, and N.H. Giles, Jr. 1964. The nature of complementation among mutants in the *histidine-3* region of *N. crassa*. *Brookhaven Symp. Biol.* **17:** 53–65.

Ames, B.N. and R.G. Martin. 1964. Biochemical aspects of genetics: The operon. *Annu. Rev. Biochem.* **33:** 235–258.

Ammerer, G. 1990. Identification, purification, and cloning of a polypeptide (PRTF/GRM) that binds to mating-specific promoter elements in yeast. *Genes Dev.* **4:** 299–312.

Anagnost, J.A. 1984. "Partial suppression of the *tup1* mutation by the *MATα2* gene product in *Saccharomyces cerevisiae*." M.S. thesis, Rutgers University, New Brunswick, New Jersey.

Anderegg, R.J., R. Betz, S.A. Carr, J.W. Crabb, and W. Duntze. 1988. Structure of *Saccharomyces cerevisiae* mating hormone **a**-factor. Identification of S-farnesyl cysteine as a structural component. *J. Biol. Chem.* **263:** 18236–18240.

Astell, C.R., L. Ahlstrom-Jonasson, M. Smith, K. Tatchell, K.A. Nasmyth, and B.D. Hall. 1981. The sequence of the DNAs coding for the mating-type loci of *Saccharomyces cerevisiae*. *Cell* **27:** 15–23.

Baba, M., N. Baba, Y. Ohsumi, K. Kanaya, and M. Osumi. 1989. Three-dimensional analysis of morphogenesis induced by mating pheromone α factor in *Saccharomyces cerevisiae*. *J. Cell Sci.* **94:** 207–216.

Barksdale, A.W. 1969. Sexual hormones of *Achlya* and other fungi. *Science* **166:** 831–837.

Bender, A. and G.F. Sprague, Jr. 1987. *MATα1* protein, a yeast transcription activator, binds synergistically with a second protein to a set of cell-type-specific genes. *Cell* **50:** 681–691.

Betz, R. and W. Duntze. 1979. Purification and partial characterization of **a**-factor, a mating hormone produced by mating type **a** cells from *Saccharomyces cerevisiae*. *Eur. J. Biochem.* **95:** 469–475.

Betz, R., V.L. MacKay, and W. Duntze. 1977. **a**-Factor from *Saccharomyces cerevisiae*: Partial characterization of a mating hormone produced by cells of mating type **a**. *J. Bacteriol.* **132:** 462–472.

Betz, R., J.W. Crabb, H.E. Meyer, R. Wittig, and W. Duntze. 1987. Amino acid sequences of **a**-factor mating peptides from *Saccharomyces cerevisiae. J. Biol. Chem.* **262:** 546–548.

Blair, L.C. 1979. "Genetic analysis of mating type switching in yeast." Ph.D. thesis, University of Oregon, Eugene.

Blinder, D., S. Bouvier, and D.D. Jenness. 1989. Constitutive mutants in the yeast pheromone response: Ordered function of the gene products. *Cell* **56:** 479–486.

Brody, S. and C. Yanofsky. 1964. Independent action of allele-specific suppressor mutations. *Science* **145:** 399–400.

Bücking-Throm, E., W. Duntze, L.H. Hartwell, and T.R. Manney. 1973. Reversible arrest of haploid cells at the initiation of DNA synthesis by a diffusible sex factor. *Exp. Cell Res.* **76:** 99–110.

Burkholder, A.C. and L.H. Hartwell. 1985. The yeast α-factor receptor: Structural properties deduced from the sequence of the *STE2* gene. *Nucleic Acids Res.* **13:** 8463–8475.

Byers, B. and L. Goetsch. 1974. Duplication of spindle plaques and integration of the yeast cell cycle. *Cold Spring Harbor Symp. Ouant. Biol.* **38:** 123–131.

———. 1975. Behaviors of spindles and spindle plaques in the cell cycle and conjugation in *Saccharomyces cerevisiae. J. Bacteriol.* **124:** 511–523.

Caplan, S. and J. Kurjan. 1991. Role of α-factor and the *MFα1* α-factor precursor in mating in yeast. *Genetics* **127:** 299–307.

Chan, R.K. and C.A. Otte. 1982. Isolation and genetic analysis of *Saccharomyces cerevisiae* mutants super-sensitive to G1 arrest by **a**-factor and α-factor pheromones. *Mol. Cell. Biol.* **2:** 11–20.

Chang, Y.Y. 1968. Cyclic 3′,5′-adenosine monophosphate phosphodiesterase produced by the slime mold *Dictyostelium discoideum. Science* **160:** 57–59.

Ciejek, E. and J. Thorner. 1979. Recovery of *S. cerevisiae* **a** cells from G1 arrest by α-factor pheromone requires endopeptidase action. *Cell* **18:** 623–635.

Ciejek, E., J. Thorner, and M. Geier. 1977. Solid phase peptide synthesis of α-factor, a yeast mating pheromone. *Biochem. Biophys. Res. Commun.* **78:** 952–961.

Crandall, M., R. Egel, and V.L. MacKay. 1977. Physiology of mating in three yeasts. *Adv. Microb. Physiol.* **15:** 307–398.

Cross, F., L.H. Hartwell, C. Jackson, and J.B. Konopka. 1988. Conjugation in *Saccharomyces cerevisiae. Annu. Rev. Cell Biol.* **4:** 429–457.

Dietzel, C. and J. Kurjan. 1987. The yeast *SCG1* gene: A $G_α$-like protein implicated in the **a**- and α-factor response pathway. *Cell* **50:** 1001–1010.

Dmochowska, A., D. Dignard, D. Henning, D.Y. Thomas, and J. Bussey. 1987. Yeast *KEX1* gene encodes a putative protease with a carboxypeptidase B-like function involved in killer toxin and α-factor precursor processing. *Cell* **50:** 573–584.

Dolan, J.W., C. Kirkman, and S. Fields. 1989. The yeast *STE12* protein binds to the DNA sequence mediating pheromone induction. *Proc. Natl. Acad. Sci.* **86:** 5703–5707.

Duntze, W. and T.R. Manney. 1968. Two mechanisms of allelic complementation among tryptophan synthetase mutants of *Saccharomyces cerevisiae. J. Bacteriol.* **96:** 2085–2093.

Duntze, W., V. MacKay, and T.R. Manney. 1970. *Saccharomyces cerevisiae*: A diffusible sex factor. *Science* **168:** 1472–1473.

Duntze, W., D. Stötzler, E. Bücking-Throm, and S. Kalbitzer. 1973. Purification and partial characterization of α-factor, a mating-type specific inhibitor of cell reproduction from *Saccharomyces cerevisiae. Eur. J. Biochem.* **35:** 357–365.

Errede, B. and G. Ammerer. 1989. *STE12*, a protein involved in cell-type-specific transcription and signal transduction in yeast, is part of protein-DNA complexes. *Genes Dev.* **3:** 1349–1361.

Field, C. and R. Schekman. 1980. Localized secretion of acid phosphatase reflects the pattern of cell surface growth in *Saccharomyces cerevisiae. J. Cell Biol.* **86:** 123–128.

Fields, S. and I. Herskowitz. 1985. The yeast *STE12* product is required for expression of two sets of cell-type-specific genes. *Cell* **42:** 923–930.

———. 1987. Regulation by the yeast mating-type locus of *STE12*, a gene required for cell-type-specific expression. *Mol. Cell. Biol.* **7:** 3818–3821.

Finegold, A.A., W.R. Schafer, J. Rine, M. Whiteway, and F. Tamanoi. 1990. Common modifications of trimeric G proteins and ras protein: Involvement of polyisoprenylation. *Science* **249:** 165–169.

Fink, G.R. 1964. Gene-enzyme relations in histidine biosynthesis in yeast. *Science* **146:** 525–527.

Gehrung, S. and M. Snyder. 1990. The *SPA2* gene of *Saccharomyces cerevisiae* is important for pheromone-induced morphogenesis and efficient mating. *J. Cell Biol.* **111:** 1451–1464.

Gething, M.-J., ed. 1985. *Current communications in molecular biology: Protein transport and secretion*, pp. 103–108. Cold Spring Harbor Laboratory, Cold Spring Harbor, New York.

Glomset, J.A., M.H. Gelb, and C.C. Farnsworth. 1990. Prenyl proteins in eukaryotic cells: A new type of membrane anchor. *Trends Biochem. Sci.* **15:** 139–142.

Hagen, D.C., G. McCaffrey, and G.F. Sprague, Jr. 1986. Evidence the yeast *STE3* gene encodes a receptor for the peptide pheromone **a**-factor; gene sequence and implications for the structure of the presumed receptor. *Proc. Natl. Acad. Sci.* **83:** 1418–1422.

Hartwell, L.H. 1980. Mutants of *Saccharomyces cerevisiae* unresponsive to cell division control of polypeptide mating hormone. *J. Cell Biol.* **85:** 811–822.

Hasek, J., I. Rupes, J. Svobodova, and E. Streiblova. 1987. Tubulin and actin topology during zygote formation of *Saccharomyces cerevisiae. J. Gen. Microbiol.* **133:** 3355–3363.

Hawthorne, D.C. 1963a. A deletion in yeast and its bearing on the structure of the mating type locus. *Genetics* **48:** 1727–1729.

———. 1963b. Directed mutation of the mating type alleles as an explanation of homothallism in yeast. *Proc. Int. Congr. Genet.* **11:** 34–35.

Hawthorne, D.C. and R.K. Mortimer. 1960. Chromosome mapping in *Saccharomyces*: Centromere-linked genes. *Genetics* **45:** 1085–1110.

Herskowitz, I. 1982. The *MATα2* gene. *Recent Adv. Yeast Mol. Biol.* **1:** 320–331.

Hicks, J.B. and I. Herskowitz. 1976. Evidence for a new diffusible element of mating pheromones in yeast. *Nature* **260:** 246–248.

Hicks, J., J. Strathern, and A.J.S. Klar. 1979. Transposable mating type genes in *Saccharomyces cerevisiae. Nature* **282:** 478–483.

Hopper, A.K. and B.D. Hall. 1975. Mating type and sporulation in yeast. I. Mutations which alter mating-type control over sporulation. *Genetics* **80:** 41–59.

Hrycyna, C.A. and S. Clarke. 1990. Farnesyl cysteine C-terminal methyltransferase activity is dependent upon the *STE14* gene product in *Saccharomyces cerevisiae. Mol. Cell. Biol.* **10:** 5071–5076.

Hrycyna, C.A., S.K. Sapperstein, S. Clarke, and S. Michaelis. 1991. The *Saccharomyces cerevisiae STE14* gene encodes a methyltransferase that mediates C-terminal methylation of **a**-factor and RAS proteins. *EMBO J.* **10:** 1699–1709.

Jacob, F. and J. Monod. 1961. Genetic regulatory mechanisms in the synthesis of proteins. *J. Mol. Biol.* **3:** 318–356.

Jarvis, E.E., K.L. Clark, and G.F. Sprague, Jr. 1989. The yeast transcription activator PRTF, a homolog of the mammalian serum response factor, is encoded by the *MCM1* gene. *Genes Dev.* **3:** 936–945.

Jenness, D.D., A.C. Burkholder, and L.H. Hartwell. 1983. Binding of α-factor pheromone to yeast **a** cells: Chemical and genetic evidence for an α-factor receptor. *Cell* **35:** 521–529.

Julius, D., L. Blair, A. Brake, G. Sprague, and J. Thorner. 1983. Yeast α-factor is processed from a larger precursor polypeptide: The essential role of a membrane-bound dipeptidyl aminopeptidase. *Cell* **32:** 839–852.

Julius, D., A. Brake, L. Blair, R. Kunisawa, and J. Thorner. 1984. Isolation of the putative structural gene for the lysine-arginine-cleaving endopeptidase required for processing of yeast prepro-α-factor. *Cell* **37:** 1075–1089.

Kassir, Y. and G. Simchen. 1976. Regulation of mating and meiosis in yeast by the mating type region. *Genetics* **82:** 187–206.

Keleher, C.A., S. Passmore, and A.D. Johnson. 1989. Yeast repressor α2 binds to its operator cooperatively with yeast protein Mcm1. *Mol. Cell Biol.* **9:** 5228–5230.

Konijn, T.M., J.G.C. van de Meene, Y.Y. Chang, D.S. Barkley, and J.T. Bonner. 1969.

Identification of adenosine-3′,5′-monophosphate as the bacterial attractant for myxamoebae of *Dictyostelium discoideum*. *J. Bacteriol.* **99:** 510–512.
Kuchler, K., R.E. Sterne, and J. Thorner. 1989. *Saccharomyces cerevisiae STE6* gene product: A novel pathway for protein export in eukaryotic cells. *EMBO J.* **8:** 3973–3984.
Kurjan, J. 1985. α-Factor structural gene mutations in *Saccharomyces cerevisiae*: Effects on α-factor production and mating. *Mol. Cell. Biol.* **5:** 787–796.
Kurjan, J. and I. Herskowitz. 1982. Structure of a yeast pheromone gene (*MF*α): A putative α-factor precursor contains four tandem copies of mature α-factor. *Cell* **30:** 933–943.
Leibowitz, M.J. and R.B. Wickner. 1976. A chromosomal gene required for killer plasmid expression, mating and spore maturation in *S. cerevisiae*. *Proc. Natl. Acad. Sci.* **73:** 2061–2065.
Lemontt, J.F., D.R. Fugit, and V.L. MacKay. 1980. Pleiotropic mutations at the *TUP1* locus that affect the expression of mating-type-dependent functions in *Saccharomyces cerevisiae*. *Genetics* **94:** 899–920.
Levi, J.D. 1956. Mating reaction in yeast. *Nature* **177:** 753–754.
Lindegren, C.C. 1949. *The yeast cell: Its genetics and cytology*. Educational Publishers, St. Louis, Missouri.
Lindegren, C.C. and G. Lindegren. 1943a. A new method for hybridizing yeast. *Proc. Natl. Acad. Sci.* **29:** 306–308.
———. 1943b. Segregation, mutation, and copulation in *Saccharomyces cerevisiae*. *Ann. Mo. Bot. Gard.* **30:** 453–469.
Lipke, P.N., A. Taylor, and C.E. Ballou. 1976. Morphogenetic effects of α-factor on *Saccharomyces cerevisiae* **a**-cells. *J. Bacteriol.* **127:** 610–618.
MacKay, V.L. 1978. Mating-type specific pheromones as mediators of sexual conjugation in yeast. In *Molecular control of proliferation and differentiation* (ed. J. Papaconstantinou and W.J. Rutter), pp. 243–259. Academic Press, New York.
———. 1983. Cloning of yeast *STE* genes in 2 μm vectors. *Methods Enzymol.* **101:** 325–343.
MacKay, V.L. and T.R. Manney. 1974a. Mutations affecting sexual conjugation and related processes in *Saccharomyces cerevisiae*. I. Isolation and phenotypic characterization of nonmating mutants. *Genetics* **76:** 255–271.
———. 1974b. Mutations affecting sexual conjugation and related processes in *Saccharomyces cerevisiae*. II. Genetic analysis of nonmating mutants. *Genetics* **76:** 273–288.
MacKay, V.L., S.K. Welch, M.Y. Insley, T.R. Manney, J. Holly, G.C. Saari, and M.L. Parker. 1988. The *Saccharomyces cerevisiae BAR1* gene encodes an exported protein with homology to pepsin. *Proc. Natl. Acad. Sci.* **85:** 55–59.
MacKay, V.L., J. Armstrong, C. Yip, S. Welch, K. Walker, S. Osborn, P. Sheppard, and J. Forstrom. 1992. Characterization of the Bar proteinase, an extracellular enzyme from the yeast *Saccharomyces cerevisiae*. In *Structure and function of the aspartic proteinases: Genetics, structures, and mechanisms* (ed. B.M. Dunn), pp. 161–172. Plenum Press, New York.
Manney, T.R. 1964. Action of a super-suppressor in yeast in relation to allelic mapping and complementation. *Genetics* **50:** 109–121.
———. 1968. Evidence for chain termination by super-suppressible mutants in yeast. *Genetics* **60:** 719–733.
Manney, T.R., W. Duntze, N. Janosko, and J. Salazar. 1969. Genetic and biochemical studies of partially active tryptophan synthetase mutants of *Saccharomyces cerevisiae*. *J. Bacteriol.* **99:** 590–596.
McGrath, J.P. and A. Varshavsky. 1989. The yeast *STE6* gene encodes a homologue of the mammalian multidrug resistance P-glycoprotein. *Nature* **340:** 400–404.
Michaelis, S. and I. Herskowitz. 1988. The **a**-factor pheromone of *Saccharomyces cerevisiae* is essential for mating. *Mol. Cell. Biol.* **8:** 1309–1318.
Miyajima, I., M. Nakafuku, N. Nakayama, C. Brenner, A. Miyajima, K. Kaibuchi, K. Arai, Y. Kaziro, and K. Matsumoto. 1987. *GPA1*, a haploid-specific essential gene, encodes a yeast homolog of mammalian G protein, which may be involved in mating factor signal transduction. *Cell* **50:** 1011–1019.
Nasmyth, K.A. and K. Tatchell. 1980. The structure of transposable yeast mating type loci. *Cell* **19:** 753–764.

Okada, T., K. Sonomoto, and A. Tanaka. 1987. Novel Leu-Lys-specific peptidase (LeuLysin) produced by gel-entrapped yeast cells. *Biochem. Biophys. Res. Commun.* **145:** 316–322.

Oshima, T. and I. Takano. 1980. Mutants showing heterothallism from a homothallic strain of *Saccharomyces cerevisiae. Genetics* **94:** 841–857.

Osumi, M., C. Shimoda, and N. Yanagishima. 1974. Mating reaction in *Saccharomyces cerevisiae.* V. Changes in the fine structure during the mating reaction. *Arch. Microbiol.* **97:** 27–38.

Petes, T.D. and W.L. Fangman. 1973. Preferential synthesis of yeast mitochondrial DNA in α-factor-arrested cells. *Biochem. Biophys. Res. Commun.* **55:** 603–609.

Pomper, S. and O.R. Burkholder. 1949. Studies on the biochemical genetics of yeast. *Proc. Natl. Acad. Sci.* **35:** 456–464.

Radin, D.N. 1976. "Genetics and physiology of mating in *Saccharomyces cerevisiae.*" Ph.D. thesis, University of California, Berkeley.

Rhodes, N., L. Connell, and B. Errede. 1990. *STE11* is a protein kinase required for cell-type-specific transcription and signal transduction in yeast. *Genes Dev.* **4:** 1862–1874.

Rine, J.D. 1979. "Regulation and transposition of cryptic mating type genes in *Saccharomyces cerevisiae.*" Ph.D. thesis, University of Oregon, Eugene.

Rine, J., G.F. Sprague, Jr., and I. Herskowitz. 1981. *rme1* mutation of *Saccharomyces cerevisiae*: Map position and bypass of mating type locus control of sporulation. *Mol. Cell. Biol.* **1:** 958–960.

Roman, H. and S.M. Sands. 1953. Heterogeneity of clones of *Saccharomyces* derived from haploid ascospores. *Genetics* **39:** 171–179.

Sakai, K. and N. Yanagishima. 1972. Mating regulation in *S. cerevisiae.* II. Hormonal regulation of agglutinability of **a** type cells. *Arch. Microbiol.* **84:** 191–198.

Sakurai, A., S. Tamura, N. Yanagishima, and C. Shimoda. 1975. Isolation of a peptide factor controlling sexual agglutination in *S. cerevisiae. Proc. Natl. Acad. Sci.* **51:** 291–294.

Sakurai, A., S. Tamura, N. Yanagishima, C. Shimoda, M. Hagiya, and N. Takao. 1974. Chemical characterization of sexual hormones in yeast. In *Plant growth substances 1973: Proceedings of the 8th International Congress of Plant Growth Substances,* pp. 185–192. Hirokawa Publishing, Tokyo.

Schafer, W.R., C.E. Trueblood, D.-C. Yang, M.P. Mayer, S. Rosenberg, C.D. Poulter, S.-H. Kim, and J. Rine. 1990. Enzymatic coupling of cholesterol intermediates to a mating pheromone precursor and to the Ras protein. *Science* **249:** 1133–1139.

Schekman, R. and V. Brawley. 1979. Localized deposition of chitin of the yeast cell surface in response to mating pheromone. *Proc. Natl. Acad. Sci.* **76:** 645–649.

Sena, E.P., D.N. Radin, and S. Fogel. 1973. Synchronous mating in yeast. *Proc. Natl. Acad. Sci.* **70:** 1373–1377.

Singh, A., E.Y. Chen, J.M. Lugovoy, C.N. Chang, R.A. Hitzeman, and P.H. Seeburg. 1983. *Saccharomyces cerevisiae* contains two discrete genes-coding for the α-factor pheromone. *Nucleic Acids Res.* **11:** 4049–4063.

Sprague, G.F., Jr. and I. Herskowitz. 1981. Control of yeast cell type by the mating type locus. I. Identification and control of expression of the **a**-specific gene, *BAR1. J. Mol. Biol.* **153:** 305–321.

Sprague, G.F., Jr., J. Rine, and I. Herskowitz. 1981. Control of yeast cell type by the mating type locus. II. Genetic interactions between *MAT*α and unlinked α-specific *STE* genes. *J. Mol. Biol.* **153:** 323–335.

Stark, H.C., D. Fugit, and D.B. Mowshowitz. 1980. Pleiotropic properties of a yeast mutant insensitive to catabolite repression. *Genetics* **94:** 921–928.

Stötzler, D. and W. Duntze. 1976. Isolation and characterization of four related peptides exhibiting α-factor activity from *Saccharomyces cerevisiae. Eur. J. Biochem.* **65:** 257–262.

Stötzler, D., H.H. Kiltz, and W. Duntze. 1976. Primary structure of α-factor peptides from *Sacchammyces cerevisiae. Eur. J. Biochem.* **69:** 397–400.

Strathern, J.N., J.B. Hicks, and I. Herskowitz. 1981. Control of cell type in yeast by the mating type locus: The α1-α2 hypothesis. *J. Mol. Biol.* **147:** 357–372.

Strathern, J.N., E. Spatola, C. McGill, and J.B. Hicks. 1980. The structure and organization of transposable mating type cassettes in *Saccharomyces* yeasts. *Proc. Natl. Acad. Sci.* **77:** 2839–2843.

Strazdis, J.R. and V.L. MacKay. 1982. Reproducible and rapid methods for the isolation and assay of **a**-factor, a yeast mating hormone. *J. Bacteriol.* **151:** 1153–1161.

———. 1983. Induction of yeast mating pheromone **a**-factor by α cells. *Nature* **305:** 543–545.

Takahashi, T. 1958. Complementary genes controlling homothallism in *Saccharomyces. Genetics* **43:** 705–714.

———. 1961. Sexuality and its evolution in *Saccharomyces. Seiken Ziho* **12:** 11–20.

Takahashi, T. and Y. Ikeda. 1959. Bisexual mating reaction in *Saccharomyces chevalieri. Genetics* **44:** 375–382.

Takano, I. and Y. Oshima. 1967. An allele specific and a complementary determinant controlling homothallism in *Saccharomyces oviformis. Genetics* **57:** 875–885.

———. 1970. Mutational nature of an allele-specific conversion of the mating type by the homothallic gene $HO_{\alpha}$ in *Saccharomyces. Genetics* **65:** 421–427.

Tan, S., G. Ammerer, and T.J. Richmond. 1988. Interactions of purified transcription factors: Binding of yeast *MATα1* and PRTF to cell type-specific, upstream activating sequences. *EMBO J.* **7:** 4255–4264.

Teague, M.A., D.T. Chaleff, and B. Errede. 1986. Nucleotide sequence of the yeast regulatory gene *STE7* predicts a protein homologous to protein kinases. *Proc. Natl. Acad. Sci.* **83:** 7371–7375.

Throm, E. and W. Duntze. 1970. Mating-type-dependent inhibition of deoxyribonucleic acid synthesis in *Saccharomyces cerevisiae. J. Bacteriol.* **104:** 1388–1390.

Tkacz, J.S. and V.L. MacKay. 1979. Sexual conjugation in yeast: Cell surface changes in response to the action of mating hormones. *J. Cell Biol.* **80:** 326–333.

Trueheart, J., J. Boeke, and G.R. Fink. 1987. Two genes required for cell fusion during yeast conjugation: Evidence for a pheromone induced surface protein. *Mol. Cell Biol.* **7:** 2329–2334.

Watzele, M. F. Klis, and W. Tanner. 1988. Purification and characterization of the inducible **a** agglutinin of *Saccharomyces cerevisiae. EMBO J.* **7:** 1483–1488.

Whiteway, M., L. Hougan, D. Dignard, D.Y. Thomas, L. Bell, G.C. Saari, F.G. Grant, P. O'Hara, and V.L. MacKay. 1989. The *STE4* and *STE18* genes of yeast encode potential β and γ subunits of the mating factor receptor-coupled G protein. *Cell* **56:** 467–477.

Wickner, R.B. 1974. Mutants of *Saccharomyces cerevisiae* that incorporate deoxythymidine-5′-monophosphate into deoxyribonucleic acid in vivo. *J. Bacteriol.* **117:** 252–260.

Wilkinson, L.E. and J.R. Pringle. 1974. Transient G1 arrest of *S. cerevisiae* cells of mating type-α by a factor produced by cells of mating type-**a**. *Exp. Cell Res.* **89:** 175–187.

Williams, F.E. and R.J. Trumbly. 1990. Characterization of *TUP1*, a mediator of glucose repression in *Saccharomyces cerevisiae. Mol. Cell. Biol.* **10:** 6500–6511.

Williams, F.E., U. Varanasi, and R.J. Trumbly. 1991. The *CYC8* and *TUP1* proteins involved in glucose repression in *Saccharomyces cerevisiae* are associated in a protein complex. *Mol. Cell. Biol.* **11:** 3307–3316.

Winge, Ø. and C. Roberts. 1949. A gene for diploidization in yeast. *C.R. Trav. Lab. Carlsberg Ser. Physiol.* **24:** 341–346.

Yanagishima, N. 1969. Sexual hormones in *Saccharomyces cerevisiae. Antonie Leeuwenhoek* (suppl.) **35:** suppl. C9.

Yanagishima, N., C. Shimoda, M. Tsuboi, M. Hagiya, N. Takao, A. Sakurai, and S. Tamura. 1974. Hormonal regulation in the life cycle of *Saccharomyces cerevisiae.* In *Plant growth substances 1973: Proceedings of the 8th International Congress on Plant Growth Substances*, pp. 173–184. Hirokawa Publishing, Tokyo.

# Homothallism, Mating-type Switching, and the Controlling Element Model in *Saccharomyces cerevisiae*

YASUJI OSHIMA
*Department of Biotechnology, Faculty of Engineering*
*Osaka University, Suita-shi, Osaka 565, Japan*

In midsummer of 1965, I returned from Southern Illinois University in Carbondale, Illinois, to the Research Laboratory of Suntory Ltd., which at that time was located at Dojima in downtown Osaka. I had just spent 2 years in Illinois studying yeast genetics as a postdoctoral research associate at the Biological Research Laboratory of Professor Carl C. Lindegren. A few years earlier, Suntory Ltd. had entered the beer business and was brewing its own beer, viewing it as a most promising beverage, and the atmosphere at the company was still somewhat excited about the new business. At that time, the beer business in Japan was dominated by three companies, Asahi, Kirin, and Sapporo, which were in strong market competition with each other. This competition became even stronger after the entry of Suntory.

Of course, I apprised my colleagues in the Research Laboratory of Suntory of my studies at Carbondale and of all my experiences in the United States and Europe where I had traveled in the summer of 1963 on the way to Carbondale. I also heard about the progress of their research. Although most of these conversations have now been mostly forgotten, I remember Isamu Takano's stories on the polyploidy of the brewing and baking yeasts and on the complicated inheritance of film formation in yeast strains used in fermentation industries. These, of course, were of interest to me, but the genetic data on film formation did not make any sense at all.

The industrial yeast strains, especially those used in brewing and baking, had a history of empirical selection even before they were even recognized as living things. Selection has been carried out for centuries in the kitchens of individual families, in manor houses of country lords, in monasteries, and in local factories. We became interested in the question of how the polyploidy of the brewing and baking yeasts (all of these yeasts are *Saccharomyces cerevisiae*) occurred and was selected. What is the advantage of polyploid strains in the brewing and baking industries? We also wondered how to construct such strains.

I was, however, most interested in another story told by Takano. Because we were so enthusiastic about the new beer business and the questions associated with polyploidy in brewing yeasts, he told me about some spore clones produced by diploids that had been constructed by crosses between haploid segregants from a strain of *Saccharomyces oviformis* and a haploid strain used in the laboratory. When these ascospores were placed on an agar plate after dissection of the asci, they could self-diploidize during their development into minute colonies. The strain of *S. oviformis*, used as a flor yeast in sherry making in Spain, was examined by Takano because it formed a characteristic thick film on the surface of wine. He had already performed some genetic analysis on this yeast, which could interbreed with genetic breeding stocks, but no definite ideas on film formation had been formulated. As just described, however, some of the segregants in the crosses showed the novel phenomenon of homothallism: Each tetrad produced by the diploid consisted of two haploid clones of **a** mating type and two nonmating diploid clones. Haploid clones of α mating type were never obtained. The diploid segregants again showed the 2 haploid **a**:2 diploid segregation when they were sporulated and the asci were dissected. In addition, there was an indication in other crosses that some of the haploid clones of **a** mating type also diploidized. These observations were compiled in our first homothallism paper published in *Genetics*; we reported that there is a single dominant gene, *HOα*, for homothallic diploidization of a haploid cell of α mating type, whereas another gene, *HM*, complementary to *HOα*, was necessary to diploidize haploid **a** cells (Takano and Oshima 1967).

In a moment of inspiration, I had the idea that homothallism, a phenomenon well known from the beginning of yeast genetics from studies of Øjvind Winge and Carl C. Lindegren, might be connected with polyploidy. The combination of circumstances at that moment—the new beer business, polyploidy in brewing yeasts, and the novel homothallism—compelled us to analyze homothallism, and we initiated the genetic analysis immediately in 1965 at the Research Laboratory of Suntory Ltd. When I moved in 1970, along with Satoshi Harashima and many other students, the analysis was continued in the Department of Fermentation Technology (now the Department of Biotechnology, since April 1, 1991) at Osaka University, the Suita campus, in parallel with Isamu Takano and his colleagues from Suntory.

## HOMOTHALLISM AND HETEROTHALLISM: BACKGROUND

Before describing our genetic study of homothallism, it is probably a good idea to summarize what was known about homothallism in *Saccharomyces* yeasts at that time. Homothallism is strictly concerned with the yeast life cycle, and thus this phenomenon had been known from the beginning of yeast genetics. In fact, yeast genetics was actually initiated with homothallic strains by Øjvind Winge at the Carlsberg Laboratory in Copenhagen in the mid 1930s. Extensive investigations had thus already been conducted on this phenomenon.

The life cycle of *Saccharomyces* yeasts includes both diploid and haploid phases. All, or essentially all, strains of *S. cerevisiae* and the related species of this yeast isolated from their natural habitat are diploids. When these diploid cells are placed on a sporulation medium, for which I first used Gorodkowa agar but now simply use sodium acetate or potassium acetate agar, vegetative growth ceases and the cells become committed to meiosis and then sporulate. In this process, each diploid cell is transformed into an ascus usually consisting of four spores. The ascospores and cells arising from them after spore germination constitute the haplophase. Then, significant developmental differences occur in the first few cell-division cycles after spore germination, depending on whether the strain is homothallic or heterothallic. In the pioneering study of yeast genetics, Winge (1935) observed homothallism in *S. cerevisiae* var. *ellipsoideus* and later in *Saccharomyces chevalieri* (Winge and Roberts 1949). (According to the recent yeast taxonomy, both of these species and *S. oviformis* are included in *S. cerevisiae*, even though these strains have differences in their abilities to ferment or assimilate some sugars.) Round oval-shaped or elliptical diploid cells, often with a single bud near one end, were observed in a culture developed from a single ascospore; these cells could sporulate but did not have mating ability. When these diploid cells were placed on sporulation medium, they sporulated, and cultivation of ascospores again gave rise to diploid cells. In contrast to the homothallic strains of Winge, Carl C. Lindegren and Gertrude Lindegren (1943) observed heterothallism in American strains in various species of *Saccharomyces*. They found that single-spore cultures grew into haploid clones having either one of two mating types, **a** and α, and that these clones were unable to sporulate. The christening of the mating types in *Saccharomyces* yeasts as **a** and α occurred in these studies. When two haploid vegetative cells of complementary mating types (one **a**, the other α) were mixed together, there arose a diploid zygote that gave rise to a diploid clone capable of sporulation but having a nonmating phenotype. When these diploids were sporulated, haploid clones of **a** and α mating types always segregated 2**a**:2α in each ascus. Haploid cells are, in general, easily distinguishable from diploid cells by their small clumped round appearance, the so-called rosette appearance (Townsend and Lindegren 1954).

The different behaviors of the strains in the laboratories of Winge and Lindegren were due to the presence of the *D* gene (for diploidization), now called *HO*, in Winge's strains; the Lindegren strains contained the *d* allele, now called *ho*, which is an inactive allele of *D* (Winge and Roberts 1949) (for genetic symbols for homothallism at phase I, see Table 1). The *D* gene proved to be epistatic to the allelic genes **a** and α, now called *MAT***a** and *MAT*α, on chromosome III and segregated independently from the *MAT* locus. A few years later, Takahashi et al. (1958) described a homothallic gene, *HO*, which is similar to the *D* gene, and subsequently suggested that homothallism is controlled by three complementary genes, $HM_1$, $HM_2$, and $HM_3$ (Takahashi 1958; Takahashi and Ikeda 1959). They considered that the *HM* gene system was different from that involving the *D* gene. Isamu Takano and I simply adopted the homothallism gene nomenclature of Takahashi and his co-authors. Later allelism tests with these homothallic strains revealed that the *D* (and *HO*) and *HM* genes were involved in the same genetic system as the *HO*α and *HM* genes (Takano and Oshima 1970a) (for genetic symbols at phase II, see Table 1).

The terminology of homothallism and heterothallism for these two developmental differences in *Saccharomyces* yeasts was adopted from that used for Mucorales by Blakeslee (1904a,b): Homothallism denotes the ability to form diploids by self-fertilization in a single-spore culture; heterothallism denotes obligatory cross-fertility, i.e., the ability to form diploids only between cells of different mating types that are derived from separate single-spore cultures. In practice, homothallism and heterothallism in yeasts are determined either by testing cells in a single-spore culture for sporulation and for mating ability or by inspecting cell shape.

Several possibilities for the mechanism of homothallic diploidization were discussed by Winge and Roberts (1958). Diploids might be formed by the ab-

*Table 1* Change of the Genetic Symbols in Homothallism

| Phenotype | Genotype | | | |
|---|---|---|---|---|
| | I[a] 1949 | II 1970 | III 1974 | IV 1980 |
| Heterothallism | | | | |
| perfect | *[b] *d* | **ho* | **ho* | **ho* |
| conditional | | | | |
| Hp | — | — | *MATa HO HMa hmα* | *MATα HO HMLα HMRα* |
| Hq | — | *MATa HOα hm* | *MATa HO hma HMα* | *MATa HO HMLa HMRa* |
| Homothallism | | | | |
| Ho; type I | — | — | **HO hma hmα* | **HO HMLa HMRα* |
| type II | **D* | **HOα HM* | **HO HMa HMα* | **HO HMLα HMRa* |
| Hp | — | — | *MATa HO HMa hmα* | *MATa HO HMLα HMRα* |
| Hq | — | *MATα HOα hm* | *MATαHO hma HMα* | *MATα HO HMLa HMRa* |

[a]Phase of nomenclature change of the genetic symbols and year of publication.
[b]Asterisk represents either *MAT*a or *MATa*.

sence of nuclear division after chromosome duplication (endomitosis), fusion of mother and daughter nuclei before cytokinesis (direct diploidization), and fusion of two separate cells. Although Winge and Laustsen (1937) suggested that direct diploidization occurred in a Danish baking yeast, Winge (1935) had proposed earlier that diploidization occurs most often by fusion between two cells in a single-spore culture of a homothallic strain. We confirmed the model proposed in 1935 by tracing cell proliferation under a microscope or by taking quick-motion pictures of cell growth in single-spore cultures of homothallic strains with a micro-cinecamera. We found that cell fusion occurs between two cells within a few generations after germination, in general, after the second cell division in single-spore culture (for results of these experiments, see Takano and Oshima 1967).

How do two cells in a spore culture of a homothallic strain fuse to produce a zygote? Winge and Roberts (1958) suggested that cell fusion occurs between cells of the same mating type, unlike matings of heterothallic strains that must be of complementary mating types. Hawthorne (1963a) suggested that the *D* gene acts by causing mutation of one mating-type allele to the other during cell growth; hence, diploidization would occur in homothallic strains just as in heterothallic strains, by fusion of cells of opposite mating types. These arguments were not, however, accompanied by any experimental evidence.

## DIPLOIDIZATION ALWAYS OCCURS BY FUSION OF **a** AND α CELLS

Prior to examining how zygotes are formed during cultivation of a homothallic spore, we investigated mating-type segregation in each ascus of homothallic diploids. To do this, we examined zygote formation by spore-to-cell fusion of each ascospore with a haploid heterothallic cell of known mating type (Oshima and Takano 1971). We observed that not more than two ascospores in each ascus could copulate directly with a given type of heterothallic cell, supporting the 2**a**:2α segregation in each ascus, even in the asci of homothallic diploids.

We next isolated a diploid segregant from a tetrad clone having the *HO*α *hm* genotype of *S. oviformis*. This diploid showed the 2 haploid **a**:2 diploid mating-type segregation, as described above, when colonies derived from its tetrad clones were examined for their thallism and mating type. However, the diploid gave a segregation of 2**a**:2α spores when the spores in each ascus were assayed directly by spore-to-cell mating. Thus, the two diploid clones in each tetrad might have originated from the two spore clones of α mating type. These two diploids segregated heterothallic haploid clones of **a** mating type in subsequent sporulation and dissection. We compared these **a** mating-type clones with the authentic heterothallic strains of **a** mating type but could not detect any significant differences in their behavior in mating and in the gene controlling their mating type. Thus, these **a** mating-type clones should have a *MAT***a** allele converted from the *MAT*α allele (Takano and Oshima 1970b). Similarly, haploid segregants of α mating type from a diploid constructed by a spore-to-cell mating between an α ascospore (derived from a homothallic diploid developed from a spore having the **a** *HO*α *HM* genotype) and an **a** *ho*α *hm* cell had phenotypes indistinguishable from that of the standard α mating-type cells. A gene responsible for α mating type was shown to map at the *MAT* locus on chromosome III (Oshima and Takano 1971). These observations led to the important conclusions that (1) diploidization in a homothallic cell is a result of cell fusion between **a** and α cells, one of which arose from a mating-type switch during the proliferation of cells right after spore germination, and (2) the mating-type switch is due to a change of the mating-type locus itself. In later studies consisting of observing changes from α-factor resistance (α cells are able to grow in the presence of α-factor) to α-factor sensitivity (cell division of **a** cells is arrested in the presence of α-factor), it was shown that mating-type change often occurs at or after the second cell division after spore germination (Hicks and Herskowitz 1976; Hicks et al. 1977b; Strathern and Herskowitz 1979). Two rules of mating-type switching were suggested by the above authors: (1) both cells of any cell division are always of the same mating type and (2) only cells that have budded at least once are competent to switch mating types. Cells that have not budded are rarely if ever observed to switch. Thus, the mating-type switch occurs at or after the second division of the cell following spore germination, and once mating type is switched, both mother and daughter cells have the same **a** or α mating type.

## HOMOTHALLISM GENES

In the first report of our study on homothallism (Takano and Oshima 1967), we described two genes, *HOα*, specific for diploidization of a haploid α mating-type cell, and *HM*, complementary to *HOα* for diploidization of a haploid **a** cell, involved in homothallism. Subsequent allelism tests with the previously known *D* gene and *HM* genes revealed that these genes belonged to the *HOα HM* system (Takano and Oshima 1970a).

A few months after the publication of our allelism tests, we were surprised by a brief report that appeared in the *Yeast News Letter* of June, 1970. J. Santa Maria described a new type of tetrad segregation involving homothallism: A diploid strain, *Saccharomyces norbensis* SBY 2535, showed two heterothallic haploid α clones and two nonmating diploid clones in each ascus. In addition, he also observed that other strains showed diploidization of all four spore clones in each ascus and yet other strains showed segregation of two diploids and two haploid clones of **a** mating type in each ascus. I immediately requested a reprint of their original paper (Santa Maria and Vidal 1970) and the strain *S. norbensis* SBY 2535 and confirmed their observations.

Santa Maria and Vidal proposed, in the same 1970 paper, a new classification of homothallic strains according to their ascus type: Homothallic strains showing a segregation of 4 diploid spore clones:0 haploid spore clone (as in the classic *D* strain) are Ho type, strains showing a segregation of 2 diploid spore clones:2 haploid spore clones of α mating type in each ascus (as in *S. norbensis* SBY 2535) are Hp type, and strains showing a 2 diploid spore clones:2 haploid spore clones of **a** mating type segregation (as in *S. oviformis*) are Hq type.

To investigate the differences in genotypes among these three different types of homothallism, Ho, Hp, and Hq, we performed genetic analyses on multigene hybrids constructed by crossing the α Hp haploid clones segregated from *S. norbensis* SBY 2535 with the **a** *hoα hm*, **a** *hoα HM*, and **a** *HOα hm* clones obtained in the previous study and by crossing spores from *S. norbensis* SBY 2535 and the *HOα HM* homothallic diploid by spore-to-spore mating. The diploids were subjected to tetrad analysis. The spore clones in each ascus were examined for their thallism and for their mating types when the spore clones showed heterothallism. Although detailed data cannot be described here, we found two types (types I and II) of the Ho type of perfect homothallism in addition to the Hp and Hq types of semihomothallism (Oshima and Takano 1972). Genotypes for these four types of homothallic strains and for the heterothallic strains were most adequately explained by postulating three kinds of homothallism genes, consisting of a single pair of alleles, *HO/ho*, *HM***a***/hm***a**, and *HMα/hmα* (for genetic symbols at phase III, see Table 1). In addition, we observed that the diploids obtained by the α Hp x **a** Hq crosses could produce tetrads consisting of four Ho-type homothallic spore clones. These asci were found to consist of two spores of each type, type I and type II (see Table 1), of Ho homothallism. An allelism test showed that type I clones should have the *HO hm***a** *hmα* genotype and that the type II clones should have the *HO HM***a** *HMα* genotype, the same as the classic *D* strain. With this and other observations, we supposed that it was possible that there were duplicate genes for *HM***a** and *HMα* function (Oshima and Takano 1972).

During investigation of the possibility of duplicate genes for the *HM***a** and *HMα* functions, G.I. Naumov and I.I. Tolstorukov, then at the Institute of Genetics and Selection of Microorganisms in Moscow, suggested that the *hmα* allele has the same function as the *HM***a** gene and that *hm***a** has the same function as the *HMα* gene. These authors kindly sent me a reprint of their paper (Naumov and Tolstorukov 1973), but, I confess, I could not understand its meaning at first, because the paper was in Russian (with a short English summary), and wrote back to them with what now looks like nonsense.

In August 20–25 of that year, 1973, the XIIIth International Congress of Genetics was held at the Berkeley campus of the University of California. At the conference, I met Isamu Takano who was, at that time, in the Laboratory of Professor Harlyn O. Halvorson at Brandeis University, Waltham, Massachusetts as a postdoctoral research associate. (Takano had received his Doctor of Science degree from Osaka University in 1970 by presentation of his thesis on homothallism in yeast.) This was the first time I had seen Takano since I had bid him good-bye at Osaka airport in August of 1972. We took the opportunity to visit Robert K. Mortimer at the Donner Laboratory. On the way to the Laboratory, as I recall, Takano began to talk about the Russian paper. While listening to him, I suddenly understood the significance of their paper, even though I could not critically examine their data. On returning to my laboratory from the conference, I asked Satoshi Harashima, who was then a graduate student, to analyze the tetrad data pooled in my laboratory on ascus types and frequencies of segregation of homothallism and mating types and to compare them with the theoretical values. The results were splendid and were well in accord with the expectation based on Naumov and Tolstorukov's idea (Harashima et al. 1974). In the meantime, I received a second letter from Naumov, dated October 15, 1973, further clearing up my confusion.

## MAPPING OF THE *HM*a AND *HMα* GENES

During the statistical analysis of the pooled genetic data, we could detect loose linkage between the *MAT* locus and the *HM***a** and *HMα* loci on chromosome III, whereas *HO* showed no linkage to any of these loci (Harashima et al. 1974). The *HO* locus was later mapped to the left arm of chromosome IV. Since no direct linkage was detected between *HM***a** and *HM*α, we supposed that these genes were separated by the *MAT* locus. It was obvious, however, that conventional mapping functions proposed for tetrad data of dihybrid diploids were not applicable for mapping the *HM*α and *HM***a** genes to their respective positions on the right and left side of the *MAT* locus, because their function in the mating-type switch is allele-specific and the diploids must be heterozygous for the mating-type alleles. These complexities made it impossible to determine allelic 2:2 segregation of the homothallism genes and the mating types in some asci. In addition, the diploids for analysis had to be trihybrids heterozygous for three loci, i.e., *HM***a**/*hm***a** (or *HM*α/*hm*α), *MAT***a**/*MAT*α, and *X*/*x* (a standard marker to be tested for linkage to the *HM***a** or *HM*α locus). It is now easy to map any gene, even *HM***a** and *HM*α, once the gene has been cloned. At that time, however, tetrad or single-spore analysis was the only way to map genes, and mapping functions for such trihybrid segregation did not exist. Thus, we were required by necessity to derive a novel algebraic procedure for three-factor analysis (Harashima and Oshima 1976, 1977). This method was based on calculating the relative frequencies of occurrence of six ascus types with respect to the distribution of homothallism/heterothallism and a standard genetic marker, such as *his4*, *leu2*, and *thr4* on chromosome III, in each ascus by placing the *HM***a** (or *HM*α) gene at an appropriate position on a chromosome. The map position of *HM***a** (or *HM*α) was determined as a site that gave the minimum Chi$^2$ value in a comparison of values for the observed and the theoretical distributions of the six ascus types. By using this procedure, we could map the *HM*α locus on the right arm of chromosome III at a site 65 cM distant from the centromere and the *HM***a** locus 64 cM from the centromere on the left arm (Fig. 1) (Harashima and Oshima 1976).

## CONTROLLING ELEMENT MODEL

During these genetic analyses of homothallism, we suggested several mechanisms as working hypotheses. Initially, we thought that the *HO*α gene produced a cytoplasmic factor, which was working as a copulating device conferring the same function as the product of the **a** mating-type allele in a certain fraction of α *HO*α *hm* cells. However, this hypothesis, which involved a cytoplasmic factor but no switch of the *MAT* gene, was negated by the subsequent observations that homothallism is caused by conversion of a mating-type allele from *MAT***a** to *MAT*α or vice versa (Takano and Oshima 1970b; Oshima and Takano 1971). We were rather surprised and a little bit bewildered at the time with the findings that the **a** and α alleles at the *MAT* locus are mutually interchangeable at high frequency in a regular manner within a few generations after spore germination and that the functions of the homothallic genes are strictly allele-specific. Before these observations, we believed that genes were absolutely stable and did not change their structure, except for the rare occurrence of mutation. We knew of a mutator gene, *mutT*, in *Escherichia coli* (Yanofsky et al. 1966; Cox and Yanofsky 1967), but the mating-type switching of yeast was obviously different and homothallism could not be explained by this analogy. Instead, I thought that there was some resemblance in the locus specificity and the mode of modulation of mating-

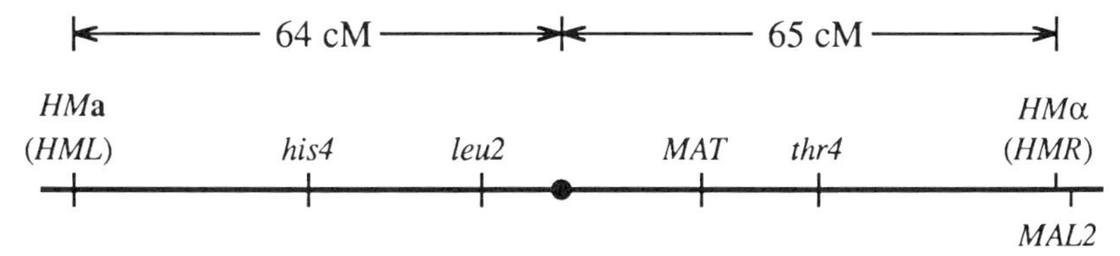


*Figure 1* Mapping of *HMα* and *HM***a** loci on chromosome III. Map positions are drawn to approximate genetic scale according to Harashima and Oshima (1976), with selected markers on the map compiled by Mortimer et al. (1989).

type switching with the specific transposable genetic controllers of mutation described by Barbara McClintock (1957) in maize. But not many transposable genetic elements were known at the beginning of 1970s. Although they were not strictly the same, we were also encouraged by the existence of mutable genes in *Drosophila melanogaster* (Green 1967, 1969a,b).

With these analogies, we proposed the controlling element model, in which the elementary structure of the mating-type locus for both **a** and α is essentially the same. The association of some kind of controlling element with this locus would cause differentiation of two mating-type alleles (Oshima and Takano 1971). One of the two homothallism genes, *HOα* or *HM*, could be concerned with the production of a controlling element; the other gene could control the association of the controlling element with its affinity site. Thus, the mutagenic event at the mating-type locus appeared to consist of several steps. In *HOα HM* cells, the event may go to completion, whereas it is interrupted at some step in haploid cells of *HOα hm*, *hoα HM*, and *hoα hm* genotypes. Activities of the homothallism genes would be blocked as soon as heterozygosity of the mating-type alleles was established by zygote formation between cells with complementary mating types.

This initial model was revised subsequently (Harashima et al. 1974) in parallel with the revision of the nomenclature of the homothallism genes from phase II to phase III (see Table 1). In the revised model, the *HM***a** and *HMα* genes produced the specific controlling elements; the association of an *HM***a** element with the mating-type locus formed the α mating-type allele and the association of an *HMα* element with the mating-type locus gave rise to the **a** mating-type allele. In this case, genetic symbols were defined as follows: *HM***a** is able to diploidize **a** cells, and *HMα* is able to diploidize α cells. In other words, *HM***a** provides α-specific information but is silent at its locus, and *HMα* is silent **a**-specific information. These genes are expressed only when their information is inserted at the *MAT* locus.

The proof that Naumov and Tolstorukov's idea was right was the observation of codominance, or functional equivalence, of the *hm***a** and *hmα* alleles with their respective counterparts; diploid cells homozygous for mating type were constructed for this experiment by reciprocal mitotic recombination (Klar and Fogel 1977), by protoplast fusion (Arima and Takano 1979), and by tetraploid segregation (Harashima and Oshima 1980). With these advances, Naumov and Tolstorukov's idea was included in the controlling element model: The *hm***a** and *hmα* alleles should have the same function as the *HMα* and *HM***a** genes but at the *HM***a** and *HMα* loci, respectively (Harashima and Oshima 1980).

Molecular mechanisms other than the controlling element model were also proposed to explain mating-type conversion. These models included the so-called flip-flop model (Hicks and Herskowitz 1977). In this case, both **a** and α information is present at the *MAT* locus; DNA modification or intramolecular recombination at a hypothetical control site at the *MAT* locus allows expression of either the *MAT***a** or *MAT*α cistron. The *HO*, *HM***a**, and *HM*α genes would code for enzymes involved in the modification or rearrangement of the DNA sequence at the control site of the *MAT* locus.

I believed, however, that the flip-flop models were not correct because we knew that the *MAT*α*-inc* allele was effectively healed during mating-type conversion. The *MAT*α*-inc* allele was discovered in a strain of *Saccharomyces diastaticus* as an α mating-type allele that is insensitive to the action of the homothallic genes; it converted to *MAT***a** only at low frequency by the function of the homothallic genes (Takano et al. 1973; Takano and Arima 1979). Strikingly, the *MAT***a** clones derived from the *MAT*α*-inc* allele behaved as a normal *MAT***a** allele and produced a normal *MAT*α allele via a subsequent mating-type switch by the homothallism genes (Takano et al. 1973). I thought that the *HM***a** element, and subsequently the *HM*α element, replaced the *MAT*α*-inc* allele. The experiment with the *MAT*α*-inc* allele is analogous to subsequent experiments with a defective α *ste* mutant allele of the *MAT* locus (Hicks and Herskowitz 1977), with a *mat*α mutation (Strathern et al. 1979a), with a *mat***a**-2 mutation (Klar et al. 1979b), and with a *MAT***a***-inc* mutation (Mascioli and Haber 1980), and many other experiments. The notion that an *HM* element replaces the *MAT* allele was further strengthened by the isolation of a class of mutants containing *mar1* (Klar et al. 1979a), *sir1* (Rine et al. 1979), or *cmt* (Haber and George 1979) mutations. These mutations confer an in situ expression of the silent mating-type information at the *HM***a** and *HM*α loci. These findings indicated that the *HM***a**/*hm***a** and *HM*α/*hm*α genes are the structural genes of the **a** and α mating types but are not necessary elements for the expression of the mating-type alleles.

With the above evidence, revision of the nomenclature of the homothallism genes was discussed by several yeast geneticists at the IXth International Conference on Yeast Genetics and Molecular Biology, held at the University of Rochester in mid June, 1978. Accordingly, the *HM***a** and *hm***a** alleles, which are located on the left arm of chromosome III (see Fig. 1) were denoted *HML*α and *HML***a**, respectively. *HML*α, for example, is the allele of homothallic genes on the left arm that bears a silent copy of α information. Similarly, the *HM*α and *hm*α alleles, which are on the right arm of chromosome III, were denoted *HMR***a** and *HMR*α, respectively. The nomenclature of *HO*/*ho* remained as it was (dominant allele, *HO*; recessive allele, *ho*) (for genetic symbols at phase IV, see Table 1).

Subsequent molecular analysis of the homothallism genes—cloning of *HML*α (Hicks et al. 1979) and *MAT*α (Nasmyth and Tatchell 1980) and DNA sequencing of the *MAT***a** and *HML*α genes (Astell et al. 1981)—showed strict conformity to the controlling element model. The arrangement of the *HML*, *MAT*, and *HMR* loci on chromosome III and their structural similarities revealed by the physical analysis of the DNAs clearly supported transposition in mating-type interconversion. The physical structures of these genes also explained the Hawthorne (1963b) deletion—an α to **a** mutation accompanied by deletion of a large chromosomal segment bearing the *THR4* locus—as due to recombination between the homologous regions of *MATα* and *HMR***a** (Herskowitz 1988). This deletion indicated that the promoter region of *MATα* might be connected with the structural region of *HMR***a**. Similar recombination between *HML* and *MAT* can produce a circular chromosome because the intervening region contains the centromere (Strathern et al. 1979b).

With the most recent revision of the genetic symbols, it is now clear that some strains in the pedigree of *S. oviformis* have only **a** information at the *HML* and *HMR* loci. Thus, the α to **a** switching can occur in the two α spore clones, whereas no mating-type switch occurs in the two **a** spore clones. This gives rise to the 2 diploid:2 haploid **a** mating-type segregation in each ascus. In contrast, *S. norbensis* SBY 2535 has only α information for both silent copies of the mating type. According to our experience, most (if not all) of the strains of *S. cerevisiae* and the related species of this yeast have α information in the *HML* locus and **a** information in the *HMR* locus. Hence, the strains of *S. oviformis* and *S. norbensis* are rare natural variants. These modifications are expected to be derived by transfer of mating-type information from *HMR* to *HML* or vice versa. This type of information transfer is known to occur in *HO* cells when the silent *HML* and *HMR* genes are expressed due to *sir* (or *mar1*) mutations (Klar et al. 1981).

## EPILOGUE

On looking back over the initial period of the studies on homothallism, it was a strict extension of the classic yeast genetics from the period of Winge and Lindegren, because we never used mutants in homothallism. The principal strains were those of *S. oviformis* and *S. norbensis* isolated from their natural habitats. Interestingly, the original isolation of both of these strains was in Spain. Most of our achievements were due to the almost simultaneous appearance of these strains and a strain of *S. diastaticus*. It was, however, not surprising, because at that time, several staff members in the Research Laboratory of Suntory Ltd. were engaged in the collection and critical examination of industrial yeasts from various sources. Substantial parts of our achievement resulted from their efforts and the willingness of J. Santa Maria and others to provide their strains.

Another characteristic feature of the study is genetic analysis. We employed most of the genetic techniques available in *Saccharomyces* yeasts: the dominance/recessiveness test, complementation test, allelism test on dominant genes, epistasis/hypostasis test, and linkage analysis of tetrad and single-spore data from trihybrids, as well as of the conventional tetrad data from dihybrids. For functional analysis of individual genes in a multigene system, a procedure based on comparison of observed and theoretical distributions of various ascus types and their frequencies of occurrence in tetrad data of multigene hybrids was extremely useful. A typical example of the multigene analysis was described in Oshima and Takano (1972). Hybrids were also constructed, along with the regular mating between **a** and α cells, by protoplast fusion between two strains of the same mating type or with nonmaters, and by direct cell-to-spore mating. Without such techniques, homothallism could not have been analyzed. In other words, elucidation of the mechanism of homothallism was possible only in *S. cerevisiae*.

Regarding the construction of polyploids, we could achieve construction of tetraploids by generating **a**/**a** or α/α cells from **a**/α diploids by mitotic recombination and by mating-type switching as in haploid cells (Takagi et al. 1983). Once such tetraploids were obtained, further construction of tetraploid and triploid hybrids was easy, because the initial tetraploids segregate diploid maters with the **a**/**a** and α/α genotypes by sporulation. We confirmed heterosis in yeast strains for alcoholic fermentation and that heterozygous polyploids are more vigorous than autopolyploids (Takagi et al. 1983). Even higher heterogeneities for genetic information are possible in polyploids than in diploids, but the reason why polyploidy is favorable in brewing and baking yeasts is still unknown. Such polyploids in industrial strains of *S. cerevisiae* are, however, more easily constructed by a protoplast fusion technique (Harashima et al. 1984; Takagi et al. 1985).

The mechanism of homothallism elucidated by this study was, however, far more significant than the initial aim. Although it is now well known that some genes are differentiated by rearrangement of DNA segments in the genome during developmental processes of the cell, we were convinced that the mating-type switch of *Saccharomyces* yeasts was one of the typical examples of this mechanism. Although our contributions in the analysis of homothallism and mating-type switching were limited to the initial phase of model building, further detailed mechanisms in homothallism were elucidated in the subsequent cellular and molecular analyses in several laboratories, especially in the laboratory of Ira Herskowitz at the University of Oregon and later at the University of California at San Francisco, and by James B. Hicks, Amar J.S. Klar, and Jeffrey N. Strathern at the Cold Spring Harbor Laboratory, and many others. With these advances, Herskowitz and his colleagues proposed a revised version of the controlling element model, the cassette model, in 1977 (Hicks and Herskowitz 1977; Hicks et al. 1977a).

## ACKNOWLEDGMENTS

I thank Isamu Takano and Satoshi Harashima for comments, and Ira Herskowitz and Reed B. Wickner for their comments and critical reading of the manuscript. This study has been supported by research grants from the Ministry of Education, Science, and Culture of Japan, and in part by Suntory Ltd.

## REFERENCES

Arima, K. and I. Takano. 1979. Evidence for co-dominance of the homothallic genes, *HM*α/*hm*α and *HM***a**/*hm***a**, in *Saccharomyces* yeasts. *Genetics* **93:** 1–12.

Astell, C.R., L. Ahlstrom-Jonasson, M. Smith, K. Tatchell, K.A. Nasmyth, and B.D. Hall. 1981. The sequence of the DNAs coding for the mating-type loci of *Saccharomyces cerevisiae*. *Cell* **27:** 15–23.

Blakeslee, A.F. 1904a. Sexual reproduction in the Mucorineae. *Proc. Am. Acad. Arts Sci.* **40:** 205–319.

———. 1904b. Zygospore formation, a sexual process. *Science* **19:** 864–866.

Cox, E.C. and C. Yanofsky. 1967. Altered base ratios in the DNA of an *Escherichia coli* mutator strain. *Proc. Natl. Acad. Sci.* **58:** 1895–1902.

Green, M.M. 1967. The genetics of a mutable gene at the *white* locus of *Drosophila melanogaster*. *Genetics* **56:** 467–482.

———. 1969a. Mapping a *Drosophila melanogaster* "controlling element" by interallelic

crossing over. *Genetics* **61:** 423–428.
———. 1969b. Controlling element mediated transpositions of the *white* gene in *Drosophila melanogaster*. *Genetics* **61:** 429–441.
Haber, J.E. and J.P. George. 1979. A mutation that permits the expression of normally silent copies of mating-type information in *Saccharomyces cerevisiae*. *Genetics* **93:** 13–35.
Harashima, S. and Y. Oshima. 1976. Mapping of the homothallic genes, *HMα* and *HM***a**, in *Saccharomyces* yeasts. *Genetics* **84:** 437–451.
———. 1977. Frequencies of twelve ascus-types and arrangement of three genes from tetrad data. *Genetics* **86:** 535–552.
———. 1980. Functional equivalence and co-dominance of homothallic genes *HMα*/*hmα* and *HM***a**/*hm***a** in *Saccharomyces* yeasts. *Genetics* **95:** 819–831.
Harashima, S., Y. Nogi, and Y. Oshima. 1974. The genetic system controlling homothallism in *Saccharomyces* yeasts. *Genetics* **77:** 639–650.
Harashima, S., A. Takagi, and Y. Oshima. 1984. Transformation of protoplasted yeast cells is directly associated with cell fusion. *Mol. Cell. Biol.* **4:** 771–778.
Hawthorne, D.C. 1963a. Directed mutation of the mating type alleles as an explanation of homothallism in yeast. *Proc. Int. Congr. Genet.* **1:** 34–35. (Abstr.)
———. 1963b. A deletion in yeast and its bearing on the structure of the mating type locus. *Genetics* **48:** 1727–1729.
Herskowitz, I. 1988. The Hawthorne deletion twenty-five years later. *Genetics* **120:** 857–861.
Hicks, J.B. and I. Herskowitz. 1976. Interconversion of yeast mating types. I. Direct observations of the action of the homothallism (*HO*) gene. *Genetics* **83:** 245–258.
———. 1977. Interconversion of yeast mating types. II. Restoration of mating ability to sterile mutants in homothallic and heterothallic strains. *Genetics* **85:** 373–393.
Hicks, J.B., J.N. Strathern, and I. Herskowitz. 1977a. The cassette model of mating-type interconversion. In *DNA insertion elements, plasmids, and episomes* (ed. A.I. Bkhari et al.), pp. 457–462. Cold Spring Harbor Laboratory, Cold Spring Harbor, New York.
———. 1977b. Interconversion of yeast mating types. III. Action of the homothallism (*HO*) gene in cells homozygous for the mating type locus. *Genetics* **85:** 395–405.
Hicks, J.B., J.N. Strathern, and A.J.S. Klar. 1979. Transposable mating type genes in *Saccharomyces cerevisiae*. *Nature* **282:** 478–483.
Klar, A.J.S. and S. Fogel. 1977. The action of homothallism genes in *Saccharomyces* diploids during vegetative growth and the equivalence of *hm***a** and *HMα* loci functions. *Genetics* **85:** 407–416.
Klar, A.J.S., S. Fogel, and K. MacLeod. 1979a. *MAR1*—A regulator of the *HM***a** and *HMα* loci in *Saccharomyces cerevisiae*. *Genetics* **93:** 37–50.
Klar, A.J.S., S. Fogel, and D.N. Radin. 1979b. Switching of a mating-type **a** mutant allele in budding yeast *Saccharomyces cerevisiae*. *Genetics* **92:** 759–776.
Klar, A.J.S., J.N. Strathern, and J.B. Hicks. 1981. A position-effect control for gene transposition: State of expression of yeast mating-type genes affects their ability to switch. *Cell* **25:** 517–524.
Lindegren, C.C. and G. Lindegren. 1943. A new method for hybridizing yeast. *Proc. Natl. Acad. Sci.* **29:** 306–308.
Mascioli, D.W. and J.E. Haber. 1980. A *cis*-acting mutation within the *MAT***a** locus of *Saccharomyces cerevisiae* that prevents efficient homothallic mating-type switching. *Genetics* **94:** 341–360.
McClintock, B. 1957. Controlling elements and the gene. *Cold Spring Harbor Symp. Quant. Biol.* **21:** 197–216.
Mortimer, R.K., D. Schild, C.R. Contopoulou, and J.A. Kans. 1989. Genetic map of *Saccharomyces cerevisiae*, edition 10. *Yeast* **5:** 321–403.
Nasmyth, K.A. and K. Tatchell. 1980. The structure of transposable yeast mating type loci. *Cell* **19:** 753–764.
Naumov, G.I. and I.I. Tolstorukov. 1973. Comparative genetics of yeast. X. Reidentification of mutators of mating types in *Saccharomyces*. *Genetika* **9:** 82–91.
Oshima, Y. and I. Takano. 1971. Mating types in *Saccharomyces*: Their convertibility and homothallism. *Genetics* **67:** 327–335.
———. 1972. Genetic controlling system for homothallism and a novel method for

breeding triploid cells in *Saccharomyces*. In *Proceedings of the 4th International Fermentation Symposium* (ed. G. Terui), pp. 847–852. Society of Fermentation Technology, Osaka, Japan.

Rine, J., J.N. Strathern, J.B. Hicks, and I. Herskowitz. 1979. A suppressor of mating-type locus mutations in *Saccharomyces cerevisiae*: Evidence for and identification of cryptic mating-type loci. *Genetics* **93:** 877–901.

Santa Maria, J. and D. Vidal. 1970. Segregacion anormal del "mating type" en *Saccharomyces. Inst. Nac. Invest. Agron.* **30:** 1–21.

Strathern, J.N. and I. Herskowitz. 1979. Asymmetry and directionality in production of new cell types during clonal growth: The switching pattern of homothallic yeast. *Cell* **17:** 371–381.

Strathern, J.N., L.C. Blair, and I. Herskowitz. 1979a. Healing of *mat* mutations and control of mating type interconversion by the mating type locus in *Saccharomyces cerevisiae. Proc. Natl. Acad. Sci.* **76:** 3425–3429.

Strathern, J.N., C.S. Newlon, I. Herskowitz, and J.B. Hicks. 1979b. Isolation of a circular derivative of yeast chromosome III: Implications for the mechanism of mating type interconversion. *Cell* **18:** 309–319.

Takagi, A., S. Harashima, and Y. Oshima. 1983. Construction and characterization of isogenic series of *Saccharomyces cerevisiae* polyploid strains. *Appl. Environ. Microbiol.* **45:** 1034–1040.

———. 1985. Hybridization and polyploidization of *Saccharomyces cerevisiae* strains by transformation-associated cell fusion. *Appl. Environ. Microbiol.* **49:** 244–246.

Takahashi, T. 1958. Complementary genes controlling homothallism in *Saccharomyces. Genetics* **43:** 705–714.

Takahashi, T. and Y. Ikeda. 1959. Bisexual mating reaction in *Saccharomyces chevalieri. Genetics* **44:** 375–382.

Takahashi, T., H. Saito, and Y. Ikeda. 1958. Heterothallic behavior of a homothallic strain in *Saccharomyces* yeast. *Genetics* **43:** 249–260.

Takano, I. and K. Arima. 1979. Evidence of the insensitivity of the α-*inc* allele to the function of the homothallic genes in *Saccharomyces* yeasts. *Genetics* **91:** 245–254.

Takano, I. and Y. Oshima. 1967. An allele specific and a complementary determinant controlling homothallism in *Saccharomyces oviformis. Genetics* **57:** 875–885.

———. 1970a. Allelism tests among various homothallism-controlling genes and gene systems in *Saccharomyces. Genetics* **64:** 229–238.

———. 1970b. Mutational nature of an allele-specific conversion of the mating type by the homothallic gene *HOα* in *Saccharomyces. Genetics* **65:** 421–427.

Takano, I., T. Kusumi, and Y. Oshima. 1973. An α mating-type allele insensitive to the mutagenic action of the homothallic gene system in *Saccharomyces diastaticus. Mol. Gen. Genet.* **126:** 19–28.

Townsend, G.F. and C.C. Lindegren. 1954. Characteristic growth patterns of the different members of a polyploid series of *Saccharomyces. J. Bacteriol.* **67:** 480–483.

Winge, Ø. 1935. On haplophase and diplophase in some Saccharomycetes. *C.R. Lab. Carlsberg Ser. Physiol.* **21:** 77–109.

Winge, Ø. and O. Laustsen. 1937. On two types of spore germination, and on genetic segregation in *Saccharomyces*, demonstrated through single-spore culture. *C.R. Trav. Lab. Carlsberg Ser. Physiol.* **22:** 99–117.

Winge, Ø. and C. Roberts. 1949. A gene for diploidization in yeast. *C.R. Trav. Lab. Carlsberg Ser. Physiol.* **24:** 341–346.

———. 1958. Yeast genetics. In *The chemistry and biology of yeasts* (ed. A.H. Cook), pp. 123–156. Academic Press, New York.

Yanofsky, C., E.C. Cox, and V. Horn. 1966. The unusual mutagenic specificity of an *E. coli* mutator gene. *Proc. Natl. Acad. Sci.* **55:** 274–281.

# CELL CYCLE

D.H. Williamson (Louvain-la-Neuve, Belgium 1980)

J.M. Mitchison and B. Stevens (Fishbourne, England 1974)

# Getting Started in the Cell Cycle

LELAND H. HARTWELL
*Department of Genetics, SK-50*
*University of Washington*
*Seattle, Washington 98195*

The most exciting time in my scientific career began in 1969 when we discovered yeast *cdc* mutants. I have therefore chosen to use this opportunity to recall the events of the first 4 years of our work with *cdc* mutants and the intellectual motivations that encouraged it. The work grew out of an intense collaboration primarily between three graduates students, Joe Culotti, Lynna Hereford, and Brian Reid, a postdoctoral fellow, John Pringle, a technician, Marilyn Culotti, and myself. Each new result generated vigorous discussions so that it is impossible to know who contributed a particular idea. By the end of 1973, most of the concepts that have guided the work of my laboratory since had been formulated (Hartwell et al. 1974), although some of the work was not published until several years later. We had characterized the functions of 32 *cdc* genes. Each of their products appeared to act at only one stage of the cell cycle. The phenotypes of the mutants revealed that the events of the cycle were organized into two pathways, one leading to the nuclear events and the other leading to budding and cytokinesis. The pathways of events exhibited an obligatory order such that early events were prerequisites for later events. The $G_1$ interval was divided into three successive steps. The first step, controlled by *CDC28* and termed Start, was essential for both dependent pathways and was also the site at which cell division was controlled by nutrition and mating pheromones. Finally, *cdc4* mutants revealed a timer that phased successive budding events.

## PRELUDE TO YEAST

My interest in eukaryotic cell division and its control came out of my postdoctoral work in the laboratory of Renato Dulbecco. During that time, under the guidance of Marguerite Vogt, I studied a variety of cell culture systems, trying to get an experimental handle on the control of cell division, but none of my attempts led anywhere. I also worked with Renato and Marguerite in their ongoing studies on the tumor virus, polyoma. I became aware of the complexity of cell division control and also intrigued by the idea that the addition of a few viral genes to a mammalian cell could overcome these elaborate controls. Discouraged with mammalian cells, but irrevocably engaged with the problem of cell division control, I left the Salk Institute in 1965 and took a job as an assistant professor at the newly created University of California at Irvine.

## YEAST BEGINNINGS

The experience with mammalian cells led me to return to my original imprinting experience, working as an undergraduate on T4 genetics in the Delbrück group under the direction of Bob Edgar. I had observed over the years how much Bob Edgar and Bill Wood had learned about T4 morphogenesis using conditional mutants. Cell division appeared to be largely a morphogenetic process as well, and it seemed like the same approach should work. I was thinking this over shortly after moving to Irvine at one of those rare times when I was forced to think because there was no way to work. I had obtained a grant to study the control of DNA replication in mammalian cells and had ordered the appropriate equipment. I had a couple of months to wait for everything to arrive and little to do in the interim except read and think. I discussed my desire to do genetics on the problem of cell division control with Dan Wulff, also a newly arrived assistant professor at Irvine, and he suggested that I find a model eukaryotic cell that would lend itself to a genetic approach. I thought that was a great idea and immediately went back to the library. Most eukaryotic genetics was being done with *Neurospora* and yeast. Since I wanted to do cell physiology along with genetics, I needed a single-celled organism. Hence, yeast was the only possible choice.

At that point, I made a list of questions about yeast that I could not find answers to in the literature. The only one I can remember now was why no one used radioactive thymidine to label yeast DNA. I called Herschel Roman, one of the prominent names in the literature and began going through my list. On about the fourth question, Herschel, in his typically insightful way, realized the depth of my ignorance and said, "I think you should come up for a visit." Soon after that I spent a couple of days with Bob Mortimer, at the University of California, Berkeley, and then with Herschel Roman at the newly formed Department of Genetics at the University of Washington, receiving what was to be my only formal training in yeast.

In Herschel's laboratory I visited with Shelly Esposito, then a graduate student, Don Hawthorne, and Herschel. Shelly provided a strain that met my requirements, having amino acid and purine auxotrophy (for the purpose of following protein and nucleic acid synthesis with radioactive labels), and being nonclumpy (for measuring cell division). Someone showed me how to dissect asci, and with incredible generosity, Herschel and Don sent me home with a micromanipulator on loan. Don is well known in the yeast community for including puzzles in the strains he sends out that are commensurate with your ability. I went home to do genetics with two *MAT***a** strains; soon I was able to obtain a *MAT*α strain as well.

## YEAST MUTANTS

I began by isolating 400 temperature-sensitive mutants of yeast and, together with the help of an undergraduate student, Pamela Michaels, characterized them all for protein synthesis, RNA synthesis, DNA synthesis, cell division, and cell morphology (Hartwell 1967). Calvin McLaughlin joined us at Irvine at about that time, and because of his expertise in protein synthesis, we set out to try to determine the molecular defects in some of the protein synthesis mutants. He identified two mutants with temperature-sensitive tRNA synthetase, and this result gave me confidence that it would be possible to trace down molecular defects in the mutants. Another important event during this period was a visit from my former undergraduate mentor, Bob Edgar, who expressed encouraging enthusiasm for the mutants and gave me some good advice. Namely, that I would learn far more about them if I shared them freely.

## CELL DIVISION MUTANTS

We recognized a few DNA replication mutants and cell division mutants in our original characterization of temperature-sensitive mutants (Hartwell 1967), but they were rare and we did not devote our attention to them until they showed up more prominently in an unexpected way.

In 1968, I accepted the invitation of Herschel Roman to join the Department of Genetics at the University of Washington, thinking that being in a genetics department would be a good influence on me since I had little formal genetics training. A year later, in the summer of 1969, an undergraduate chemistry major, Brian Reid, serendipitously discovered a rapid way to detect *cdc* mutants. Brian's project was to see if yeast exhibited the phenomenon of cortical inheritance discovered by Tracy Sonneborn with *Paramecium*. Sonneborn had found that if the cortex of a *Paramecium* cell was surgically rotated, then the progeny of that cell also had a patch of rotated cilia. We knew that some of our yeast mutants formed odd shapes at the restrictive temperature, and we wondered if they would continue to propagate the distorted shape after a return to the permissive temperature. Brian quickly discovered that clones of mutant cells grown up at the permissive temperature after the cells had been exposed to the restrictive temperature were normal in shape. Not willing to leave it there, Brian decided to look at the first few divisions and record his observations by photomicroscopy. When Brian and I examined his pictures, we realized how much information was there about cell cycle behavior. Because yeast cells bud, one could order the cells in the cell cycle at early time points by bud size, and then one could see how each cell developed at the restrictive temperature from later time points.

Brian began screening many temperature-sensitive mutants photographically, spending most of the summer in a bathing suit in the warm room. Although some potential candidates turned up, we were not sure they were cell cycle mutants until a new graduate student, Joe Culotti, joined the laboratory in the fall of 1969 and succeeded in visualizing the nucleus by cytological techniques (Hartwell et al. 1970). We contacted C.F. Robinow, who had pioneered yeast cytology, for advice, and he generously offered to visit the laboratory and show us how to stain nuclei since he was coming to Seattle soon for another meeting. Joe had collected all the reagents that Robinow required but discovered to his chagrin on the appointed morning that he had forgotten to grow any cells with which to work! Joe scurried around the building and eventually found a culture growing in another laboratory. Over the next year, Mary Ashton-Hill, an undergraduate student, Joe Culotti, and myself exploited photomicroscopy to find 148 cell cycle mutants in 32 genes and define their execution points and terminal phenotypes (Hartwell et al. 1973). Ironically, the first *cdc* gene so designated, *cdc1*, turned out not to be a good cell cycle mutant, since the cells arrested heterogeneously, some with small buds and others without, some having replicated DNA and others not; we were a little too anxious to find *cdc* mutants.

## SYNCHRONOUS CULTURES

Assays for most of the cell cycle events, DNA replication, nuclear migration, nuclear division, cytokinesis and cell division, were at hand or easily developed. However, we also needed a rapid technique for synchronizing mutant strains so that we could unambiguously determine which of these events stopped first. The best technique at the time was one described, ironically, as "A rapid method for synchronizing division" that required several days of media shifts (Williamson and Scopes 1962). I think the idea of using Renografin gradients came from the fact that they had been used to separate different organelles or cell types, but I do not think they had been used to fractionate cells within the cell cycle. In any case, we were lucky; when dividing yeast cells were banded in Renografin gradients, they formed a broad band containing cells with small buds at the top and cells with large buds at the bottom. Either fraction yielded large quantities of sufficiently synchronized cultures to meet our needs for mutant analysis (Hartwell 1970). Dick Shulman, a postdoc who arrived in 1970, also exploited this technique for analyzing the time of synthesis of components during the cell cycle.

## DEPENDENT PATHWAYS

In addition to defining the stage of action of each gene, the results of these experiments revealed that the events in the cell cycle were ordered into dependent pathways. When a mutant arrested at one early step, it usually did not complete steps that normally came later. This behavior paralleled exactly what had been learned from the genetic analysis of T4 morphogenesis and reinforced my conviction that cell division was more of a morphogenetic problem than one of gene regulation. We made predictions of the phenotypes of double mutants based on this model and tested many double mutant combinations; all conformed to our expectations. Two dependent pathways, one leading to DNA replication and nuclear division and the other to bud emergence and cytokinesis, were established by 1971. In contrast to the rapid dissection of the dependent pathways, it was 18 more years before we gained any insight into how the dependence of events was attained (Hartwell and Weinert 1989).

## MAP LOCATIONS

All respectable geneticists know that they must map their genes. I had put one yeast gene on the map, *ils1*, and I knew it was a tedious and chancy business. In those days, if a gene was not centromere-linked, one kept doing crosses hoping that the gene showed linkage to some random marker in the cross. Fortunately, Bob Mortimer spent the 1970–1971 academic year on sabbatical in Seattle working with Don Hawthorne. Bob was interested in developing new mapping strategies using mitotic recombination and aneuploidy methods to locate new genes. All I had to do was supply Bob with the *cdc* mutants, and he placed 14 on the map (Hartwell et al. 1973).

## *CDC28*

The *CDC28* gene of *Saccharomyces cerevisiae* and the homologous gene, *CDC2*, of *Schizosaccharomyces pombe* are currently seen as playing a major role in cell cycle control. The identification of these two genes had a precarious history. Only one allele of *CDC28* was detected among the 150 *cdc* mutants; it was easily recognized since the cells arrested cleanly as mononucleate, unbudded cells at the restrictive temperature. However, the first mutant was a double mutant containing both *cdc28* and *cdc1* alleles. When the *cdc1* mutation was removed, the *cdc28* mutant became morphologically aberrant quickly at the restrictive temperature, and it is likely that we would not have recognized it as a cell cycle mutant without this secondary mutation. Ironically, the homologous gene of *S. pombe* apparently had an equally tenuous beginning. Of 25 small mutants, 24 were in the *wee1* gene and only 1 identified a *wee* allele of the *cdc2* gene (Thuriaux et al. 1978). However, this allele appears to have been critical in focusing attention on the *cdc2* gene as an important element in mitotic control.

## α-FACTOR

In 1970, Duntze, MacKay, and Manney published a paper showing that purified mating pheromone differentially arrested DNA replication relative to RNA and protein synthesis (Duntze et al. 1970). This was a clear sign that α-factor was having a cell cycle effect. Tom Manney sent us the protocol for preparing α-factor, and I invited him to visit so that together we could investigate the cell cycle effects of α-factor. Lynna Hereford spent a month preparing α-factor from a large fermentor run and had finally obtained 100 ml of concentrated factor the day before Manney's arrival. When Lynna went to retrieve it the next morning, she found a huge mold growing in the flask that had been substituted as a joke by Joe and Brian. We quickly found that α-factor arrested the cell cycle in $G_1$ and that the cells had the phenotype of *CDC28* mutants. Furthermore, if *cdc* mutants were first synchronized at the α-factor block and then released at high temperature, they all arrested in the first cell cycle, indicating that none of the *CDC* steps could be completed at the α-factor block (Bücking-Throm et al. 1973).

## **a**-FACTOR

It seemed likely that two conjugating cells would need to synchronize their cell cycles and that α-factor was one part of the regulatory system to achieve this end. However, the complementary **a**-factor had not been detected. I decided to try to determine whether the two cells were both at the same stage of the cell cycle at the time of cell fusion. This was somewhat complicated to determine because cells mated under the same conditions that promoted division, and in a mating mixture, cell fusion and cell division were going on at the same time. A *cdc* mutant solved this dilemma. By carrying out the mating at the restrictive temperature for a nuclear division mutant, I could limit events to one cycle. The analysis clearly showed that both cells were unbudded at the time of fusion (Hartwell 1973). Linda Wilkinson, a graduate student who joined the laboratory in 1971, then set out with the help of John

Pringle to find the elusive **a**-factor; together they showed that concentrated culture medium from *MAT***a** cells caused $G_1$ arrest at the *cdc28* step of *MAT*α cells (Wilkinson and Pringle 1974).

## CONJUGATION

Brian Reid had gone to the Massachusetts Institute of Technology for graduate school, but after a year returned to the laboratory in the fall of 1970 to do his thesis in my laboratory. We had learned that cells normally synchronize themselves in $G_1$ at the *CDC28* step for cell fusion during conjugation. Brian decided to ask whether cell fusion was restricted to that stage of the cell cycle. By synchronizing *cdc* mutants at other stages and then challenging them to mate, he was able to show that cells could conjugate from the *CDC28* step but not at any other step (Reid and Hartwell 1977). Thus, the *CDC28* step marked not only the position of arrest for mating, but also a point beyond which fusion was not permitted.

## RECIPROCAL SHIFTS

The finding that α-factor prevented budding in the *cdc4* and *cdc7* mutants suggested that the α-factor block preceded the *cdc4* and *cdc7* blocks in a pathway. These results led Joe and I to devise a method for determining the order of gene product function that underlies the events themselves. In thinking this through, we realized that there were four possible orders of two events. Epistasis experiments using double mutants were capable of resolving only two of the alternatives, whereas the new method, called reciprocal shifts, could resolve all four. This method, which was independently developed by Jarvik and Botstein (1973), required two different conditional blocks. We used heat-sensitive mutants with stage-specific cell cycle inhibitors; Botstein and Jarvik used heat-sensitive mutants in combination with cold-sensitive mutants in phage P22.

This method was used by Lynna Hereford to map the order of gene functions in $G_1$ relative to the α-factor block (Hereford and Hartwell 1974). Lynna's results showed that *CDC28*, *CDC4*, and *CDC7* functioned in that order in $G_1$, with the *CDC28* step being interdependent with the α-factor-sensitive step. The order deduced by Lynna was supported by the facts that *cdc28* mutants were unbudded and *cdc4* and *cdc7* mutants were budded and that mutants arrested at the *cdc28* or *cdc7* step required protein synthesis before they could replicate DNA and *cdc7* mutants did not.

## SPINDLE POLE BODIES

Around 1971 or 1972, I was trying to talk Dick McIntosh into reconstructing the ultrastructures of all of the *cdc* mutants by electron microscopy of serial thin sections. I did not have anything specific in mind but thought that something interesting would come out of it. I discussed this with Breck Byers who was working on the ultrastructure of ribosome crystals at the time and he pointed out, in his perspicacious way, that the only structure of interest was the spindle pole body. This was the beginning of a long and fruitful collaboration. Breck defined the morphogenesis of the spindle pole body during mitosis and conjugation as well as in all of the *cdc* mutants (Byers and Goetsch 1975). He showed that it was a single pole in *cdc28* mutants, a side-by-side double pole in *cdc4* mutants, and formed separated poles with a complete spindle in *cdc7* mutants. These results were congruent with Lynna's on the order of function and further strengthened my view of the cell cycle as a dependent series of morphogenetic events.

## GROWTH

John Pringle discovered through a bit of serendipity that stationary-phase yeast cells were also arrested in $G_1$, at or before the *CDC28* step. We routinely inoculated several dilutions of cells the night before an experiment and used one that was not dense the next day to be sure that cells were in exponential growth. John utilized a culture that had gotten too dense and noticed that the cells were all unbudded. Upon questioning Herschel Roman about this phenomenon, he was told that some strains did this and some did not. John examined a variety of strains and found that if they were sonicated to separate individual cells, then all strains arrested as unbudded cells in stationary phase. Although I was disposed to expect cell cycle control in $G_1$ as a result of my familiarity with the animal cell literature, John recalls having a hard time convincing me of the significance of this result.

John found that although nitrogen-starved cells were arrested homogeneous in the cell cycle at $G_1$, they were very heterogeneous in size. However, the small cells in starved populations all grew to about the same size as the large cells before they initiated a new cell cycle after nutrient addition. These results suggested that there was a cell size requirement for the *CDC28* step (Johnston et al. 1977).

## START

Clearly, the step executed by the *CDC28* product was the key regulatory step in the cell cycle. It was the first event in the cell cycle, was the point from which two pathways of dependent events diverged, and was the point at which control occurred by both mating factors and by growth. Completion of this step committed the cell to complete the cell cycle even if nutrients were withdrawn or if a mating partner was presented. We decided that this step needed a special designation and spent some time debating what to call it. Lynna suggested that we call it "G-whiz" to mock our enthusiasm and to mimic $G_0$, $G_1$, and $G_2$; unfortunately, we opted for the term "Start."

## ACKNOWLEDGMENTS

In addition to the people mentioned above, I wish to acknowledge the contributions of a number of undergraduate students during this period, Trudy Holland, Susan Purrington, Susan Healey, Richard Gutierrez, and Theresa Naujack. Constant and invaluable advice was provided by my colleagues, Calvin McLaughlin, Walton Fangman, Breck Byers, Larry Sandler, Don Hawthorne, and Herschel Roman. The beginnings of the cell cycle work was supported by a grant from the National Science Foundation and later by grants from the National Institutes of Health.

## REFERENCES

Byers, B. and L. Goetsch. 1975. Behaviors of spindles and spindle plaques in the cell cycle and conjugation in *Saccharomyces cerevisiae*. *J. Bacteriol.* **124:** 511–523.

Bücking-Throm, E., W. Duntze, L.H. Hartwell, and T. R. Manney. 1973. Reversible arrest of haploid yeast cells at the initiation of DNA synthesis by a diffusible sex factor. *Exp. Cell Res.* **76:** 99–110.

Duntze, W., V. MacKay, and T. Manney. 1970. *Saccharomyces cerevisiae*: A diffusible sex factor. *Science* **168:** 1472–1473.

Hartwell, L.H. 1967. Macromolecule synthesis in temperature-sensitive mutants of yeast. *J. Bacteriol.* **93:** 1662–1670.

———. 1970. Periodic density fluctuation during the yeast cell cycle and the selection of synchronous cultures. *J. Bacteriol.* **104:** 1280–1285.

———. 1973. Synchronization of haploid yeast cell cycles, a prelude to conjugation. *Exp. Cell Res.* **76:** 111–117.

Hartwell, L.H. and T.A. Weinert. 1989. Checkpoints: Controls that ensure the order of cell cycle events. *Science* **246:** 629–634.

Hartwell, L.H., J. Culotti, and B. Reid. 1970. Genetic control of the cell division cycle in yeast I. Detection of mutants. *Proc. Natl. Acad. Sci.* **66:** 352–359.

Hartwell, L.H., J. Culotti, J.R. Pringle, and B.J. Reid. 1974. Genetic control of the cell division cycle in yeast. *Science* **183:** 46–51.

Hartwell, L.H., R.K. Mortimer, J. Culotti, and M. Culotti. 1973. Genetic control of the cell division cycle in yeast. V. Genetic analysis of mutants. *Genetics* **74:** 267–286.

Hereford, L.M. and L.H. Hartwell. 1974. Sequential gene function in the initiation of *Saccharomyces cerevisiae* DNA synthesis. *J. Mol. Biol.* **84:** 445–461.

Jarvik, J. and D. Botstein. 1973. A genetic method for determining the order of events in a biological pathway. *Proc. Natl. Acad. Sci.* **70:** 2046–2050.

Johnston, G.C., J.R. Pringle, and L.H. Hartwell. 1977. Coordination of growth with cell division in the yeast *Saccharomyces cerevisiae*. *Exp. Cell. Res.* **105:** 79–98.

Reid, B.J. and L.H. Hartwell. 1977. Regulation of mating in the cell cycle of *Saccharomyces cerevisiae*. *J. Cell Biol.* **75:** 355–365.

Thuriaux, P., P. Nurse, and B. Carter. 1978. Mutants altered in the control coordinating cell division with cell growth in the fission yeast *Schizosaccharomyces pombe*. *Mol. Gen. Genet.* **161:** 215–220.

Wilkinson, L.E. and J.R. Pringle. 1974. Transient G1 arrest of *S. cerevisiae* cells of mating type alpha by a factor produced by cells of mating type. *Exp. Cell Res.* **89:** 175–187.

Williamson, D.H. and A.W. Scopes. 1962. A rapid method for synchronizing division in the yeast, *Saccharomyces cerevisiae*. *Nature* **193:** 256–257.

# The Cell Cycle of Fission Yeast

J. MURDOCH MITCHISON
*Institute of Cell, Animal, and Population Biology*
*University of Edinburgh, Edinburgh EH9 3JT, Scotland*

Modern work on the fission yeast *Schizosaccharomyces pombe* began with two people. The first was Urs Leupold, who took up the genetics in the autumn of 1946 (see Leupold, this volume) on the advice of Øjvind Winge at the Carlsberg Laboratory in Copenhagen (Leupold 1950). I was the second person. In the early 1950s, I began physiological experiments and worked very enjoyably for some years with Michael Swann on the mechanism of cleavage in sea urchin eggs. I then moved to the Zoology Department of Edinburgh University with Michael when he was appointed to a Chair there. By the mid 1950s, we had progressed as far as I thought we could with our existing ideas and techniques, and it was time to move to another aspect of the cell cycle. We would try to answer the question of how cells grew between one division and the next. This was largely virgin territory at the time, and evidence was only just appearing that DNA synthesis was periodic with gaps between it and the divisions ($G_1$ and $G_2$). My familiar material of sea urchin eggs was obviously unsuitable because, although they synthesize DNA each cycle, they do not show overall growth. Bacteria were too small for the light microscope, my favorite tool, and mammalian cells in culture grew too slowly for one accustomed to experiments that could be completed in a few hours. Yeast seemed to be a good material—fairly large, fast growing, and easy to handle. I was, however, put off by the unusual mode of budding that occurs in the cycle of budding yeast. I did not know much about yeast, and it was rather by chance that I came upon fission yeast while thumbing through the classical book on yeast taxonomy by Lodder and Kreger-Van Rij (1952). *S. pombe* seemed to be a good choice, since not only did it split in two at division like most other growing cells, but it also grew only in length so that a cell could be positioned in the cell cycle by a length measurement. I thus started to work with fission yeast, and probably the first time it was shown to a wide audience was in a time-lapse film I had made that was incorporated in a popular BBC television program on "The Cell." For several years afterward in the late 1950s, I would get small checks from the BBC for repeat programs.

Research on fission yeast has spread out enormously since those early days, and a mark that *S. pombe* has "come of age" is the recent book *Molecular Biology of the Fission Yeast* (Nasim et al. 1989). When I started, the state of its chromosomes and nuclear cytology was still murky, and it could not be called a simple eukaryote as the distinction between prokaryotes and eukaryotes was not made until the next decade. I suppose I hoped it would serve as a model for other cells, and thus it has been a pleasure for me in recent years to see that discoveries in fission yeast have linked up with those in eggs and other cells to reveal a "universal mechanism" for the early molecular stages of mitosis (Nurse 1990).

*The Early Days of Yeast Genetics*

## MACROMOLECULAR SYNTHESIS

I had been working with a shearing interference microscope, so my first work with fission yeast was to use this instrument to measure the growth of single cells in total dry mass and in volume. Dry mass in a single optical section can be derived from the light retardation, but more sophisticated equipment is needed to find the total dry mass of a single living and growing cell. I and two colleagues (one of whom, F.H. Smith, was the inventor of this type of interference microscope) spent several months in developing a technique for measuring the total integrated dry mass of a single cell at one reading (Mitchison et al. 1956). The equipment, largely homemade, was spread over a large laboratory table and involved a spinning disk, an oscilloscope, and a photomultiplier, as well as a "dissected" interference microscope.

I used this equipment on single cells of fission yeast growing in a gelatin medium to reduce the refractive index difference between the cell and the medium (Mitchison 1957). Perhaps the most striking result was the appearance of a "linear pattern" in the total dry mass. The dry mass increased at a constant rate until the end of the cycle. The rate then doubled just before the mother cell divided into two daughters. This was the first demonstration of the linear pattern since shown to occur in many of the components of growth in fission yeast (for review, see Mitchison 1989). Volume growth followed a different pattern, so the specific gravity of the cell varied during the cycle.

Are these linear patterns peculiar to fission yeast? I think not. I used the same equipment on single cells of budding yeast and found the same linear pattern in total dry mass (Mitchison 1958). As with fission yeast, there were changes in specific gravity that were later exploited by Hartwell (1970) to generate synchronous cultures by equilibrium centrifugation in density gradients. Linear patterns in total protein were also found by Williamson and Scopes (1961), although these authors were cautious in their interpretation since it is difficult to distinguish linear patterns from exponential patterns without rate measurements. Linear patterns in acid phosphatase activity were also found later in both budding and fission yeast (Creanor et al. 1983). Why worry about these subtle and difficult to establish patterns? There are two answers: (1) Doubling the rate of production is likely to be an important event for cell metabolism and (2) this doubling indicates that a signal comes to growth from the DNA-division cycle (Mitchison et al. 1992).

The late 1950s and 1960s were a time in which a good deal of the cell cycle work with a variety of cells was concerned with the patterns of synthesis of the main macromolecules, and in particular DNA (for review, see Mitchison 1971). For example, $G_1$ was found to be much more variable in length than S and $G_2$. There was extensive use of tritiated thymidine with several ingenious techniques, but work with yeast was delayed because tritiated thymidine is not incorporated into the DNA of these cells. I spent some time in the late 1950s trying to detect such incorporation in fission yeast without realizing that yeasts lack thymidine kinase. The problem was solved in budding yeast by Williamson and his colleagues using bulk measurements in synchronous cultures and autoradiography of single cells, where the DNA had been labeled together with the RNA, and the RNA had then been extracted (Williamson 1966). For fission yeast, bulk measurements were made by Bostock et al. (1966), and autoradiography was used by Nasmyth et al. (1979). Unlike budding yeast, fission yeast has a short $G_1$ and S phase and a long $G_2$ phase (for review, see Mitchison 1989).

After the initial measurements of total dry mass and volume, it was obvious that we should look at some of the major cell components such as protein and RNA. These components could not be measured in living cells, so instead we adopted an autoradiographic technique first suggested to me in the late 1950s by Ole Maaløe in Copenhagen: Growing cells were pulse-labeled with the appropriate radioactively labeled precursor, and autoradiographs were made after fixation. Grain counts were made over individual cells to give the rate of incorporation, and the cells were also positioned in the cell cycle by a length measurement. This was done for protein and carbohydrate (Mitchison and Wilbur 1962), for RNA (Mitchison and Lark 1962), and for the uptake of bases into RNA (Mitchison et al. 1969) using a method developed by Mitchison and Cummins (1964) for retaining the acid-soluble pool. The results did not conflict seriously with later results on synchronous cultures (for review, see Mitchison 1989), but the method had limitations. It is basically an "age-fractionation" technique (see section below on Synchronous Cultures) that has an inherent variation, since cells of the same length are at varying stages in the cycle. The method is also least effective when trying to detect changes at the beginning and end of the cycle. In addition, grain numbers over cells must be kept low if they are to be resolved, and this increases the variation due to the statistics of random radioactive decay. Except in special cases, I now regard the method as obsolete, and it has been largely superseded by synchronous cultures.

## SYNCHRONOUS CULTURES

Only limited information can be obtained from single cells, either living or dead, so there has always been a strong case for developing synchronous cultures that open up the cell cycle to biochemical assays of the events of the cycle. Even though the degree of synchrony is always imperfect in such cultures, and it decays because of the natural variation in cycle times, synchronous cultures are powerful tools. The first synchronous cultures were made in the early 1950s by E. Zeuthen and O.H. Scherbaum with *Tetrahymena* and by H. Tamiya and his colleagues with *Chlorella*. These were "induction" methods in which an initially asynchronous growing culture was synchronized by environmental changes (sublethal heat shocks or changes in illumination). These induction methods are still useful, but they have the disadvantage that normal cell cycle growth is often distorted.

The first successful method for producing synchronous cultures in budding yeast was that developed by Williamson and Scopes (1960). This method involved starvation, cell separation, and regrowth. Although there is an initial lag, there is much less distortion of the cycle than the induction methods applied to growing cultures, and, like all induction methods, it has the merit of a large yield of cells.

Another method for producing synchronous cultures is "selection." In this case, cells at one stage of the cycle are selected from an asynchronous culture and separated off so that they can be grown up as a synchronous culture. In principle at least, there should be less distortion of the cycle, although the yield is much less than with induction synchrony.

Various methods had been used to select the cells. Two successful techniques were to select small cells of *Escherichia coli* by elution from a membrane (Helmstetter and Cummings 1963) or to select mitotic mammalian cells by "wash off" (Terasima and Tolmach 1961). I tried several of these techniques on fission yeast in the early 1960s. Starvation and regrowth did not work, and I was unable then to stick cells to a membrane and elute off the young cells. Perhaps others may be more successful in the future. However, Walter Vincent and I did find a satisfactory way of selecting small cells by centrifuging the packed cells from an asynchronous culture through a sucrose gradient (velocity centrifugation) and taking off the top layer to generate a synchronous culture (Mitchison and Vincent 1965). This worked for fission yeast, budding yeast, and *E. coli*. Our rationale at the time was that what was being used for separating molecules could also be used for separating whole cells. It has proved to be an important method that has been widely used for a variety of cell types. One important point is that an asynchronous control culture can be established by growing up all of the cells in the gradient. This tests for possible perturbations caused by cell collection and the effect of the gradient.

Under certain situations, there are perturbations with fission yeast, and these are even more marked with *E. coli*. A later and less-perturbing variant of this technique which we developed was to carry out the selection in an elutriating rotor where the cells are growing in a continuous flow of warm medium (Creanor and Mitchison 1979). We used what I believe was the first such rotor available in this country.

Another variant of this method is to use "age-fractionation." This method relies on the fact that the stages of the cell cycle should be spread through the gradient so that successive samples down the gradient should give successive stages of the cycle. This technique does not work reliably for fission yeast, but it has been widely used for budding yeast with a large gradient set up in a zonal rotor. The yield is large, and perturbations can be reduced by chilling the gradient, but serious problems exist both in the degree of selection of the stages, particularly at the end of the cycle, and in the analysis of the data and deducing the real cell cycle pattern (Mitchison 1988). In addition, it is impossible to run a satisfactory control.

Selection synchrony by elutriation is the best technique at present for generating synchronous cultures of fission yeast, but the results that have emerged are beyond the time scale of this book. Some of them have been reviewed previously (Mitchison 1989).

## ENZYME ACTIVITY

Once it became possible to carry out biochemical assays on synchronous cultures or age-fractionated cells, an obvious step forward was to measure enzyme activity to follow the behavior of individual proteins through the cell cycle. This was an active field in the late 1960s, with many measurements being made on a wide variety of cells from *E. coli* to mammalian cells in culture. It excited a lot of interest, and when I wrote a short review about it at that time (Mitchison 1969), I received more than a 1000 reprint requests, to the understandable irritation of the University of California at Berkeley, where I had been working, who had to forward them all to Scotland.

A considerable proportion of the enzymes appeared to be "step enzymes," in which the activity doubled over a limited period of the cycle, like DNA. This period of activity increase differed from one enzyme to another so I concluded my review with, "These patterns imply that the cell cycle (and cell growth) is an ordered sequence of syntheses, with a continued change in the chemical composition of the cell." I still hold this view, although I now believe that the changes are much more subtle than I imagined at the time. There were several theories to account for these activity patterns of which the one most relevant to budding yeast was that of "linear reading" put forward by Halvorson and his colleagues (for review, see Halvorson et al. 1970; Halvorson 1977). Their work had shown that most of the enzymes in budding yeast were step enzymes, and they suggested that genes are transcribable for only a restricted part of the cycle, that they are transcribed in order, and that this order is the same as the sequence of the appropriate genes on the chromosomes. This was a very interesting theory, and there was experimental evidence to back it up. For example, the step timings of nine enzymes were correlated with the positions of their genes on the genetic map. Their timing was consistent with linear transcription of the chromosomes from end to end (Tauro et al. 1968).

I was not one to be left out of this mainstream activity of the period so we started to look at enzyme activity in fission yeast. Our most careful study was on sucrase, acid phosphatase, and alkaline phosphatase (Mitchison and Creanor 1969). In view of the results with budding yeast, I had expected to find step patterns of activity, but what actually emerged from this work was three of the linear patterns mentioned above. I had listed in my 1969 review five other enzymes with step patterns in fission yeast, but three of them came from overinterpreted work by a research student that was never published.

I continued to be interested in patterns of enzyme activity (and still am). We did a survey in the early 1970s of the activity of 19 enzymes in fission yeast and found that 18 of them showed continuous increases in activity through the cycle. There were no step patterns, but we did not go into sufficient detail to establish whether these continuous increases followed a linear pattern or an exponential one (Mitchison 1977). We also used the new elutriation technique to show the absence of step activity patterns in budding yeast in acid phosphatase, α-glucosidase, and β-galactosidase (with *Kluyveromyces lactis*) (Creanor et al. 1983). Not that there are no step enzymes in yeast. They certainly exist and as yet appear mostly to be concerned with DNA synthesis, but this came later.

Looking back, I am slightly surprised how rapidly the wave of enthusiasm for activity measurements subsided, especially since the in vivo activity of an enzyme is an important part of cell metabolism. Work in this field largely stopped in the early 1970s with some of the differences unresolved, especially in the yeasts. I suspect rumor got around that it was a difficult and controversial field that would not yield much profit. Certainly, some of the interpretation was overoptimistic, and steps were sometimes drawn too easily through the data points. But most of the problems were technical ones. People did not fully realize that synchronizing methods could produce marked perturbations that sometimes looked like cell cycle patterns and that it was vitally important to run controls for this. Age-fractionation, despite its attractively high yield, also had the serious problems mentioned above.

## GENETICS AND CELL CYCLE MODELS

A major advance in the work on the yeast cell cycle was made in the early 1970s by Hartwell's isolation of cell cycle (*cdc*) mutants in budding yeast (described elsewhere in this volume). We followed this up in the mid 1970s when Paul Nurse went to learn fission yeast genetics in Bern from Urs Leupold and then came to my laboratory. This was a "Golden Age" here, with a number of very bright and effective biologists working on the cell cycle of fission yeast, including Paul Nurse, Peter Fantes, Kim Nasmyth, Ronny Fraser, Pierre Thuriaux, and my long-term colleague Jim Creanor. It was not a large group by modern standards, but it was of high quality.

One of the first results was the isolation of 27 *cdc* mutants in 14 unlinked genes (Nurse et al. 1976). Although we were following in Hartwell's footsteps, it is worth mentioning that the method of isolating these mutants was different. Whereas Hartwell had found mutants to accumulate with terminal cell cycle phenotypes, the first fission yeast mutants were isolated as long cells that had continued to grow at the restrictive temperature, and the position of their cell cycle block was then established by cytology and DNA measurements. These mutants have proved to be of great importance in defining the genetic elements of the mitotic control. By the end of the 1970s, the three main mutants were *cdc2*, *cdc25*, and *wee1* (although more have been discovered since). *cdc2* is unusual in having two cell cycle block points, and the *cdc2* gene product, the protein kinase $p34^{cdc2}$, is central in the mitotic control (Nurse 1990).

In addition to the *cdc* mutants, another important group of cell cycle mutants in fission yeast are *wee* mutants. Their phenotypic expression is reduced size at division, although they grow normally and are not blocked in the cell cycle. They were called "wee" because of their small size and their isolation in Scotland. A temperature-sensitive allele *wee1-50* (originally called *cdc9-50*) was particularly important. Paul Nurse had the perception to realize that what was happening in *wee* mutants was an advance of mitosis so that the cells divided earlier and at a smaller size. Cell size appeared to be an important trigger for division, and temperature shifts with *wee1-50* revealed that size was monitored in the cycle close to mitosis (Nurse 1975). A similar conclusion was reached from experiments using nutritional shifts with wild-type cells (Fantes and Nurse 1977).

It is worth carrying the story a little further since the "true" molecular biology of fission yeast did not start until the early 1980s. The late 1960s and the 1970s were very much a time for cell cycle control models in a variety of cells. These models sought to explain the timing of mitosis and of the S period, and cell size or mass was an important component of many of the models. Hence, there was much discussion of "sizers" and "timers," even though their molecular basis was unknown. The first of these models was for *E. coli* (Cooper and Helmstetter 1968), and the most recent model is for algal cells (Donnan et al. 1985; McAteer et al. 1985). The flavor of this period and a good description of a number of the models can be found in John (1981).

The model for fission yeast was developed in the late 1970s (for review, see Fantes and Nurse 1981; Nurse and Fantes 1981). Briefly, the model involved two size controls. One of these controls monitored size just before mitosis and initiated mitosis when a critical size was reached. The other control initiated

DNA synthesis at a critical size. In wild-type cells, the mitotic control was the important one. The DNA control was cryptic since the cells after division were larger than the critical size for the $G_1$/S transition; hence, the very short $G_1$. In *wee* mutant cells, however, $G_1$ was extended until the cells grew to the critical size for the initiation of DNA synthesis. It was uncertain what determined the mitotic timing, although there appeared to be a minimum time for the events of $G_2$ to be completed. Budding yeast was like the *wee* mutants of fission yeast, with the main size control at the $G_1$/S boundary or "START" (Carter 1981).

When the methods of molecular biology were applied in the 1980s to the problems of the yeast cell cycle, they produced striking advances in our knowledge of the molecules involved in the controls. But the questions of timing and the nature of the size controls as yet have been elusive. I have no doubt that these problems will be resolved and that the earlier models will be provided with a molecular basis. But I have the feeling, which is common enough in older biologists, that some of the younger biologists pay less attention than they should to the earlier experiments and models. I wonder how many of the new generation of molecular biologists are aware of the evidence for the instability of the "mitogen" that comes from the amputation experiments on *Amoeba* which were first done by Hartmann (1928).

## REFERENCES


Bostock, C.J., W.D. Donachie, M. Masters, and J.M. Mitchison. 1966. Synthesis of enzymes and DNA in synchronous cultures of *Schizosaccharomyces pombe*. *Nature* **210:** 808–810.

Carter, B.L.A. 1981. The control of cell division in *Saccharomyces cerevisiae*. *Soc. Exp. Biol. Semin. Ser.* **10:** 99–117.

Cooper, S. and C.E. Helmstetter. 1968. Chromosome replication and the division cycle of *Escherichia coli* B/r. *J. Mol. Biol.* **31:** 519–540.

Creanor, J. and J.M. Mitchison. 1979. Reduction of perturbations in leucine incorporation in synchronous cultures of *Schizosaccharomyces pombe*. *J. Gen. Microbiol.* **112:** 385–388.

Creanor, J., S.G. Elliott, Y.C. Bisset, and J.M. Mitchison. 1983. Absence of step changes in activity of certain enzymes during the cell cycle of budding and fission yeasts in synchronous cultures. *J. Cell Sci.* **61:** 339–349.

Donnan, L., E.P. Carvill, T.J. Gilliland, and P.C.L. John. 1985. The cell cycles of *Chlamydomonas* and *Chlorella*. *New Phytol.* **99:** 1–40.

Fantes, P. and P. Nurse. 1977. Control of cell size at division in fission yeast by a growth modulated size control over nuclear division. *Exp. Cell Res.* **107:** 377–386.

———. 1981. Division timing: Controls, models and mechanisms. *Soc. Exp. Biol. Semin. Ser.* **10:** 11–33.

Halvorson, H.O. 1977. The cell division cycle. A review of current models on temporal gene expression in *Saccharomyces cerevisiae*. In *Cell differentiation in microorganisms, plants and animals* (ed. L. Nover and K. Mothes), pp. 361–376. Fischer, Jena.

Halvorson, H.O., B.L.A. Carter, and P. Tauro. 1970. Synthesis of enzymes during the cell cycle. *Adv. Microb. Physiol* **6:** 47–106.

Hartmann, M. 1928. Über experimentelle Unsterblichkeit von Protozoen-Individuen. Ersatz der Fortpflanzung von *Amoeba proteus* durch fortgesetzte Regeneration. *Zool. Jahrb.* **45:** 973–987.

Hartwell, L.H. 1970. Periodic density fluctuation during the yeast cell cycle and the selection of synchronous cultures. *J. Bacteriol.* **104:** 1280–1285.

Helmstetter, C.E. and D.J. Cummings. 1963. Bactrial synchronisation by selection of cells at division. *Proc. Natl. Acad. Sci.* **50:** 767–774.

John, P.C.L. 1981. *The cell cycle* (ed. P.C.L. John). Society for Experimental Biology Seminar Series 10. Cambridge University Press, Cambridge, England.

Leupold, U. 1950. Die Vererbung von Homothallie und Heterothallie bei *Schizosaccharomyces pombe. C.R. Trav. Lab. Carlsberg. Ser. Physiol.* **24:** 381–480.
Lodder, J. and N.J.W. Kreger-Van Rij. 1952. *The yeasts. A taxonomic study*. North Holland, Amsterdam.
McAteer, N., L. Donnan, and P.C.L. John. 1985. The timing of division in *Chlamydomonas. New Phytol.* **99:** 41–56.
Mitchison, J.M. 1957. The growth of single cells. I. *Schizosaccharomyces pombe. Exp. Cell Res.* **13:** 244–262.
———. 1958. The growth of single cells. II. *Saccharomyces cerevisiae. Exp. Cell Res.* **15:** 214–221.
———. 1969. Enzyme synthesis in synchronous cultures. *Science* **165:** 657–663.
———. 1971. *The biology of the cell cycle*. Cambridge University Press, Cambridge, England.
———. 1977. Enzyme synthesis during the cell cycle. In *Cell differentiation in microorganisms, plants and animals* (ed. L. Hover and K. Mothes), pp. 377–401. Fischer, Jena.
———. 1988. Synchronous cultures and age fractionation. In *Yeast: A practical apnroach* (ed. J.H. Duffus and I. Campbell), pp. 51–63. I.R.L. Press, Oxford.
———. 1989. Cell cycle growth and periodicities. In *Molecular biology of the fission yeast* (ed. A. Nasim et al.), pp. 205–242. Academic Press, San Diego.
Mitchison, J.M. and J. Creanor. 1969. Linear synthesis of sucrase and phosphatases during the cell cycle of *Schizosaccharomyces pombe. J. Cell Sci.* **5:** 373–391.
Mitchison, J.M. and J.E. Cummins. 1964. A method of making autoradiographs of yeast cells which retain pool components. *Exp. Cell Res.* **34:** 406–409.
Mitchison, J.M. and K.G. Lark. 1962. Incorporation of $^{3}$H-adenine into RNA during the cell cycle of *Schizosaccharomyces pombe. Exp. Cell Res.* **28:** 452–455.
Mitchison, J.M. and W.S. Vincent. 1965. Preparation of synchronous cell cultures by sedimentation. *Nature* **205:** 987–989.
Mitchison, J.M. and K.M. Wilbur. 1962. The incorporation of protein and carbohydrate precursors during the cell cycle of fission yeast. *Exp. Cell Res.* **26:** 144–157.
Mitchison, J.M., J. Creanor, and B. Novak. 1992. Coordination of growth and division during the cell cycle of fission yeast. *Cold Spring Harbor Symp. Quant. Biol.* **56:** 557–565.
Mitchison, J.M., L.M. Passano, and F.H. Smith. 1956. An integration method for the interference microscope. *Q. J. Microsc. Sci.* **97:** 287–302.
Mitchison, J.M., J.E. Cummins, P.R. Gross, and J. Creanor. 1969. The uptake of bases and their incorporation into RNA during the cell cycle of *Schizosaccharomyces pombe* in normal growth and after a step-down. *Exp. Cell Res.* **57:** 411–422.
Nasim, A., P. Young, and B.F. Johnson, eds. 1989. *Molecular biology of the fission yeast.* Academic Press, San Diego.
Nasmyth, K., P. Nurse, and R.S.S. Fraser. 1979. The effect of cell mass on the cell cycle timing and duration of the S-phase in fission yeast. *J. Cell Sci.* **39:** 215–233.
Nurse, P. 1975. Genetic control of cell size at cell division in yeast. *Nature* **256:** 547–551.
———. 1990. Universal control mechanism regulating onset of M-phase. *Nature* **344:** 503–508.
Nurse, P. and P.A. Fantes. 1981. Cell cycle controls in fission yeast: A genetic analysis. *Soc. Exp. Biol. Semin. Ser.* **10:** 85–98.
Nurse, P., P. Thuriaux, and K. Nasmyth. 1976. Genetic control of the cell division cycle in the fission yeast *Schizosaccharomyces pombe. Mol. Gen. Genet.* **146:** 167–178.
Tauro, P., H.O. Halvorson, and R.L. Epstein. 1968. Time of gene expression in relation to centromere distance during the cell cycle of *Saccharomyces cerevisiae. Proc. Natl. Acad. Sci.* **59:** 277–284.
Terasima, T. and L.J. Tolmach. 1961. Changes in X-ray sensitivity of HeLa cells during the division cycle. *Nature* **190:** 1210–1211.
Williamson, D.H. 1966. Nuclear events in synchronously dividing yeast cultures. In *Cell synchrony* (ed. I.L. Cameron and G.M. Padilla), pp. 81–101. Academic Press, New York.
Williamson, D.H. and A.W. Scopes. 1960. The behaviour of nucleic acids in synchronously dividing cultures of *Saccharomyces cerevisiae. Exp. Cell Res.* **20:** 338–349.

———. 1961. Protein synthesis and nitrogen uptake in synchronously dividing cultures of *Saccharomyces cerevisiae. J. Inst. Brew* **67:** 39–42.

# Circles and Cycles: Early Days at Nutfield

DONALD H. WILLIAMSON
*Parasitology Division*
*National Institute for Medical Research*
*Mill Hill, London, NW7 1AA, United Kingdom*

## BREWING, BREAD, AND BUTTER

My first encounter with the yeast cell, as an Edinburgh undergraduate, was not very encouraging. My bacteriology lecturer was keen on the fact that the bacterial nucleoid had been found fairly recently and that bacteria were therefore not exceptions to the general rule that genetic information was carried on DNA. However, he said, darkly, that this might not be the case for the brewer's yeast cell; no one had actually found DNA in this organism, and someone called Lindegren had actually shown that its nucleus was a rather large and empty looking object some people had quite wrongly regarded as a vacuole.

This was an unpromising start for a long-term relationship with such an anomalous little creature, and for the rest of my stay in Edinburgh, finally gaining a Ph.D. for studying "lipid droplets" (poly-β-hydroxybutyrate inclusions) in *Bacillus* rods (Williamson and Wilkinson 1958), I gave yeast little thought. However, by chance, my first paid job on leaving Edinburgh was with the Brewing Industry Research Foundation at Nutfield in Surrey, an environment where I was bound sooner or later to run into the organism that was to become my bread and butter.

This choice of employment was not the result, as it might be today, of careful thought about future career prospects. Nor was it perception of the future importance of yeast as a eukaryotic model, because in 1956, the great divide of the biological world into eukaryotes and prokaryotes (Stanier and van Neil 1962) had not surfaced. Nor yet was it the genetic versatility of the yeast cell that appealed, because the awful truth is that I was a genetic ignoramus. My courses at Edinburgh had involved very little formal genetics, and although I had a superficial grasp of bacterial and phage genetics, nobody had really opened my eyes to the power of the discipline or to its central role in cell biology. This sad state of affairs was partly due to the teaching system then operating at Edinburgh, where courses were taught very much as isolated modules, with relatively little interaction. Students could choose from a wide range of modules to build up to a final honors year, and the student's choice had more to do with expediency than with the requirements of a broad education.

My choice of career owed much more to the pleasant location of the Foundation in a country house near the little Surrey village of Nutfield, coupled with excellent working conditions, a reasonable salary, and the promise that although I would have to do a certain amount of "bread and butter" work on brewing bacteriology, I would be free to dabble in something more fundamental, as I wished.

*The Early Days of Yeast Genetics*

The Brewing Industry Research Foundation (then known to its inmates as "BIRF," perhaps the reason they later changed its name to the Brewing Research Foundation), seemed like one of the last bastions of the British Empire: an elegant country mansion set in tranquil landscaped grounds and equipped with an ornamental fish pond (tame carp as big as your foot), as well as an outdoor swimming pool (one of my major memories of Nutfield). A small group of us became habitués of the swimming pool at lunchtime from May to October, when the frogs took over. These days I doubt if anyone would think of starting up a research institution in such isolation or for that matter of maintaining such a splendid swimming pool. However, the laboratories were beautifully equipped (wonder of wonders, they even had their own home-built analytical ultracentrifuge with a 5-yard light path and a rotor shielded by brick walls; I hope it's still there), and the library and small-scale pilot brewery were all that could be desired.

When I joined the laboratory in 1956, BIRF was under the command of Sir Ian Heilbronn, a chemist with a considerable reputation. Sir Ian established the basic philosophy of BIRF, giving individuals considerable freedom to follow their academic noses, provided at the same time they maintained some more applied project "to keep the brewers happy." Under Sir Ian, this policy seemed to work very well, and the arrangement was pretty stable. A year or two later, when Sir Ian retired and his deputy Dr. Arthur Cook took over his job, it was not quite as good. The paymasters, in the form of the Brewers Society, coughing up a farthing (about £0.001) on every barrel of beer sold by their members, always seemed to be chopping and changing their attitudes. One moment BIRF was doing "*Great fundamental work for the good of the nation,*" while the next, it was "*Why the hell aren't they doing something useful down there?*" At least that is how it seemed to us on the shop floor.

## SNAILS AND CIRCLES

As it turned out, the then head of the Microbiology Section, Alan Eddy, was an academically minded scientist who was very keen on yeast as an experimental organism. He was particularly interested in the physicochemical nature of its surface and its role in flocculation, an age-old concern of brewers. He wanted to be able to remove the yeast cell wall with enzymes and had already dug out an ancient observation by Giaja (1922) that yeasts could be digested by treatment with the gut juice of the Roman (edible) snail *Helix pomatia*. (In checking this reference for the purpose of this article, I came across two earlier notes by Giaja [1914, 1919] which reveal that he first observed the degradation of the yeast cell wall by snail enzyme in 1914. He also noted that the enzyme had no proteolytic activity and that yeast cells treated with it in the presence of sucrose retained their "fermentatif" power for several hours. The details are sparse, and the nature of his preparations is not totally clear, but he evidently recognized the potential value of his wall-less material for answering "nombreux problems qui se rattachent a la presence de la membrane.")

This treatment with snail gut juice was to be the humble beginning of the yeast protoplast as a subject for scientific curiosity and a basis for the first procedures for yeast transformation. Knowing of my experience with bacterial protoplasts, Alan suggested that I should try to use snail enzyme to make protoplasts from yeast, in the same way that Weibull (1953) had first used lysozyme to isolate them from bacilli. I was a bit surprised, because it seemed so straightforward that I thought he must have already done it. However, to my astonishment, he showed me cells lysing in snail enzyme in the absence of an osmotic stabilizer and said "look, they're rather funny protoplasts, aren't they?" I tactfully suggested that an osmotic stabilizer might be desirable, and the search for suitable compounds then became the main focus of my activities.

Snail enzyme in those days was not something that came in a bottle, and we would freeze-dry gram quantities obtained from live snails. We were lucky to be housed not far from chalky areas of Surrey where these beautiful creatures abounded. The snails were not so lucky; one of our technicians knew a good spot where, on a humid summer morning, preferably just after a thunderstorm, he could collect a few hundred snails in a couple of hours. So it was that snail hunting became a laboratory ritual. Some laboratories fish for sea urchins in the Mediterranean, others go for salmon sperm; we grubbed around for snails. It sounds very primitive, and we tried to avoid winter shortages by buying edible snails commercially and reviving them in a hot room on a diet of cabbage leaves, and even at times incubating their eggs and raising their young. However, the yields of the precious enzyme were never very good from our domesticated livestock so, sadly, the slaughter of their wild cousins had to go on. It was a relief when glusulase became commercially available.

In the course of this work, I became obsessed with circles. I chose to monitor different protocols microscopically rather than turbidimetrically, and to help this, I chose a strain of what was then called *Saccharomyces carlsbergensis* (NCYC 74), was later renamed *Saccharomyces uvarum*, but is now regarded as a variety of *Saccharomyces cerevisiae*. This strain, originally from a brewing source and subsequently used for different purposes by a number of laboratories, was chosen by us for two outstanding properties: its exquisite sensitivity to snail enzyme and the cylindrical shape of its cells. These properties were most important because of the ease with which the transition from cylindrical cell to precisely spherical protoplast could be observed. My whole enthusiastic aim was to fill the microscope field with these beautiful circles (Fig. 1). They were to be a symbol of our ultimate success (Eddy and Williamson 1957), and I began to dream about circles and see them wherever I went.

Alan Eddy and I both thought that protoplasts would be a scientific cornucopia, opening the door to all sorts of goodies like organelle isolation, cell wall synthesis, and the like. I am not sure that this early promise has really been fulfilled, although protoplasts have been used in various ways. Perhaps their main claim to fame (and of satisfaction to Alan and myself) was their central role in the development of the first transformation procedures (Beggs 1978; Hinnen et al. 1979). I suppose, in another quantum mechanical universe, Giaja would not have looked to see what snail gut juice did to yeast cells and/or Alan Eddy would not have seen the reference. This might have delayed the achievement of transformation of yeast protoplasts, but probably not by much; the Japanese (or someone) would have discovered zymolyase or novazyme (or lithium or electroporation) sooner than was the case. It is tempting to think of one's tiny contribution to science as a vital factor in human

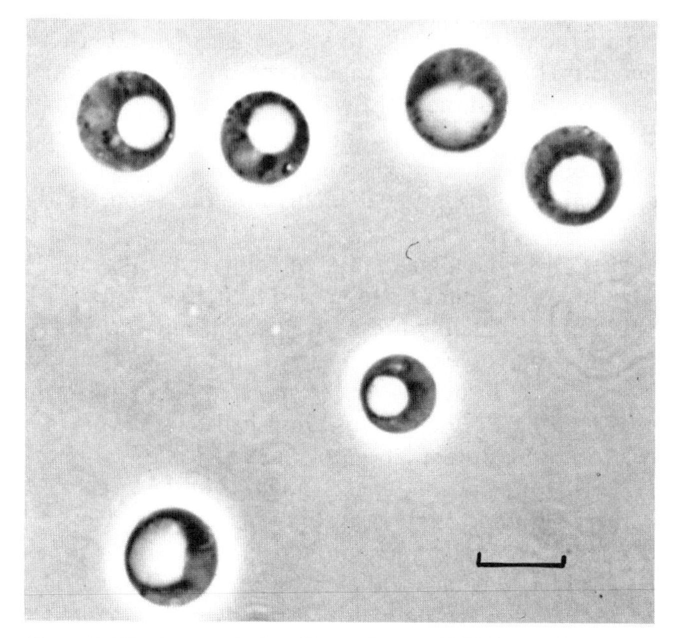

*Figure 1* The target circles: Protoplasts of NCYC 74. Bar, 5 μm.

progress, but the brutal truth is that if you did not do it someone else would, or the end would have been reached by a different route. However, there might now have been a few more snails on the Surrey Downs.

In any event, Alan and I both played quite a lot with protoplasts, looking for the fruits of the cornucopia; Alan was very keen on isolating nuclei, but did not get very far (Eddy 1959), and I did not really do anything very constructive. We both wanted to get protoplasts to regenerate whole cells but were totally without success. Given an osmotically stabilized nutrient medium, they would generate enormous fibrous sacs that we called sausage skins, but it seemed that without a bit of wall as template, protoplasts were unable to remember what size and shape whole cells were supposed to be (Eddy and Williamson 1959). Nĕcas (1956) had recorded that he could regenerate whole cells from some sort of preparation of a protoplast-like nature, but we could not repeat his trick with our protoplasts, and I suspect that Nĕcas' preparations were more in the nature of spheroplasts—cells with seriously thinned walls—rather than true protoplasts totally devoid of wall. Our lack of success in this direction left me rather surprised by the initial reports of successful transformation of protoplasts, dependent as that process is on regeneration of cells. I now suppose, however, that these transformation protocols, like the Nĕcas preparations, involved spheroplasts rather than the true protoplasts that Alan Eddy and I were used to, and efficient regeneration probably requires a little bit of wall.

## ESCAPE

I got on well with Alan Eddy. He was an impressive person, both physically and intellectually, and not too long after I went to Nutfield, he took the Chair of Biochemistry at Manchester's UMIST. However, he took a proprietary interest in just about every aspect of yeast biology one could think of, to the point where I was beginning to chafe, feeling I would never develop something I could call my own. Our relationship became a bit strained one day when I tackled him about this, asking if there was anything about yeast he was not interested in, a little corner I could regard as mine. "No" he said, "nothing." I knew then that I needed an escape route.

This came more easily than I might have hoped. It happened that in my last months at Edinburgh that I came across an article by Dan Mazia in a journal called *The American Scientist* on "The Life History of the Cell" (Mazia 1956). This article was to me quite amazing. Perhaps because of my bacteriological background, it had never occurred to me that microorganisms had a life other than that of bulk cultures in lag, log, or stationary phases. The idea of even *thinking* about what went on in a single individual cell traversing the division cycle was totally stunning. Moreover, Mazia said there were several ways of tackling this problem, and he outlined not only Murdoch Mitchison's elegant studies on single cells (Mitchison 1957, 1958), but also the use of synchronously dividing cultures, pioneered at that time mainly by Eric Zeuthen (1964b) on *Tetrahymena*. With these, *one could actually get enough cells at particular stages of the cell cycle to allow bulk biochemical analyses!*

Wow! This really blew my mind, and under other circumstances, I would have dropped everything to work in this new world. However, there was no way I could do so at that embryonic stage of my career, so the concept had to be put on the back burner. It remained there even in my first months at Nutfield, because the budding yeast cell, with its lop-sided mode of division, hardly seemed a prime candidate for synchronization. It was only because of my quest for independence that the idea of synchronously dividing cultures resurfaced as something that just might be worth trying, and I was relieved when Alan Eddy readily agreed that this unpromising little corner was something he had no designs on!

But how to synchronize yeast cells? Fortunately, the field of division synchronization had progressed considerably during the dog years of the 1950s, and a number of successful approaches with various organisms had been published, most of them based on exposure to repeated changes of temperature (heat shock), nutrition (feeding and starving) or, for green cells, light intensity. However, these procedures often generated abnormally large, division-inhibited cells. During their subsequent "free-wheeling" recovery period, they would go through several synchronized divisions, but with little growth, making their behavior somewhat more akin to that of developing embryonic cells than "normal" cycling individuals. (The basic philosophy of synchrony, the pros and cons of induction methods versus selection ones, and the biological relevance of unbalanced growth are discussed in Zeuthen [1964a]. The volume referred to also has a number of interesting articles describing some of the early synchronous systems.)

For this reason, these systems were not very attractive, but it occurred to me that there might be a way of avoiding such gross artifacts with yeast. The basic idea was to apply the synchronizing stresses to resting cells rather than growing ones. This looked really feasible because Lindegren and Haddad (1953) had shown that unbudded yeast cells in fresh medium normally do not grow in size until they produce a bud (only true for full-grown parents, but near enough right) and were therefore clearly "resting." All we needed to do then was to use unbudded cells, which of course accumulate in stationary-phase cultures.

This approach in any case looked hopeful since Maurice Ogur, who was originally from the Lindegren laboratory and had pioneered the estimation of DNA in various organisms including yeast (Ogur and Rosen 1950), had noted that reinoculated stationary-phase cells went through a moderately synchronous burst of DNA synthesis as they produced their first buds (Ogur et al. 1953). The degree of synchrony was not too exciting, and there were no data on cell division, as opposed to bud formation, but this signaled the possible value of such cells as starting material. This was underlined in a paper by Sylvén et al. (1959), who described a moderate degree of synchrony obtained using cells that had been starved in potassium succinate. Their process seemed very fiddly, and the synchronous rounds did not look too much like doublings, but this paper gave another push in the direction of the resting cells. So the idea was born.

## HELP FROM THE BREWERS

One bothersome little difficulty, however, was the tendency of many yeast strains to grow in chains, something that would obviously impede cell counting. Fortunately, however, we were lucky enough at Nutfield to have the National Collection of Yeast Cultures, then being ably run by my good friend Beryl (Kitty) Brady. In those days, the collection housed very few respectably marked haploids, but it had a great selection of brewing strains. Kitty helped us find a bunch of strains that tended to divide cleanly, and out of these, I settled on NCYC 239, originally donated to the Collection by Whitbread's brewery. It is hard to overestimate the value of this strain to the development of our synchronous system. It never formed chains, and occasional adherent daughter cells could easily be knocked off by a touch of ultrasonics. Initially, we counted cells with hemocytometers, and although we learned how to distinguish divided pairs from nondivided ones microscopically (Williamson and Scopes 1960), a chain-forming strain would have made the whole business impossibly tedious. The clean-living habit of NCYC 239 solved this problem. It became even more important when, a little later on, we bought our first Coulter Counter; for all its electronic wizardry, this admirable machine has no way of identifying divided but adherent pairs, and clean separation at cell division, aided if necessary by a touch of ultrasonics, is essential to accurate electronic counting.

Brewing strains like NCYC 239 are not all sunshine and light incidentally. Although nominally diploids, many of them carry homoeologous chromosomes (Kielland-Brandt et al. 1983; Petersen et al. 1987). The components of these homoeologous chromosome pairs are evidently functional (at least for brewing purposes!) but have very different ancestries, and it seems that they are sufficiently different in sequence to prevent proper meiotic recombination. As a consequence, sporulation is usually defective, and viable haploids are rare. This of course prevents conventional genetic analysis and seriously limits the scope of molecular genetic procedures. This last point was of no concern to us at that time, because transformation and all that follows from it had not then been invented. I am ashamed to say, however, that the classical genetic shortcomings of these strains also never entered my head, simply because of my ignorance of the power and value of genetics. Perversely, however, the failure of NCYC 239 to sporulate was probably essential to the success of the feed-starve procedure. Subsequent experience with genetically respectable diploids (at least of the usual **a**/α variety) has demonstrated that these often do not respond well to the procedure, possibly because starvation signals the cells to attempt entry into meiosis, and this is probably inimicable to mitotic synchronization by this route. An appropriately marked *a/a* diploid would be a much better bet; this could be synchronized by all currently available procedures including the feed-starve one, would be transformable, and by comparison with NCYC 239 would be relatively amenable to genetic analysis.

## FEEDING AND STARVING: EUREKA!

As it was, I was oblivious to these niceties. Having chosen a good strain, I set to work with more enthusiasm than logic. It was predictable that the few stationary cells that still retained small buds would be a problem, but these were reduced to an acceptable level simply by using relatively old stationary-phase cultures. A more pressing problem was that the individuals in these cultures fell into two classes, large and small. The large ones were cells that had previously had one or more daughters, and the small ones were the last generation of daughters produced in the final nutrient-limited divisions preceding stationary phase. In fresh medium, the parental cells (about 60% of the population) showed no apparent growth before they produced their first bud, about 1 hour after inoculation. The smaller daughter cells, on the other hand, were very slow to bud and needed to grow to a parent-like size before doing so. This clearly affected the overall synchrony of the population (Fig. 2) and led me to the tedious necessity of eliminating them by repeated rounds of differential centrifugation. Although laborious (but for a shortcut, see Williamson 1964b), this strikingly improved the synchrony of the population (Fig. 2). In fact, many people working with α-hormone-induced or rotor-selected synchronous populations might be happy with the degree of synchrony attained by this simple selection process. However, as shown in Figure 2, there were still considerable improvements to be made by feeding and starving the selected cells.

The aim of this part of the process was to maintain the cells in their nongrowing (unbudded) state during repeated cycles of feeding and starving, and the first experiments were aimed at seeing just how long a feed could be used without generating buds. One day, I ran out of time and had to leave the cells in growth medium in the refrigerator overnight. I still remember my annoyance the next day at finding that they had all produced a bud (Fig. 3) and would have to be discarded; but as I gazed at them I realized that the young buds were all the same size! Synchrony!

It did not take long to work that method up into a reasonably reproducible procedure, at first incorporating feeding overnight at 4°C, and then repeating the whole process at least seven times! Still, "gee whiz it worked," and Tony Scopes and I published a paper on nucleic acid synthesis in cells synchronized in this way (Williamson and Scopes 1960), which reported the first reasonably accurate determination of the timing of the S phase in *S. cerevisiae*. I still remember the intense satisfaction of walking home one night after working late

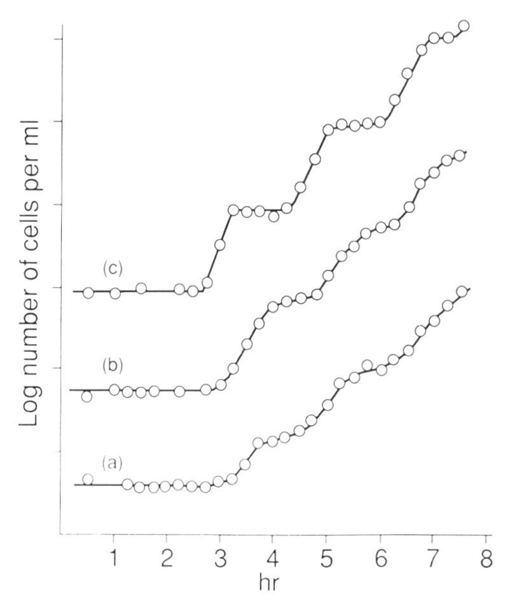


*Figure* 2 Synchronous cell divisions of NCYC 239. The intervals of the ordinate correspond to doublings in cell numbers, estimated using a Coulter Counter. All three cultures were inoculated with approximately 1.5 x $10^6$ cells/ml and incubated at 25°C in a semidefined medium (Williamson 1966). (*a*) Stationary-phase cells, untreated; (*b*) "large" cells selected from same population; (*c*) large cells after feed/starve synchronization. The improved synchrony brought about by first selecting the large cells and then feeding and starving them is evident.

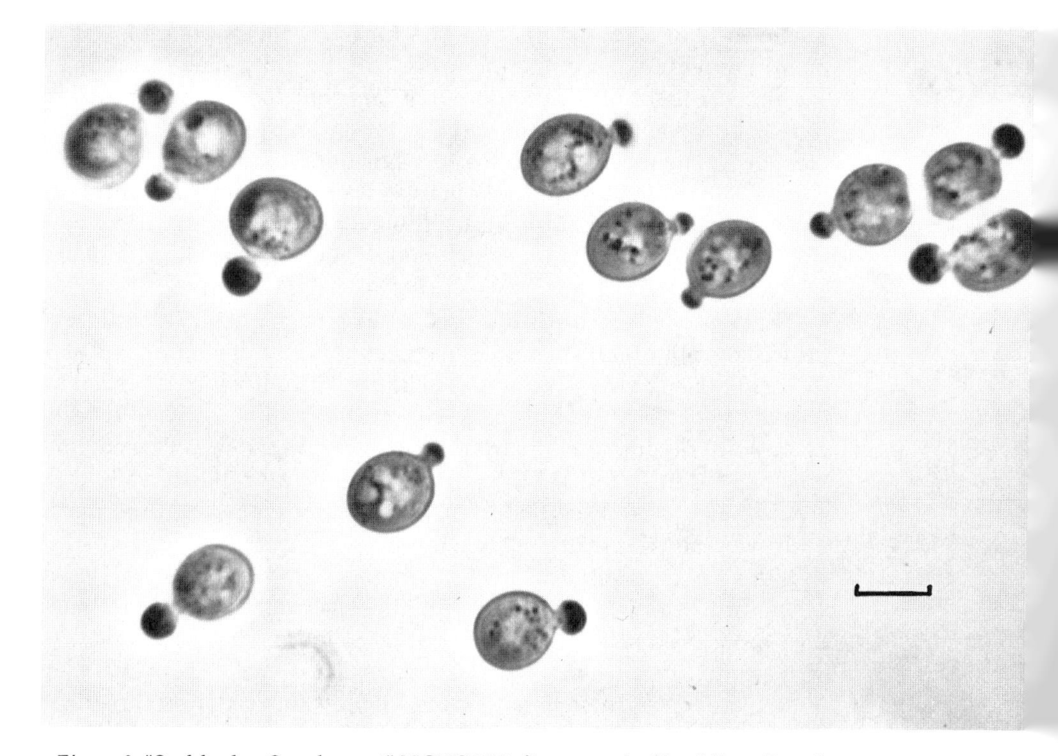

*Figure* 3 "Suddenly—Synchrony!" NCYC 239, first round of budding. Bar, 5 µm.

to get the first results, the only person in the world who knew when the yeast cell replicated its chromosomes in the cell cycle. Wow!

This was a great pleasure, but a greater and more long-lasting one arising from that fairly trivial bit of work was that it induced Murdoch Mitchison, at that time to me an icon for his beautiful studies on single cells of both *Schizosaccharomyces pombe* and *S. cerevisiae*, to write a long and friendly letter saying how much he had enjoyed our paper. What a nice guy! We have remained good friends ever since.

Of course the lengthy time scale of that primordial synchronization process was impossible to live with, and it was not long before we found that the original overnight feed in the cold could be replaced with a 40-minute feed at 25°C. In addition, cycles of feeding and starving could be repeated in quick succession, as long as the starving period was at least 2 hours, so that in principle the whole process could be completed in 1 day (Williamson and Scopes 1962). That sounds like a long time, but when the process can provide enough cells (storable for a couple of weeks at 4°C) for several good-sized cultures, it begins to sound more attractive. We also found that the synchrony of the cultures was greatly enhanced by improving the growth medium, and armed with this, we followed cell cycle changes of various basic parameters—protein and nucleic acids, dry weight, volume, respiration, nitrogen uptake (Williamson and Scopes 1960, 1961b; Scopes and Williamson 1964; for review, see Williamson 1964a), and, somewhat later, the timing of mitosis (Williamson 1966).

## WHO SAID ARTIFACTS?

This all sounds pretty pedestrian to modern ears, but at that stage of the game, we were obliged to be natural historians. Apart from a delightful study of volume growth of the budding yeast cell by Bayne-Jones and Adolph (1932), Murdoch Mitchison's paper on *S. cerevisiae* (Mitchison 1958), and the two yeast synchrony papers (Ogur et al. 1953; Sylvén et al. 1959), there was very little literature on the overall growth processes of the cell. So we had to look at every basic parameter we could. (Both Sylvén et al. [1959] and Williamson and Scopes [1961b] observed that the total nitrogen content of synchronized cells increased discontinuously in the cell cycle in a single discrete step, probably in $G_2$. This was quite unlike the continuous pattern of synthesis of protein and RNA, and implied that there was a sharp cell-cycle-regulated uptake of nutrients soon after budding. I do not think this has been examined by anyone since.)

To tell the truth, in those far off BC (before cloning) days, there was precious little else we could do, but in any case, we had another powerful motive for this descriptive approach. This was the lingering doubt that our system, like so many other artificially synchronized culture populations, might be grossly "abnormal." The lack of comparable data from single cells or other synchronized yeast systems did not make this easy. We did a fair amount of hand-waving (see, e.g., Williamson and Scopes 1961a) and were keen on detecting balanced growth, i.e., doublings per cycle, to contrast with the clearly unbalanced behavior of, for instance, Zeuthen's heat-shock-synchronized *Tetrahymena* system (Zeuthen 1964b). In fact, the first cell cycle was decidedly out of balance in various respects, but subsequent ones were a lot better, and this is probably normal for resting cells transferred into fresh growth medium.

In contrast, DNA behaved exactly as expected, showing a clear doubling in each cycle, including the first. However, it seemed important to try to find out if the timing of this doubling (in this strain coincident with the emergence of the bud) was the same as that in unsynchronized individuals, and this I set out to do when I moved in 1962 to Henry Harris's newly formed Cell Biology group at the John Innes Institute.

The approach was simple; all that was needed was to pulse-label growing cells with a DNA-specific precursor, make whole-cell autoradiographs, and determine the ages of labeled cells by measuring buds; one of the great advantages of using brewer's yeast for single-cell studies is the ease with which the position of a cell in the division cycle can be determined in this way. The bad news was that no yeast strains could be labeled specifically with thymidine, the usual DNA label for other systems, and an alternative labeling method was required. The method that eventually evolved, using tritiated adenine and rigorous elimination of non-DNA label, was a bit cumbersome, but very specific for DNA (Williamson 1965). Given that, the estimation of cell ages on the basis of bud size was a snip, and it soon became clear that log-phase cells of NCYC 239 entered S phase at just about the same time as their synchronized bretheren, i.e., at the time of bud initiation.

In my eyes, this gave the feed-starve procedure its *Good Housekeeping* seal, and in a sense rounded off my formative years at Nutfield. I am happy that the method has been successfully used by a number of people, most notably in recent years by my colleague Lee Johnston in his group's elegant studies on the cell cycle regulation of DNA synthesis enzymes (for review, see Johnston 1990). It is very satisfying to find that all of the important molecular controls seem to be just the same in feed-starve synchronized cells as in those synchronized by other techniques.

## GIF AND SEATTLE

The study of DNA synthesis in single cells acted as a springboard for later interests. Confident of the specificity of the labeling protocol, I realized that a low level of labeling of cells throughout the cell cycle, outside of the S phase, might be due to replication of mitochondrial DNA. Thus, it was that when Herschel Roman invited me to spend a year at the Genetics Department of the University of Washington at Seattle, it seemed a good idea to spend the time there exploring mitochondrial DNA. It was there that I met Ethel Moustacchi, who became a lifelong friend. She was then at Orsay, but like me was on a year's sabbatical in Seattle also at Hersch's invitation. Like me, she wanted to learn physical techniques for handling DNA, and we joined forces in a very fruitful collaboration learning from Ben Hall how to use the analytical ultracentrifuge and finding both mitochondrial DNA and what we called the $\gamma$-satellite DNA, i.e., the yeast cell's reiterated ribosomal genes (Moustacchi and Williamson 1966). Ultimately, our work at Seattle enabled us to show that mitochondrial DNA in yeast is indeed synthesized throughout the cell cycle (Williamson and Moustacchi 1971).

I cannot end this paper without paying tribute to the memory of Herschel Roman, whose untimely death was a sad blow to his many friends and admirers. Since his name will surely crop up throughout these memoirs, it cannot be out of place to record my indebtedness to him, for he more than anyone else introduced me to the joys of genetics. He it was who, having seen the early papers from Nutfield on protoplasts and synchronous division, kindly invited Alan Eddy and me to the Second International Conference on Yeast Genetics, which was held at Piotr Slonimski's place at Gif-sur-Yvette near Paris in 1963. This meeting was only the second in the series that has since developed into the 800-strong biannual International Conference on Yeast Genetics and Molecular Biology (the last one being in Vienna in 1992), and the 1974 version of which I was to coorganize with Alan Bevan and Robin Holliday, at Brighton (see group photo elsewhere in this volume).

This early meeting at Gif was great fun—there were only 50 or so participants—and I made many lasting friendships among the worldwide yeast community. But the most important aspect of that meeting, for me, was that I began to learn about tetrad analysis, recombination, gene conversion (a very hot topic), cytoplasmic inheritance (very rudimentary then), and similar wonders that previously had barely touched my universe. It was at this meeting that Herschel asked me to spend a year at Seattle, incidentally following in the footsteps of David Wilkie and Robin Holliday. Hersch's generous nature started me, and I suspect many others, along the path to genetic literacy and for that I shall always remain truly grateful.

## REFERENCES


Bayne-Jones, S. and E.F. Adolph. 1932. Growth in size of micro-organisms measured from motion pictures. I. Yeast *Saccharomyces cerevisiae*. *J. Cell. Comp. Physiol.* **1:** 387–407.

Beggs, J.D. 1978. Transformation of yeast by a replicating hybrid plasmid. *Nature* **275:** 104–109.

Eddy, A.A. 1959. The probable nuclear origin of certain of the bodies released from yeast protoplasts by ultrasonic treatment. *Exp. Cell Res.* **17:** 447–464.

Eddy, A.A. and D.H. Williamson. 1957. A method of isolating protoplasts from yeast. *Nature* **179:** 1252–1253.

———. 1959. Formation of aberrant cell walls and of spores by the growing yeast protoplast. *Nature* **183:** 1101–1104.

Giaja, J. 1914. Sur l'action de quelqes ferments sur les hydrates de carbone de la levure. *C.R. Soc. Biol.* **77:** 2–4.

———. 1919). Emploi des ferments dans les études de physiologie cellulaire: Le globule de levure dépouillé de sa membrane. *C.R. Soc. Biol.* **82:** 719–720.

———. 1922. Sur la levure dépouillée de membrane. *C.R. Soc. Biol.* **86:** 708–709.

Hinnen, A.A., J.B. Hicks, and G.R. Fink. 1979. Transformation of yeast. *Proc. Natl. Acad. Sci.* **75:** 1929–1933.

Johnston, L.H. 1990. Periodic events in the cell cycle. *Curr. Opin. Cell Biol.* **2:** 274–299.

Kielland-Brandt, M.C., T. Nilsson-Tillgren, J.G.L. Peterson, S. Holmberg, and C. Gjermansen. 1983. Approaches to the genetic analysis and breeding of brewer's yeast. In *Yeast genetics: Fundamental and applied aspects* (ed. J.F.T. Spencer et al.), pp. 421–437. Springer-Verlag, Berlin.

Lindegren, C.C. and S.A. Haddad. 1953. The control of nuclear and cytoplasmic synthesis by the nucleocytoplasmic ratio in *Saccharomyces*. *Exp. Cell Res.* **5:** 549–550.

Mazia, D. 1956. The life history of the cell. *Am. Sci.* **44:** 1–32.

Mitchison, J.M. 1957. The growth of single cells. I. *Schizosaccharomyces pombe*. *Exp. Cell Res.* **13:** 244–262.

———. 1958. The growth of single cells. II. *Saccharomyces cerevisiae*. *Exp. Cell Res.* **15:** 214–221.

Moustacchi, E. and D.H. Williamson. 1966. Physiological variations in satellite components of yeast DNA detected by density gradient centrifugation. *Biochem. Biophys. Res. Commun.* **23:** 56–61.

Nečas, O. 1956. Regeneration of yeast cells from naked protoplasts. *Nature* **177:** 898–899.
Ogur, M. and G. Rosen. 1950. The nucleic acids of plant tissues. I. The extraction and estimation of desoxypentose nucleic acid and pentose nucleic acid. *Arch. Biochem. Biophys.* **25:** 262–276.
Ogur, M., S. Minckler, and D.O. McClary. 1953. Desoxyribonucleic acid and the budding cycle in the yeast. *J. Bacteriol.* **66:** 642–645.
Petersen, J.G.L., T. Nilsson-Tillgren, M.C. Kielland-Brandt, C. Gjermansen, and S. Holmberg. 1987. Structural heterozygosis at genes ILV2 and ILV5 in *Saccharomyces carlsbergensis. Curr. Genet.* **12:** 167–174.
Scopes, A.W. and D.H. Williamson. 1964. The growth and oxygen uptake of synchronously dividing cultures of *Saccharomyces cerevisiae. Exp. Cell Res.* **35:** 361–371.
Stanier, R.Y. and C.B. van Niel. 1962. The concept of a bacterium. *Arch. Mikrobiol.* **42:** 17–35.
Sylvén, B., C.A. Tobias, H. Malmgren, R. Ottoson, and B. Thorell. 1959. Cyclic variations in the peptidase and catheptic activities of yeast cultures synchronised with respect to cell multiplication. *Exp. Cell Res.* **16:** 75–87.
Weibull, C. 1953. The isolation of protoplasts from *Bacillus megaterium* by controlled treatment with lysozyme. *J. Bacteriol.* **66:** 688–695.
Williamson, D.H. 1964a. Division synchrony in yeasts. In *Synchrony in cell division and growth* (ed. E. Zeuthen), pp. 351–390. Wiley, London.
———. 1964b. Techniques for synchronizing yeast cells. In *Synchrony in cell division and growth* (ed. E. Zeuthen), pp. 589–591. Wiley, London.
———. 1965. The timing of deoxyribonucleic acid synthesis in the cell cycle of *Saccharomyces cerevisiae. J. Cell Biol.* **25:** 517–528.
———. 1966. Nuclear events in synchronously dividing yeast cultures. In *Cell synchrony—Studies in biosynthetic regulation* (ed. I.L. Cameron and G.L. Padilla), pp. 88–101. Academic Press, New York.
Williamson, D.H. and E. Moustacchi. 1971. The synthesis of mitochondrial DNA during the cell cycle in the yeast *Saccharomyces cerevisiae. Biochem. Biophys. Res. Commun.* **42:** 195–201.
Williamson, D.H. and A.W. Scopes. 1960. The behaviour of nucleic acids in synchronously dividing cultures of *Saccharomyces cerevisiae. Exp. Cell Res.* **20:** 338–349.
———. 1961a. Synchronization of division in cultures of *Saccharomyces cerevisiae* by control of the environment. *Symp. Soc. Gen. Microbiol.* **11:** 217–242.
———. 1961b. Protein synthesis and nitrogen uptake in synchronously dividing cultures of *Saccharomyces cerevisiae. J. Inst. Brew.* **67:** 39–42.
———. 1962. A rapid method for synchronizing division in the yeast, *Saccharomyces cerevisiae. Nature* **193:** 256–257.
Williamson, D.H. and J.F. Wilkinson. 1958. The isolation and estimation of the poly-β-hydroxybutyrate inclusions of *Bacillus* species. *J. Gen. Microbiol.* **19:** 198–209.
Zeuthen, E., ed. 1964a. Introduction. In *Synchrony in cell division and growth*, pp. 1–8. Interscience Publishers, New York.
———. 1964b. The temperature-induced division synchrony in *Tetrahymena*. In *Synchrony in cell division and growth*, pp. 99–158. Interscience Publishers, New York.

# GENE STRUCTURE AND EXPRESSION

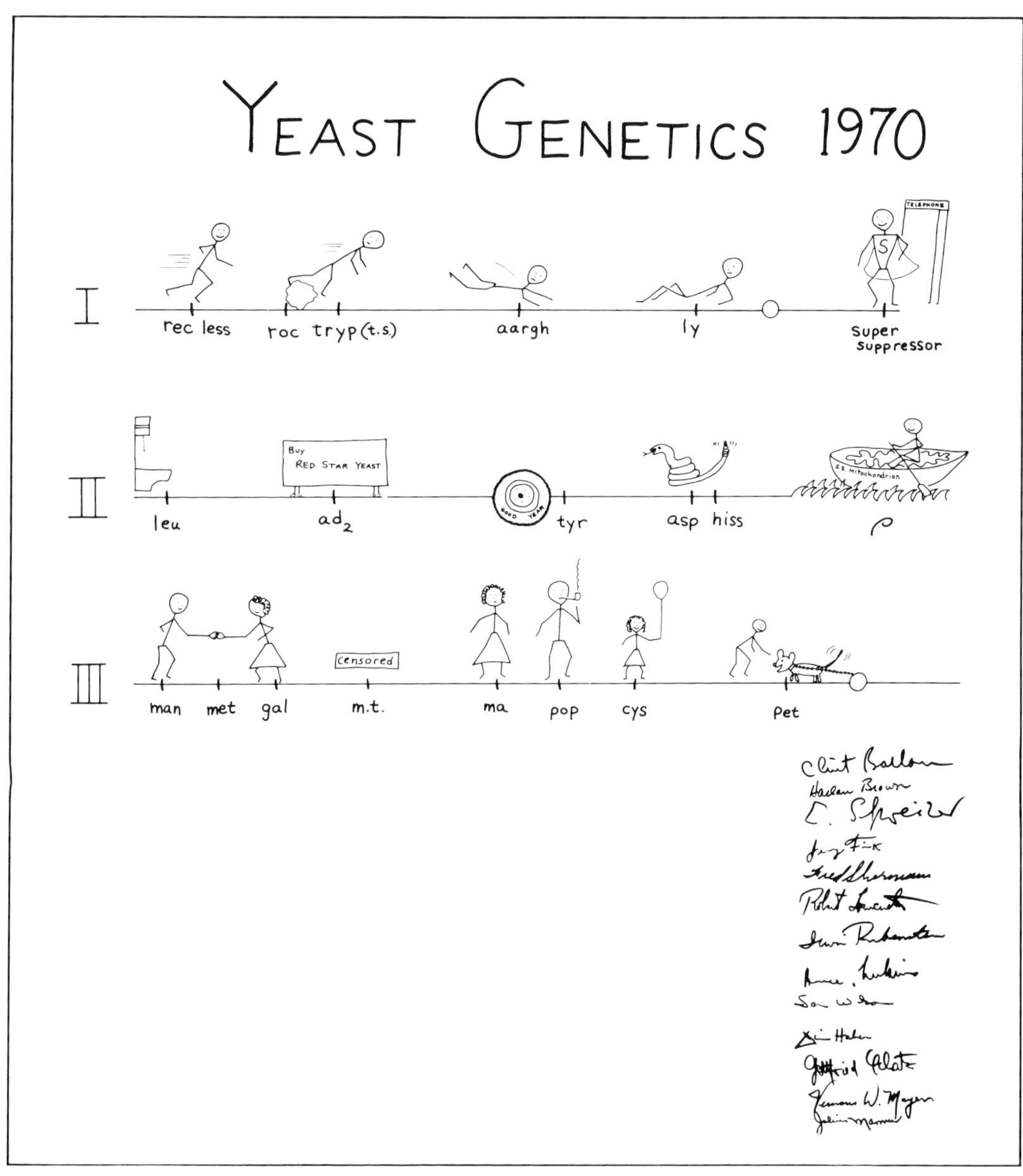


Map conceived by participants of 1970 Yeast Course:
C. Ballou, H. Brown, E. Schweizer, G. Fink,
F. Sherman, R. Lowenstein, I. Rubinstein,
H. Lukins, S. Wilson, J. Haber, G. Schatz,
V. Mayer, and J. Marmur

# Genetic Roots*

ELIZABETH W. JONES
*Carnegie Mellon University*
*Pittsburgh, Pennsylvania 15213*

I entered the University of Washington with the intent of becoming an organic chemist and then working in industry. Real life wasted little time in confronting me with two facts: Chemistry was completely hostile to the presence of women within the discipline, and I was not very good at it at the depth required anyway. I began a random walk through zoology, achieving little satisfaction. To supplement my parents' and godparents' contributions, I obtained, in what proved to be the luckiest day in my life, a part-time job washing dishes in Herschel Roman's research laboratory in the Botany Department. With time, I was *allowed* to make media—even Don Hawthorne's (I *think* I was the only one he ever allowed to do so!)—and my curiosity began to be aroused about what was going on in the laboratory. A year later, I took a genetics course and was hooked. By the time I graduated in 1960 with a B.S. in chemistry, I had taken seven genetics courses, six of which were substituted for chemistry requirements. I have never been sure whether allowing these substitutions reflected enlightened educational policy on the part of the Chemistry Department or indifference to the education of women.

Upon graduation, I ventured as far as Brooklyn for the summer, working for Sy Fogel and Don Hurst. My most vivid memories of that summer are the size of the cockroaches (we didn't have them in Seattle), my first try at inventing the wheel (devising the *best* and *universal* sporulation medium), and our joint discovery of those seductively beautiful asci, derived from what proved to be a triploid strain, that were so big and beautiful and gave so few viable progeny.

I returned to Seattle to enroll for graduate school in the Genetics Department at Washington, where I worked with Herschel once again, obtaining, in 1964, the first Ph.D. degree granted by that illustrious Department. The problem on which I worked arose from observations Roman had made on mutation in red *ade2* strains (Roman 1956). Red strains accumulate white derivatives that have acquired a second mutation that blocks purine nucleotide biosynthesis (and, hence, secondarily red pigment formation) at a step prior to formation of phosphoribosyl-5-amino-4-imidazole (otherwise known as AIR). The secondary mutations fell into six different groups that ranged in frequency from 11% for *ade3* to 36% for *ade6*. When larger samples of mutants were obtained, the mutations fell into seven groups; 4% were *ade3* alleles and 35% were *ade6* alleles (Jones 1964). The initial question I addressed was whether these differences in frequency reflected differences in gene size. What was not entirely obvious at that time was how one might measure gene size.

Seymour Benzer had shown that one could define a unit of genetic function by complementation and that such a functional unit contained many different

*I dedicate this chapter to the memory of Herschel Roman, 1914–1989, who introduced me to research and provided unfailing encouragement as I joined this most perfect profession.

mutable sites within it (Benzer 1955). Results similar to those in phage T4 were being reported for yeast (Leupold 1955; Roman 1955, 1957), *Neurospora* (Giles 1955; St. Lawrence 1956), and *Aspergillus* (Pritchard 1955), and attempts were being made to map alleles within a locus. David Catcheside had visited Herschel's laboratory in 1958 or 1959, and together they had attempted to map alleles using mitotic recombination frequencies and fluctuation tests. The effort failed. Of course, X-ray-induced recombination as a basis for mapping had not yet been devised (Manney and Mortimer 1964). I began to try allelic mapping using meiotic progeny, plating whole asci. This method was moderately successful, although the data were not very pretty (Jones 1964, 1972a). I journeyed to Bob Mortimer's laboratory in 1964 to try to obtain an independent estimate of the relative sizes of the two loci using X-ray mapping. The results were qualitatively in agreement with the meiotic results. The take-home message from both sets of experiments was clear. *ADE6* was genetically twice as large as *ADE3*, but it was clearly not seven to ten times as large. The size difference alone could not account for the difference in recovery of mutations for the two loci.

During this period, gene conversion was being studied intensively in a number of laboratories. One recurring question was whether intragenic recombination was always nonreciprocal or could yield reciprocal products. The high rate of intragenic recombination seen in meiotic progeny in yeast, for these *ade* loci at least, allowed a search for the double mutant. I dissected a few hundred asci and found reciprocal products among the progeny for widely separated *ade6* alleles (Jones 1964). Several of these mapped *ade6* alleles and the double mutants were used by Herschel in further studies of recombination (see, e.g., Fabre and Roman 1977; Roman and Ruzinski 1990).

The Genetics Department at Washington was started in 1959; I enrolled in 1960. Although the exact chronology of the development of the Department escapes me, I do remember that the faculty consisted in the early days of Herschel Roman, Dave Stadler, and Don Hawthorne. Larry Sandler, Jon Gallant, and Ben Hall joined soon thereafter. The human genetics group in the Department of Medicine, Arno Motulsky and Stan Gartler initially, was also developing, and the two groups interacted. Both Herschel and Arno welcomed visitors and ensured that students met and interacted with them. A frequent visitor was Bob Mortimer, who often stayed for a week at a time while he and Don Hawthorne wrote papers together. We often saw Frank Stahl and George Streisinger after they moved to the University of Oregon. One memorable lecture given in tandem by these two researchers introduced us to two bull snakes (or such like), each biting the other's tail, as a metaphor for describing circular maps from linear genomes (for a related illustration, see Buckley 1987). Other visitors included Boris Ephrussi (he talked about bigs and littles; we, of course, called them *grandes* and *petites*), Harry Harris, Harlyn Halvorson, Dan Lindsley, David Catcheside, David Perkins, and Robert Wagner. Sabbatical visitors included Enrico Gandini and Robin Holliday. Early postdoctoral fellows were Charles Epstein, Jørgen Friis, Freddie Sherman (before he changed his name to Fred), Rita (Vicky) Colwell, Alberta Herman, and Toshiaki Takahashi.

During this period, Don Hawthorne described mating-type switching as a mutational event (he called it gene perversion), he and Mortimer discovered and described super (nonsense) suppressors in yeast (Hawthorne and Morti-

Pictured from left to right are Beth Jones, Dave Stadler, Rita Colwell, Herschel Roman, and David Catcheside (ca. 1960).

mer 1963), and he and Jørgen Friis discovered osmotic remedial alleles (Hawthorne and Friis 1964); Fred Sherman discovered that gene conversion occurred at two times during sporulation (Sherman and Roman 1963); and genetic polymorphism in humans was a hot topic. Half the department worked on recombination. I had begun to become bored with recombination and was becoming enamored of biochemistry.

During my graduate training, I had made a hobby of taking biochemistry courses (eight total) and elected to do postdoctoral work with Boris Magasanik in order to gain practical experience in biochemistry. I later learned that I owed this opportunity to Jon Gallant, who promised Boris that I would not do something stupid like get married and quit. I carted my *ade3* mutants across country to Boris's laboratory at the Massachusetts Institute of Technology. I wanted to determine the biochemical basis of the dual adenine and histidine requirement of *ade3* mutants. Since the two pathways are, in essence, sequential linear pathways, i.e., PRPP→AMP, ATP→histidine, it was not clear how the dual requirement might arise. I had very little success in finding the block initially. I assayed a goodly number of enzymes of histidine and purine nucleotide biosynthesis, only to find all of them present at normal levels. We demonstrated early on that *ade3* mutants were not defective in phosphoribosyl-5-amino-imidazole formyltransferase (Jones and Magasanik 1967a) and thus were not comparable to the *ade10* mutants of *Schizosaccharomyces pombe* (Whitehead et al. 1966). I surfaced the quite misleading fact that *ade3* mutants, when starved of histidine, excrete copious quantities of what I finally tracked down to be anthranilic acid (had I but known it, of course, this probably presaged cross-pathway control). After a year without cracking the problem, Boris suggested I consider another project. I persuaded him to allow me to continue. After a second year, I volunteered to move to a different project. He persuaded me to continue. In the third year, in desperation, I decided to assay the enzymes involved in interconverting the various single carbon derivatives of tetrahydrofolate (THFA), even if I could not see how this analysis could conceivably provide the answer, since there were two routes of synthesis to each of the two derivatives of tetrahydrofolate, namely, 5,10-methenyl THFA and 10-formyl THFA, that are required for IMP biosynthesis. And, of course, the *ade3* mutants proved to be deficient for three THFA interconversion activities (Jones and Magasanik 1967a,b; Lazowska and Luzzatti 1970a,b; Lam and Jones 1973). Nagy et al. (1969) later showed that *ade9* mutants of *S. pombe,* which also require adenine and histidine, are defective for the same enzymatic activities. I obtained soft evidence that 5,10-methylene THFA, which might accumulate in *ade3* mutants, inhibited the fourth enzyme of histidine biosynthesis and that 5,10-methenyl THFA activated the same enzyme. These two effects, if true, could account for the histidine requirement in the mutants and, conceivably, for the finding that mutations mapping to the left end of the locus do not require histidine (Jones 1977a) since they might lack only synthetase activity in vivo (Staben and Rabinowitz 1986; Whitehead and Rabinowitz 1988). I never succeeded in solidifying that evidence. It remained puzzling why all of the mutants lacked all of the activities.

It is unclear why it never occurred to me that a single mutation might eliminate several enzymatic activities at once, especially if the activities were part of an enzyme complex or encoded in a single polypeptide. I suspect it was in part because we yeast people were mesmerized by the beauty of the operon model (Jacob and Monod 1961) and thought primarily in terms of single activities encoded by contiguous, related genes; we were not thinking of monocistronic mRNAs and polyfunctional proteins. We viewed yeast as a large bacterium rather than a small cow, to quote Boris Magasanik.

Once I had my own laboratory at Case Western Reserve University, I investigated the nature of the *ade3* mutations. I found that a very large fraction of the mutations (70%) were nonsense mutations (UAA or UAG) (Jones 1972b). Several were what we called "multisite" mutations (Jones 1972a); they did not recombine with mutations that recombined with each other. These multisite mutations, of course, were probably deletion mutations. Because multisite mutations in the *r*II region of T4 did not result in T4 DNA that was smaller than wild-type DNA (Nomura and Benzer 1961), the nature of multisite mutations was obscure. Of course, these multisite *r*II mutants did carry deletion mutations; they had headfuls of DNA of unaltered size that carried longer terminal redundancies in their headfuls of DNA (Streisinger et al. 1967).

The classification of *ade3* point mutations as nonsense mutations was based on segregational analysis (Jones 1972b). In the manuscript I submitted, some of the expected ratios were incorrect. Fortunately, a reviewer caught the errors (Bob Mortimer later kidded me gently about them). I had never really been taught tetrad analysis and realized that I did not know how to determine segregation ratios systematically. I began to teach myself as I was learning to teach undergraduate students. I ran all of my deductions past two very bright M.D.-Ph.D. students, Bruce Cohen and Peter Harris. My benighted genetics classes also were exposed. Once I had taught polyploid segregation in plants to my classes I was even able to do tetraploid tetrad analysis (Roman et al. [1955] was a marvelous guide), and I now put our graduate students through this exercise every few years whether they need it or not!

The finding that all of the *ade3* mutations recovered were probably "polar" in nature, meaning that they prevented synthesis of a gene product downstream from the mutation, suggested that mutations resulting in loss of only one of three interconversion activities might not have the typical *ade3* mutant phenotype of requiring adenine and histidine. I devised a way to isolate the mutants deficient for a single activity by making use of the fact that yeast could use the combination of formate plus glycine to satisfy a requirement for serine. The formate is activated as 10-formylTHFA and then converted to 5,10-methyleneTHFA through the agency of the interconversion enzymes. This latter derivative donates a single carbon to glycine to yield serine. Any block in any activation or interconversion step will prevent serine synthesis whether or not an adenine requirement results. A number of mutants were obtained that lacked a single interconversion activity or two activities and that required neither adenine nor histidine. The mutations proved to be alleles of the *ADE3* locus and to map among the previously described nonsense mutations. We obtained biochemical evidence that the three activities cofractionated and that point mutations which affected one activity could alter the properties of all remaining activities. Evidence for a second set of interconversion activities was also obtained (McKenzie and Jones 1977).

All of this evidence was in hand by the time the 1973 International Congress of Genetics took place in Berkeley. I attended and made a point of introducing myself to Jesse Rabinowitz in the Biochemistry Department at Berkeley, one of the world's experts on THFA interconversion enzymes in bacteria. I told him all about our results in this eukaryote and about our suspicions that this was probably a multifunctional polypeptide chain just ripe for biochemical investigation and, as it proved, persuaded him to do the biochemical followup (Paukert et al. 1977; Staben and Rabinowitz 1983, 1986; Appling and Rabinowitz 1985; Whitehead and Rabinowitz 1988). Despite my long stay in Boris's laboratory, I had not become a card-carrying biochemist but preferred to walk the boundary between genetics and biochemistry. Indeed, I did not publish in a biochemical journal until 1991–1992 (Jones 1991; Nebes and Jones 1991; Manolson et al. 1992).

The finding that a substantial fraction of mutations at the *ADE3* locus go undetected in the color screen because they do not block purine biosynthesis, together with the finding of a size difference between the two loci, can probably in large measure account for the difference in recovery of "*ade3*" and *ade6* mutations as whites. However, I confess that the journey through *ADE3* has left me with a profound skepticism about finding simple or single explanations for differences (or absences) in mutant recovery. The "genome paradox," according to which the number of essential genes identified by temperature-sensitive lethal mutations is far less than the number expected, leaves me utterly unmoved!

Another by-product of that 1973 International Congress was my exposure to Ames's use of overlay tests containing liver extract (Ames et al. 1973). This had a remarkably liberating effect on me when I came to design agar overlays, as I did when I took up the study of proteases. If you can put *liver extract* in agar overlays, you can put *anything* in them! And so we did, particularly in HPA overlays for detecting protease B activity (Zubenko et al. 1979).

During 1972–1973, I had become interested in yeast proteases after talking with Charlie Miller, who studied *Salmonella* proteases. I worked out a plate test that I thought would detect proteases. The substrate was acetylphenylalanine β-naphthyl ester (APE). Previous workers had reported cleavage of acetyltyrosine ethyl ester, the closest comparable substrate to APE, by both carboxypeptidase Y (CpY) and protease B (PrB). I was uncertain initially whether one or the other or both enzymes would catalyze cleavage of APE and whether I would be able to obtain mutants unable to cleave APE. I did obtain mutants, of course (Jones 1977b). However, I thought it necessary to test for both CpY and PrB activity in all of the mutants (protease A was assayed as well for no particularly good reason). The initial screen yielded many mutants and resulted in identification of many genes. The pleiotropic nature of most of the mutations revealed by the enzyme assays (and growth tests) was immediately exciting. Little did I know that these mutants would provide an entry into the problems of vacuolar hydrolase precursor processing and vacuole biogenesis, problems that have occupied those of us in my laboratory every since.

## ACKNOWLEDGMENTS

I am grateful for the support I have received throughout my graduate education and research career from the National Institutes of Health, first through a traineeship and then through research grants. Since moving to Carnegie Mellon University, support has been from research grants DK-18090 and GM-29713. I have greatly enjoyed the undergraduate and graduate students and postdoctoral fellows who have worked in my laboratory or taken my courses or both. I have been very lucky.

## REFERENCES

Ames, B.N., W.E. Durston, E. Yamasaki, and F.D. Lee. 1973. Carcinogens as mutagens: A simple test system combining liver homogenates for activation and bacteria for detection. *Proc. Natl. Acad. Sci.* **70:** 2281–2285.

Appling, D.R. and J.C. Rabinowitz. 1985. Evidence for overlapping active sites in a multifunctional enzyme: Immunochemical and chemical modification studies on $C_1$-tetrahydrofolate synthase from *Saccharomyces cerevisiae*. *Biochemistry* **24:** 3540–3547.

Benzer, S. 1955. Fine structure of a genetic region in bacteriophage. *Proc. Natl. Acad. Sci.* **41:** 344–354.

Buckley, R. 1987. *The greedy python*. Picture Book Studio, Saxonville, Massachusetts.

Fabre, F. and H. Roman. 1977. Genetic evidence for inducibility of recombination competence in yeast. *Proc. Natl. Acad. Sci.* **74:** 1667–1671.

Giles, N.H. 1955. Forward and back mutation at specific loci in *Neurospora*. *Brookhaven Symp. Biol.* **8:** 103–123.

Hawthorne, D.C. and J. Friis. 1964. Osmotic remedial mutants, a new classification for nutritional mutants in yeast. *Genetics* **50:** 829–839.
Hawthorne, D.C. and R.K. Mortimer. 1963. Super-suppressors in yeast. *Genetics* **48:** 617–620.
Jacob, F. and J. Monod. 1961. Genetic regulatory mechanisms in the synthesis of proteins. *J. Mol. Biol.* **3:** 318–356.
Jones, E.W. 1964. "A comparative study of two adenine loci in *Saccharomyces cerevisiae.*" Ph.D. thesis, University of Washington, Seattle.
———. 1972a. Fine structure analysis of the *ADE3* locus in *Saccharomyces cerevisiae. Genetics* **70:** 233–250.
———. 1972b. Nonsense mutations in the *ADE3* locus of *Saccharomyces cerevisiae. Genetics* **71:** 217–232.
———. 1977a. Bipartite structure of the *ade3* locus of *Saccharomyces cerevisiae. Genetics* **85:** 209–233.
———. 1977b. Proteinase mutants of *Saccharomyces cerevisiae. Genetics* **85:** 23–33.
———. 1991. Three proteolytic systems in the yeast *Saccharomyces cerevisiae. J. Biol. Chem.* **266:** 7963–7966.
Jones, E.W. and B. Magasanik. 1967a. Phosphoribosyl-5-amino-4-imidazole carboxamide formyltransferase activity in the adenine-histidine auxotroph of *Saccharomyces cerevisiae. Biochem. Biophys. Res. Commun.* **29:** 600–604.
———. 1967b. Genetic block in the interconversion of folic acid coenzymes in *Saccharomyces cerevisiae. Bacteriol. Proc.* p. 127.
Lam, K.B. and E.W. Jones. 1973. Mutations affecting levels of tetrahydrofolate interconversion enzymes in *Saccharomyces cerevisiae. Mol. Gen. Genet.* **123:** 199–208.
Lazowska, J. and M. Luzzatti. 1970a. Biochemical deficiency associated with *ade3* mutations in *Saccharomyces cerevisiae.* I. Levels of three enzymes of tetrahydrofolate metabolism. *Biochem. Biophys. Res. Commun.* **39:** 34–39.
———. 1970b. Biochemical deficiency associated with *ade3* mutations in *Saccharomyces cerevisiae.* II. Separation of two forms of methylenetetrahydrofolate dehydrogenase. *Biochem. Biophys. Res. Commun.* **39:** 40–45.
Leupold, U. 1955. Versuche zur genetischen klassifizierung adenin-abhängiger mutanten von *Schizosaccharomyces pombe. Arch. Julius Klaus-Stift. Vererbungsforsch. Sozialanthropol. Rassenhyg.* **30:** 506–516.
Manney, T.R. and R.K. Mortimer. 1964. Allelic mapping in yeast using X-ray induced mitotic reversion. *Science* **143:** 581–582.
Manolson, M.F., D. Proteau, R.A. Preston, A. Stenbit, B.T. Roberts, M.A. Hoyt, D. Preuss, J. Mulholland, D. Botstein, and E.W. Jones. 1992. The *VPH1* gene encodes a 95-kDa integral membrane polypeptide required for *in vivo* assembly and activity of yeast vacuolar $H^+$-ATPase. *J. Biol. Chem.* **267:** 14294–14303.
McKenzie, K.Q. and E.W. Jones. 1977. Mutants of the formyltetrahydrofolate interconversion pathway in yeast. *Genetics* **86:** 85–102.
Nagy, M., H. Heslot, and L. Poirier. 1969. Conséquences d'une mutation affectant la biosynthèse des coenzymes foliques chez le *Schizosaccharomyces pombe. C.R. Acad. Sci.* **269:** 1268–1271.
Nebes, V.L. and E.W. Jones. 1991. Activation of the proteinase B precursor of the yeast *Saccharomyces cerevisiae* by autocatalysis and by an internal sequence. *J. Biol. Chem.* **266:** 22851–22857.
Nomura, M. and S. Benzer. 1961. The nature of the "deletion" mutants in the rII region of phage T4. *J. Mol. Biol.* **3:** 684–692.
Paukert, J.L., G.R. Williams, and J.C. Rabinowitz. 1977. Formyl-methenyl-methylenetetrahydrofolate synthetase (combined): Correlation of enzymic activities with limited proteolytic degradation of the protein from yeast. *Biochem. Biophys. Res. Commun.* **77:** 147–154.
Pritchard, R. 1955. The linear arrangement of a series of alleles of *Aspergillus nidulans. Heredity* **9:** 343–371.
Roman, H. 1955. Mutations studies in yeast. *Genetics* **40:** 592.
———. 1956. A system selective for mutations affecting the synthesis of adenine in yeast. *C.R. Trav. Lab. Carlsberg Ser. Physiol.* **26:** 299–314.
———. 1957. Studies of gene mutation in *Saccharomyces. Cold Spring Harbor Symp.*

*Quant. Biol.* **21:** 175–185.

Roman, H. and M.M. Ruzinski. 1990. Mechanisms of gene conversion in *Saccharomyces cerevisiae. Genetics* **124:** 7–25.

Roman, H., M.M. Phillips, and S.M. Sands. 1955. Studies of polyploid *Saccharomyces.* I. Tetraploid segregation. *Genetics* **40:** 546–561.

Sherman, F. and H. Roman. 1963. Evidence for two types of allelic recombination in yeast. *Genetics* **48:** 255–261.

St. Lawrence, P. 1956. The *q* locus of *Neurospora crassa. Proc. Natl. Acad. Sci.* **42:** 189–194.

Staben, C. and J.C. Rabinowitz. 1983. Immunological crossreactivity of eukaryotic $C_1$-tetrahydrofolate synthase and prokaryotic 10-formyltetrahydrofolate synthetase. *Proc. Natl. Acad. Sci.* **80:** 6799–6803.

———. 1986. Nucleotide sequence of the *Saccharomyces cerevisiae ADE3* gene encoding $C_1$-tetrahydrofolate synthase. *J. Biol. Chem.* **261:** 4629–4637.

Streisinger, F., J. Emrich, and M.M. Stahl. 1967. Chromosome structure in phage T4. III. Terminal redundancy and length determination. *Proc. Natl. Acad. Sci.* **57:** 292–295.

Whitehead, E., M. Nagy, and H. Heslot. 1966. Interactions entre la biosynthèse des purines nucléotides et celle de l'histidine chez le *Schizosaccharomyces pombe. C.R. Acad. Sci.* **263:** 819–821.

Whitehead, T.R. and J.C. Rabinowitz. 1988. Nucleotide sequence of the *Clostridium acidiurici* ("*Clostridium acidi-urici*") gene for 10-formyltetrahydrofolate synthetase shows extensive amino acid homology with the trifunctional enzyme $C_1$-tetrahydrofolate synthase from *Saccharomyces cerevisiae. J. Bacteriol.* **170:** 3255–3261.

Zubenko, G.S., A.P. Mitchell, and E.W. Jones. 1979. Septum formation, cell division, and sporulation in mutants of yeast deficient in proteinase B. *Proc. Natl. Acad. Sci.* **76:** 2395–2399.

# My Life with Cytochrome *c*

FRED SHERMAN
*Departments of Biochemistry and Biophysics*
*University of Rochester School of Medicine and Dentistry*
*Rochester, New York 14642*

## THE BEGINNING

I have devoted almost my entire scientific career to using cytochrome *c* for investigating many diverse problems in biology. In this short chapter, I describe why I chose and continued to use this system for more than 30 years. I was introduced to yeast as a graduate student with Robert K. Mortimer at Berkeley, where I studied the induction of $\rho^-$ mitochondrial mutants by elevated temperature. After receiving my degree, I investigated recombination of yeast as a postdoctoral fellow at Seattle with the late Herschel Roman. Subsequently in 1960, I entered a second postdoctoral position in the laboratory of the late Boris Ephrussi at Gif-sur-Yvette, France, where I initiated a research program that led to the discovery of the *CYC1* gene.

After arriving in Boris Ephrussi's laboratory, Piotr P. Slonimski and I began a study of *pet* mutants, i.e., those mutants that have mutations of nuclear genes and that are unable to grow on media having nonfermentable carbon sources as a sole energy source (Sherman and Slonimski 1964). These *PET* nuclear genes encode essential components of mitochondria that are required for aerobic metabolism but are distinct from the $\rho^+$ determinant that was eventually shown to correspond to mitochondrial DNA. I began this study by collecting mutants that were already demonstrated or suspected to have nuclear defects. For this purpose, I contacted David Pittman, Maurice Ogur, Donald C. Hawthorne, and Robert K. Mortimer, who constituted the majority of yeast geneticists at that time. One of the strains, 662.8, obtained from Maurice Ogur, turned out to be of considerable importance. The strain consisted of a mixture of haploid and diploid cells and, as expected, did not grow on nonfermentable substrates. Genetic analysis revealed that the strain contained two mutations: *pet4-1*, which prevented growth on nonfermentable substrates, and *cyc1-1*, which still allowed growth on nonfermentable substrates but caused a 95% diminution of cytochrome *c*. Thus, the *cyc1-1* mutation was uncovered only because it was fortuitously in the same strain with *pet4* and because cytochrome spectra of the meiotic segregants were examined. At that time, it was indeed puzzling why the *cyc1-1* mutation was in the strain. Furthermore, *cyc1-1* was also sensitive to UV light and hypertonic media (Sherman et al. 1965), phenotypes eventually shown to be due to deletion of an approximately 12-kb segment that encompassed the *RAD7* and *OSM1* loci, as well as the *CYC1* locus (Singh and Sherman 1978; Stiles et al. 1981a; Melnick and Sherman 1990). Some 5 years later, a second deletion (*cyc1-237*) with the same seemingly pleiotropic phenotypes was unexpectedly uncovered among meiotic segregants. The origin of this type of deletion was not understood until Liebman et al. (1979) noted that certain laboratory strains spontaneously gave rise to high frequencies of deletions encompassing the *CYC1*, *OSM1*, and *RAD7* genes and that the deletions are flanked by Ty1 elements (Liebman et al. 1981; Stiles et al. 1981a). Thus, *cyc1-1*, the first cytochrome-*c*-deficient mutant, was uncovered because of the rare occurrence of a spontaneous deletion in a strain containing an unrelated *pet* mutation.

## IMPORTANCE OF CYTOCHROME *C*

Finding a cytochrome *c* yeast mutant was of considerable importance in 1960. At that time, when DNA sequencing was in the realm of science fiction, all information on gene structure was inferred from mutationally altered proteins. A major effort was to decipher the genetic code on the bases of amino acid replacements and mutagenic specificity. Early in 1960, only three proteins were amenable to mutational analysis: tryptophan synthetase from *Escherichia coli* (Yanofsky et al. 1962), lysozyme from bacteriophage T4 (Streisinger et al. 1967), and tobacco mosaic virus coat protein (Tsugita and Fraenkel-Conrat 1963). Cytochrome *c* was one of the few proteins that could be easily purified. Its low molecular weight allowed easy diagnosis of altered sequences by peptide mapping and amino acid compositional analysis. Furthermore, the 12-amino-acid segment encompassing the heme group was already sequenced (Tuppy 1958), and entire cytochromes *c* from various species were being sequenced in several laboratories. Thus, investigating cytochrome *c* appeared to be an ideal project, especially because yeast had the advantage of being a eukaryotic microorganism with a well-defined genetic system. The advantange of using yeast cytochrome *c* became even more evident a few years later when the complete amino acid sequence of iso-1-cytochrome *c* was reported by Narita et al. (1963).

## *CYC1* ENCODES ISO-1-CYTOCHROME *C*

Because the *cyc1-1* mutant contained a minor form of a chromatographic distinct cytochrome *c*, I initiated experiments after arriving at the University of Rochester early in 1960 that led to the finding of two forms of cytochrome *c* in yeast, iso-1-cytochrome *c* and iso-2-cytochrome *c* (Sherman et al. 1965). A similar study was carried out at Gif-sur-Yvette by Slonimski et al. (1965), who also suggested the intriguing but fallacious hypothesis that the apo form of the minor species, iso-2-cytochrome *c*, was the repressor of iso-1-cytochrome *c*.

In addition, a major effort was to devise methods to detect cytochrome-*c*-deficient mutants (see below). The first systematic screen involving spectroscopic examinations of large number of strains resulted in uncovering *cyc1-2*, the second mutation at the *CYC1* locus, as well as mutations of the *CYC2* and *CYC3* loci (Sherman 1964). To establish that *CYC1* encodes iso-1-cytochrome *c* and was not a regulatory gene, it was necessary to demonstrate that an allele encodes an altered form of iso-1-cytochrome *c* with a change in the primary sequence. Because *cyc1-1* was a deletion and did not produce intragenic revertants, the *cyc1-2* mutant was critical for establishing in 1966 that the *CYC1* gene encodes the primary structure of iso-1-cytochrome *c* (Sherman et al. 1966). An intragenic revertant, *CYC1-2-A*, was shown to have a Gln-21→Tyr-21 replacement within the heme peptide.

It is also of interest to note that the *CYC3* gene, uncovered in the initial screen, was shown 23 years later to encode heme lyase, the enzyme catalyzing the covalent attachment of the heme group to the iso-cytochromes *c* (Dumont et al. 1987). More recently, the *CYC2* gene has been cloned, sequenced, and shown to encode a mitochondrial protein required for normal mitochondrial import of cytochrome *c* (Dumont et al. 1992).

## ISOLATION AND CHARACTERIZATION OF *cyc1* MUTANTS

In the early 1960s, studies of gene structure and gene expression were almost entirely dependent on the isolation and characterization of a large number of mutants. As mentioned above, the first methods to detect cytochrome-*c*-deficient mutants were developed more than 25 years ago (Sherman 1964). The first method involved low-temperature (–196°C) spectroscopic examination of large numbers of strains on the surfaces of nutrient agar plates. Colonies, derived from mutagenized cells, were inoculated on square plastic petri dishes, each containing 36 strains. After incubation and growth of the strains, the dishes were frozen in liquid nitrogen. The dishes that did not explode were placed on a rack under a simple spectroscope and were moved by hand to center each strain in the light path. Although altered absorption spectra were just barely perceptible, I was eventually able to examine more than 2000 strains per day after the method was perfected. However, I was hardly able to see anything the next day. More importantly, one critical single mutant, *cyc1-2*, described above, was isolated by this technique after examining approximately 14,800 strains.

A few years later, we developed a more expedient procedure that relied on the staining of colonies with benzidine reagents (Sherman et al. 1968). The staining of colonies required a short exposure to a $H_2O_2$ solution, followed by exposure to a benzidine solution. Because the transfer and removal of solutions disrupted the colonies, a search was made for a method to fix the colonies but still allow effective contact with the solutions. After testing numerous agents, we discovered that gently spraying the surfaces of petri plates with ordinary hair spray was effective. Conditions were worked out with a low-priced brand of hair spray, which we used for several years. However, one day this brand was no longer available from our usual vender. I immediately wrote to the company asking who were the local distributors. I received a letter informing me that their product was no longer available in Rochester. Furthermore, the company told me that they had sent copies of my letter to all local wholesalers, as evidence that their hair spray was in great demand. Fortunately, other, more expensive, brands were equally effective.

Eventually, we developed an even more expedient procedure for isolating Cyc$^-$ mutants that depended on either the absence or lack of function of cytochrome *c* (Sherman et al. 1974). This method was based on the finding that mutants partially deficient in cytochrome *c*, containing approximately 5% of the total normal level, are defective in the utilization of lactate but are still able to utilize other nonfermentable substrates such as glycerol or ethanol. These partially deficient mutants that are unable to utilize lactate are resistant to the toxic action of the analog chlorolactate. Thus, chlorolactate medium, which contains chlorolactate and the nonfermentable carbon source glycerol, can be used to enrich for mutants partially defective in cytochrome *c*. As expected and fortunately for us, the major class of cytochrome-*c*-deficient mutants arising on chlorolactate medium were *cyc1* mutants, which lacked iso-1-cytochrome *c* but retained the normal low-level amount of iso-2-cytochrome *c*.

The determination of the sites of the mutations in the *CYC1* gene was also critical for their characterization. At that time before DNA sequencing, genetic mapping was the only means to estimate the relative positions of point mutations without resorting to protein sequencing. Furthermore, the only reliable mapping scheme required a combination of both deletion mapping and two point crosses. Deletion mapping unambiguously established the order, whereas two point crosses established the identity of sites. However, deletions were extremely rare and almost unknown in yeast. Fortunately, a novel procedure for generating deletions at the *CYC1* locus was developed. Deletions were recovered from crosses that contained extensive dissimilarities of sequences in homologous regions of two *CYC1* alleles. These alleles encoded iso-1-cytochromes *c* that were functional but contained two different sequences, respectively, in the dispensable amino-terminal region of the protein. The diploids were sporulated and plated on chlorolactate medium, and *cyc1* mutants deficient in iso-1-cytochrome *c* were selected from the meiotic progeny. More than 25% of the *cyc1* mutants contained deletions of various lengths, from those covering adjacent codons to those encompassing the entire *CYC1* locus and flanking genes. A total of 60 different lengths of deletions were uncovered among the 104 deletions obtained by this procedure (Sherman et al. 1975). Although it is still unclear exactly how these deletions arose, the mechanism may be related to heteroallelic mispairing. Nevertheless, these deletions proved to be invaluable for mapping point mutations, especially after calibration with sites defined by amino acid replacement of iso-1-cytochrome *c* from intragenic revertants. Years later, DNA sequencing revealed that the sites of *cyc1* point mutations were generally within a codon or two of the sites estimated by genetic mapping.

## GENERATING ALTERED GENES AND PROTEINS: THEN AND NOW

Two major classes of revertants are distinguished after high densities of *cyc1* cells are plated on synthetic medium containing lactate as the sole carbon source (Sherman et al. 1968). The first class usually forms visible colonies after 5–7 days of incubation, whereas colonies of the second class usually arise after 10 days. The first class constitutes intragenic revertants with normal or altered iso-1-cytochromes *c*, together with the low amount of iso-2-cytochrome *c* characteristic of normal strains. In contrast, most of the second type of revertants contain only iso-2-cytochrome *c*, but usually in amounts higher than normal. Thus, a large number of altered iso-1-cytochromes *c* were uncovered in early studies by analyzing the series of mutations of the type $CYC1^{+}$→*cyc1-x*→*CYC1-x-y*, where $CYC1^{+}$ denotes the wild-type gene that encodes iso-1-cytochrome *c*, *cyc1-x* denotes mutations that cause deficiency or nonfunction of iso-1-cytochrome *c*, and *CYC1-x-y* denotes intragenic reversions that restore at least partial activity and give rise to either the normal or altered iso-1-cytochrome *c*. More than 500 *cyc1-x* mutants were isolated and characterized, and more than 100 different iso-1-cytochrome *c* sequences were obtained from *CYC1-x-y* revertants (Hampsey et al. 1988).

In more recent times, numerous altered iso-1-cytochromes *c* have also been generated using the standard method of site-directed mutagenesis, which relies on single-stranded *Escherichia coli* vectors containing the target sequence and a short synthetic oligonucleotide containing the desired alterations. Of more importance, we have described a more convenient procedure for producing specific alteration of genomic DNA by transforming yeast directly with synthetic oligonucleotides (Moerschell et al. 1988, 1991; Yamamoto et al. 1992). This procedure is easily carried out by transforming a defective *cyc1* mutant and selecting for revertants that are at least partially functional. The oligonucleotide used for transformation contains a sequence that corrects the defect and produces additional alterations at nearby sites. This technique is ideally suited for producing a large number of specific alterations that change a completely nonfunctional allele to at least a partially functional form. The selection procedure used with *cyc1* mutants allows recovery of altered iso-1-cytochromes *c* with less than 1% of the normal activity (Moerschell et al. 1988, 1991). By using various *cyc1* mutants having alterations along the gene, all 20 amino acid replacements can be conveniently generated at almost any site by simply transforming the strain with sets of oligonucleotides. For example, all possible amino acid residues were conveniently introduced adjacent to the initiator methionine residue for systematically investigating amino-terminal processing of iso-1-cytochrome *c* (Moerschell et al. 1990). In addition, transformation directly with degenerate oligonucleotide, followed by DNA sequencing the pertinent polymerase chain reaction (PCR)-amplified region, has been used to produce iso-1-cytochromes *c* with all 20 replacements of Gly-11 (L. Linske-O'Connell and F. Sherman, unpubl.).

In a sense, transformation directly with oligonucleotide is the achievement of an ultimate goal to uncover specific mutagens. In early studies, many mutagens were investigated with the hope of controlling specific base-pair changes or at least distribution of base-pair changes. Only a few of the more commonly used mutagens were reported, but the bulk of the studies was disappointing and not published. Although the procedure with synthetic oligonucleotides is a form of DNA transformation, I cannot help but secretly look upon this process as a dream come true, the ultimate specific and controllable mutagen.

## DEDUCING DNA SEQUENCES FROM PROTEIN SEQUENCES

Of critical importance in these early studies was the collaboration with John Stewart, in which DNA sequences were deduced from the amino acid alterations in revertant proteins (Sherman and Stewart 1971, 1978). During the course of experiments that spanned over two decades, Stewart analyzed more than 3000 altered forms of iso-1-cytochromes *c*. These early studies covered diverse topics, including nonsense codons and suppressors, initiation of translation, mutagenesis, recombination, and structure-function relationships of iso-1-cytochrome *c*. For example, the nucleotide sequences of chain-terminating codons were deduced in the early 1970s from the finding that almost all of the revertant proteins contained single replacements of amino acids whose codons differed from TAA or TAG by single bases (Stewart and Sherman 1972; Stewart et al. 1972).

One of the major highlights of these early studies was the identification of the ATG initiator codon by mutationally altered iso-1-cytochromes *c*. In the first report by Stewart et al. (1971), 9 of 210 *cyc1* mutants were shown to be deficient in iso-1-cytochrome *c* due to alterations of the ATG codon that is required for initiation of protein synthesis. Structural analysis of 64 revertant

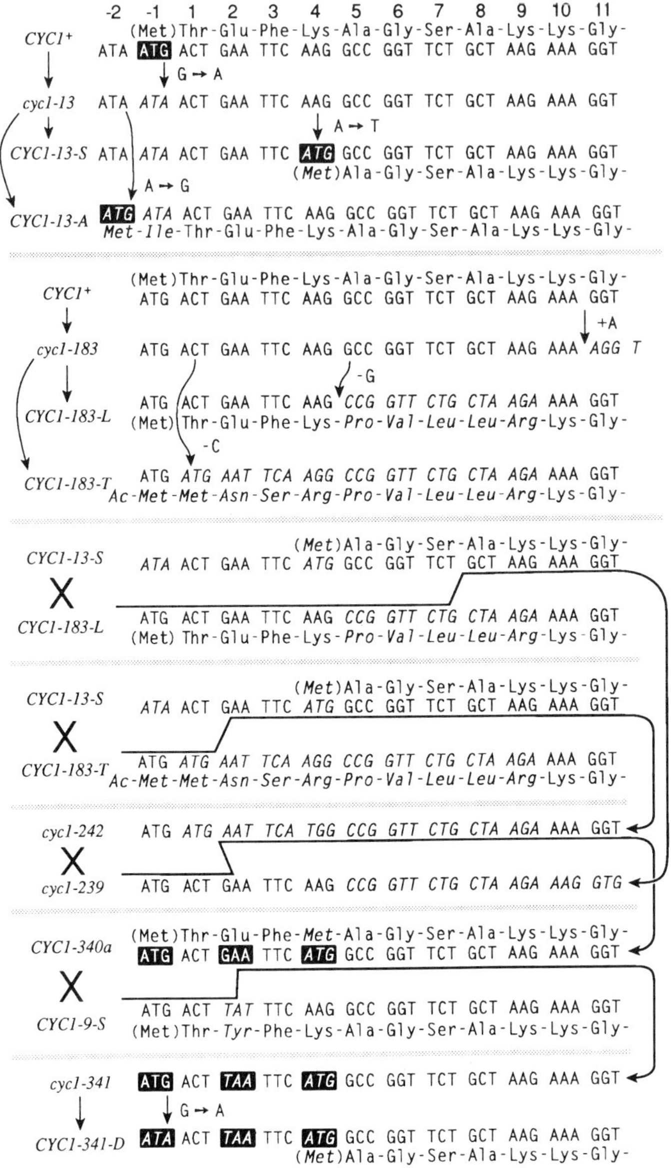


*Figure 1* Mutational events leading from the normal gene *CYC1*+ to the initiator mutant *cyc1-13* and the frameshift mutant *cyc1-183*, and the mutational events giving rise to intragenic revertants with altered iso-1-cytochromes *c*. Specific *cyc1* and *CYC1* meiotic recombinants were obtained from the indicated crosses. The amino acid residues and codons that differ from the normal are shown in italics. The methionine residues shown in parentheses are excised from the iso-1-cytochromes *c*. The formation of the *cyc1-13* mutation and to the revertants containing the long (*CYC1-13-A*) and short (*CYC1-13-S*) forms of iso-1-cytochrome *c* is shown at the top of the figure. These and other results (Stewart et al. 1971) indicated that ATG was the initiation codon, protein synthesis can initiate at several sites without any obvious ribosomal binding site, and a methionine aminopeptidase cleaves amino-terminal residues of methionine from some but not all termini. Recombination was used to generate sequences with ATG triplets at positions –1 and 4 and TAA triplets at position 2, as shown at the bottom of the figure. These results, as well as other results summarized in Sherman and Stewart (1982), established some of the basic properties of translation (see text).

-1 1 2 3 4 5 6 7 8 9 10 11 12 13 14
(Met) Thr - Glu - Phe - Lys - Ala - Gly - Ser - Ala - Lys - Lys - Gly - Ala - Thr - Leu -
ATG ACT GAA TTC AAG GCC GGT TCT GCT AAG AAA GGT GCT ACA CTN
CGA TTC TTT CCA CGA
Synthetic probe

*Figure 2* The amino acid sequence of the amino-terminal region of iso-1-cytochrome *c* and the corresponding sequence of the 44 nucleotides that was deduced from frameshift and initiator mutants (Sherman and Stewart 1973; Stewart and Sherman 1974). Synthetic probe denotes the 15-mer (Szostak et al. 1979) used to clone the *CYC1* gene by hybridization (Stiles et al. 1981b).

proteins from 17 of these *cyc1* intitiator mutants indicated that some of the reverse mutations introduced initiator codons at new sites as illustrated at the top of Figure 1 (Sherman and Stewart 1982). Each of the *cyc1* mutants gave rise to revertant iso-1-cytochromes *c* that had one of the following amino-terminal additions: Met-Ile-; Met-Leu-; Met-Arg-; Met-Lys-; and Met-Val-. These results were explained by the mutational pathways presented in Figure 1, illustrating the formation and reversion of the *cyc1-13* mutant. Further rules governing initiation of translation were deduced from specific sequences that were generated by recombination between *cyc1* mutations in vivo, long before the existence of site-directed mutagenesis (see below) (Sherman and Stewart 1975, 1982).

Early in 1970, we also attempted to isolate altered forms of revertant proteins that could be used to deduce the DNA sequence of the *CYC1* gene. The 44-base-pair sequence at the 5′-translated region of the gene, shown in Figure 2, was actually deduced from altered iso-1-cytochromes *c* from frameshift and initiator revertants (Sherman and Stewart 1973; Stewart and Sherman 1974).

## "SITE-DIRECTED" MUTAGENESIS THE OLD-FASHIONED WAY

A variety of mutagens were employed in these early studies in order to recover all possible single-base-pair changes and also in attempts to obtain certain specific changes. Although mutagenic treatments alone gave rise mainly to single-base-pair changes that could not be predicted, except G·C→A·T transitions, specific sequences were constructed by elaborate and systematic mutational and recombinational steps in vivo. Using techniques similar to those developed for selecting *cyc1* mutants, it was possible to obtain *cyc1* recombinants in vivo with desired sequences by crossing certain *CYC1* mutants that contained altered iso-1-cytochromes *c* and plating the sporulated cross on chlorolactate medium. The resistant colonies were analyzed genetically for *cyc1* defects, and the sites of the lesions were determined by fine-structure mapping with defined *cyc1* tester strains. Likewise, *CYC1* recombinants could be constructed from two *cyc1* mutants by crossing, sporulating, and plating on lactate medium. The initial *cyc1* mutants and *CYC1* revertants served as a resource for designing the desired sequences. This approach of using mutation and recombination was especially useful for generating single and multiple ATG triplets in the 5′ region of the gene (Sherman and Stewart 1975, 1982). For example, sequences containing various combinations of ATG triplets at positions −1 and 4 and TAA triplets at position 2 were generated by the steps outlined in Figure 1. Some of the basic conclusions derived from these studies were that initiation occurred only at ATG codons, translation can occur at any site within a 37-nucleotide region, translation initiates only at the most 5′ ATG codon, and translation does reinitiate after a terminating codon (Sherman and Stewart 1982).

## CLONING *CYC1* AND LATER DEVELOPMENTS

The development of recombinant DNA procedures in the mid 1970s obviously eliminated the critical need of altered protein sequences to deduce DNA sequences. However, prior to the reports of Struhl et al. (1976) and especially Hinnen et al. (1978), the only practical procedure for identifying a DNA yeast clone was hybridization to nucleic acid probes. Other than the tRNA and rRNA genes, the *CYC1* gene was the only yeast gene with at least a partially known DNA sequence. With a knowledge of the 44-base-pair sequence (Fig. 2), Szostak et al. (1977, 1979) and Montgomery et al. (1978) synthesized oligonucleotides 13 and 15 residues long (shown in Fig. 2), respectively, that were used to identify the clone containing the *CYC1* gene (Montgomery et al. 1978; Stiles et al. 1981b).

The availability of the cloned *CYC1* gene, and soon after the *CYC7* gene (Montgomery et al. 1980), stimulated investigations of transcriptional regulation in numerous laboratories, including those of Michael Smith, Leonard Guarente, Benjamin Hall, and Richard Zitomer, as well as in my laboratory. Detailed analysis of the promoter region, primarily by Guarente and coworkers, revealed upstream activation sites, UAS1 and UAS2 (Guarente 1987). These upstream regions were instrumental for identifying and isolating proteins required for *CYC1* transcription, including HAP1, which activates UAS1, and HAP2 and HAP3, which activate UAS2 (Guarente 1987). Furthermore, the multiple TATA elements (α and β) and their rules for transcription initiation have been investigated in great detail (Li and Sherman 1991). In addition, the transcription termination region, originally identified with the *cyc1-512* mutant obtained in vivo (Zaret and Sherman 1982), has been systematically investigated for mRNA cleavage (Butler and Platt 1988) and termination properties (Russo et al. 1991).

## THE ISO-CYTOCHROMES *C* SYSTEM TODAY

Recently, provocative developments have occurred in the areas of posttranslational modification, heme attachment, mitochondrial import, and structure, stability, and folding of the iso-cytochromes *c* (Hickey et al. 1992; Sherman et al. 1992). The three-dimensional structures of iso-1-cytochrome *c* (Louie et al. 1988) and iso-2-cytochrome *c* (G.D. Brayer, unpubl.) have been determined, allowing better interpretations of altered forms (Hampsey et al. 1988; Hickey et al. 1992). These studies are revealing a role for protein stability in the posttranslational regulation of the iso-cytochromes *c*. Apparently, partially repressed yeast uses both differential transcription (Laz et al. 1984; Pillar and Bradshaw 1991) and differential stability (Dumont et al. 1990) of the apocytochromes *c* as regulatory mechanisms for maintaining elevated proportions of iso-2-cytochrome *c*.

Reflecting a basic philosophy, we are still trying to use the iso-cytochrome *c* system to address diverse problems. Nuclear localization signals and genes involved in nuclear import have recently been investigated by mistargeting iso-1-cytochrome *c* and by using the techniques developed for this system (Gu et al. 1992). Altered forms of the iso-cytochromes *c* are being used for such studies as protein-protein interactions, electron transfer, electrochemistry, immunology, ubiquitination, etc. It is satisfying to know that the use of this beautiful iso-cytochrome *c* system has not been exhausted, even though it started 30 years ago.

## ACKNOWLEDGMENT

The work reported in this chapter and carried out by the author was supported by U.S. Public Health Service research grant RO1-GM-12702.

## REFERENCES

Butler, I.S. and T. Platt. 1988. RNA processing generates the mature 3′end of yeast *CYC1* messenger RNA *in vitro*. *Science* **242:** 1270–1274.

Dumont, M.E., J.F. Ernst, and F. Sherman. 1987. Identification and sequence of the gene encoding cytochrome *c* heme lyase in the yeast *Saccharomyces cerevisiae*. *EMBO J.* **6:** 235–241.

Dumont, M.E., J.B. Schlichter, T.S. Cardillo, M.K. Hayes, and F. Sherman. 1992. *CYC2* encodes a factor involved in mitochondrial import of yeast cytochrome *c*. *Mol. Cell. Biol.* (in press).

Dumont, M.E., A.J. Mathews, B.T. Nall, S.B. Baim, D.C. Eustice, and F. Sherman. 1990. Differential stability of two apo-isocytochromes *c* in the yeast *Saccharomyces cerevisiae*. *J. Biol. Chem.* **265:** 2733–2739.

Guarente, L. 1987. Regulatory proteins in yeast. *Annu. Rev. Genet.* **21:** 425–452.

Gu, Z., R.P. Moerschell, F. Sherman, and D.S. Goldfarb. 1992. *NIP1*, a gene required for nuclear transport in yeast. *Proc. Natl. Acad. Sci.* **89:** 10355–10359.

Hampsey, D.M., G. Das, and F. Sherman. 1988. Yeast iso-1-cytochrome *c*: Genetic analysis of structure-function relationships. *FEBS Lett.* **231:** 275–283.

Hickey, D.R., G. McLendon, and F. Sherman. 1992 Increased stability of mutant forms of yeast cytochrome *c*. In *Stability of protein pharmaceuticals:* In vivo *pathways of degradation and strategies for protein stabilization* (ed. T.M. Ahem and M.C. Manning). Plenum, New York. (In press.)

Hinnen, A., J.B. Hicks, and G.R. Fink. 1978. Transformation of yeast. *Proc. Natl. Acad. Sci.* **75:** 1929–1933.

Laz, T.M., D.F. Pietras, and F. Sherman. 1984. Differential regulation of the duplicated iso-cytochrome *c* genes in yeast. *Proc. Natl. Acad. Sci.* **81:** 4475–4479.

Li, W.-Z. and F. Sherman. 1991. Two types of TATA elements for the *CYC1* gene of the yeast *Saccharomyces cerevisiae*. *Mol. Cell. Biol.* **11:** 666–676.

Liebman, S.W., P. Shalit, and S. Picologlous. 1981. Ty elements are involved in the formation of deletions in *DEL1* strains of *Saccharomyces cerevisiae*. *Cell* **26:** 401–409.

Liebman, S.W., A. Singh, and F. Sherman. 1979. A mutator affecting the region of the iso-1-cytochrome *c* gene in yeast. *Genetics* **92:** 783–802.

Louie, G.V., W.L.B. Hutcheon, and G.D. Brayer. 1988. Yeast iso-1-cytochrome *c*; A 2.8 Å resolution three-dimensional structure determination. *J. Mol. Biol.* **199:** 295–314.

Melnick, L. and F. Sherman. 1990. Nucleotide sequence of the COR region: A cluster of six genes in the yeast *Saccharomyces cerevisiae*. *Gene* **87:** 157–166.

Moerschell, R.P., G. Das, and F. Sherman. 1991. Transformation of yeast directly with synthetic oligonucleotides. *Methods Enzymol.* **194:** 362–369.

Moerschell, R.P., S. Tsunasawa, and F. Sherman. 1988. Transformation of yeast with synthetic oligonucleotides. *Proc. Natl. Acad. Sci.* **85:** 524–528.

Moerschell, R.P., Y. Hosokawa, S. Tsunasawa, and F. Sherman. 1990. The specificities of yeast methionine aminopeptidase and acetylation of amino-terminal methionine *in vivo*: Processing of altered iso-1-cytochromes *c* created by oligonucleotide transformation. *J. Biol. Chem.* **265:** 19638–19643.
Montgomery, D.L., B.D. Hall, S. Gillam, and M. Smith. 1978. Identification and isolation of the yeast cytochrome *c* gene. *Cell* **14:** 673–680.
Montgomery, D.L., D.W. Leung, M. Smith, P. Shalit, G. Faye, and B.D. Hall. 1980. Isolation and sequence of the gene coding for iso-2-cytochrome *c*. *Proc. Natl. Acad. Sci.* **77:** 541–545.
Narita, K., K. Titani, Y. Yaoi, and H. Murakami. 1963. The complete amino acid sequence in baker's yeast cytochrome *c*. *Biochim. Biophy. Acta* **77:** 688–690.
Pillar, T.M. and R.E. Bradshaw. 1991. Heat shock and stationary phase induce transcription of the *Saccharomyces cerevisiae* iso-2 cytochrome *c* gene. *Curr. Genet.* **20:** 185–188.
Russo, P., W.-Z. Li, D.M. Hampsey, K.S. Zaret, and F. Sherman. 1991. Distinct *cis*-acting signals enhance 3′endpoint formation of *CYC1* mRNA in the yeast *Saccharomyces cerevisiae*. *EMBO J.* **10:** 563–571.
Sherman, F. 1964. Mutants of yeast deficient in cytochrome *c*. *Genetics* **49:** 39–48.
Sherman, F. and P.P. Slonimski. 1964. Respiration-deficient mutants of yeast II. Biochemistry. *Biochim. Biophys. Acta* **90:** 1–15.
Sherman, F. and. W. Stewart. 1971. Genetics and biosynthesis of cytochrome *c*. *Annu. Rev. Genet.* **5:** 257–296.
———. 1973. Mutations at the end of the iso-1-cytochrome *c* gene of yeast. In *The biochemistry of gene expression in higher organisms* (ed. J.K. Pollak and J.W. Lee), pp. 56–86. Australian and New Zealand Book Co., Sydney.
———. 1975. The use of iso-1-cytochrome *c* mutants of yeast for elucidating the nucleotide sequences that govern initiation of translation. *FEBS Proc. Meet.* **38:** 175–191.
———. 1978. The genetic control of yeast iso-1 and iso-2-cytochrome *c* after 15 years. In *Biochemistry of genetics of yeast. Pure and applied aspects* (ed. M. Bacila et al.), pp. 273–316. Academic Press, New York.
———. 1982. Mutations altering initiation of translation of yeast iso-1-cytochrome *c*: Contrast between the eukaryotic and prokaryotic process. In *Molecular biology of the yeast* Saccharomyces: *Metabolism and gene expression* (ed. J.N. Strathern et al.), pp. 301–333. Cold Spring Harbor Laboratory, Cold Spring Harbor, New York.
Sherman, F., H. Taber, and W. Campbell. 1965. Genetic determination of iso-cytochromes *c* in yeast. *J. Mol. Biol.* **13:** 21–39.
Sherman, F., M. Jackson, S.W. Liebman, A.M. Schweingruber, and J.W. Stewart. 1975. A deletion map of *cyc1* mutants and its correspondence to mutationally altered iso-1-cytochromes *c* of yeast. *Genetics* **81:** 51–73.
Sherman, F., R.P. Moerschell, S. Tsunasawa, R. Sternglanz, and M.E. Dumont. 1992. Co- and post-translational processes and mitochondrial import of yeast cytochrome *c*. In *Translational regulation of gene expression* (ed. J. Ilan), vol. 2. Plenum Press, New York. (In press.)
Sherman, F., J.W. Stewart, M. Jackson, R.A. Gilmore and J.K Parker. 1974. Mutants of yeast defective in iso-1-cytochrome *c*. *Genetics* **77:** 255–284.
Sherman, F., J.W. Stewart, E. Margoliash, J. Parker, and W. Campbell. 1966. The structural gene for yeast cytochrome *c*. *Proc. Natl. Acad. Sci.* **55:** 1498–1504.
Sherman, F., J.W. Stewart, J.H. Parker, E. Inhaber, N.A. Shipman, G.J. Putterman, R.L. Gardisky, and E. Margoliash. 1968. The mutational alteration of the primary structure of yeast iso-1-cytochrome *c*. *J. Biol. Chem.* **243:** 5446–5456.
Singh, A. and F. Sherman. 1978. Deletions of the iso-1-cytochrome *c* and adjacent genes of yeast: Discovery of the *OSM1* gene controlling osmotic sensitivity. *Genetics* **89:** 653–665.
Slonimski, P.P., R. Acher, G. Péré, A. Sels, and M. Somio. 1965. Éléments du système respiratoire et leur régulation: Cytochromes et iso-cytochromes. *Colloq. Int. CNRS* **124:** 435–461.
Stewart, J.W. and F. Sherman. 1972. Demonstration of UAG as a nonsense codon in bakers' yeast by amino-acid replacements in iso-1-cytochrome *c*. *J. Mol. Biol.* **68:**

429–443.
———. 1974. Yeast frameshift mutations identified by sequence changes in iso-1-cytochrome *c*. In *Molecular and environmental aspects of mutagenesis* (ed. L. Prakash et al.), pp. 102–127. C.C. Thomas, Springfield, Illinois.

Stewart, J.W., F. Sherman, N.A. Shipman, and M. Jackson. 1971. Identification and mutational relocation of the AUG codon initiating translation of iso-1-cytochrome *c* in yeast. *J. Biol. Chem.* **246:** 7429–7445.

Stewart, J.W., F. Sherman, M. Jackson, F.L. Thomas, and N. Shipman. 1972. Demonstration of the UAA ochre codon in bakers yeast by amino-acid replacements in iso-1-cytochrome *c*. *J. Mol. Biol.* **68:** 83–96.

Stiles, J.I., L.R. Friedman, and F. Sherman. 1981a. Studies on transposable elements in yeast. II. Deletions, duplications and transpositions of the COR segment that encompasses the structural gene of yeast iso-1-cytochrome *c*. *Cold Spring Symp. Quant. Biol.* **45:** 602–607.

Stiles, J.I., J.W. Szostak, A.T. Young, R. Wu, S. Consaul, and F. Sherman. 1981b. DNA sequence of a mutation in the leader region of the yeast iso-1-cytochrome *c* mRNA. *Cell* **25:** 277–284.

Streisinger, G., Y. Okada, J. Emrich, J. Newton, A. Tsugita, E. Terzaghi, and M. Inouye. 1967. Frameshift mutations and the genetic code. *Cold Spring Harbor Symp. Quant. Biol.* **31:** 77–84.

Struhl, K., J.R. Cameron, and R.W. Davis. 1976. Functional genetic expression of eukaryotic DNA in *Escherichia coli*. *Proc. Natl. Acad. Sci.* **73:** 1471–1475.

Szostak, J.W., J.I. Stiles, C.P. Bahl, and R. Wu. 1977. Specific binding of a synthetic oligonucleotide to yeast cytochrome *c* mRNA. *Nature* **265:** 61–63.

Szostak, J.W., J.I. Stiles, B.K. Tye, P. Chiu, F. Sherman, and R. Wu. 1979. Hybridization with synthetic oligonucleotides. *Methods Enzymol.* **68:** 419–428.

Tsugita, A. and H. Fraenkel-Conrat. 1963. Contributions from TMV studies to the problem of genetic information transfer and coding. In *Molecular genetics* (ed. J.H. Taylor), part I, pp. 477–520. Academic Press, New York.

Tuppy, H. 1958. Über die Artspeziftät der Proteinstrukur. In *Proceedings of Symposium on Protein Stucture* (ed. A. Neuberger), pp. 66–67. Wiley, New York.

Yamamoto, T., R.P. Moerschell, L.P. Wakem, D. Ferguson, and F. Sherman. 1992. Parameters affecting the frequencies of transformation and co-transformation with synthetic oligonucleotides in yeast. *Yeast* **8:** (in press).

Yanofsky, C., D.H. Helinski, and B.D. Maling. 1962. The effects of mutation on the composition and properties of the A protein of *Escherichia coli* tryptophan synthetase. *Cold Spring Harbor Symp. Quant. Biol.* **26:** 11–23.

Zaret, K.S. and F. Sherman. 1982. DNA sequence required for efficient transcription termination in yeast. *Cell* **28:** 563–573.

# MOLECULAR BIOLOGY

G. Fink and D. Botstein (CSHL 1975)

R. Davis (Louvain-la-Neuve, Belgium 1980)

# A Phage Geneticist Turns to Yeast

DAVID BOTSTEIN
*Department of Genetics*
*Stanford University School of Medicine*
*Stanford, California 94305*

## MIGRATING FROM PHAGE TO YEAST

I came to Cold Spring Harbor Laboratory in the summer of 1971 to take the Yeast Course taught by Gerry Fink and Fred Sherman. Many (if not quite all) of the early yeast molecular geneticists were, like me, "immigrants," in the sense that we learned molecular genetics in bacterial systems and then chose to apply this way of thinking to *Saccharomyces cerevisiae*. Possibly the best way to convey the spirit and the ideas of the early years of yeast molecular biology is to describe why and how I started to work with yeast.

In 1971, all of my research was focused on the temperate *Salmonella* phage P22. I identified completely with the phage community. Zinder and Lederberg (1952) had discovered generalized transduction in *Salmonella typhimurium*; P22 is the phage vehicle that carries the bacterial DNA from host to host, and it featured a wide variety of interesting biological properties that one could study. Despite this, very few laboratories were working on P22. The opportunity to study such a diversity of phenomena suited perfectly my eclecticism. Then as now, study sections would regularly comment on my "lack of focus." Although small at the time, my laboratory studied DNA replication, transduction and recombination (soon to branch into the study of transposons), morphogenesis of the virion, and of course the complicated genetic regulation of lysogeny. A very interactive and productive research community was interested in these issues, most of them working with the coliphage λ. I had some of the best scientific interactions in my career with this community and I learned from them. My work was interesting, it was challenging, it was productive, and it was great fun.

Why, then, did I even think about doing something different? The reasons derived from a general perception that the end of the road was near for phage molecular genetics. The mechanistic and regulatory paradigms, it seemed, were already all on the table, and many believed that no new principles might be found in these relatively simple systems. Articles appeared in prominent journals trumpeting the "end" of molecular genetics, the more amazing in retrospect because this was well before recombinant DNA technology was conceived. Those of us without tenure were earnestly warned that we might not be perceived as having a future were we to stick with our prokaryotic intellectual game for too long. It was time to get eukaryotic, to look to the future.

I took this advice very seriously. I am glad that I did, even though I believe now as I did then that the intellectual case against phage and bacterial genetics was entirely specious. However, the fact remained that most of my peers and betters thought the end was near, and, unfortunately, it was clear that their

*The Early Days of Yeast Genetics*

thinking it being so made it so. Tenure committees, deans, and study sections, then as now, were impressed mainly by conventional wisdom. It is now clear that the whole eukaryotic/prokaryotic argument was silly. The intervening 20 years, especially the revolution wrought by recombinant DNA methods, showed prokaryotic biology to be full of new ideas, discoveries, and principles, many of which apply to eukaryotes as well. Nevertheless, there was a big rush of people and money out of the prokaryotic fields. The impressive progress that has changed the face of prokaryotic biology since was accomplished by what can only be called a supremely talented, courageous, and tenacious skeleton crew. Bacterial biology remains underfunded and underpopulated even today, showing the power of conventional wisdom even in the teeth of contravening facts.

Feeling the need to branch out into something eukaryotic, I determined to find a minimalist solution. I sought a field that would allow me to do sophisticated genetics in real time, as with phage; provide access to biochemical tests of genetic insights, as with phage; encompass a breadth of accessible biological phenomena, as with phage; be in an open and interactive community, as with phage; be relatively cheap to set up and bootleg if grants were not immediately forthcoming; and finally, be compatible with continuing work with phage and bacteria in the same laboratory. I considered several alternatives, which I then narrowed down to two: animal virology and yeast. I studied them both seriously, and favored yeast slightly, partly because of the expense associated with animal tissue culture. At about this time I met Gerry Fink, always an effective recruiter for *Saccharomyces*. His arguments and enthusiasm clinched the deal for me. I decided to add yeast to the portfolio of my already perilously "unfocused" phage laboratory and to continue with both. And so I did, although the strictly prokaryotic work has diminished to a low level in the last decade.

I was not alone in my way of thinking. Many other phage and bacterial geneticists traveled the same road. Almost all of us began by taking the Yeast Course at Cold Spring Harbor. In 1971, three bacterial geneticists came to Cold Spring Harbor from the Massachusetts Institute of Technology (MIT); the others were my laboratory partner for the course, Gerry Smith (then a student with Boris Magasanik), and Ira Herskowitz (then just starting a short stint in my laboratory as a postdoc). The Yeast Course, our first introduction to *S. cerevisiae*, was an intellectually wonderful experience. It was also fun. I learned a great deal more than just yeast genetics. By the most direct method, I learned that Ira Herskowitz is a consummate teacher. I am no athlete, yet in 3 weeks Ira taught me enough table tennis to hold my own with all but the most dedicated players. In subsequent years, I even became one of the Ping-Pong set at Gordon Conferences. I got far enough to play at the same table with David Freifelder (who was once national junior champion) and Frank Stahl (who takes his Ping-Pong very seriously indeed). Ira taught me by stressing the fundamentals, so that even though I could not always execute a shot correctly, I knew right from wrong. Genetics is like table tennis in that sense: The many different organisms upon which we practice genetics present diverse difficulties and opportunities in execution, but underneath the fundamentals remain always the same. That is why we geneticists are so easily able to recognize those of our peers who are "real geneticists," regardless of whether they work with phage, worms, flies, mice, or people. At Cold Spring Harbor, I became convinced that Ira, who had then just finished his Ph.D., would become an uncommonly influential teacher and intellectual leader in genetics. I now believe that Ira was the most effective of all of the phage converts to yeast in applying the fundamentals of molecular genetic thinking to the biology of yeast.

The Cold Spring Harbor Yeast Course was central, in my opinion, in making the yeast molecular biology community what it is today. Gerry Fink and Fred Sherman in 1971 were leading yeast researchers who carried a conviction that the future lay with the melding of two schools of thought in genetics: the formally genetic and the molecular. Fred came to this through his long-term interest in cytochrome *c* and suppression (see, e.g., Sherman 1964; Sherman et al. 1973), and Gerry came to this through his devotion to histidine biosynthesis, the study of which he had undertaken both in yeast and in *Salmonella* (see, e.g., Fink 1966; Fink and Martin 1967). Gerry thus had his education divided between the two traditions he and Fred undertook to join and teach at Cold Spring Harbor. We, the prokaryote geneticists, brought with us an already sophisticated molecular thought process, following the intellectual paths illuminated by Jacob and Monod on the one side and Luria, Delbrück, and Hershey on the other. We brought with us an obsession with DNA and the central dogma, especially with the analysis of regulation at the level of transcription and, as we often hoped but rarely found, translation. We worried about messengers, ribosomes, and tRNA. Our favorite genetic tools were the selective cross and the *cis-trans* test.

Fred and Gerry offered their students formal diploid genetics, especially the marvel of tetrad analysis. We, being familiar mainly with bacteria and phage genetics, were impressed by the power of being able to make stable diploids at will and to carry out literally hundreds of complementation tests at once. We revelled in the simplicity with which tetrads and complementation testing on replica plates allows one to construct double mutants, even when the doubles cannot be distinguished from the single mutant by phenotype. We learned that there is genetics beyond the selective cross. We chafed, of course, at the difficulty in doing fine-structure genetics, the phage geneticist's bread and butter. Most of all, however, we were impressed with the garden of phenomena to which little, if any, molecular thinking and analysis had been applied.

All of this was of course already known to Fred and Gerry. But the differences in the styles of genetics between phage and yeast were a constant source of new ideas nevertheless. The prokaryotic geneticists proposed a great variety of schemes, only some of which were crackpot ideas that Fred and Gerry could shoot down at once. Other ideas reflected important insights that led to interesting science, some of which is now quite well-known. We were fascinated with the eukaryotic novelties: diploidy, chromosomes, centromeres, mitosis, mating, meiosis. We were somewhat taken aback at the relatively unsophisticated way (compared to bacteria) in which metabolic and especially regulatory studies were then still being done. The yeast community, beginning with our instructors, were just beginning to learn how truly powerful the molecular genetic tools could be. We immigrants had found, in yeast, a proper eukaryotic system in which we could practice molecular genetics, until then a discipline applicable only to bacteria and their viruses.

It is worth recalling some of the issues and ideas that date from those days.

One that preoccupied me was isogenicity. There was then no way to transfer small segments of the genome from one yeast strain to another. Thus, there was no good way to make isogenic strains. Yeast geneticists used strains of many different backgrounds that were crossed indiscriminately. From our experience with bacteria, we knew that this practice would compromise severely studies of regulation. Gerry Fink, with his hybrid background, already understood this, and following Gerry's lead, many of the immigrants in the early years spent much time backcrossing interesting mutations into our favorite strains. Our concern proved entirely justified when, 15 years later, the first pulsed-field gels showed massive and unpredictable differences in gross sizes of chromosomes among the laboratory strains! Much of the early work in my own laboratory began with programs of backcrossing; having decided upon S288C as our standard wild type, we crossed useful markers (such as Lacroute's uracil auxotrophs and Hartwell's standard *cdc* alleles) into this background, producing "congenic" derivatives, before working extensively with them. Only with the advent of DNA transformation did it become possible to construct rigorously isogenic sets of yeast strains with which to study subtle differences among mutants.

Many of us were interested in knowing which features of prokaryotic molecular biology are essentially the same in yeast and which are basically different. Sometimes we wanted to know for technical reasons; it seemed important to know, for instance, whether suppression by the dominant allele-specific "supersuppressors" worked via mutant tRNAs that have mutated to read stop codons, as in bacteria. We hoped, in this case, that the mechanism of suppression would be much the same, because then we could exploit nonsense mutations in similar ways. At that time, for example, there was no good way to find null mutations other than nonsense mutations that map near the beginning of a gene. Yet we always had a deeply ambivalent feeling when we found close similarity between bacteria and yeast. After all, if something turned out to work in yeast just as it did in bacteria, it would not be the kind of novel "eukaryotic" principle that conventional wisdom dictated we pursue. Out of this ambivalence, this balance between searching for what was truly new, on the one hand, and what we could understand and manipulate, on the other, we immigrant molecular geneticists were drawn to studies of suppressors, DNA replication, protein synthesis, and, most of all, regulation of enzyme synthesis. We all wanted to find the model regulatory system—the "Lac operon" of yeast. I was quite typical in this way; my earliest work in yeast included work on amber suppressors (Brandriss et al. 1975, 1976) and later the inducible secreted enzyme invertase (Carlson et al. 1981a,b). Ira began to work right away on mating type (Hicks and Herskowitz 1976; Hicks et al. 1977), which already that summer he perceived was really a regulatory system that controls cell type. This meant, as we never tired of telling whomever would listen, that there must be a great hierarchy of morphogenetic and regulatory genes that execute all the functions that differ among cells with the α, **a**, and **a**/α phenotypes (see Herskowitz and Oshima [1981] for a definitive statement of this view, first clearly articulated by MacKay and Manney [1974]). I believe that the mating-type system, as a regulatory paradigm, is the closest analog in yeast to the Lac operon.

My major interest, however, lay elsewhere. I was struck with the simplicity with which one could probe essential cellular functions in yeast, taking advantage of the haploid vegetative state and the simplicity of complementation analysis. The great innovator in this arena was Lee Hartwell, who with Cal McLaughlin had isolated and characterized many hundreds of temperature-sensitive mutants (Hartwell and McLaughlin 1968). Their explicit intention had been to use mutations to investigate any and all aspects of cell function; they would do for yeast what Edgar, Epstein, and colleagues had done for T4 (Epstein et al. 1964). Hartwell and McLaughlin applied very simple approaches lifted from phage and bacterial physiology (Hartwell had studied with Magasanik at MIT) to characterize the mutant phenotypes. They were able to identify genes specifically affecting RNA and protein metabolism and, of course, the progress of the cell cycle per se (for a fine early review, see Hartwell 1974). I recognized that this work was well advanced beyond anything that had been done with bacteria. It was comparable only to the attempts to saturate the genomes of phages (T4, λ, and, of course, P22) with mutations in essential genes. Indeed, this was the work I was engaged in with phage P22; I was already well on the way to identifying, using amber, heat-sensitive, and cold-sensitive mutations, every one of its essential genes (I missed only two or three in the end) and trying to determine what each of their products did for the organism (Botstein et al. 1972; for a comprehensive review of my work during this period, see Susskind and Botstein 1978).

Hartwell and McLaughlin's work was thus the single most important intellectual reason I decided to study yeast. I shared with the phage group the idea that the great use of genetics is to probe biological mechanisms. Hartwell's cell cycle work made it already clear that molecular genetics, properly applied to *Saccharomyces*, could lay bare the mechanisms underlying phenomena like eukaryotic DNA replication and mitosis, which already were known to be quite different from their prokaryotic counterparts. It was in this direction that I resolved that my laboratory should go with yeast.

## GETTING TOGETHER THE MOLECULAR TECHNOLOGY

The arguments for yeast as a eukaryotic system in which one could study molecular genetics in the style of phage and bacteria had some weaknesses. These concerned molecular technology, which was, by bacterial standards, very primitive. During this period, bacterial geneticists had found ways to obtain genes of interest as pieces of DNA, mainly as λ specialized transducing phages. With these tools, it was possible to measure mRNAs by hybridization. Without them, one could not convincingly distinguish regulation at the level of transcription from that at the level of translation. Only by being able to measure mRNA level could we hope to study regulation in a meaningful way. The genes about which we knew a lot were, in the main, those that either were phage genes to begin with (e.g., λ*c*I, encoding the λ repressor) or those that could be manipulated onto specialized transducing phage (notably, the *gal*, *lac*, and *ara* operons; for a good summary of the technology of the time, see Miller and Reznikoff 1978). Gerry Smith, my laboratory partner at Cold Spring Harbor, had produced a significant breakthrough in Boris Magasanik's laboratory just by isolating a λ phage that carried the *hut* (histidine utilization) operon from *Salmonella* (Smith 1971), with which he and his successors in Boris' laboratory figured out much about nitrogen regulation in bacteria (cf. Magasanik 1978).

During the summer of 1971, a group of us (including Ira, Gerry Fink, and myself) decided to look for viruses that might infect *S. cerevisiae*. Our idea was that if we could find viruses, some of them might allow us to make the equivalent of λ transducing phages for yeast. We were determined to find a "yeast phage," without which we would find it difficult to bring molecular genetics to full flower in yeast. We called breweries, associations of breweries, yeast taxonomists, and the like in the hope of learning whether batches of beer sometimes went bad because the yeast lysed. In our minds, we had visions of fermenters full of *Escherichia coli* lysing as a result of phage contamination (usually the dreaded T1). This was indeed a scenario that regularly plagued our biochemist friends. We thought that the brewing industry might therefore even be interested in funding our search; after all, had not Pasteur helped the vintners?

All of our respondents told us we were crazy—there were no yeast viruses. Fred Sherman was particularly downbeat. He understood our wish to find the yeast phage; this was by no means a new idea. Indeed, he related that his first project in Bob Mortimer's laboratory was to find yeast viruses! He had looked in all kinds of exotic places (including the zoo, in materials that apparently smelled not so nice) but never found anything. He thought it a waste of our time. The brewers were considerably more emphatic: Not only had they never observed viruses, but they did not want to know about it if one were found. They feared the public relations consequences of the news that Americans might be imbibing viruses along with their beer! In the aftermath of the recombinant DNA debacle, I look back with respect at the brewers' understanding of the American public psyche. In the midst of all this negative feedback, one brewer did answer the call. Jaime Conde, a brewmaster we soon affectionately called the "brewer from Seville," told anecdotes of failed beer batches in Spanish breweries involving lysis. There was correspondence, Jaime sent strains, and he eventually joined the Fink laboratory in Ithaca.

Nobody has yet found a yeast virus, at least not in the sense of something that makes an overtly infectious particle. Interesting science came out of the search, however. Both Gerry and I found quickly that the Seville strains harbored a very efficient killer to which most of the laboratory strains were susceptible. The Fink laboratory used these to define the killer phenomenon more accurately; they learned much about the particle and the RNA, and, probably most significantly, they isolated mutations of yeast that affected the maintenance of the killer (Fink and Styles 1972; Vodkin et al. 1974). In my laboratory, Ira and I found many killers in many different strains of yeast, some from laboratories and many more from the wild, although we stopped well short of the zoo. Jaime spent his time in Ithaca isolating the first karyogamy mutants with Gerry, although he did use the *kar1* mutant to show that killers can be transferred by cytoplasmic mixing (Conde and Fink 1976). After all this effort, we were technologically no better off than before.

In the fall of 1973, I persuaded Gerry Fink and John Roth (a mutual friend and colleague who had worked with Gerry on *Salmonella*) to go on sabbatical (due for all three of us the following year). We would work together on yeast, with an emphasis on genetic and molecular techniques. John suggested and vigorously pursued San Diego as a venue, but this fell through because our hosts were afraid that our yeast would contaminate their tissue culture. Like biochemists, who feared all phage because of their problems with phage T1, cell culturists to this day fear all yeast because of their contamination problems with *Candida* species. Despite all our learned and perfectly sound arguments that *Saccharomyces* and *Candida* are as different as cats and bats, reason did not prevail, and so we had to look elsewhere. Jim Watson suggested we come to Cold Spring Harbor instead and use the Davenport Laboratory (since renamed in honor of Max Delbrück). Davenport had been used for courses in the summer (including the Yeast Course) but not in the winter. The building had just been renovated, heating had been installed, and, best of all, the equipment we needed would be available from the summer courses. To get fellowships and some modest grant support, we proposed a big inbreeding scheme to get Hartwell's mutants into the S288C background so that we could properly do pseudoreversion genetics, a collaboration with Ray Gesteland (then permanent staff at Cold Spring Harbor) to prove that yeast suppressors are alterations in tRNA, and of course the continued search for yeast viruses. We were fortunate enough to get a small grant that covered supplies and even a technician.

The sabbatical at Cold Spring Harbor in the academic year 1974–1975 was possibly the most important year of my career. Although we did not do all that much useful experimental work, we did think about what we were doing in genetics, what others were doing, the relative importance of genetic, biochemical, and other molecular tools, and quite generally what the future might hold. I think we stopped to question these general issues at a critical moment in the history of genetics. We debated the future of genetics, and the relative virtues of yeast and animal viruses, the great strength of the research program at Cold Spring Harbor. Joe Sambrook was the senior virologist, and his group included as postdocs such embryonic luminaries as Phil Sharp and Mike Botchan. One of the direct consequences of our proximity to Sambrook's laboratory was that we learned about restriction enzyme technology from the pioneers. One discussion that had a particularly powerful effect on me was Mike Botchan's seminar on his then planned use of the gel-transfer hybridization technique (Southern 1975 [then still unpublished]) to determine the site(s) of integration of SV40 (Botchan et al. 1976); we discussed for weeks how one might use Southern's blotting method to study isolated yeast genes someday. Most of these ideas have actually been reduced to practice since. Researchers at Cold Spring Harbor thought hard about each others' work, but we three did it non-stop. John and I talked a lot about transposition in bacteria, discussions that led to several years worth of experiments we later pursued in our own laboratories (cf. Kleckner et al. 1977). We became excited about the usefulness of transposons as genetic tools, which no doubt influenced Gerry to pursue the insertion mutations he encountered several years later in yeast. But most importantly, we recognized that the yeast virus might now not be necessary—there might be much better ways to get at the problem of gene isolation.

## FROM YEAST DNA TO RECOMBINANT DNA

The key experiments establishing the recombinant DNA technology had just been done—the papers were already published or in press. I had played a small part in one of the experiments: Dale Kaiser had asked me whether I knew of a simple way to get a large quantity of DNA of uniform size with flush ends that have a 5′ phosphate and a 3′ hydroxyl. He wanted this DNA for his student Peter Lobban's experiments. I suggested the P22 genome which Charley Thomas had shown (and I had later confirmed) had such a structure. This was why the seminal Lobban-Kaiser experiment (historically the true beginning of recombinant DNA technology) was done with P22 DNA (Lobban and Kaiser 1973). Gerry Fink and I were particularly impressed with the reports from David Hogness' laboratory at Stanford, where Pieter Wensink, using the Lobban-Kaiser technique, had established "banks" of *E. coli* clones each bearing an insert of randomly sheared genomic DNA from *Drosophila* (Wensink et al. 1974).

Gerry and I decided to make such a bank for yeast, whereas John put most of his effort into the tRNA collaboration (ultimately successful; Gesteland et al. 1976) and the backcrosses. Making recombinant DNA banks was, at that time, a laborious procedure, requiring not only considerable effort, but also expertise not generally available. We made a deal with Pieter Wensink (who had just moved to Brandeis' new Rosenstiel building) to collaborate. All we needed was to bring some yeast DNA, some vector DNA, and an offering of a useful enzyme. This offering was a common practice in the days before commercial enzymes: If you used somebody's reagent, you tried to repay with another scarce reagent. I provided Tom Maniatis (then at Cold Spring Harbor doing his famous cDNA cloning of the globin genes; Maniatis et al. 1976) with some semi-skilled labor for a preparation of the 5′ exonuclease of λ, some of which we later gave to Pieter. That was the easy part, now we needed the yeast and vector DNAs.

Nobody had published a good DNA preparation from yeast, let alone one that gave good yields of high-molecular-weight DNA of high purity. I tried the best one and found that it yielded only about 15% of the nuclear DNA, most of it in small fragments on the order of 10 kb. Fortunately, I had some relevant experience. My Ph.D. thesis had involved recovering unsheared DNA preparations from phage-infected bacteria (cf. Botstein 1968). So Gerry and I set about learning how to lyse yeast thoroughly without degrading the DNA and then how to purify that DNA. For the latter step, we needed an ultracentrifuge. Our hosts at Cold Spring Harbor graciously agreed to let us move one of theirs into Davenport. We thought all was lost when the centrifuge fell off the bulldozer that was used to carry it cross-country over the lawn from the Demerec building. Fortunately, Beckman's Spinco division made sturdy machines, and the dented centrifuge worked fine after the Spinco repairman had ministered to it. Beckman didn't even charge us extra, even though it was obvious that the damage was the result of the failure of our exotic transport system. We learned to make DNA, but in pitifully small amounts. The good news about *Saccharomyces*, after all, is its very small genome; the corresponding bad news is that one gets very little DNA per gram of cells. After several months, we pooled three of our best preparations, which yielded barely enough to bother visiting Brandeis.

The vector DNA was not much simpler to obtain. In those days, there was much controversy concerning the plasmid vectors. We chose to work with pSC101, the earliest Cohen-Boyer vector, which we obtained from Stan Cohen (Cohen et al. 1973). There was considerable competitive excitement among different laboratories about how to grow the cells containing various plasmids in order to achieve good yield and, even more important, how best to separate the plasmid DNA from the chromosomal DNA of the bacteria. Gerry and I were confused by the conflicting advice and claims, so we started on a comprehensive study of parameters such as cell density, presence or absence of chloramphenicol, and the myriad variables in the extraction and purification procedures themselves. We made some progress but were depressed by the reality that we were still making tenfold less DNA per liter of cells than the published claims of some researchers. One day, Gerry called one of these claimants and explained our distress, providing some indication of the results of our study of the problem. After Gerry got to the punch line—that after all this effort we had recovered "only so many micrograms per liter"—there was silence on the line. Finally, our colleague asked "tell me again how much you got per liter?" Gerry repeated the number, and then was astonished to be asked most urgently to send full details of our protocol. It turned out that our claimant friend had published his "best" result, which he could no longer repeat. Our amateur effort was threefold better than his current efforts! We might even, at that moment, have held a world's record for yield and purity of pSC101 DNA! We eventually were able to make adequate, if not yet massive, amounts of pure pSC101 DNA.

The truly wonderful part of our sabbatical year was that we three were equals, thinking and working together on problems of mutual interest. We had all day to think and do experiments. Each of us had a large and active laboratory in Cambridge, Ithaca, and Berkeley, where our postdocs and graduate students were all working hard. Although we were at Cold Spring Harbor also working hard (probably thinking harder about science there than at home), we were not doing nearly as many of the non-science chores involved in running a laboratory and being its guru. We were reminded of this by the almost daily calls from our home laboratories. For example, 1974 was the year of the agar crisis; hard on the heels of the oil crisis, microbiologists were subjected to the unavailability of agar for plates from the normal sources. Each of these august suppliers rationed their customers; like most of the world, they did not understand that genetics uses considerable amounts of plates. One of us discovered that a German company (Fluka) had agar for sale in quantity but that it was not quite pure enough for minimal medium. So we each instructed our laboratories back home to buy some, wash it, and use it for most experiments, saving the rationed Difco agar for critical experiments. One day, one of Gerry's postdocs called Davenport in a panic. They had decided to wash all of their Fluka agar (several kilos) in what they regarded as a large pot. When they had added the water, the agar began to swell and quite quickly to overflow, at which point the calls to Cold Spring Harbor began. What was to be done? What, indeed, could Gerry tell them? New garbage cans were found, the agar was washed in gallons of water with considerable losses, but the day was saved. We got some amusement from the picture of the sorcerer's apprentices (Gerry made a fine sorcerer) dealing with the overflowing agar, but we also were reminded that we were truly enjoying science in a way we could not responsibly do at home. My postdocs were very conscious of this as well; I encouraged them to be self-reliant in my absence and they more than rose to the challenge. In the spring of 1975, the time came for first-year MIT students to choose a thesis advisor. Three of my postdocs (Susan Gottesman, Nancy Kleckner, and Mimi Susskind) conspired to accept one of the students they liked (Fred Winston) without my knowledge. One day they called, each on an extension, and presented their case to me as a *fait accompli*, complete with a sad tale of the injustice that Fred's entry into my laboratory would avert. What could I do? I went along, thinking all the time how much agonizing had been avoided and hoping that the choice was a good one. Once I got to meet Fred in person, I was of course most happy with their choice. So even in this most important of academic activities I got the year off.

I attended only one meeting that year, the famous Asilomar meeting called with much fanfare to consider the potential dangers of the recombinant DNA technology. This is not the place to recount the history associated with that event. In the present context, however, it is worth noting that I went with the conviction that this technology was the way of the future for yeast and with a real understanding of the magnitude of the opportunity. I will admit that at that time the possibility of danger from recombinant DNA seemed more real to me than it does now. Even then, however, I thought of danger only in terms of accidental inoculation of a laboratory worker with massive numbers of harmful organisms; such an event might occur with an industrial-scale preparation of, let us say, an oncogene-bearing plasmid. I was as surprised as anyone when the process of risk estimation by worst-case scenario began to run amok, as it did during the Asilomar meeting and even more afterward. I never thought yeast could pose a danger and, as a bacterial geneticist, was truly amazed that genes from *Salmonella*, which can cross with *E. coli* naturally, would seriously be considered a potential hazard. It was at Asilomar, with the real possibility of a moratorium that might inhibit our work with yeast, that I realized how much I had come to accept that recombinant DNA was the technical solution we were seeking and that our decision to work with yeast was going to depend on our ability to make that technology work. (A footnote to Asilomar: Much later, when we finally had useful clone banks in microtiter plates sitting in a liquid nitrogen freezer at MIT, a serious effort was undertaken to have them destroyed. This was a consequence of the notorious Cambridge City Council hearings, for at that time our yeast bank was the only existing collection of eukaryotic clones in bacterial hosts in the city. We sent a copy of the bank to Cold Spring Harbor, where it all began, to preserve it. Another copy went to Rochester, to Fred Sherman's laboratory. Whatever Fred may have thought about the wisdom of our search for the yeast phage, he was never in doubt about the usefulness of gene isolation.)

Later in the spring, after Asilomar, Gerry and I went to Pieter Wensink's laboratory where together we made the first yeast recombinants in plasmids. We had too little DNA and had made a minor mess of some of the experiments, but we showed it could be done, and done well enough to saturate the yeast genome. Tom Petes, who joined my laboratory at MIT as a postdoc just as I returned from sabbatical, repeated our DNA preparations and, with Pieter Wensink's help, made the first big plasmid bank (Petes et al. 1978). With it he characterized the ribosomal DNA (the most prominent feature in the DNA of yeast) and used a restriction fragment difference to show genetically that all of the rDNA maps to a single location (Petes and Botstein 1977). This experiment was our first published use of the Southern procedure, and it is the direct precursor of the use of restriction-fragment-length polymorphisms to map the human genome (Botstein et al. 1980). This is a story for another occasion, and I mention it only to reemphasize that genetics is a unitary discipline (cf. Botstein 1990). Yeast had a role very early in the development of a molecular genetics of the human, just as phage had a role in the development of the molecular genetics of yeast.

## MODERN TIMES BEGIN

The beginning of the modern era in yeast molecular genetics for me was the first Molecular Biology of Yeast meeting at Cold Spring Harbor. Gerry Fink and I, both of us still on sabbatical, organized the meeting, which included 167 participants. The program is interesting today both because so many of the major laboratories were already represented and because the influence of the alumni of the Yeast Course (by then 6 years old) was already clear. Every 2 years, the number of people at this meeting doubled, until the meeting outgrew the capacity of the Cold Spring Harbor facilities. In 1975, we adopted the strictly egalitarian phage meeting style—everyone who submitted an abstract gave a talk, and every talk was 15 minutes long. It was very like the phage meetings organized by Luria and Delbrück about 30 years before, and we were proud of that comparison. Indeed, the yeast community was beginning to resemble, in scientific impact as well as egalitarian style, the phage community. Everybody seemed to feel this sense of community, that our science was moving ahead, and that the yeast group was somehow going to make it into the vanguard of molecular biology.

The private discussions at the meeting, of course, concentrated on the possibility of isolating individual genes with recombinant DNA technology. Already many of us had made plans. We knew that our friend Ron Davis (who did not attend the meeting) was in the midst of isolating genes using bacteriophage λ vectors; his student Kevin Struhl would soon isolate the *HIS3* gene on the basis of its ability to complement a bacterial mutation in the gene specifying the corresponding enzyme, and John Carbon would do the same for *LEU2* (Ratzkin and Carbon 1977; Struhl and Davis 1977). I met François Lacroute at this 1975 meeting, and we decided to collaborate on isolating the *URA3* gene in the same way. In the end, we would use these clones to demonstrate transcriptional regulation of the gene in yeast (Bach et al. 1979). Although the really powerful DNA technologies (transformation, cloning by complementation, gene replacement, gene disruption, and gene fusions) were still far in the future, we could see already that the new technology would mean that yeast molecular genetics would flourish.

I remember clearly, at that meeting in 1975, being confident for the first time that my gamble (i.e., investing my effort in yeast instead of animal viruses) was going to pay off. I even fancied that others might also be "coming around" on the possibility that yeast might become a significant model system for eukaryotic molecular biology. The previous winter, Jim Watson had joined Gerry, John, and me for a memorable lunch at Blackford Hall during which he cross-examined us very thoroughly about yeast, especially our claims for its prospects as a model system for eukaryotes. We argued at length, told him about the future we envisioned, and illustrated with examples the actual and potential power of genetic analysis combined with the hoped-for recombinant DNA technology. We came away from the discussion thinking that we had made no impression whatsoever on Jim, who remained openly skeptical. But that summer, Jim came to many of the sessions of that first Molecular Biology of Yeast meeting. For 15 years I have believed that something must have impressed him, taking as proof the famous little essay suggesting that yeast should become the *E. coli* for eukaryotic cell biology that I found in the next (third) edition of *Molecular Biology of the Gene* (Watson 1976). The joke is on me, of course, because the editors of this volume have pointed out to me that the essay first appeared in the *second* edition published in 1970! Of course this makes sense, because it was after all the Cold Spring Harbor Yeast Course that introduced most of us to yeast, and it had to be Jim Watson who invited Fred Sherman and Gerry Fink to teach such a course in the first place. So I no longer have evidence that Jim was impressed by anything that happened in 1975; instead, I have more evidence for Jim's ability to foresee, sometimes even prescribe, the scientific future. No matter, although the concept is clearly much older, I still believe that 1975 was the year in which yeast molecular biology became a reality.

## REFERENCES

Bach, M.L., F. Lacroute, and D. Botstein. 1979. Evidence for transcriptional regulation of OMP-decarboxylase in yeast by hybridization mRNA to the yeast by structural gene cloned in *E. coli. Proc. Natl. Acad. Sci.* **76:** 386–390.

Botchan, M., W. Topp, and J. Sambrook. 1976. The arrangement of simian virus 40 in the DNA of transformed cells. *Cell* **9:** 269–287.

Botstein, D. 1968. Synthesis and maturation of phage P22 DNA. I. Identification of intermediates. *J. Mol. Biol.* **34:** 621–643.

———. 1990. 1989 Allen Award Address: The American Society of Human Genetics Annual Meeting, Baltimore. *Am. J. Hum.Genet.* **47:** 887–891.

Botstein, D., R.K. Chan, and C.H. Waddell. 1972. Genetics of bacteriophage P22. II. Gene order and gene function. *Virology* **49:** 268–282.

Botstein, D., R.L. White, M. Skolnick, and R.W. Davis. 1980. Construction of a genetic linkage map in man using restriction fragment length poymorphisms. *Am. J. Hum. Genet.* **32:** 314–331.

Brandriss, M.C., L. Soll, and D. Botstein. 1975. Recessive lethal amber suppressor in yeast. *Genetics* **79:** 551–560.

Brandriss, M.C., J.W. Stewart, F. Sherman, and D. Botstein. 1976. Substitution of serine caused by a recessive lethal suppressor in yeast. *J. Mol. Biol.* **102:** 467–476.

Carlson, M., B.C. Osmond, and D. Botstein. 1981a. Mutants of yeast defective in sucrose utilization. *Genetics* **98:** 25–40.

———. 1981b. Genetic evidence for a silent SUC gene in yeast. *Genetics* **98:** 41–54.

Cohen, S.N., A.G.Y. Chang, H.W. Boyer, and R.B. Helling. 1973. Construction of biologically functional bacterial plasmids *in vitro. Proc. Natl. Acad. Sci.* **70:** 3240–3244.

Conde, J. and G.R. Fink. 1976. A mutant of *Saccharomyces cerevisiae* defective for nuclear fusion. *Proc. Natl. Acad. Sci.* **73:** 3651–3655.

Epstein, R.H., A. Bolle, C.M. Steinberg, E. Kellenberger, E. Boy de la Tour, R. Chevallay, R.S. Edgar, M. Susman, G.H. Denhardt, and A. Lielausis. 1964. Physiological studies of conditional mutants of bacteriophage T4D. *Cold Spring Harbor Symp. Quant. Biol.* **28:** 375–394.

Fink, G.R. 1966. A cluster of genes controlling three enzymes in histidine biosynthesis in *Saccharomyces cerevisiae. Genetics* **53:** 445–459.

Fink, G.R. and R.G. Martin. 1967. Polarity in the histidine operon. *J. Mol. Biol.* **30:** 97–107.

Fink, G.R. and C.A. Styles. 1972. Curing of a killer factor in *Saccharomyces cerevisiae. Proc. Natl. Acad. Sci.* **69:** 2846–2849.

Gesteland, R.F., M. Wolfner, P. Grisafi, G. Fink, D. Botstein, and J.R. Roth. 1976. Yeast suppressors of nonsense codons work efficiently *in vitro* via tRNA. *Cell* **7:** 381–390.

Hartwell, L. 1974. *Saccharomyces cerevisiae* cell cycle. *Bacteriol. Rev.* **38:** 164–198.

Hartwell, L. and C. McLaughlin. 1968. Temperature-sensitive mutants of yeast exhibiting a rapid inhibition of protein synthesis. *J. Bacteriol.* **96:** 1664–1671.

Herskowitz, I. and Y. Oshima. 1981. Control of cell type in *Saccharomyces cerevisiae*: Mating type and mating-type interconversion. In *The molecular biology of the yeast* Saccharomyces: *Life cycle and inheritance* (ed. J.M. Strathern et al.), pp. 181–209. Cold Spring Harbor Laboratory, Cold Spring Harbor, New York.

Hicks, J.B. and I. Herskowitz. 1976. Interconversion of yeast mating types. I. Direct observations of the action of the homothallism (*HO*) gene. *Genetics* **83:** 245–258.

Hicks, J.B., J.N Strathern, and I. Herskowitz. 1977. The cassette model of mating-type interconversion. In *DNA insertion elements, plasmids and episomes* (ed. A.I. Bukhari et al.), pp. 457–462. Cold Spring Harbor Laboratory, Cold Spring Harbor, New York.

Kleckner, N., J. Roth, and D. Botstein. 1977. Genetic engineering *in vivo* using translocatable drug-resistance elements: New methods in bacterial genetics. *J. Mol. Biol.* **116:** 125–159.

Lobban, P.E. and A.D. Kaiser. 1973. Enzymatic end-to-end joining of DNA molecules. *J. Mol. Biol.* **78:** 453–471.

MacKay, V. and T.R. Manney. 1974. Mutations affecting sexual conjugation and related processes in *Saccharomyces cerevisiae*. Isolation and phenotypic characterization of non-mating mutants. *Genetics* **76:** 255–271.

Magasanik, B. 1978. Regulation in the *hut* system. In *The operon* (ed. J.H. Miller and W.S. Reznikoff), pp. 373–387. Cold Spring Harbor Laboratory, Cold Spring Harbor, New York.

Maniatis, T., S.G. Kee, A. Efstratiadis, and F.C. Kafatos. 1976. Amplification and characterization of a β-globin gene synthesized *in vitro*. *Cell* **8:** 163–182.

Miller, J.H. and W.S. Reznikoff, eds. 1978. *The operon*. Cold Spring Harbor Laboratory, Cold Spring Harbor, New York.

Petes, T.D. and D. Botstein. 1977. Simple Mendelian inheritance of the reiterated ribosomal DNA of yeast. *Proc. Natl. Acad. Sci.* **74:** 5091–5095.

Petes, T.D., J.R. Broach, P.C. Wensink, L.M. Hereford, G.R. Fink, and D. Botstein. 1978. Isolation and analysis of recombinant DNA molecules containing yeast DNA. *Gene* **4:** 37–49.

Ratzkin, B. and J. Carbon. 1977. Functional expression of cloned yeast DNA in *E. coli*. *Proc. Natl. Acad. Sci.* **74:** 487–491.

Sherman, F. 1964. Mutants of yeast deficient in cytochrome *c*. *Genetics* **49:** 39–48.

Sherman, F., S.W. Liebman, J.W. Stewart, and M. Jackson. 1973. Tyrosine substitution resulting from the suppression of amber mutants of iso-1-cytochrome *c* in yeast. *J. Mol. Biol.* **78:** 157–168.

Smith, G.R. 1971. Specialized transduction of the *Salmonella hut* operons by coliphage λ: Deletion analysis of the *hut* operons employing λ*phut*. *Virology* **45:** 208–223.

Southern, E. 1975. Detection of specific sequences among DNA fragments separated by gel electrophoresis. *J. Mol. Biol.* **98:** 503–517.

Struhl, K. and R.W. Davis. 1977. Production of a functional eukaryotic enzyme in *Escherichia coli*: Cloning and expression of the yeast structural gene for imidazole glycerol phosphate dehydratase (*his3*). *Proc. Natl. Acad. Sci.* **74:** 5255–5259.

Susskind, M.M. and D. Botstein. 1978. Repression and immunity in *Salmonella* phages P22 and L: Phage L lacks a functional secondary immunity system. *Virology* **89:** 618–622.

Vodkin, M., F. Katterman, and G.R. Fink. 1974. Yeast killer mutants with altered ribonucleic acid. *J. Bacteriol.* **117:** 681–686.

Watson, J.D. 1970. *Molecular biology of the gene*, 2nd edition, pp. 519–520. W.A. Benjamin, New York.

———. 1976. *Molecular biology of the gene*, 3rd edition, pp. 506–507. W.A. Benjamin, Menlo Park, California.

Wensink. P.C., D.J. Finnegan, J.E. Donelson, and D.S. Hogness. 1974. A system for mapping DNA sequences in the chromosomes of *Drosophila melanogaster*. *Cell* **3:** 315–325.

Zinder, N.D. and J. Lederberg. 1952. Genetic exchange in *Salmonella*. *J. Bacteriol.* **64:** 679–699.

# Genes, Replicators, and Centromeres: The First Artificial Chromosomes

JOHN CARBON
*Department of Biological Sciences*
*University of California*
*Santa Barbara, California 93106*

## LIBRARIES AND GENES

My decision to work on the molecular biology of yeast was born in the midst of a typical mid-life crisis. In 1972, I was 41 years old and poised for important changes in my life, including my field of research. For several years, my laboratory at the University of California, Santa Barbara, had concentrated on the biochemistry and genetics of bacterial transfer RNAs. We had been quite successful, having shown that the molecular mechanism of genetic suppression of missense and frameshift mutations involved various tRNAs with mutationally altered anticodons. I was particularly proud of the experiments on frameshift suppression carried out with a very talented postdoctoral fellow, Don Riddle (now on the faculty of the University of Missouri). Don showed that suppression of a particular class of +1 frameshift mutations in *Salmonella* was caused by a mutationally altered glycine tRNA containing an extra base in the anticodon loop; the four-base anticodon recognized a four-base codon in the message, showing that tRNA was responsible for framing the translation of the messenger RNA (Riddle and Carbon 1973). In other experiments, we had obtained the first evidence that the tRNA anticodon can serve as a recognition site for the activating enzyme responsible for "loading" an amino acid onto the tRNA (Carbon and Fleck 1974; Roberts and Carbon 1974). However, I had become somewhat bored with the tRNA field; the rewards seemed relatively small for the tremendous amount of work that was required to purify a specific wild-type or mutationally altered $^{32}$P-labeled tRNA from in-vivo-labeled cells and to determine the nucleotide sequence by the Sanger "fingerprint" method. This was before DNA sequencing and recombinant DNA techniques were invented, of course; life is much easier now in the tRNA field.

Sometime in 1972, I became aware of the fascinating experiments on in vitro recombinant DNA under way in the laboratories of Herb Boyer at the University of California, San Francisco, and Stanley Cohen and Paul Berg at Stanford. We invited Dave Jackson, a postdoc in Berg's laboratory, to UCSB, to give a "job" seminar talk on his work. Dave had successfully joined two different DNA molecules together in vitro. He had converted the circular SV40 and bacteriophage λ*dvgal* DNA molecules to full-length linears using the new *Eco*RI restriction endonuclease recently purified in Herb Boyer's laboratory and had covalently joined the two molecular species by using poly(dA) and poly(dT) "connectors" (Jackson et al. 1972). Dave described how in principle this method could be used to insert and maintain foreign DNA sequences in living cells by joining the foreign DNA to "vectors" (SV40 in his experiment) capable of replicating in the host cells. The experimental opportunities opened up by the new recombinant DNA techniques seemed boundless. I made the decision to jump headfirst into this new field as soon as possible.

Fortunately, I knew Paul Berg very well, since I had spent a very enjoyable and fruitful year in his laboratory in 1965–1966 working with him and Charlie Yanofsky on the mechanism of genetic suppression of missense mutations in *Escherichia coli* (Carbon et al. 1966, 1967). Paul had launched me into the tRNA field 7 years before; why not spend another sabbatical year at Stanford learning how to work with SV40 and recombinant DNA? Paul immediately agreed to give me space in his laboratory, and I applied for and received a National Science Foundation grant to help support my research at Stanford. It was one of those rare times in life when several nonparallel lines seem to converge at a common point. My marriage had recently disintegrated, I had become romantically involved with Louise Clarke, a beautiful and talented graduate student in my laboratory, and I had the opportunity and funds to spend an entire year in one of the best laboratories and biochemistry departments in the world, learning a fascinating new discipline.

In the early summer of 1973, Louise and I piled her two cats and innumerable potted plants into our cars and moved to the San Francisco Bay area, she to do postdoctoral research in Howard Goodman's laboratory at UCSF, and I to plunge happily into my new experiments on DNA at Stanford. Although I did not work with yeast that year, the techniques that I learned and developed for the manipulation of DNA were absolutely invaluable and provided the real groundwork for our isolation of the yeast *LEU2* gene just 2 years later. In Paul's laboratory, I purified some restriction enzymes (no commercial sources in those early days!), ran the first agarose slab gels at Stanford, and developed methods for the construction of DNA insertions and deletions at specific restriction sites in the SV40 circular double-stranded DNA molecule. I stumbled onto a basic principle of DNA deletion analysis by sheer serendipity. I had a preparation of SV40 DNA that had been linearized by cleavage at the single *Hpa*II recognition site and trimmed back at each end with an exonuclease in preparation for poly(dA) "3′-tailing" by a terminal transferase reaction. When, as a control experiment, this exonuclease-treated linear DNA was used to infect animal cells in culture, plaques were obtained! Viral DNAs isolated from these clones were circular and contained deletions centered around the *Hpa*II site. Recircularization had occurred in vivo, and since by good fortune the restriction site we had cleaved was in a nonessential region of the viral genome, infectious virus was recovered. Tom Shenk, a postdoc in Paul's laboratory at that time (and now a highly respected member of the Molecular Biology faculty at Princeton University), helped me develop the method into a general procedure to construct deletions at specific sites in the circular DNA genome and continued to pursue the project after I returned to UCSB (Carbon et al. 1975; Shenk et al. 1976).

Immediately upon returning to UCSB in the late summer of 1974, Louise and I embarked upon an ambitious project. I wanted to design a definitive experiment to determine if eukaryotic genes would be expressed in bacterial cells. At the time, little or nothing was known about eukaryotic gene promoters and transcription, since recombinant DNA techniques were in their infancy and, other than some ribosomal RNA genes, no eukaryotic structural genes had yet been isolated. Thus, it seemed quite possible that transcription and translation control signals would turn out to be universal, and thus eukaryotic genes might be expressed freely in *E. coli* host cells. Aside from the intrinsic scientific interest of this question, the potential commercial applications did not escape me! The overall plan for the experiment was relatively simple, but first we needed to construct complete genomic libraries, or "gene banks" as we called them in those days. The idea was to look for complementation of various auxotrophic mutations in *E. coli* by expression of foreign genes introduced into the bacterial cell on DNA segments sealed into bacterial plasmid vectors by using the new recombinant DNA technology. Since most metabolic and catabolic biochemical reactions were known to be catalyzed by similar enzymes in eukaryotic and prokaryotic cells, we reasoned that functional expression of the appropriate eukaryotic gene in *E. coli* should lead to complementation of a genetic defect in the analogous bacterial gene. Thus, by using various auxotrophic bacterial strains unable to grow on selective media as recipients for the recombinant DNAs, we could select directly for bacterial transformants in which the genetic defect had been corrected by expression of a plasmid-borne foreign gene.

Unfortunately, however, several difficulties had to be overcome before this experiment could be carried out. The most obvious problem was that a negative result, i.e., inability to complement a particular bacterial mutation with foreign DNA, could simply mean that the genomic library was incomplete and did not contain a cloned DNA segment including the desired eukaryotic gene. Clearly, we needed some independent evidence that our genomic libraries were reasonably complete, otherwise a negative result in the complementation experiments would be meaningless. Although everyone seemed to be talking about the possibility of constructing complete genomic libraries, as far as I knew, no one at the time had really made one and determined how much of the foreign genome was represented in the collection of bacterial clones. A second problem was more sociopolitical than scientific; a nationwide panic was developing over the presumptive dangers of recombinant DNA technology, and it was not clear whether we would even be allowed to do the experiments without very elaborate safety facilities then unavailable at UCSB. Therefore, my decision was to construct the first genomic library using fragments of *E. coli* DNA, as a control to determine the effectiveness of our technology in terms of generating complete libraries. Strains bearing auxotrophic mutations in genetic loci scattered rather uniformly around the *E. coli* circular genome were readily available, and thus the library could be tested for completeness by looking for hybrid plasmids capable of complementing these mutations. The second library was to be constructed from genomic DNA of common baker's yeast (*Saccharomyces cerevisiae*), because the genome size was known to be relatively small as compared to that of most eukaryotes, one hundredth that of animal cell genomes and only about three times larger than that of *E. coli*. In addition, a great deal was known about metabolic reactions and enzymes in yeast, and it appeared that most or all of the amino acid biosynthetic enzymes in bacteria would have a yeast counterpart (wild-type yeast can grow on media containing only sugar, biotin, and various inorganic salts and thus can synthesize all of the amino acids). Somehow I felt that genes from this relatively simple eukaryote would have a higher probability of being expressed in bacterial cells than would genes from animal or plant cells. In addition, if we could obtain some yeast genes by these procedures, it could be a starting point for the investigation of gene expression and regulation in a simple model eukaryotic cell. Fortunately, I made the right decision in choosing yeast for these experiments; little did I know at the time that the yeast genome was relatively free of introns, which were discovered a couple of years later as common interruptions in genes from most eukaryotic cells. Finally, I could foresee little danger in introducing either endogenous bacterial or yeast DNA into *E. coli* and hoped that high-containment laboratories would not be necessary.

As a vector for construction of our genomic libraries, we decided to use the *E. coli* colicinogenic factor ColE1, a circular double-stranded DNA plasmid present in quite high copy number in many naturally occurring bacterial strains. Years before recombinant DNA techniques were invented, Charlie Yanofsky had suggested to me that the high-copy-number ColE1 plasmid would make an excellent vehicle to obtain overexpression of gene products in bacterial cells, and in the early 1970s, Craig Squires and I had attempted without success to insert some tRNA genes into ColE1 by in vivo recombination. In 1974, Vicky Hershfield and Don Helinski at the University of California, San Diego, collaborated with Charlie and the Boyer laboratory to covalently seal a DNA segment containing the *E. coli trpE* gene (excised from transducing phage DNA) into the single *Eco*RI site of plasmid ColE1 and subsequently establish the chimeric plasmid in *E. coli* cells (Hershfield et al. 1974). The hybrid plasmid with its new *trpE* gene was maintained intact in *E. coli*, and the *trpE* enzyme was overexpressed to very high levels in the transformed cells. The main disadvantage of ColE1 was the selection, which had to be carried out on media containing the ColE1 protein toxin, laboriously purified from cells carrying the intact ColE1 plasmid. Unfortunately, the convenient Boyer "pBR" cloning vectors conferring antibiotic resistance were not available until about 1977 (Bolivar et al. 1977).

In late 1974, Louise and I used the Stanford poly(dA:dT) "connector" method to join segments of *E. coli* genomic DNA to *Eco*RI-cleaved ColE1 DNA and then transformed various *E. coli* mutants with the hybrid DNA circles, selecting for complementation as evidenced by growth on selective media. After a day or two in the incubator, we were overjoyed to find a few colonies growing on the selection plates. Subsequent work indicated that these transformants contained large hybrid ColE1 plasmids with DNA inserts containing specific regions of the bacterial genome. By selecting for complementation of *trpE* and *araC* mutations, we had cloned the *trp* and *ara* gene clusters directly from the bacterial genome (Clarke and Carbon 1975). Then we had what seemed at the time to be a terrific idea, but which later caused us no end of trouble with the new Recombinant DNA Advisory Committee (RAC), formed at the National Institutes of Health to regulate the use of recombinant DNA and enforce safety procedures. A talented and mathematically minded graduate student in the laboratory, Curt Covey (now a well-known planetary physicist), had calculated that if we inserted genomic DNA fragments 15,000 base pairs (15 kb) in length into the plasmid vector, it would require a collection of only about 1500 bacterial colonies to reach the 99% probability level of having the entire *E. coli* genome represented in the collection. To reach the same 99% probability level for the genomic yeast library, we would need only 4500 bacterial plasmid clones! Our idea was to prepare these colony "banks" in an *E. coli* strain containing the F plasmid, a so-called "conjugative plasmid" that is capable of transferring a copy of itself from the $F^+$ host cell to an $F^-$ bacterial cell during cell-cell mating (conjugation). This F plasmid was well known to have the ability to mobilize the ColE1 plasmid to transfer also into the recipient $F^-$ cell. We could prepare complete libraries of *E. coli* and yeast genomic DNAs in the ColE1 vector replicating in an $F^+$ host, store the collections as separate clones in the ultra-cold freezer (–80°C), and then score the clones for plasmids capable of complementing any *E. coli* mutation by simple bacterial "plate mating" techniques. The colonies in the library were grown in grid-like arrays on agar media and then replica-plated onto uniform "lawns" of various mutant $F^-$ recipients. A patch of growth on selective plates indicated transfer of a plasmid capable of correcting the genetic defect in the recipient.

This method worked beautifully! The first genomic library of *E. coli* DNA consisted of 2100 colonies that we stored in plastic microtiter dishes in the deep-freeze. Nearly every *E. coli* mutant strain that we could find could be complemented by a hybrid ColE1 plasmid from a colony in this collection. In 1976, Louise and I described this "Clarke-Carbon library" in *Cell*, reporting that at least 80% of the bacterial genome was represented (Clarke and Carbon 1976). Immediately, we were deluged with requests from laboratories all over the world for copies of the library. Fortunately, Barbara Bachmann, curator of the *E. coli* Genetic Stock Center at Yale University, agreed to preserve and distribute clones from the library. Even today, after 16 years, I still get occasional requests for clones from this library. As of 1987, more than 300 of the library clones had been characterized in terms of identifying the exact region of the *E. coli* genome contained on the hybrid plasmids (Phillips et al. 1987). The Clarke-Carbon library has been put to good use by innumerable laboratories; but for us, it was only a control experiment to see if a complete genomic library could be made. We were anxious to use the same technology to determine if genes from yeast would complement the mutations in *E. coli*.

In 1975–1976, we prepared similar genomic libraries in an $F^+$ host strain using total DNAs prepared from baker's yeast and from the fruit fly *Drosophila melanogaster* (4300 and 10,000 colonies, respectively, were picked and stored). Postdoc Barry Ratzkin (now in the research division of Amgen, an enterprising biotechnology firm) prepared the yeast DNA libraries. The *E. coli* host strain that we were using (JA199) happened to contain an inactivating mutation in the *leuB* gene. The mutation, known as *leu-6*, was originally isolated by Edward Tatum at Stanford in 1945 and was one of the earliest genetic markers used in *E. coli* research; in fact, Tatum and Lederberg used *leu-6* as one of the markers in their classical experiments that first demonstrated genetic recombination in bacteria (Tatum 1945; Tatum and Lederberg 1947). The bacterial *leuB* gene is analogous to the yeast *LEU2* gene; both express a β-isopropylmalate dehydrogenase activity, an enzyme in the leucine biosynthetic pathway. In fact, *leu2* was a well-known centromere-linked genetic marker, in common use at the time by yeast geneticists. Barry and I discussed a preliminary experiment to be done with the yeast genomic library—why not pool a large number of ColE1-resistant bacterial transformants and plate them out selecting directly for complementation of the *leuB* mutation in the library host strain? If the yeast *LEU2* gene was expressed in the bacterial cells, then we should get complementation and growth on the selective media. Barry plated out the pooled yeast library on media lacking leucine and stuck the plates away in the incubator. In fact, a few fairly slow-growing colonies were obtained on these selection plates, but only after they had sat for several days in the incubator and we had just about given up hope. A plasmid, designated pYe*leu*10, isolated from one of these colonies, complemented the bacterial *leu-6* mutation most efficiently (Ratzkin and Carbon 1977). We first established by hybridization reassociation kinetics that the DNA insert in pYe*leu*10 was in fact yeast DNA, since we had been fooled in a similar experiment by a plasmid containing a cloned segment of contaminating bacterial DNA! Fortunately, pYe*leu*10 did contain yeast DNA, but in the absence of a transformation system for yeast, we were at a loss for a method to actually prove that the *LEU2* gene was on the plasmid (the Maxam-Gilbert DNA sequencing method was not in common use until after 1977). At about the same time that we isolated pYe*leu*10, Kevin Struhl, a graduate student in Ron Davis' laboratory at Stanford, told me that he had also been able to complement an *E. coli* auxotrophic mutation (a deletion of the *hisB* gene) with cloned genomic yeast DNA. They had also succeeded in constructing a complete genomic library of yeast DNA, using their newly developed bacteriophage λ cloning vectors. Presumably, Struhl and Davis had cloned the yeast *HIS3* gene, which specifies the same enzymatic activity as does bacterial *hisB*, although again final proof was lacking until a DNA transformation system for yeast became available. Barry Ratzkin used our plasmid-based yeast genomic library to isolate a hybrid plasmid capable of complementing a bacterial *hisB* deletion and obtained a plasmid containing a DNA insert that hybridized with the Struhl-Davis clone. It seemed clear that the genomic libraries were reasonably complete, at least some yeast genes were being expressed in *E. coli*, and the methods were quite reproducible; the same yeast DNA segment had been isolated in different laboratories from two independently constructed genomic libraries (Struhl et al. 1976; Ratzkin and Carbon 1977).

Gerry Fink, then at Cornell University, suggested a clever hybridization experiment that they could do to determine if the DNA segment cloned in pYe*leu*10 was derived from yeast chromosome III, the known location of the *LEU2* gene. The experiment was based on the availability in his laboratory of strains of yeast aneuploid for chromosome III, i.e., with two copies of III but only one of all other chromosomes (N + 1) or with one copy of III and two copies of the others (2N – 1). Quantitative hybridizations of a labeled pYe*leu*10 DNA probe to genomic DNAs isolated from these aneuploid strains should correlate with the relative copy numbers of chromosome III, if the clone in fact contained DNA from the *LEU2* region. I sent the clone to Gerry, and Jim Hicks, then a postdoc in Gerry's laboratory and later a member for many years of the first Cold Spring Harbor yeast group, carried out the experiment. The results were nearly exactly those predicted if the cloned DNA originated from chromosome III, giving us further assurance that we had indeed cloned the yeast *LEU2* gene (Hicks and Fink 1977).

In the next year or two (1976–1978), several other yeast genes were cloned from yeast genomic libraries by selecting for complementation of *E. coli* mutations. These included the yeast *ARG4*, *TRP1*, *TRP5*, and *URA3* genes, all commonly used now as selectable markers on yeast cloning vectors (for review, see Botstein and Davis 1982). Approximately one in four of the heterologous complementations that were attempted in that early work was fruitful (Carbon et al. 1977b), an amazing success rate when one considers the considerable differences now known to exist between prokaryotic and eukaryotic transcriptional control signals. Apparently, DNA sequences fortuitously recognized as weak promoter signals by the *E. coli* RNA polymerase are reasonably common on the cloned yeast DNA segments, resulting in low-level expression of the foreign genes. Later, we showed that expression of yeast genes in the library bacterial colonies also could be detected by screening the colonies with antisera directed against a particular yeast protein, thus avoiding the troublesome in vivo complementation, which requires synthesis of a functional protein (Clarke et al. 1979). In this way, postdoc Ron Hitzeman isolated the yeast phosphoglycerokinase gene (*PGK*), a highly expressed glycolytic enzyme gene whose "hot" promoter has been put to good use by the biotechnology industry (Hitzeman et al. 1980).

Before leaving this section, I should point out that all of this early work on genomic libraries was conducted in an atmosphere charged with tension and doubts, generated by the controversy regarding the presumptive dangers of this type of experimentation. I shall never forget presenting a talk on our research at a conference on "Recombinant Molecules: Impact on Science and Society," held at the Massachusetts Institute of Technology in mid 1976 at the height of the debate (Carbon et al. 1977a). At the time, Cambridge was a center for the activists, led by a small but highly vocal group of Harvard and MIT faculty members, who were convinced that molecular biologists were out to destroy the world. After every scientific presentation, someone from the audience would rise and present a long talk, obviously prepared ahead of time, on the dangers inherent in the research. Unfortunately, my results demonstrating expression of eukaryotic genes in bacteria added considerable fuel to the fire, since it proved to some that the dangers were not imaginary. To make matters worse, we had prepared libraries in $F^+$ hosts, capable of transferring the recombinant plasmids to other bacteria in the intestines and sewers of the world. When the official National Institutes of Health Guidelines for Recombinant DNA Research were issued in mid 1976, the use of bacterial hosts containing conjugative plasmids (such as the F plasmid) was prohibited. I wrote several letters to influential members of the Recombinant DNA Advisory Committee (RAC) suggesting that use of our $F^+$ yeast genomic libraries should be permitted under P3 conditions (a high-containment laboratory under negative air pressure with work conducted in a biosafety hood). However, this plea was dismissed, and in an official letter, dated October 22, 1976, from the head of the NIH Office of Recombinant DNA Activities, I was instructed to destroy all of our genomic libraries in $F^+$ host cells (except those constructed from *E. coli* genomic DNA). The offending libraries were autoclaved in late 1976. This strange episode is described more fully, along with copies of relevant letters and news items, in the excellent book by James Watson and John Tooze, which documents the history of gene cloning (Watson and Tooze 1981).

## YEAST TRANSFORMATION AND DNA REPLICATORS

Shortly after we had isolated pY*eleu*10 in 1976, the idea of using the presumptive *LEU2* cloned DNA to transform a yeast *leu2* mutant to $Leu^+$ surfaced. The dogma at the time was that yeast could not be transformed by exogenous DNA; many had tried using genomic DNA preparations and had failed. I remember clearly telling a visitor from a well-known French yeast laboratory that we would like to use our cloned yeast genes to transform yeast. He gave me a patronizing smile and told me that it was common knowledge in yeast laboratories that yeast could not be transformed by exogenous DNA. The idea that several orders of magnitude amplification of transformation efficiency would be obtained by using a cloned DNA segment containing a known gene for the transformation escaped many people at the time. But again we ran into a bureaucratic nightmare with the NIH RAC, which had to approve formally each new species proposed as a host to receive recombinant plasmids. We could not attempt to transform yeast with any of our cloned yeast genes until the RAC approved *S. cerevisiae* as a host for transformation with recombinant DNA. In the early fall of 1977, I submitted a written proposal to the RAC, asking for permission to attempt to transform yeast with our ColE1 plasmids containing the cloned yeast genes. I pointed out the great advantages that would ensue if the transformation was successful: "development of suitable transformation and vector systems in yeast (*S. cerevisiae*) cells would eventually permit the isolation of any yeast gene system by complementation of various yeast mutations by the cloned yeast DNA." I alluded to the innocuous nature of baker's yeast, but suggested that the experiments could be done under P2 conditions.

Of course, weeks went by without a reply from the RAC. However, at about this time I received a telephone call from an excited Jim Hicks at Cornell. He told me that Albert Hinnen in the Fink laboratory had successfully transformed a yeast *leu2* double mutant to $Leu^+$ with the pYe*leu*10 plasmid that I had sent them. The transforming DNA had integrated into the yeast genome at the *leu2* locus (Hinnen et al. 1978). I was pleased that we finally had evidence that we really had cloned the *LEU2* gene and our hopes for a yeast transformation system were to be realized, but at the same time I was puzzled. How had Gerry obtained permission to carry out this "dangerous" experiment? It turned out that the research was supported by the National Science Foundation and Gerry had obtained their approval without difficulty. I felt very frustrated by being caught in this bureaucratic vise; God knows when the NIH RAC would approve yeast as a recombinant DNA host. I had NIH funding, and it would take me months to obtain NSF funding. To make matters worse, in early January, 1978, nearly 2 months after I submitted my request, I received a letter from the RAC spokeswoman at NIH telling me that the committee had met, had not yet approved yeast as a host, and had asked: "If these particular yeast strains are pathogenic or if they are eaten by humans?" I couldn't believe it! Did the committee really not know that baker's or brewer's yeast is nonpathogenic and in fact is eaten by humans? Obviously, they were unable or unwilling to make a decision and had resorted to a commonly used bureaucratic strategy to gain time: ask for additional information. I knew that Ron Davis had submitted the same request to the RAC and was anxiously awaiting approval. Ron and I had a long telephone conversation to discuss the problem, and we came to the only possible solution. Gerry Fink had found the only way out; we also would use non-NIH funds to do the experiments. Fortunately, both of us had small grants from non-NIH sources (not surprisingly, this loophole was closed by the NIH at a later date).

Within a day or two, I had moved one of my postdocs (Chulai Hsiao) onto non-NIH funds, and we used the Hinnen procedure to transform the appropriate yeast mutants with various plasmid DNAs containing our cloned yeast genes. However, the approval from the RAC to transform yeast with recombinant plasmids did not materialize until March 23, 1978, nearly 5 months after I had submitted the request for approval. Even then, approval was only granted to introduce DNA from yeast or from nonpathogenic prokaryotes into yeast; cloning of other DNAs into yeast was still not permitted. A letter from Gerry Fink to Don Frederickson, the Director of the NIH, pointing out the absurdity of the situation must have helped considerably to break the deadlock. In fact, both the NSF and the American Cancer Society had approved yeast as a recombinant DNA host months earlier.

Chulai soon was able to use the Hinnen procedure to transform the appropriate yeast mutants to wild type with our plasmids containing the putative *LEU2*, *ARG4*, and *TRP5* genes. However, he made a surprising discovery. The plasmid containing the *ARG4*-complementing DNA segment (termed pYe[*ARG4*]1) was amazingly efficient in transforming yeast *arg4* mutants to $Arg^+$; Chulai obtained hundreds of $Arg^+$ transformant colonies per plate, rather than only one or two as Hinnen had obtained with pYe*leu*10. Shortly after Ron Davis and I had decided to proceed with the yeast transformation experiments, I heard from Dan Stinchcomb, a graduate student in Ron's laboratory at Stanford, that a plasmid containing the yeast *TRP1* gene transformed yeast *trp1* mutants to $Trp^+$ at very high frequency, again about a thousandfold higher frequency than Hinnen had reported. Dan had discovered that the *TRP1* plasmid was maintained autonomously in the transformed cells and apparently was replicating independently without integrating into the yeast genome (Stinchcomb et al. 1979; Struhl et al. 1979). Chulai observed the same phenomenon with pYe(*ARG4*)1; this plasmid also was capable of autonomous replication in yeast cells (Hsiao and Carbon 1979). Subsequent work in both laboratories showed that the genomic DNA segments containing *ARG4* and *TRP1* also included a DNA sequence region capable of sustaining autonomous replication of the plasmid in yeast cells. Ron's laboratory named their sequence *ARS1* (for *A*utonomously *R*eplicating *S*equence) (Stinchcomb et al. 1979; Tschumper and Carbon, 1980) and Chulai and I followed suit by naming the *ARG4*-associated replicator *ARS2* (Hsiao and Carbon 1981; Tschumper and Carbon 1982). Subsequent research in many laboratories has verified Ron's original suggestion that these *ARS* elements are origins of DNA replication in the yeast genome.

At about this same time, Jean Beggs in England had covalently inserted a DNA fragment containing an *E. coli* plasmid vector plus the yeast *LEU2* gene into the 2-micron plasmid, an endogenous plasmid that occurs in many laboratory strains of yeast. This hybrid yeast–*E. coli* plasmid also would transform yeast *leu2* mutants to $Leu^+$ with high frequency and was maintained autonomously in high copy number (Beggs 1978). Thus, by late 1978, early 1979, both 2-micron- and *ARS*-based plasmid "shuttle" vectors were available. These vectors would replicate autonomously in either *E. coli* or yeast; they contained selectable genetic markers for use in both organisms, and thus could be recovered conveniently from total yeast DNA and prepared in quantity by introduction into *E. coli* and amplification in the bacterial host. The essential tools were now available to begin the molecular dissection of a model eukaryotic cell; the golden era of yeast molecular biology had begun.

## CENTROMERES AND ARTIFICIAL CHROMOSOMES

By some strange coincidence, several of the yeast genes we had cloned from genomic libraries by selecting for functional expression in *E. coli* happened to be centromere-linked; i.e., each is located relatively close to a centromere on a yeast chromosome. This is true for *TRP1*, *ARG1*, and *LEU2*. In fact, for a time after *ARS1* and *ARS2* were discovered, the possibility that these sequences were actually centromeres was entertained in both the Davis and Carbon laboratories. After all, both were located in the genome adjacent to centromere-linked genes (*TRP1* and *ARG4*), and they supported autonomous replication of plasmids in the yeast nucleus. However, this theory was discarded before too long; the properties of the *ARS* elements just did not match with what would be expected of centromeres. The *ARS* plasmids are very unstable mitotically, they are lost with high frequency in meiosis, and when integrated into the genome, they do not cause any problems typical of dicentric chromosomes. Moreover, *ARS* elements were soon discovered that were located on the chromosome arms at some distance from a centromere.

Shortly after the *LEU2* gene was cloned, it occurred to us that it should be possible to use that clone as a stepping stone to isolate a yeast centromere on a cloned segment of genomic DNA. At an international yeast symposium held in Sao Paulo, Brazil in December of 1977, I described the experimental strategy that we were using to isolate a yeast centromere (Carbon et al. 1978). The plan was first to clone genes known to occur on opposite sides and closely adjacent to a centromere (as established by standard genetic mapping techniques) and then to isolate all of the DNA in between these loci as a set of plasmid clones containing overlapping genomic DNA inserts. Obviously, the centromere would have to be located somewhere on one or more of these plasmids. However, at the time, it was not completely clear how we would identify the centromere, but I was confident that a genetic element as important as a centromere would contain some obvious structural feature that would give some hint as to its location. For example, since there are 16 yeast chromosomes, most likely we would find a sizeable DNA sequence repeated 16 times in the yeast genome. Or, since highly repetitive DNA sequences were known to occur in the centromere regions of higher eukaryotic chromosomes, perhaps we would find a stretch of repetitive sequences. How wrong I was!

We chose the centromere of chromosome III as our goal, because this centromere is flanked fairly closely by the *LEU2* gene on the left arm and by the *PGK* (3-phosphoglycerokinase) gene on the right arm. We already had the *LEU2* gene on hand and were confident that the *PGK* gene could be cloned without too much difficulty. Craig Chinault, a newly arrived postdoc in the laboratory, worked out a laborious procedure to identify plasmid clones with overlapping DNA inserts in our yeast genomic libraries (Chinault and Carbon 1979). The procedure is now known as "chromosome walking," but then I insisted on the fancier name, "overlap hybridization screening." It involved isolating small DNA fragments from the two ends of the pY*eleu*10 insert and then using these as hybridization probes to find other library clones containing these same DNA sequences. Since the recombinant plasmids in the library were constructed using randomly cleaved genomic DNA, clones containing plasmids with overlapping inserts could thus be identified. By serial repetitions of this procedure, Craig "walked" in both directions from the *LEU2* region cloned in pYe*leu*10 and obtained a set of four plasmids with overlapping inserts that encompassed about 30 kb of genomic DNA surrounding the *LEU2* locus. Unfortunately, at the time, we had no way of knowing which direction from *LEU2* was toward the centromere, since Craig's set of clones from the *LEU2* region did not connect up with the *PGK* clone isolated by Ron Hitzeman in 1978. We were blissfully unaware in 1978-early 1979 that the centromere was in fact on one of Craig's "walk" clones. Since, as we found out later, the budding yeast centromeres do not cross-hybridize with each other, and the region contains no repetitive DNA or other easily recognized structural features, my structural criteria for identification of the centromere were useless.

A tantalizing false lead occupied much of our time in late 1978-early 1979. Craig had found a genomic DNA segment from the *LEU2* region that hybridized quite strongly to many genomic DNA restriction fragments; on Southern blots, it looked like it could be hybridizing to about 16–17 different restriction fragments. I was very hopeful that this repeated sequence was the yeast centromere. Alan Kingsman, now a professor at Oxford University, isolated several genomic clones containing other copies of this repeated sequence and carefully prepared restriction maps of the cross-hybridizing regions. The repeat was about 6 kb in length and occurred at least 16–20 times in the genome. But then we heard that Ron Davis' laboratory had isolated a repeated DNA sequence with similar properties from yeast. This repeat had been called Ty1 (for *t*ransposon *y*east 1), and, as the name implies, it seemed to be a transposon capable of moving about the yeast genome (Cameron et al. 1979). I obtained a clone of Ty1 from Ron and, to our great disappointment, soon determined that our repeat from the *LEU2* region was not a centromere at all; it was definitely a type of Ty1. Alan extended these Ty1 studies after taking up his new faculty position at Oxford (Kingsman et al. 1981), and later, both the Kingsman and Fink laboratories made the important finding that Ty1 is a type of retrovirus that can package into particles in vivo, but it apparently has no extracellular existence (Garfinkel et al. 1985; Mellor et al. 1985).

Meanwhile, Louise Clarke was heading for the centromere by a slightly different route. Gerry Fink and Bob Mortimer had informed me that one of Lee Hartwell's temperature-sensitive cell division cycle mutations, *CDC10*, had recently been genetically mapped to a location very tightly linked to the centromere of chromosome III; much closer than either *LEU2* or *PGK*. Louise set out to clone the *CDC10* gene by transforming a yeast *cdc10* mutant with pooled plasmid DNAs from a yeast genomic library that she had constructed in a *TRP1-ARS1* shuttle vector, selecting for complementation of the temperature-sensitive *cdc10* mutation. Using this powerful new approach, it was now possible to clone almost any yeast gene; the only requirement was a strain containing an inactivating or conditional mutation in the gene of interest. The plasmid (pYe[*CDC10*]1) that Louise obtained by this procedure contained the wild-type *CDC10* gene, and, in addition, the yeast DNA insert overlapped with one of the genomic clones that Craig Chinault had obtained by walking from *LEU2*, and thus extended the walk (Clarke and Carbon 1980a). Since we knew that *CDC10* was closer to the centromere than is *LEU2*, finally we had correlated the physical and genetic maps in the *LEU2-CDC10-PGK* region. The clones from the walk now extended about 30 kb from *LEU2* in the direction of the centromere, but we still had not reached the *PGK* locus (only

10 map units [centiMorgans] from *LEU2*). Because the yeast genetic map at that time incorrectly placed *CDC10* on the *LEU2* side of the centromere, we thought for a time that we had not yet reached the centromere. With the exception of the Ty1 element located near *LEU2*, all of the genomic DNA uncovered by the walk scored as unique sequences by hybridization analysis. My preconception still was that the centromeric DNA would be recognizable as a repeated sequence.

In working with pYe(*CDC10*)1 in the fall of 1979, Louise made some interesting observations. The plasmid DNA was capable of transforming yeast with an incredibly high efficiency, at least one order of magnitude higher than we had ever seen before. Even more interesting, the yeast transformants maintained the plasmid quite stably, even in the absence of selection, for many generations, whereas the parent *ARS* vector was lost at high frequency, as is typical of all *ARS* plasmids. Even after growth in complete media overnight, more than 95% of the cells still retained pYe(*CDC10*)1 in an autonomous state, whereas under the same growth conditions, the parent *ARS* vector was almost completely lost. We began to entertain the possibility that pYe(*CDC10*)1 contained the centromere and that it was functioning as a mitotic stabilizing element on the plasmid. We now know, of course, that the presence of a complete centromeric DNA sequence on a plasmid converts it into an "artificial chromosome"; in effect, the cell accepts the new centromere-bearing replicating DNA molecule as one of its own chromosomes, and the imposter is segregated during cell divisions on the mitotic apparatus right along with the "real" chromosomes (for review, see Carbon and Clarke 1990; Clarke 1990). In 1979, however, this was a new idea; it was not at all obvious that a relatively short DNA sequence was all that would be necessary to convert a replicating circular DNA molecule into a stably segregating chromosome. We needed some way to determine if the "stabilizing element" on pYe(*CDC10*)1 in fact was the chromosome III centromere.

Fortunately, two well-known yeast researchers, Bob Mortimer and Terry Cooper, spent their sabbatical leaves during the late 1970s in my laboratory to learn the new recombinant DNA technology, and Louise and I benefited greatly from their presence. I had been trained in organic chemistry and biochemistry, and, in fact, after high school had never even taken a formal course in biology. From conversations with Bob and Terry, I learned how yeast tetrad analysis could be used to determine the centromere linkage of a genetic marker. It seemed quite likely that we could use meiotic tetrad analysis to ascertain if pYe(*CDC10*)1 was segregating through meiosis as an independent centromere-bearing linkage group (a chromosome). However, it seemed like a long shot—Would a small circular plasmid containing only a couple of yeast genes, a DNA replicator, and a centromere be recognized as a chromosome and segregate properly through the two complicated meiotic divisions? Previously, Bob had spent considerable time teaching us his streamlined method for micromanipulation and separation of the four spores in yeast asci. The biochemists in the laboratory did not take to the finicky and nerve-wracking dissection procedure; one postdoc described that it was "like pushing peas around with a telephone pole!" After a bit of practice with the micromanipulator, I decided that I was ready to put a yeast haploid-containing pYe(*CDC10*)1 through a cross and analyze the meiotic segregation pattern of the plasmid *TRP1* and *CDC10* genes.

Fortunately, the 1979 Christmas recess was upon us, giving me time for uninterrupted work in the laboratory. Louise set up the "Xmas cross," as we called it, and I did the tetrad dissections during the 3-week holiday. The idea was to cross a haploid-containing pYe(*CDC10*)1 with a haploid of the opposite mating type containing several centromere-linked genetic markers, select the diploid, sporulate it, and look at the genotypes of the haploid progeny. I was hoping to see stable maintenance of the plasmid through meiosis (*ARS* plasmids are usually lost and are rarely present in meiotic progeny). We also hoped that the plasmid genes would segregate as true centromere-linked markers, appearing in the so-called "sister" spores, produced by the second meiotic division. If the centromere-bearing plasmid was segregating as an independent linkage group, it should appear in only two of the four spores, and these should be sister spores; in our case, against the centromere-XI-linked marker *met14*, plasmid pYe(*CDC10*)1 should segregate in either the two *MET14* spores or the two *met14* spores. However, as stated above, this seemed to be asking far too much of this relatively simple plasmid construct. In the first experiment I laboriously dissected 33 tetrads, only 16 of which produced four healthy progeny. However, when these "Xmas tet's," as my laboratory book describes them, were analyzed for genotypes, we were absolutely amazed at the results. The plasmid segregated to two of the four spores in 10 of the 16 complete tetrads, and, in those cases, it always appeared in sister spores!

Other experiments proved that the plasmid was not integrated into any of the yeast chromosomes, but it was replicating autonomously. Thus, pYe(*CDC10*)1 was segregating as a "real" chromosome and, albeit circular, was recognized as the first "artificial" chromosome containing a functional centromere (Clarke and Carbon 1980b).

Subsequent experiments quickly showed that the segment of DNA containing the full centromere activity (termed *CEN3*) was surprisingly short (now established to be only about 130 bp). In 1982, with the identification by Jack Szostak and Elizabeth Blackburn of small DNA segments capable of functional telomere formation in yeast (Szostak and Blackburn 1982), all of the DNA elements necessary for construction of linear artificial chromosomes became available (Murray and Szostak 1983). These essential building blocks of a eukaryotic chromosome include the origins of DNA replication (*ARS* elements), telomeres to ensure proper replication of linear chromosome ends, centromeres (*CEN*) to attach to the mitotic apparatus and move the replicated chromatids to the daughter cells, and, of course, the *raison d'être* for the chromosome, the genes that carry genetic information. These yeast artificial chromosomes (known as YACs) have become useful not only as model systems to study the mechanism of chromosome segregation (Carbon and Clarke 1990), but also as vectors to clone ultra-large heterologous segments of DNA, literally millions of base pairs in length (Burke et al. 1987). The ability to clone megabases of DNA in a single vector drastically reduces the number of clones necessary to include the vast amount of genetic information present in animal cell genomes. Thus, the latter technological advance is being put to good use in the current concerted effort to map and eventually sequence the human genome.

## EPILOGUE

In thinking back over the path that our research followed from the construction of complete *E. coli* and yeast genomic libraries to the isolation of genes, DNA replicators, centromeres, and the construction of artificial chromosomes, I am impressed by how direct the path seems, as if the idea of artificial chromosomes was already there in 1974 when Louise made the genomic libraries (or even in 1945 when Edward Tatum isolated the *E. coli leuB-6* mutation that Barry Ratzkin later used to clone the yeast *LEU2* gene). Obviously, however, we had no idea in 1974 that, just 5 years later, we would construct a functional artificial chromosome. For us at the time, the driving force was the isolation of important components of the eukaryotic chromosome, and genes were the most obvious place to start. Everything that followed flowed naturally from that initial event. This is a prime illustration of the unforeseen profits that very often arise from pure basic research carried out in small independent laboratories funded by extramural grants. Certainly "big" science, highly organized projects with many laboratories supported by federal funds and working on a common technological goal, has its place, but I hope never at the expense of my kind of science.

## ACKNOWLEDGMENTS

A large part of this chapter describes the research accomplishments of Louise Clarke, to whom, as always, I am greatly indebted. I thank Louise also for help with editing the manuscript and for jogging my memory about distant events. I also owe a great deal to the many talented graduate students and postdoctoral fellows who contributed to the work described here. The National Cancer Institute of the National Institutes of Health has generously and continuously supported my research for the past 24 years (NIH grant CA-11034). Finally, I thank the American Cancer Society for support in the form of an ACS Research Professorship.

## REFERENCES

Beggs, J.D. 1978. Transformation of yeast by a replicating hybrid plasmid. *Nature* **275:** 104–109.

Bolivar, F., R.L. Rodriguez, P.J. Greene, M.C. Betlach, H.L. Heyneker, and H.W. Boyer. 1977. Construction and characterization of new cloning vehicles. II. A multipurpose cloning system. *Gene* **2:** 95–113.

Botstein, D. and R.W. Davis. 1982. Principles and practice of recombinant DNA research with yeast. In *The Molecular biology of the yeast* Saccharomyces: *Metabolism and gene expresion* (ed. J.N. Strathern et al.), pp. 607–636, Cold Spring Harbor Laboratory, Cold Spring Harbor, New York.

Burke, D.T., G.F. Carle, and M.V. Olson. 1987. Cloning of large segments of exogenous DNA into yeast by means of artificial chromosome vectors. *Science* **236:** 806–812.

Cameron. J.R., E.Y. Loh, and R.W. Davis. 1979. Evidence for transposition of dispersed repetitive DNA families in yeast. *Cell* **16:** 739–751.

Carbon, J. and L. Clarke. 1990. Centromere structure and function in budding and fission yeasts. *New Biol.* **2:** 10–19.

Carbon, J. and E.W. Fleck. 1974. Genetic alteration of structure and function in glycine transfer RNA of *Escherichia coli*: Mechanism of suppression of the tryptophan synthetase A78 mutation. *J. Mol. Biol.* **85:** 371–391.

Carbon, J.A., P. Berg, and C. Yanofsky. 1966. Studies on missense suppression of the

tryptophan synthetase A-protein mutant A36. *Proc. Natl. Acad. Sci.* **56:** 764–771.
———. 1967. Missense suppression due to a genetically altered RNA. *Cold Spring Harbor Symp. Quant. Biol.* **31:** 487–497.
Carbon, J., T.E. Shenk, and P. Berg. 1975. Biochemical procedure for production of small deletions in simian virus 40 DNA. *Proc. Natl. Acad. Sci.* **72:** 1392–1396.
Carbon, J., L. Clarke, C. Ilgen, and B. Ratzkin. 1977a. The construction and use of hybrid plasmid gene banks in *Escherichia coli*. In *Recombinant molecules: Impact on science and society* (ed. by R.F. Beers, Jr. and E.G. Bassett), pp. 355–378, Raven Press, New York.
Carbon, J., B. Ratzkin, L. Clarke, and D. Richardson. 1977b. Genetic interaction and gene tranfser. *Brookhaven Symp. Biol.* **29:** 277–296.
Carbon, J., L. Clarke, C. Chinault, B. Ratzkin, and A. Walz. 1978. The isolation and characterization of specific gene systems from the yeast, *Saccharomyces cerevisiae*. In *Biochemistry and genetics of yeasts* (ed. M. Bacila et al.), pp. 425–443, Academic Press, New York.
Chinault, A.C. and J. Carbon. 1979. Overlap hybridization screening; Isolation and characterization of overlapping DNA fragments surrounding the *leu2* gene on yeast chromosome III. *Gene* **5:** 111–126.
Clarke, L. 1990. Centromeres of budding and fission yeasts. *Trends Genet.* **6:** 150–154.
Clarke, L. and J. Carbon. 1975. Biochemical construction and selection of hybrid plasmids containing specific segments of the *Escherichia coli* genome. *Proc. Natl. Acad. Sci.* **72:** 4361–4365.
———. 1976. A colony bank containing synthetic ColE1 plasmids representative of the entire *E. coli* genome. *Cell* **9:** 91–99.
———. 1980a. Isolation of the centromere-linked *CDC10* gene by complementation in yeast. *Proc. Natl. Acad. Sci.* **77:** 2173–2177.
———. 1980b. Isolation of a yeast centromere and construction of functional small circular chromosomes. *Nature* **287:** 504–509.
Clarke, L., R. Hitzeman, and J. Carbon. 1979. Selection of specific clones from colony banks by screening with radioactive antibody. *Methods Enzymol.* **68:** 436–442.
Garfinkel, D.J., J.D. Boeke, and G.R. Fink. 1985. Ty element transposition: Reverse transcriptase and virus-like particles. *Cell* **42:** 507–517.
Hershfield, V., H.W. Boyer, C. Yanofsky, M.A. Lovett, and D.R. Helinski. 1974. Plasmid ColE1 as a molecular vehicle for cloning and amplification of DNA. *Proc. Natl. Acad. Sci.* **71:** 3455–3459.
Hicks, J. and G.R. Fink. 1977. Identification of chromosomal location of yeast DNA from hybrid plasmid *pYeleu10*. *Nature* **269:** 265–267.
Hinnen, A., J.B. Hicks, and G.R. Fink. 1978. Transformation of yeast. *Proc. Natl. Acad. Sci.* **75:** 1929–1933.
Hitzeman, R.A., L. Clarke, and J. Carbon. 1980. Isolation and characterization of the yeast 3-phosphoglycerokinase gene (*PGK*) by an immunological screening technique. *J. Biol. Chem.* **255:** 12073–12080.
Hsiao, C.-L. and J. Carbon. 1979. High frequency transformation of yeast by plasmids containing the cloned yeast *ARG4* gene. *Proc. Natl. Acad. Sci.* **76:** 3829–3833.
———. 1981. Characterization of a yeast replication origin (*ars2*) and construction of stable mini-chromosomes containing a cloned yeast centromere (*CEN3*). *Gene* **15:** 157–166.
Jackson, D.A., R.H. Symons, and P. Berg. 1972. Biochemical method for inserting new genetic information into DNA of simian virus 40: Circular SV40 DNA molecules containing lambda phage genes and the galactose operon of *Escherichia coli*. *Proc. Natl. Acad. Sci.* **69:** 2904–2909.
Kingsman, A.J., R.L. Gimlich, L. Clarke, A.C. Chinault, and J. Carbon. 1981. Sequence variation in dispersed repetitive sequences in *Saccharomyces cerevisiae*. *J. Mol. Biol.* **145:** 619–632.
Mellor, J., N.H. Malim, K. Gull, M.F. Tuite, S. McCready, T. Dibbayawan, S.M. Kingsman, and A.J. Kingsman. 1985. Reverse transcriptase activity and Ty RNA are associated with virus-like particles in yeast. *Nature* **318:** 583–586.
Murray, A.W. and J.W. Szostak. 1983. Construction of artificial chromosomes in yeast. *Nature* **305:** 189–193.

Phillips, T.A., V. Vaughn, P.L. Bloch, and F.C. Neidhardt. 1987. Gene-protein index of *Escherichia* K-12. In Escherichia coli *and* Salmonella typhimurium, *cellular and molecular biology*, 2nd edition (ed. F.C. Neidhardt et al.), vol. 2, pp. 919–966. American Society for Microbiology, Washington, D.C.

Ratzkin, B. and J. Carbon. 1977. Functional expression of cloned yeast DNA in *Escherichia coli. Proc. Natl. Acad. Sci.* **74:** 487–491.

Riddle, D.L. and J. Carbon. 1973. Frameshift suppression: A nucleotide addition in the anticodon of a glycine transfer RNA. *Nat. New Biol.* **242:** 230–234.

Roberts, J.W. and J. Carbon. 1974. Molecular mechanism for missense suppression in *E. coli. Nature* **250:** 412–414.

Shenk, T.E., J. Carbon, and P. Berg. 1976. Construction and analysis of viable deletion mutants of simian virus 40. *J. Virol.* **18:** 664–671.

Stinchcomb, D.T., K. Struhl, and R.W. Davis. 1979. Isolation and characterization of a yeast chromosomal replicator. *Nature* **282:** 39–43.

Struhl, K., J.R. Cameron, and R.W. Davis. 1976. Functional genetic expression of eukaryotic DNA in *Escherichia coli. Proc. Natl. Acad. Sci.* **73:** 1471–1475.

Struhl, K., D.T. Stinchcomb, S. Scherer, and R.W. Davis. 1979. High-frequency transformation of yeast: Autonomous replication of hybrid DNA molecules. *Proc. Natl. Acad. Sci.* **76:** 1035–1039.

Szostak, J.W. and E.H. Blackburn. 1982. Cloning yeast telomeres on linear plasmid vectors. *Cell* **29:** 245–255.

Tatum, E.L. 1945. X-ray induced mutant strains of *Escherichia coli. Proc. Natl. Acad. Sci.* **31:** 215–219.

Tatum, E.L. and J. Lederberg. 1947. Gene recombination in the bacterium *Escherichia coli. J. Bacteriol.* **53:** 673–684.

Tschumper, G. and J. Carbon. 1980. Sequence of a yeast DNA fragment containing a chromosomal replicator and the *TRP1* gene. *Gene* **10:** 157–166.

———. 1982. Delta sequences and double symmetry in a yeast chromosomal replicator region. *J. Mol. Biol.* **156:** 293–307.

Watson, J.D. and J. Tooze. 1981. *The DNA story*, pp. 82–89. W.H. Freeman, San Francisco.

# Starting to Probe for Yeast Genes

BENJAMIN D. HALL
*Department of Genetics, SK-50*
*University of Washington*
*Seattle, Washington 98195*

This "early days" perspective recounts the excitement as well as some of the frustrations of getting started in yeast molecular genetics in the early to mid 1970s. Yeast work in my laboratory at the University of Washington began with the characterization of yeast nuclear RNA polymerases (Adman et al. 1972; Schultz and Hall 1976) and genetic studies of sporulation control and of a conditional RNA synthesis mutant (Hopper and Hall 1975a,b; Andrew et al. 1976). At that time (1973–1975), my co-workers and I had the ambition to study the details of RNA synthesis in yeast nuclei, and we began to realize that we lacked certain tools needed to bring genetic and biochemical approaches into conjunction. Although RNA polymerases I, II, and III could be separated from one another and their separate functions could be demonstrated (Schultz and Hall 1976; Sentenac et al. 1976), there seemed little likelihood that in vitro transcription of total yeast DNA by these RNA polymerases would occur with specificity. Early reports (Van Keulen et al. 1975) of selective polymerase I transcription of rDNA within total yeast DNA have gone unconfirmed.

The conviction that studies with *individual* isolated genes would offer the way out of this transcription dilemma was one influence that motivated me to think about isolating yeast genes. Also important were the various rumblings emanating from bacteriophage genetics and nucleic acid enzymology (Lobban and Kaiser 1973) suggesting that isolated nucleic acid sequences might before long be amplified and propagated. Terms such as molecular cloning and polymerase chain reaction (PCR) did not yet exist; however, I recall several discussions, one being with Tom Broker in about 1970, that foreshadowed these developments.

Viewed from a present-day perspective, against the background of 1980s yeast research, cloning genes by complementation of yeast mutant strains would have been the obvious way to bridge between plasmid vectors and yeast genetics in order to isolate genes. This was not apparent in 1970–1975, because the prospects for a reliable yeast transformation system appeared to be very bleak. There were occasional reports of successful yeast transformation, but none of them inspired confidence that the results could be repeated in another laboratory. More promising, it seemed to us, would be the approach of enriching for the yeast gene of interest, using specific nucleic acid hybridization to a genetically defined probe. Initially, we envisioned using these probes to screen through large quantities of restriction-nuclease-cleaved DNAs to enrich for interesting sequences directly. In fact, by the time our specific hybridization probes were actually ready, sophisticated methods for hybridization screening of plaques or transformant colonies in libraries (Grunstein and Hogness 1975; Benton and Davis 1977) had made physical enrichment for the gene of interest unnecessary.

From extensive reading, listening to seminars, and discussing yeast genes with experts, particularly with Don Hawthorne, I could find only two examples of genetically defined hybridization probes that could be obtained, made, or inferred for specific yeast genes. One such probe was the single species of yeast tyrosine tRNA, sequenced by Madison and co-workers (Madison and Kung 1967), a probe for a family of nonsense suppressor loci. The other was the structural gene for yeast iso-1-cytochrome *c*, identified by Sherman and Slonimski (Sherman 1963; Sherman and Slonimski 1964) and intensively analyzed by the *cyc1* mutant analyses and protein sequencing studies of Sherman and Stewart (Stewart et al. 1971; Sherman et al. 1974). In fact, the knowledge that suppressor genes *SUP2* through *SUP8* and *SUP11* encoded yeast $tRNA^{Tyr}$ also depended on the amino acid insertion specificity demonstrated for ochre alleles of *CYC1* (Stewart et al. 1972). These eight genes, then called super-suppressors, had been mapped to specific chromosomal locations (Mortimer and Hawthorne 1973). Because tRNA fractionation and sequence analysis indicated that yeast contained only one $tRNA^{Tyr}$ species, we suspected that all eight genes might encode an identical gene product and that their sequence would be that determined by Madison and co-workers.

The coding sequence of a section of the mRNA for the *CYC1* gene had been deduced by Sherman and Stewart (Sherman et al. 1974) by protein sequencing of suppressed frameshift mutations near the beginning of the gene. The known response of the iso-1-cytochrome *c* level to carbon source changes and to $O_2$ induction made this an attractive gene for studies of in vivo transcriptional regulation. Toward this end, we sought to isolate a gene-specific DNA fragment for *CYC1* mRNA hybridization studies.

The means for actualizing these hypothetical gene-specific probes for *CYC1* and for suppressor tRNA genes involved essential collaborative interactions with Michael Smith and with other scientists at the University of British Columbia in Vancouver. Shortly after coming to the Pacific Northwest in 1963, I met both Mike Smith and Gordon Tener. During the next 10 years, Mike and I had frequent contacts, and, in the early 1970s, we actively discussed ways whereby the DNA probes he was able to make by chemical synthesis might be used to assay for and to isolate specific mRNAs. I recall that this discussion reached a crucial juncture around 1974, when Richard Zitomer and I met with Mike at Deception Pass, Washington—the halfway point between Seattle and Vancouver. After the families had a picnic, Richard and I began twisting Michael's arm very hard to convince him that yeast *CYC1* offered greater potential scientific benefits for synthetic DNA studies than the T4 lysozyme-specific oligonucleotide studies he had already initiated. A vigorous effort that Cor Hollenberg made the next year to learn nucleotide chemistry and to synthesize a *CYC1* probe during a short stay in Vancouver was premature. As this project eventually developed, it was necessary to have *both* improved chemical synthesis procedures and a quick convenient assay for probes hybridizing to genomes (i.e., Southern blots) before the *CYC1* project could really gather steam.

For the other genes of interest, *SUP2* through *SUP8* and *SUP11*, the necessary probe existed both in nature and in several laboratories devoted to tRNA biochemistry—this probe was yeast tyrosine tRNA. Maynard Olson, Anita

Hopper, and I became convinced that working with $tRNA^{Tyr}$ probes might lead us in some useful direction, if only to provide practice to use later in isolating "real genes." Other reasons for our interest were the proximity of *SUP4* to *CYC1* on the linkage map of yeast chromosome X (Lawrence et al. 1975) and a genuine interest in learning about the transcription and processing of yeast suppressor tRNAs.

These embryonic plans to probe for yeast tRNA genes with labeled $tRNA^{Tyr}$ were greatly aided by Gordon Tener and his group at the University of British Columbia in Vancouver. Gordon was at this time heavily involved in a *Drosophila* tRNA project, but since he and Ian Gillam had previously worked extensively with yeast tRNAs (Gillam et al. 1967), they gave us general advice as well as enriched yeast tRNA fractions. Gordon also instructed Maynard in a method he was then perfecting to make amino-acid-specific labeled tRNA probes. He had developed an elegant general procedure for coupling an easily iodinated target molecule (the Bolton-Hunter reagent) onto the free $\alpha$-$NH_2$ group of aminoacylated tRNA. This method enabled Gordon to label specifically a single charged tRNA species existing in a crude mixture of other uncharged tRNAs.

To initiate this project, Anita Hopper and I began some work on yeast tRNA column fractionation, and Maynard Olson, together with Anita, made a brief visit to Vancouver to learn the condensation-iodination procedure. When he returned and applied the method, it was impressively successful, giving labeled $tRNA^{Tyr}$ molecules. Only later did we learn, from Guy Page's experiments, that the $^{125}I$ label incorporated into tRNA was entering a modified base and not the acylation-specific Bolton-Hunter iodination target. Because only the $tRNA^{Tyr}$ molecules in our tRNA fractions contained this modified base, Maynard succeeded by serendipity in making a very reliable and clean $^{125}I$-labeled tRNA probe. Had he followed Tener's protocol exactly, and carried out the mock iodination step prior to aminoacylation, the probes would have been less stable and perhaps less hot as well.

Having made such a $tRNA^{Tyr}$ probe and believing on genetic grounds that it should hybridize to eight regions of the yeast genome, we faced the problem of how to scan across a restriction digest of yeast DNA fragments with the $^{125}I$-labeled tRNA to find the genes. We attempted two pre-Southern approaches to this problem in genome analysis (Fig. 1, top). The most encouraging data came from a laborious gel-slicing experiment done on an *Eco*RI digest of yeast DNA in the summer of 1975. Sections from a large slab gel were dissolved in salt solution, and each extract was then passed through a warm hydroxyapatite column to purify DNA. The DNA eluted from each slice was loaded on an individual nitrocellulose filter disk and then hybridized to the $^{125}I$-labeled tRNA probe. Because of these primitive methods, it took three people most of a week to analyze a single digest. However, the results gave a clear indication that multiple fragments in the yeast genome were hybridizing to the tRNA probe. This approach reached its zenith and abruptly came to an end in the summer of 1975, just prior to the Cold Spring Harbor Yeast Meeting. Anita Hopper, who was to present our paper, had left Seattle for the University of Massachusetts 6 months before the meeting. A flat out last minute effort by the rest of us, including Maynard Olson, Gayle Lamppa, and Donna Montgomery, produced the genomic hybridization results shown in Figure 1 (bottom). One day prior to Anita's talk, these fresh data were handed to her, as

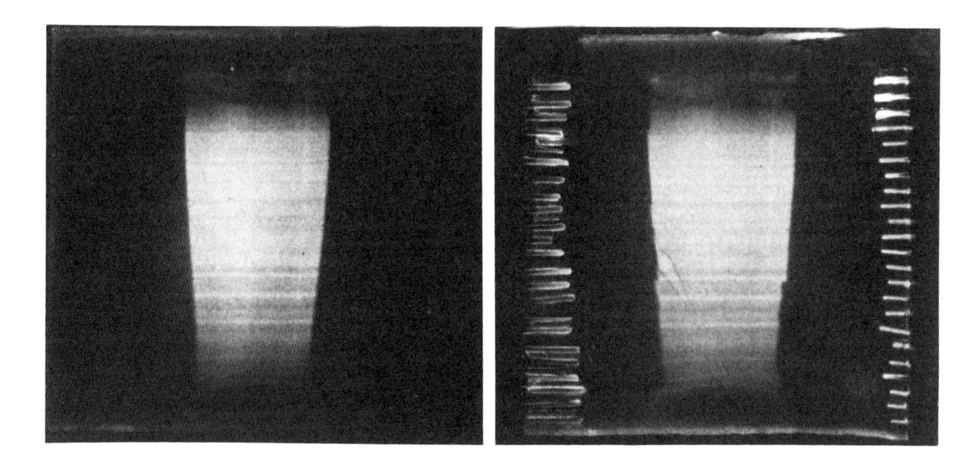

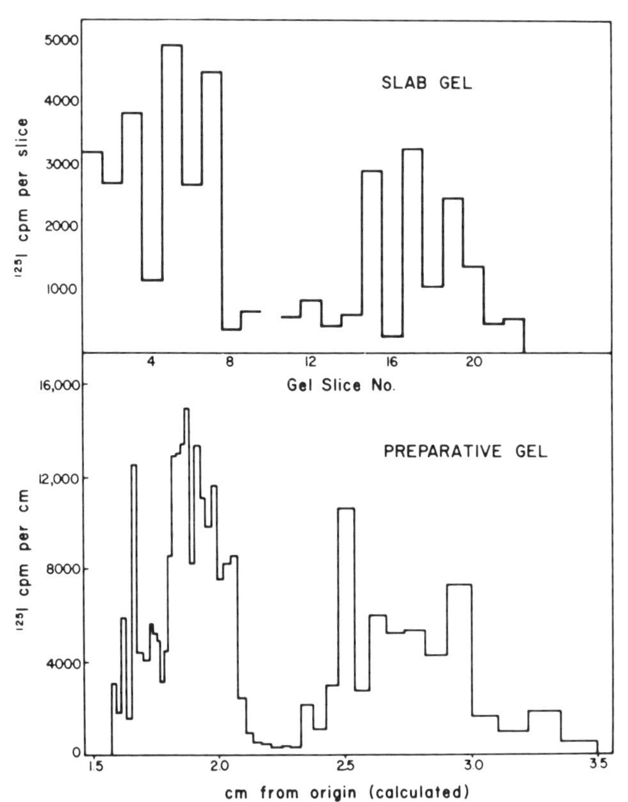


*Figure 1* (*Top*) Agarose slab gel fractionation of *Eco*RI-cut yeast chromosomal DNA. An *Eco*RI digest of total yeast DNA (10 mg) was loaded on a 10 x 25-cm 0.7% agarose gel and electrophoresed for 20 hr. (*Right*) Separation of fragments was achieved by cutting transverse slices of equal width with a sharp blade. Here, the slices are shown with toothpicks as spacers. (*Left*) After ethidium bromide staining, many bands were visible. (*Bottom*) Identification of yeast $tRNA^{Tyr}$ genes in gel slices. To recover DNA, the agarose slices were dissolved in warm (60°C) perchlorate solution. The solution from each gel fraction was loaded on an individual hydroxyapatite column to purify DNA and then eluted with 0.5 M phosphate buffer. The eluate was heated to 100°C and then passed through a nitrocellulose membrane filter to immobilize the DNA. The set of filters was hybridized to $^{125}$I-labeled $tRNA^{Tyr}$ to assay for the positions of tRNA genes.

printed tape output from the scintillation counter, in the library of the Cold Spring Harbor Laboratory. Despite the crude appearance of these data (Fig. 1, bottom), we can now say with hindsight that the pattern of appearance of the hybridizing fragments was real—a set of five genes (*SUP3* through *SUP6* and *SUP8*) on large *Eco*RI fragments and three genes (*SUP2*, *SUP7*, and *SUP11*) on small *Eco*RI fragments, with a blank area between DNA fragment sizes of 1.7 and 5 kb (Olson et al. 1979b).

At this same Cold Spring Harbor meeting, Donna Montgomery took the very important step of visiting Michael Mathews' laboratory to learn the Southern hybridization technique (not yet published at that time). The Southern procedure (Southern 1975) proved to be an immensely valuable tool for both the *CYC1* and $tRNA^{Tyr}$ projects. Upon her return to Seattle after the meeting, Donna obtained excellent results by $^{125}I$-labeled $tRNA^{Tyr}$ hybridization to a Southern blot of *Eco*RI-cut yeast DNA. For this she used a tRNA probe $^{125}I$-labeled by Siwo de Kloet during his brief stay in Seattle several weeks before. We were all very excited to find that the Southern pattern very clearly gave eight bands coinciding with the expectation from suppressor genetics that there should be eight different loci for $tRNA^{Tyr}$ (Olson et al. 1977). This was the first convincing indication that our attempts to link up molecular hybridization probes with yeast genetics had some prospect of being successful.

Over the course of the next year, several important events made possible the fruition of the tRNA and *CYC1* projects in 1977–1979. Foremost among these was the return of Maynard Olson to my laboratory in early 1976. During the fall, he had completed some teaching obligations in the Chemistry Department at Dartmouth College; he then began a new career as a postdoctoral fellow in yeast molecular genetics. A second important event was the successful chemical synthesis of a *CYC1*-specific oligonucleotide by Shirley Gillam in Michael Smith's laboratory. This made it possible for Donna Montgomery to apply her expertise in yeast genomic Southern hybridization (gained with tRNA probes) to the analysis of the *CYC1* gene. Third, there began a brief but highly productive collaboration between Maynard Olson and John Cameron, a graduate student working at Stanford University with Ron Davis. They applied the powerful λgt cloning vector technology, developed by Davis, to cloning the yeast genomic restriction fragments that carried $tRNA^{Tyr}$ genes (Olson et al. 1979a).

During this period, when the main analytical tool used in our laboratory was Southern analysis, a persistent but ultimately incorrect assumption served as a logical bridge between the $tRNA^{Tyr}$ and *CYC1* projects. Since the *SUP4* gene was close to *CYC1* on the yeast chromosomal map, I supposed that Southern observations of *CYC1* mutant alleles might yield some clue as to which of the eight $tRNA^{Tyr}$ bands corresponded to *SUP4*. Yeast strains bearing two of the *cyc1* mutant alleles seemingly provided a test of this concept. The ochre mutation *cyc1-9* consisted of a single-base-pair substitution that destroyed a predicted *Eco*RI site present in the wild-type *CYC1* gene (Sherman et al. 1974), whereas *cyc1-1* appeared, by genetic criteria, to consist of a deletion of the entire *CYC1* locus (Sherman et al. 1975). We reasoned that if the physical map distance between *CYC1* and *SUP4* were as short as the recombination distance indicated, one of the eight bands might be shifted or deleted in yeast strains bearing these *CYC1* alleles. The initial Southern analysis with

$^{125}$I-labeled tRNA$^{Tyr}$ compared the patterns of *cyc1-9* strain B596 and *cyc1-1* strain D234-4D with that of the unrelated wild-type strain S288C. The results showed an interesting difference among the three strains, with regard to the largest hybridization band. A 14-kb band that was present in B596 was replaced by a 12-kb band in wild type and by a blank space (no band) in deletion strain D234-4D. The exciting conclusion that *CYC1* and *SUP4* lay on the same *Eco*RI fragment seemed irresistible to me, but, fortunately, Maynard Olson was unconvinced. He did the additional experiments necessary to show that both B596 and its wild-type parent, D311-3A, differed from S288C by a length polymorphism in the largest tDNA$^{Tyr}$ fragment and that the missing band in D234-4D DNA was due to a slight DNA degradation during extraction that selectively removed the largest DNA fragments from the Southern pattern. In fact, as later work showed (Shalit et al. 1981), *SUP4* and *CYC1* are separated by 21 kb of DNA. Ironically, we found out shortly after *CYC1* had been cloned using an oligonucleotide (see below) that the clone had been in our laboratory already for more than 1 year because it was on the same *Eco*RI fragment with a tRNA$^{Ser}$ gene that Guy Page had cloned (Page and Hall 1981) using the tRNA as probe. Clearly, good luck is often more powerful than reasoning.

## FURTHER ANALYSIS OF THE YEAST tRNA$^{Tyr}$ GENES

In 1976, Maynard Olson began in a concerted way to study the yeast tRNA$^{Tyr}$ genes. His project grew to include (1) the cloning of chromosomal sequences encoding these genes, (2) the identification of individual cloned fragments with tyrosine-inserting suppressors on the genetic map, and (3) the sequence analysis of the genes themselves. Later, in collaboration with Andre Sentenac, Peter Piper, and Guy Page, Maynard characterized a smaller family of serine-inserting suppressor tRNA genes (Olson et al. 1981). Cloning of seven of the eight tRNA$^{Tyr}$ genes was accomplished by Olson and Cameron by plaque hybridization screening of λgt-yeast recombinant clones using the $^{125}$I-labeled

*Table 1* Summary of Assignments of tRNA$^{Tyr}$-hybridizing Restriction Fragments to Tyrosine-inserting Suppressor Loci; Sizes of the Restriction Fragments

| | | Fragment size (base pairs x $10^{-3}$) | |
|---|---|---|---|
| Fragment | Suppressor | *Eco*RI D311-3A | *Eco*RI AB350 |
| A | *SUP4* | 14 ($A_1$) | 12 ($A_2$) |
| B | *SUP5* | 9.9 | 9.9 |
| C | *SUP8* | 6.7 ($C_1$) | 12 ($C_2$) |
| D | *SUP3* | 6.3 | 6.3 |
| E | *SUP6* | 5.2 | 5.2 |
| F | *SUP2* | 1.3 ($F_1$) | 1.7 ($F_2$) |
| G | *SUP11* | 1.2 | 1.2 |
| H | *SUP7* | 0.9 ($H_1$) | 1.8 ($H_2$) |

The tRNA$^{Tyr}$-hybridizing restriction fragments are named as *Eco*RI fragments in order of decreasing size. The sizes of the *Eco*RI fragments in D311-3A are from Olson et al. (1979a; these authors analyzed the strain B596, which is virtually isogenic with D311-3A and has an indistinguishable tRNA$^{Tyr}$ hybridization pattern). These fragments were used as size markers for determining the sizes of the remaining fragments. When more than one variant of a particular fragment was analyzed, its designation is given after the size in parentheses. (Reprinted, with permission, from Olson et al. 1979a.)

$tRNA^{Tyr}$ probe (Olson et al. 1979a). Determination of the size of the yeast chromosomal *Eco*RI fragment in each isolated clone made possible its assignment to a particular yeast genomic fragment (*Eco*RI-A = 14 kb, *Eco*RI-B = 9.9 kb, etc.) (Table 1). These early cloning and genomic hybridization studies indicated that two of the eight $tRNA^{Tyr}$ bands occurred at different locations in Southern hybridization of $tRNA^{Tyr}$ to *Eco*RI-digested DNA from strain D311-3A, as compared to S288C. It then became evident that these band-size differences are genetically determined, their variation being due either to localized polymorphisms affecting cutting sites or to insertion/deletion variation. Indistinguishable from S288C in both *Eco*RI and *Hin*dIII $tRNA^{Tyr}$ Southern patterns was yeast strain W87, an extensively back-crossed diploid made by Rodney Rothstein. W87 had the very useful feature that each of the eight tyrosine-inserting ochre suppressors (*SUP2* through *SUP8* and *SUP11*) had been reisolated by Rothstein in this background (Rothstein 1977). Available as a source of tDNA restriction variants was D311-3A, with two alterations in the *Hin*dIII pattern, and two *Eco*RI variants and the related yeast *Saccharomyces diastaticus*. Two variant $tRNA^{Tyr}$ hybridization bands had been discovered in *S. diastaticus* during a summer course in 1977 by a student project to screen distant but interfertile relatives of *Saccharomyces cerevisiae* for variations.

With this set of physical and genetic markers, Maynard Olson carried out the first applications of restriction-fragment-length polymorphisms (RFLPs) to gene identification, first for *SUP4* (Goodman et al. 1977) and then for the remainder of the set of $tRNA^{Tyr}$ genes (Table 1; Fig. 2) (Olson et al. 1979b). At the risk of straying beyond the time frame of this article, I will mention several follow-up consequences of the identification of the 14-kb *Eco*RI fragment that hybridized $tRNA^{Tyr}$ with the *SUP4* genetic locus. These are the first wild-type versus mutant sequence comparison for a eukaryotic chromosomal gene (*SUP4-o* vs. *sup4*$^+$), the demonstration that introns occur in certain tRNA genes (Goodman et al. 1977), and a short time later, the mapping and sequencing of many different *SUP4* loss-of-suppression mutations by Janet Kurjan (Kurjan et al. 1980). In retrospect, an amusing feature of the sequence comparison between *SUP4-o* and *sup4*$^+$ (Fig. 3) was its interplay with the fast-breaking discovery in 1977 that introns are ubiquitous in the genes of higher organisms. The existence of an intron in a mutant gene bearing a dominant allele (i.e., *SUP4-o*) provided strong evidence that introns were compatible with gene expression. In 1977, there was a real concern that most or all of the intron-containing genes then being isolated by physical methods were pseudogenes. A short time later, the prevailing wisdom evolved to the view that higher organism genes *without* introns are the presumptive pseudogenes (Sharp 1983). The initial sequencing studies on the $tRNA^{Tyr}$ genes were greatly assisted by a sabbatical visit in the spring and summer of 1977 by Howard Goodman, whose laboratory at the University of California, San Francisco, was one of the first on the West Coast to implement Maxam-Gilbert DNA sequencing.

## CLONING AND MOLECULAR CHARACTERIZATION OF THE YEAST *CYC1* GENE

On the basis of the *CYC1* mRNA sequence inferred from Sherman's genetic studies, Shirley Gillam and Michael Smith (University of British Columbia)

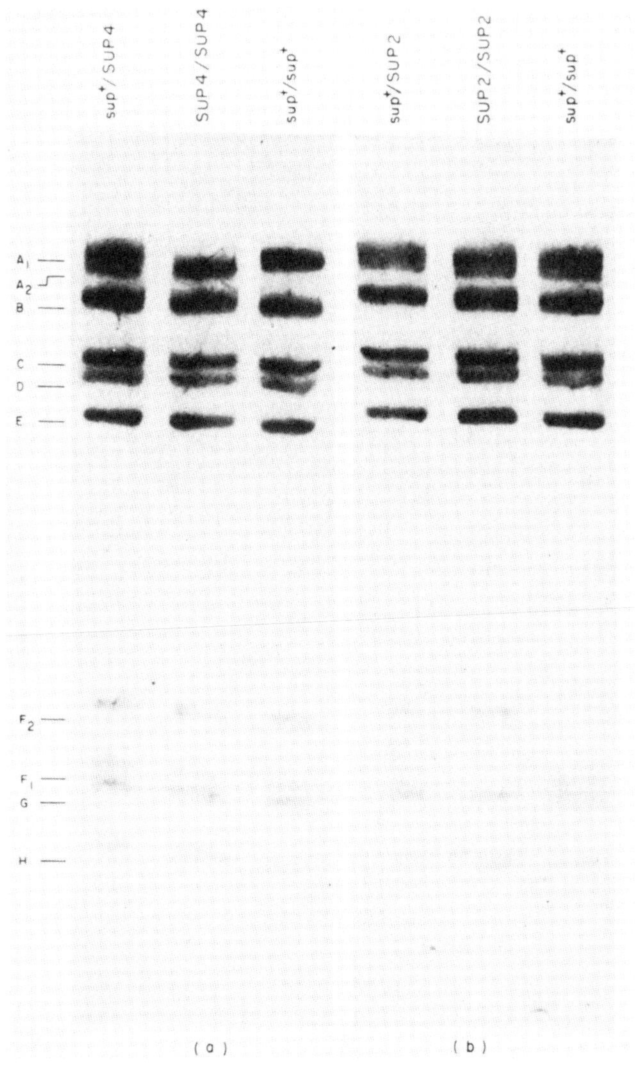


*Figure 2* Demonstration of mitotic linkage between *SUP4* and *Eco*RI-A variant and *SUP2* and an *Eco*RI-F variant. (*a*) The three strains represent a pink SUP4 heterozygote and the white and red sides of a sectored colony derived from it: AB81 (*SUP4/sup*$^+$ $A_1/A_2$ $F_1/F_2$); AB81-1W (*SUP4/SUP4* $A_1/A_2$ $F_1/F_2$); AB81-1R (*sup*$^+$*/sup*$^+$ $A_1/A_2$ $F_1/F_2$). (*b*) The three strains represent a pink SUP2 heterozygote and the white and red sides of a sectored colony derived from it: AB51 (*SUP2/sup*$^+$ $A_1/A_2$ $F_1/F_2$); AB51-1W (*SUP2/SUP2* $A_1/A_2$ $F_1/F_2$); AB51-1R (*sup*$^+$*/sup*$^+$ $A_1/A_2$ $F_1/F_2$). (Reprinted, with permission, from Olson et al. 1979a.)

synthesized a complementary oligodeoxynucleotide of length 13. At that time, when neither the solid-phase automated synthesis technology nor the phosphotriester synthetic method was available, synthesizing an oligonucleotide of this length in quantity was a substantial accomplishment. Moreover, this 13-mer as far as we could tell contained only molecules with the correct sequence. Donna Montgomery immediately used the *CYC1* molecular hybridization probe when it became available in 1976 to analyze nuclear DNA from

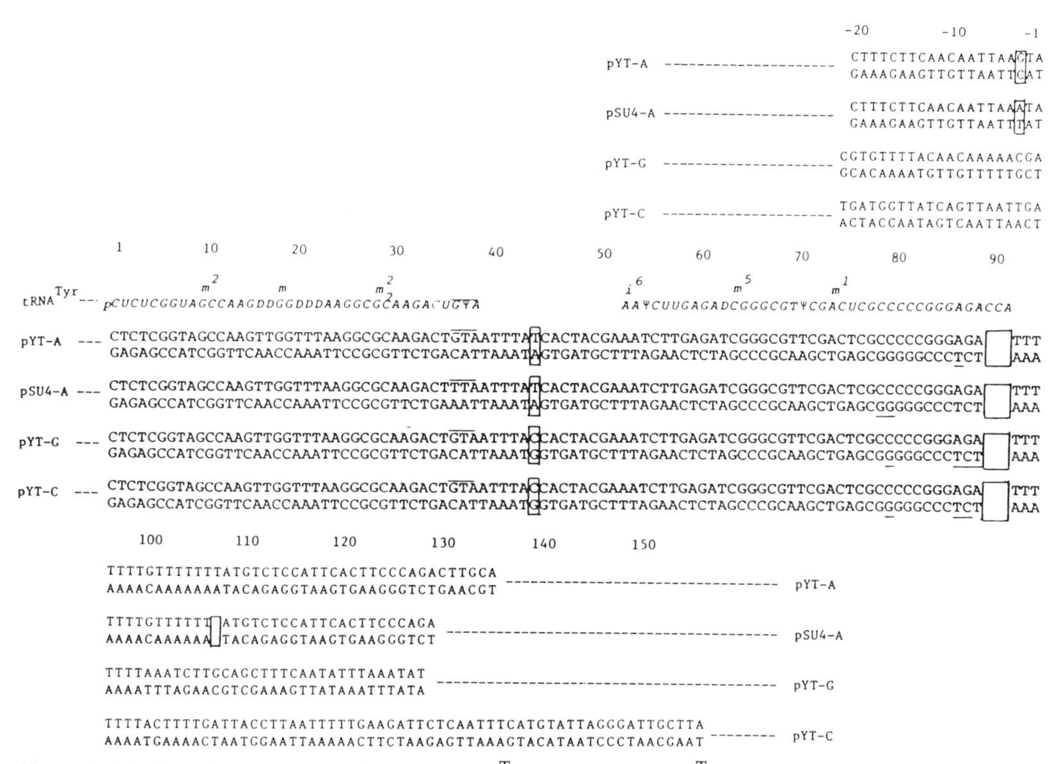

*Figure 3* Nucleotide sequences of yeast tRNA$^{Tyr}$ and four tRNA$^{Tyr}$ genes. The anticodon and its coding triplet are overlined in both the tRNA and the gene sequences. The empty boxes in all four genes at positions 90–92 indicate the absence of a coding triplet for the terminal C-C-A. Also enclosed in boxes are the variable base pair within the 14-bp "insert" and two points of divergence between the genes on pYT-A and pSU4-A. The position of the insert, as displayed here, is arbitrary in that it could equally well be displaced one base pair in either direction. (Reprinted, with permission, from Goodman et al. 1977.)

yeast carrying various *cyc1* alleles. She compared Southern patterns with the 13-mer probe for *Eco*RI-cut DNA from strains D311-3A (wild-type at *CYC1*), B596 (lacking the internal *Eco*RI site), and D234-4D (*cyc1-1* deletion). We were discouraged by the initial results, which showed a great number of bands, the same for all three DNAs. The initial explanation surmised for this—that most of the bands were due to poor sequence matching—seemed not to be the case, for there was no apparent temperature dependence of the pattern at 25–50°C. With more experience, it dawned on us that at the high molar concentration of probe being used, hybridization occurred very rapidly, occurring even during the time needed to fill and empty the hybridization vessel. As Donna became increasingly successful in achieving high specific activity of kinase end labeling of the probe and in controlling the hybridization kinetics, the Southern bands became clearer and fewer in number, decreasing to about seven (Montgomery et al. 1978). A careful comparison of the 13-mer hybridization patterns of *Eco*RI-cut DNA from strains D311-3A and B596 showed a band difference. A unique 2.6-kb band present in wild type was absent from DNA of yeast strain B596, replaced there by an 8.8-kb hybridization band. Neither of these bands was observed in DNA from the *cyc1-1* deletion strain.

To carry over this information to gene cloning, it then was of great interest to test the same 13-mer hybridization probe on a library of yeast-λ recombinant phage as a means of identifying a clone carrying *CYC1*. In fact, the desired clone was obtained fairly quickly. To recognize this clone for what it was, Donna Montgomery received help from an unexpected source. At the Cold Spring Harbor Yeast Meeting in 1977, Jack Szostak presented his Southern hybridization results with a similar oligonucleotide (Szostak et al. 1977). His data helped us to realize that the candidate *cyc1-9* clone we had that (1) gave plaque hybridization with 13-mer and (2) yielded an 8.8-kb *Eco*RI fragment on Southern blots with 13-mer in all likelihood contained the *cyc1* gene. Partial sequencing of the clone confirmed the presence of a *cyc1* open reading frame (Montgomery et al. 1978). Subsequent complete sequencing of the gene by Michael Smith showed that, as expected, the entire *CYC1* locus had been cloned (Smith et al. 1979).

The isolation and molecular analysis of *CYC1* (Smith et al. 1979; Stiles et al. 1981) have made possible many subsequent studies of this gene, including the characterization of its promoter (Faye et al. 1981; Guarente and Ptashne 1981) and regulatory sequences (Guarente and Mason 1983), the measurement of *CYC1* mRNA made under various conditions, and the cloning of related cytochrome-*c*-encoding genes (Zitomer et al. 1979; Russell and Hall 1982; Limbach and Wu 1985; Swanson et al. 1985). Because of the small size and high degree of evolutionary conservation of this gene and particularly because of the excellent genetic fine-structure analysis that had been done, *CYC1* proved to be an excellent choice for initiating detailed studies of gene action, using specific DNA of the cloned gene as a hybridization reagent.

## CONCLUSION

Viewing the work described here retrospectively, I see that in the relatively short period between 1974 and 1977, the approach to the study of transcription in my laboratory shifted from rather imprecise holistic methods to an approach much more like that used today. Direct studies of transcription, replication, or other DNA transactions in yeast are no longer made by measuring overall size or size distributions of DNA or RNA molecules or by transcribing total yeast DNA. Once the DNA of individual cloned genes became available, these could be transcribed individually in vitro (Klekamp and Weil 1982; Koski et al. 1982; Lue and Kornberg 1987) or used as labeled probes to measure the level or determine the structure of in vivo transcripts. Because shortly after the period I have described, yeast transformation provided a general method for isolating the DNA of nearly any yeast gene (and PCR takes care of all the others!), the reductionist approach to studies of gene action in yeast has since become the standard approach. During the studies described here, my co-workers and I certainly did not imagine that yeast genes of all kinds would soon be available. However, we gave considerable thought to what might be done with the few genes we could lay our hands on.

## ACKNOWLEDGMENTS

I express my gratitude for the work of all those who participated in this research, both those I have mentioned and those I have not mentioned in this account. Going back over the experiments done nearly two decades ago and seeing how the projects somehow succeeded despite unforeseen difficulties makes me appreciate how important the ability and dedication of these colleagues proved to be. As I recall, it was generally great fun when we at last identified the band or clone being sought—more exciting than in these days of ubiquitous kits, better techniques, and modern apparatuses, when everything is expected to succeed automatically. Patrick Linder's gentle persistence and good humor caused me to enjoy writing this article. I thank Anita Hopper, Donna Montgomery, and Maynard Olson for helpful comments on this manuscript and Andre Sentenac for the hospitality of his laboratory while this account was being written. I am also indebted to Jean-Marie Buhler, Michel Werner, and Sylvie Hermann for the crucial help they gave with computer problems.

## REFERENCES

Adman, R., L.D. Schultz, and B.D. Hall. 1972. Transcription in yeast: Separation and properties of multiple RNA polymerases. *Proc. Natl. Acad. Sci.* **69:** 1702–1706.

Andrew, C., A.K. Hopper, and B.D. Hall. 1976. A yeast mutant defective in the processing of 27S r-RNA precursor. *Mol. Gen. Genet.* **144:** 29–37.

Benton, W.D. and R.W. Davis. 1977. Screening λgt recombinant clones by hybridization to single plaques in situ. *Science* **196:** 180–182.

Faye, G., D.W. Leung, K. Tatchell, B.D. Hall, and M. Smith. 1981. Deletion mapping of sequences essential for in vivo transcription of the iso-l-cytochrome *c* gene. *Proc. Natl. Acad. Sci.* **78:** 2258–2262.

Gillam, I., S. Millward, D. Blew, M. von Tigerstrom, E. Wimmer, and G.M. Tener. 1967. The separation of soluble ribonucleic acids on benzoylated DEAE cellulose. *Biochemistry* **6:** 3043–3056.

Goodman, H., M. Olson, and B.D. Hall. 1977. Nucleotide sequence of a mutant eucaryotic gene: The yeast tyrosine-inserting ochre suppressor SUP4-o. *Proc. Natl. Acad. Sci.* **74:** 5453–5457.

Grunstein, M. and D.S. Hogness. 1975. A colony hybridization method for isolation of cloned DNAs that contain a specific gene. *Proc. Natl. Acad. Sci.* **72:** 3961–3965.

Guarente, L. and T. Mason. 1983. Heme regulates transcription of the *CYC1* gene of *S. cerevisiae* via an upstream activation site. *Cell* **32:** 1279–1286.

Guarente, L. and M. Ptashne. 1981. Fusion of *Escherichia coli* lacZ to the cytochrome *c* gene of *Saccharomyces cerevisiae. Proc. Natl. Acad. Sci.* **78:** 2199–2203.

Hopper, A.K. and B.D. Hall. 1975a. Mating-type and sporulation in yeast. I. Mutations which alter mating-type control over sporulation. *Genetics* **80:** 41–59.

———. 1975b. Mutation of a heterothallic strain to homothallism. *Genetics* **80:** 77–85.

Klekamp, M.S. and P.A. Weil. 1982. Specific transcription of homologous class III genes in yeast soluble cell-free extracts. *J. Biol. Chem.* **257:** 8432–8441.

Koski, R.A., D.S. Allison, M. Worthington, and B.D. Hall. 1982. An in vitro RNA polymerase III system from *S. cerevisiae*: Effects of deletions and point mutations upon *SUP4* gene transcription. *Nucleic Acids Res.* **10:** 8127–8143.

Kurjan, J., B.D. Hall, S. Gillam, and M. Smith. 1980. Mutations at the yeast SUP4 $tRNA^{Tyr}$ locus: DNA sequence changes in mutants lacking suppressor activity. *Cell* **20:** 701–709.

Lawrence, C.W., F. Sherman, M. Jackson and R.A. Gilmore. 1975. Mapping and gene conversion studies with the structural gene for iso-1-cytochrome *c* in yeast. *Genetics* **81:** 615–629.

Limbach, K.J. and R. Wu. 1985. Characterization of a mouse somatic cytochrome *c* gene and three cytochrome *c* pseudogenes. *Nucleic Acids Res.* **13:** 617–630.

Lobban, P.E. and A.D. Kaiser. 1973. Enzymatic end-to-end joining of DNA molecules. *J. Mol. Biol.* **78:** 453–471.

Lue, N.F. and R.K. Kornberg. 1987. Accurate initiation at RNA polymerase II promoters

in extracts from *Saccharomyces cerevisiae*. *Proc. Natl. Acad. Sci.* **84:** 8839–8843.
Madison, J.T. and H.K. Kung. 1967. Large oligonucleotides isolated from yeast tyrosine transfer ribonucleic acid after partial digestion with ribonuclease T1. *J. Biol. Chem.* **242:** 1324–1330.
Montgomery, D.L., B.D. Hall, S. Gillam, and M. Smith. 1978. Identification and isolation of the yeast cytochrome *c* gene. *Cell* **14:** 673–680.
Mortimer, R.K. and D.C. Hawthorne. 1973. Genetic mapping in *Saccharomyces*. IV. Mapping of temperature-sensitive genes and use of disomic strains in localizing genes. *Genetics* **74:** 33–54.
Olson, M.V., K. Loughney, and B.D. Hall. 1979a. Identification of the yeast DNA sequences that correspond to specific tyrosine-inserting nonsense suppressor loci. *J. Mol. Biol.* **132:** 387–410.
Olson, M.V., B.D. Hall, J.R. Cameron, and R.W. Davis. 1979b. Cloning of the yeast tyrosine transfer RNA genes in bacteriophage lambda. *J. Mol. Biol.* **127:** 285–295.
Olson, M.V., D.L. Montgomery, A.K. Hopper, G.S. Page, F. Horodyski, and B.D. Hall. 1977. Molecular characterization of the tyrosine tRNA genes of yeast. *Nature* **267:** 639–641.
Olson, M.V., G.S. Page, A. Sentenac, P.W. Piper, M. Worthington, R.B. Weiss, and B.D. Hall. 1981. Only one of two closely related yeast suppressor tRNA genes contains an intervening sequence. *Nature* **291:** 464–469.
Page, G.S. and B.D. Hall. 1981. Characterization of the yeast $tRNA^{Ser}{}_2$ gene family: Genomic organization and DNA sequence. *Nucleic Acids Res.* **9:** 921–934.
Rothstein, R. 1977. A genetic fine structure analysis of the suppressor 3 locus in *Saccharomyces*. *Genetics* **85:** 55–64.
Russell, P.R. and B.D. Hall. 1982. Structure of the *Schizosaccharomyces pombe* cytochrome *c* gene. *Mol. Cell. Biol.* **2:** 106–116.
Schultz, L.D. and B.D. Hall. 1976. Transcription in yeast: α-amanitin sensitivity and other properties which distinguish between RNA polymerases I and III. *Proc. Natl. Acad. Sci.* **73:** 1029–1033.
Sentenac, A., S. Dézelée, F. Iborra, J.M. Buhler, J. Huet, F. Wyers, A. Ruet, and P. Fromageot. 1976. Yeast RNA polymerases. In *RNA polymerase* (ed. R. Losick and M. Chamberlin), p. 763. Cold Spring Harbor Laboratory, Cold Spring Harbor, New York.
Shalit, P., K. Loughney, M.V. Olson, and B.D. Hall. 1981. Physical analysis of the CYCl-sup4 interval in *Saccharomyces cerevisiae*. *Mol. Cell Biol.* **1:** 228–236.
Sharp, P.A. 1983. Conversion of RNA to DNA in mammals. Alu-like elements and pseudogenes. *Nature* **301:** 471–472.
Sherman, F. 1963. Respiration-deficient mutants of yeast. I. Genetics. *Genetics* **48:** 375–385.
Sherman, F. and P.P. Slonimski. 1964. Respiration-deficient mutants of yeast. *Biochim. Biophys. Acta* **90:** 1–15.
Sherman, F., M. Jackson, S.W. Liebman, M. Schweingruber, and J.E. Stewart. 1975. Deletion map of *CYC1* mutants and its correspondence to mutationally altered iso-1-cytochrome *c* of yeast. *Genetics* **81:** 51–73.
Sherman, F., J.W. Stewart, M. Jackson, R.A. Gilmore, and J.H. Parker. 1974. Mutants of yeast defective in iso-1-cytochrome *c*. *Genetics* **77:** 255–284.
Smith, M., D.W. Leung, S. Gillam, C.R. Astell, D.L. Montgomery, and B.D. Hall. 1979. Sequence of the gene for iso-l-cytochrome *c* in *Saccharomyces cerevisiae*. *Cell* **16:** 753–761.
Southern, E. 1975. Detection of specific sequences among DNA fragments separated by gel electrophoresis. *J. Mol. Biol.* **98:** 503–517.
Stewart, J.W., F. Sherman, N.A. Shipman, and M. Jackson. 1971. Identification and mutational relocation of the AUG codon initiating translation of iso-1-cytochrome *c* in yeast. *J. Biol. Chem.* **246:** 7129–7145.
Stewart, J.W., F. Sherman, M. Jackson, F.L.X. Thomas, and N. Shipman. 1972. Demonstration of the UAA ochre codon in baker's yeast by amino-acid replacements in iso-1-cytochrome *c*. *J. Mol. Biol.* **68:** 83–96.
Stiles, J.I., J.W. Szostak, A.T. Young, R. Wu, S. Consaul, and F. Sherman. 1981. DNA sequence of a mutation in the leader region of the yeast iso-1-cytochrome *c* mRNA.

*Cell* **25:** 277–284.
Swanson, M.S., S.M. Zieminn, D.D. Miller, E.A. Garber, and E. Margoliash. 1985. Developmental expression of nuclear genes that encode mitochondrial proteins: Insect cytochromes *c. Proc. Nat. Acad. Sci.* **82:** 1964–1968.
Szostak, J.W., J.I. Stiles, C.P. Bahl, and R. Wu. 1977. Specific binding of a synthetic oligodeoxyribonucleotide to yeast cytochrome *c* mRNA. *Nature* **265:** 61–63.
Van Keulen, H., R.J. Planta, and J. Retel. 1975. Structure and transcription specificity of yeast RNA polymerase A. *Biochim. Biophys. Acta* **395:** 179–190.
Zitomer, R., D.L. Montgomery, D.L. Nichols, and B.D. Hall. 1979. Transcriptional regulation of the yeast cytochrome *c* gene. *Proc. Natl. Acad. Sci.* **76:** 3627–3631.

# From Caterpillars to Yeast

FRANÇOIS LACROUTE
*Centre de Génétique Moléculaire*
*CNRS, 91190 Gif sur Yvette, France*

As early as I can remember, I was fascinated like many children by small living organisms and notably by their ability to give rise to huge progeny. I recall having borrowed for a full week my mother's clothes-washing basin to keep toads that I had collected on their way to the pond. The toads would mate and lay long translucent strings dotted by small black eggs. After transfer to a more appropriate aquarium, tadpoles developed on a boiled-spinach diet to yield finally some small adult toads. I also remember an unexpected hatching of a *Bombyx neustria* laying on a bough that I had collected with my father during a winter trip to Burgundy to treat the orchard fruit trees. As the branch had been incubated in the warmth of my pocket, the hatching was far ahead of season, and to save the small hungry caterpillars, I quickly had to cut leafless branches of an apple tree and let the shoots develop in a flower vase inside the house. The nice orange-blue caterpillars had another surprise for me. They were very easy to keep in my bedroom, staying without any containment on the apple branches that I would replace once their leaves had been eaten. However, when I came home from high school one afternoon, I found that they had spread throughout the room and were beginning to weave their cocoons on the walls, the closet, and my desk. In the few hours since I had left, the catepillars' climbing tropism had completely inverted. Contrary to the caterpillars, the adult butterflies were of a very dull gray. I had many other opportunities to stay in contact with nature at that time because the garden of the Astronomical Observatory of Strasbourg, which was headed by my father, was left largely neglected and was close to an old pond of the Botanical Garden.

When the time arrived to choose a career, I decided to take Natural Sciences courses at Strasbourg University, although I had obtained better results in mathematics and physics. I must say that I soon regretted my choice as the majority of the lectures were still devoted to systematics, with only a few interesting lectures by Henry Maresquelle, Alice Gagneu, and Pierre Joly on genetics and by Jean Vivien on vertebrate evolution. After my graduation in 1957, as I was interested only in what was then called "physiological genetics," my father consulted some Parisian colleagues who advised him that the main French laboratories corresponding to my interests were the laboratories of Boris Ephrussi, Jacques Monod, and Pierre Chouard. They also told him that a new curriculum completely focused on advanced genetics had been created at the University of Paris in Gif sur Yvette, "Le troisième cycle de génétique." I therefore decided to spend a year taking courses before trying to join a laboratory. It was a wonderful year. The laboratory courses were prepared by Georges Prevost who was always with us discussing all the exciting current papers of Seymour Benzer, Charles Yanofsky, Francis Crick, and many others. The professors were Boris Ephrussi, Philippe l'Héritier, and Georges Teissier. It was also during a very fortunate period when the French government was emphasizing research and university teaching. At the end of the year, a contractual position was offered to each student who had satisfied the examination; we were free to chose a laboratory in which to prepare our Ph.D. thesis from among the laboratories at Gif. I hesitated between the laboratories of Harriet Ephrussi-Taylor and Boris Ephrussi. In the Ephrussi-Taylor laboratory, I would have worked on *Diplococcus pneumoniae* transformation and the naive idea of purifying a gene by direct affinity to its encoded protein (at this time, the Francis Crick adaptor hypothesis had not yet been presented). The laboratory of Boris Ephrussi was working on the mechanism of *petite* production in yeast. I was eventually swayed by the strong personality and Slav charisma of Boris Ephrussi, the interesting genetics of yeast, and the beauty of yeast under the microscope. I worked only 1 year and only half time with Boris Ephrussi. Since I had not received a good biochemical background in Strasbourg, Boris required quite rightly that I follow the Biochemistry course at Paris University, which occupied about half of my time, and at the end of the year my military deferment was cancelled due to the Algerian war. I joined the French NATO unit in Germany where I spent about 3 years working with United States Nike anti-airplane missiles.

The short period I spent with Boris Ephrussi was fruitless if one considers only my experimental results. I mainly carried out control experiments to assess the degree of clumping of yeast cells and their survival to UV irradiation; my eventual intention was to induce chromosomal *petites*. However, this period was very important for my formation. Boris Ephrussi was extremely strict on the use of appropriate controls and on pushing as far as possible the deductions from experimental results. The first trait was very important for I still have the tendency to rush ahead and to think only afterward about controls.

When I came back from the army in the spring 1962, I was very disappointed. I had chosen genetics because there was this great mystery of how DNA could encode a protein. Now, everything was there: the triplet codon, the genetic code, the tRNA adaptor system, and a regulatory mechanism for transcription. However, I felt that it was too late for me to form an adequate attack on what was for me the last great problem of biology: the mechanism by which the vertebrate brain reaches the state of consciousness. I stayed in genetics.

As Boris Ephrussi had left Gif for Cleveland, I naturally went to work with his successor and former student Piotr Slonimski. Despite some very good advice given to me by Ephrussi, I also chose to change my research position to the teaching position of an Assistant at Paris University, as this seemed to be more secure. Since then, I have always, like many of us, fought to find enough time for research between the many hours devoted to teaching. I was mainly interested in the ways genes controlled cellular functions, and Piotr suggested that I should look in the literature for biochemical pathways that were well known and for which a maximum of intermediates and inhibitors were commercially available and that I then study one of them in yeast. However, this is not how I ultimately settled on the pyrimidine pathway. Had we followed Piotr's reasoning, we would have most likely chosen the tryptophan pathway so well studied by Charlie Yanofsky in *Escherichia coli*. What happened was that in a search for dominant negative regulatory mutations analogous to the $i^S$ mutations of the *E. coli* lactose operon, we obtained, after mutagenesis of a wild-type diploid strain of *Saccharomyces cerevisiae* and nystatin selection, numerous auxotrophic strains. Unfortunately, it quickly became evident that these strains contained homozygous auxotrophic recessive mutations due to the high mitotic conversion rate following the mutagenic treatments. By chance, I found in an issue of the *Proceedings of the National Academy of Sciences* a paper by Seymour Benzer describing the phenotypic curing of some T4 rII phages by 5-fluorouracil (5-FU) (Champe and Benzer 1962). I decided to see whether my useless collection of *S. cerevisiae* auxotrophs contained a similar phenotypic curing by 5-FU. I obtained only a very weak effect with an arginine auxotroph that, I assumed, was caused not by the ambiguity due to the incorporation of 5-FU in the mRNA but rather by the interrelation between the arginine and the uracil pathways via carbamoyl phosphate. Nevertheless, the 5-FU revealed itself to be a very powerful inhibitor of yeast growth, and, moreover, resistant pyrimidine-secreting mutants were easily obtained. So, we decided to study the regulation of the pyrimidine pathway. This was a good choice. The following 3 years were exciting and rewarding, revealing successively the induction of cytoplasmic *petites* by 5-FU, the existence of a bifunctional protein encoded by the *URA2* gene for which the two activities (carbamoyl phosphate synthetase and aspartate transcarbamylase) were feedback-inhibited by UTP, the repression of *URA2* by an excess of pyrimidines, the regulation by dihydrorotic induction of some of the intermediate steps of the pathway, and finally the precise half level of enzymatic activity found in heterozygous diploids (Lacroute 1964a,b, 1968). A very pleasant and fruitful collaboration was established in 1964 with Marcelle Grenson and André Piérard from the laboratory of Jean-Marie Wiame in Brussels. In fact, Jean-Marie Wiame visited Piotr in Gif just at about the time I obtained a double auxotrophic mutant with simultaneous requirements for arginine and pyrimidines. Tetrad analysis showed that the phenotype was due to two mutations, each giving separately a wild-type phenotype on minimal medium; however, one of the mutations gave a phenotype of growth inhibition by arginine and the other gave a full block of growth by uracil. As Marcelle Grenson on her side had also obtained a mutant with a uracil-inhibited phenotype, we thought that we had mutations in isoenzymes involved in the synthesis of carbamoyl phosphate, an intermediate common to the arginine pathway, studied in Brussels, and the uracil pathway. We decided to collaborate on the subject. Direct measurements of enzymatic activities done with André Piérard in Brussels revealed that the isoenzymes idea was indeed the case and that each isoenzyme was regulated by its end product. Moreover, it appeared that the mutation found by Marcelle Grenson, despite having my same phenotype of inhibition by uracil, was in a different gene and we named the two genes *cpa1* and *cpa2*. We ended up with the paradoxical situation that two genes were necessary for one enzymatic activity (the carbamoyl phosphate synthetase) in the arginine pathway, whereas only one gene encoded two enzymatic activities (the carbamoyl phosphate synthetase and the aspartic transcarbamylase) in the uracil pathway (Lacroute et al. 1965). A similar situation was published a little earlier by Rowland Davis on *Neurospora crassa* (Davis 1963), but in the case of *Neurospora*, there is a strict channeling of carbamoyl phosphate which is not exchangeable between the two pathways, leading to uracil auxotrophy for mutations in the pyrimidine isoenzyme and to arginine auxotrophy for mutations in the arginine isoenzyme. During 1965 and 1966, the last 2 years of my Ph.D. thesis in Gif, I began to work with Pierre Meuris, a new student of Piotr's, on mutations that conferred a growth-inhibition phenotype by normal metabolites on otherwise prototrophic strains. The purpose of this study was to investigate regulatory interactions between pathways (Meuris et al. 1967). I also collaborated with Michèle Denis and with Gordin Kaplan, who was in Gif for a sabbatical year, to study the mechanism of UTP feedback inhibition on the cabarmoyl phosphate synthetase–aspartic transcarbamylase enzymatic complex. In light of the results of Richard Epstein on phage T4, I also started a collection of yeast temperature-sensitive mutants to obtain mutants impaired in indispensable functions (I was looking mainly for RNA transcription mutants) and to reveal yet unsuspected cellular functions. When I met Lee Hartwell and Cal McLaughlin the following year at an Asilomar genetics meeting, I decided to stop this approach momentarily, since their characterization of such a collection was much more advanced. I presented my Ph.D. thesis in June 1966 in front of a jury composed of Jacques Monod, Georges Cohen, Pierre Schaeffer, and Piotr Slonimski. I remember being disappointed because Jacques Monod, for whom I have great admiration, had not had the time to read the thesis.

I left Gif just after my thesis to work for 1 year in Berkeley. After some letter exchanges, Piotr Slonimski had found a place for me in Gunther Stent's laboratory to test an iconoclastic model for the regulation of the *E. coli* lactose operon. The biochemical part of the model was derived from the Cline and Bock model for translational control of gene expression. All of the regulatory mutants known for the lactose operon were also explained by this model. In summary, regulation occurred during translation by a masking of the aminoacyl tRNA acceptor site on the ribosome by the nascent β-galactosidase polypeptide. The inducer bound and changed the conformation of the nascent polypeptide, thereby unveiling the acceptor site and allowing completion of the protein. The differences in the transcription rates found between induced and noninduced cells were explained by the coupling of translation and transcription cherished by Gunther. In our model, the *i* gene encoded an enzyme of an auxiliary sugar pathway that transformed sugar I (able to induce) into sugar NI (unable to induce). Moreover, this enzyme was subject to feedback inhibition by NI, by lactose (or a derivative), and by IPTG (isopropyl-β-D-thiogalactopyranoside). The $i^-$ strains were induced due to the accumulation of I. The $i^S$ strains had lost the feedback inhibition, which gave a dominant phenotype, and never accumulated I. The $O^c$ mutations were either small amino-terminal deletions shortening the β-galactosidase polypeptide or amino acid changes giving an inducer-bound conformation in the absence of the inducer. According to this model, in a state of partial induction, there must be a longer delay between the synthesis of the amino-terminal part of β-galactosidase and its completion than in fully induced cells. We thus followed the kinetics of appearance of radioactivity in the amino terminus of completed β-galactosidase in 30% induced cultures and in fully induced cultures. We found no difference. This ruled out the model (Lacroute and Stent 1968). This was in fact not a surprise because in the middle of the experiments, Wally Gilbert had presented a seminar describing his isolation of the lactose repressor. At least we had had the pleasure of obtaining the first direct measurement of the growth rate of an individual polypeptide chain that fit the induction kinetics data of Adam Kepes. Despite the lack of success of our model, my stay in Gunther's laboratory was a delight. Discussions with Gunther were always challenging due to his unconventional ideas in molecular biology and his strong sense of humor. There were also many friendly postdocs in the laboratory: Pierca Donini, "Doc" Edlin, Felix Wettstein, and Paul Broda. Paul, notably, helped me improve my poor English and organized laboratory ski trips to the nearby Sierra Nevada mountains and visits to National Parks.

In the spring of 1967, during my stay in Gunther Stent's laboratory, I gave a seminar in Boris Magasanik's laboratory where I met David Botstein for the first time. I still remember his enthusiasm about yeast and the possibilities it offered as a eukaryotic model for molecular biology, which comforted me in my intention to continue working on yeast. As I knew I would get a position as an assistant professor in Strasbourg University after my postdoc work at Berkeley, I attended all of Gunther's lectures on genetics and molecular biology to have a framework for my first year of teaching. This was indeed very useful for me the next year.

I arrived in Strasbourg in September 1967 to teach a unit of genetics to third- and fourth-year students. The setting up of my laboratory was not too difficult. I was helped by Professor Leon Hirth who headed a very active group of plant virologists and was responsible for my return to Strasbourg. The University gave me a space of about 400 square meters in the new Botanical Institute, created in 1965, and enough money to buy basic instrumentation. Moreover, there were enough grants at this time for people doing molecular genetics. A small group joined my laboratory in the middle of 1968 from a CNRS institute on macromolecules whose biological part had been dissolved. Among them were Marie Renée Chevallier, who studied and cloned the pyrimidine permeases, and Marie-Louise (Marlyse) Bach, who cloned the *URA3* gene in David Botstein's laboratory and carried out the first studies on its transcription. Thanks to the understanding of a CNRS commission, a high-level technician, Francine Exinger, joined the laboratory in 1968. This was the beginning of a long, trusting, and fruitful collaboration that was ended only by my departure to Gif in 1987. A teaching assistant, Michel Aigle, and later on a second one, Francis Karst, were appointed by the university for the teaching of genetics.

Michel Aigle developed the genetics of a non-Mendelian mutation (*ure3*) that allowed the incorporation of ureidosuccinic acid by yeast cells cultivated on ammonium salts through a deregulation of nitrogen metabolism (Aigle and Lacroute 1975). Francis Karst obtained the first significant collection of sterol auxotrophic mutants that are still the main tool for cloning the genes of this pathway (Karst and Lacroute 1977). A former student of Richard Snow, Christopher Korch, came to the laboratory for postdoc work on the relationship between purine biosynthesis and growth inhibition by ureidosuccinic acid, and he showed that it is a complex phenotype partly due to competition for 5-PRPP (phosphoribosyl pyrophosphate) by the pyrimidine and the purine pathways. The CNRS accepted the candidature of Richard Jund to prepare a thesis on the regulation of the pyrimidine pathway. This was very fortunate for the laboratory because, besides his own work on pyrimidines (in which he characterized *URA5*, the main gene encoding OMP pyrophosphorylase, and many of the genes involved in nucleoside and nucleotide conversions), he was always a source of expertise for enzymatic problems (Jund and Lacroute 1970, 1972).

In 1972, Ole Maaløe invited me as a visiting professor for a stay of 6 months in Copenhagen. He wanted to build a small group on yeast molecular biology in his laboratory and expected me to introduce yeast techniques. I was not very convincing since nobody decided to abandon *E. coli*, but I had interesting discussions on the intercorrelation of growth rates and biosynthesis of stable RNAs, which was a major interest of Ole. An assay in collaboration with Niels Fiil and Kaspar von Meyenburg to find an equivalent of the *E. coli* "magic spot" (ppGpp) during the stringent response of yeast cells was negative. I took advantage of the good expertise of Ole's laboratory in RNA separation to measure directly in yeast the elongation rate of RNA molecules and to evaluate indirectly the elongation rate of proteins. I was surprised to find that despite the impossibility of a direct coupling between RNA and protein biosynthesis in eukaryotes, the ratio of the RNA versus protein elongation rate like in bacteria was still 3 to 1 (Lacroute 1973). I left Copenhagen in November to be back in Strasbourg in time for the teaching of the new academic year.

Marlyse Bach, who had returned in 1970 from a postdoctoral stay in Denver, decided to quantify the amount of *URA2* mRNA under different physiological conditions to show the repression of transcription by the UTP pool. She chose *URA2* because the suspected great size of the gene would increase the sensitivity of the measurements and because of its broader range of regulation compared to the other pyrimidine genes. We thought at this time to use a subtractive approach with an exhaustive hybridization of all mRNAs except for *URA2* mRNA to the DNA of a *URA2*-deleted strain, followed by hybridization to the DNA of a wild-type strain; background would have been estimated with a parallel experiment done with the RNAs isolated from the *URA2*-deleted strain. We were aware of the extreme difficulty of the task but hoped that the continuous improvement of hybridization techniques would allow us to obtain significant results. Obtaining deletions was difficult despite the development of a specific selection for *ura2* mutants (Bach and Lacroute 1972). In fact, it was impossible to get any in the genetic background of our laboratory strain FL100; to obtain them, Marlyse had to shift to a strain in which Beth Jones had already obtained deletions in adenine genes (Jones 1972).

Some deletions were found with this new strain, and two of them were complete, giving no wild-type recombinant with any *ura2* mutants. First attempts to use the DNA of one of the full deletions for mRNA subtraction were not satisfactory. However, at about this time, efficient methods for *E. coli* transformation had appeared, as well as a paper by Ariane Toussaint showing that the bacteriophage Mu was able in a dimeric state to promote integration of completely unrelated linear DNAs into the *E. coli* genome via nonhomologous transposition (Toussaint and Faelen 1973). We therefore decided to try to transform nonreverting pyrimidine mutants of *E. coli*, obtained by Mu lysogeny, with yeast DNA in the hope that after partial induction of the phage, the corresponding yeast gene would be integrated and expressed. In a second step, we thought to get specialized transductants with the yeast gene giving us a positive hybridization probe. Marlyse constructed a full set of *E. coli* strains with an insertion of Mu in the different pyrimidine genes starting from strain DB6656 sent to us by David Botstein. But the strains were not used the way we had planned. In vitro recombinant DNA techniques were just being developed in 1974, and in early 1976, Kevin Struhl isolated the *HIS3* wild-type gene from a yeast DNA bank in bacteriophage λ by complementation of the corresponding *E. coli* mutant. At the 1975 yeast Cold Spring Harbor summer meeting, I met Tom Petes and David Botstein, who had just prepared a bank of yeast DNA in pMB9. After I talked with them, David offered to receive Marlyse in his laboratory to try to complement her pyrimidine-less collection with the bank prepared by Tom. When I went back to Strasbourg and told Marlyse about the project, she was very excited and immediately took steps to join David's laboratory. Her first successful complementation was for PyrF (*pyrF* is the *E. coli* equivalent of *URA3*) at the end of 1976.

Marlyse showed that the *URA3* gene was completely encoded in a 1.2-kb *Hin*dIII restriction fragment, which was subsequently sequenced in David Botstein's laboratory (Rose et al. 1984). Once back in Strasbourg, she measured the half-life of the *URA3* mRNA with a purified hybridization probe and quantified its amount in normal and induced cells, conclusively demonstrating for the first time transcriptional regulation in yeast (Bach et al. 1979). Soon after, with the same tools, Regine Losson and I were able to show the strong destabilizing effect of N-proximal nonsense mutations on mRNA half-lives (Losson and Lacroute 1979). The next few years were very exciting and rewarding for us and others involved in yeast molecular biology. Albert Hinnen, Jim Hicks, and Gerry Fink developed yeast transformation. Soon after the demonstration in 1978 by Jean Beggs of high-efficiency yeast transformation, we were able to prepare good yeast DNA banks in 2-μm pBR322 yeast–*E. coli* shuttle vectors, thanks to Jim Hicks' kind gift of plasmids bearing 2-μm fragments. During a visit in Strasbourg, Ben Hall told us about the use of partial *Sau*3A digestions to prepare gene banks, a trick that had not yet been spread around; this allowed us to be reasonably sure that the genome was well represented in our banks. The small size of the *URA3* gene made it a candidate of choice for both plasmid constructs and gene disruptions. This was reinforced by the observation I presented at a yeast Cold Spring Harbor Workshop that *ura3* mutants are resistant to growth inhibition by 5-fluoro-orotic acid, thus providing a positive selection for *ura3*. This method was more general and practical than a previous method based on growth inhibition by ureidosuccinic acid that Marlyse and I had developed.

A Ph.D. student, Gérard Loison, found a regulatory mutant that was constitutive for all the inducible enzymes of the pyrimidine pathway; we named the affected gene *PPR1* (Loison et al. 1980). Unfortunately, the mutation gave no phenotype that could be used for cloning the wild-type allele in a mutant strain, or vice versa. A few months later, during the summer 1980, I found with Regine Losson recessive uninducible mutations in the same gene that gave a phenotype of highly increased sensitivity to growth inhibition by 6-azauracil. The first clones obtained by gene complementation of the mutant always gave us the *URA3* structural gene, which in high copy number relieved the inhibition since it encodes OMP decarboxylase, the direct target of 6-azauracil UMP inhibition. Nevertheless, we did not have to prepare a new bank starting from a *ura3* mutant strain to get the gene, but only to recover plasmids from some poorly growing transformants obtained during the initial selection and that we had, wrongly, rejected. What happened was that our *PPR1* gene in high copy number slowed down the growth of transformants due to a not yet understood effect of Ppr1 protein overproduction. This effect was even more pronounced in our case because the bank we had used was prepared in Jean Beggs' pJDB207 plasmid, an exceptionally high-copy-number plasmid due to an impaired *LEU2* promoter. Finally, we were among the first to clone a yeast regulatory gene (Losson and Lacroute 1981), to study its transcription, and to show the high instability of its mRNA (Losson et al. 1983). However, we were less lucky with the study of the regulatory mechanism itself due to our poor biochemical technology. We also had the pleasure of finding that the *PPR1* regulatory gene still regulated the *URA3* structural gene when both were introduced into the completely unrelated species *Schizosaccharomyces pombe*, for which we had developed a very efficient vector (Losson and Lacroute 1983).

New Ph.D. students, teaching assistants, and assistant professors joined the laboratory and completed the characterization, cloning, and sequencing of the different genes of the pyrimidine pathway. Among the Ph.D. students, Frederique Pelsy established the meiotic map of the *URA1* gene and showed that ocher nonsense mutations in the amino-terminal end had the same destabilizing effect as amber mutations on mRNA half-lives. Martine Nguyen cloned and characterized the *URA4* and *URA1* genes. Patricia Liljelund developed the meiotic map of *PPR1*, characterized mutants for uridine monophosphokinase, and cloned the corresponding gene (Liljelund et al. 1984). Armel Gyonvarch and Benoît Kammerer sequenced the *PPR1* and *PPR2* genes and prepared antibodies against the corresponding proteins (Kammerer et al. 1984). Elizabeth Weber, in collaboration with Marie-Renée Chevallier and Richard Jund, cloned, sequenced, and characterized the uracil permease and the adenine-cytosine permease (Weber et al. 1988). Among the teaching assistants and assistant professors, Jean-Claude Hubert participated in the majority of the gene-sequencing projects and showed the asymmetrical transcription of the *URA3* gene. Serge Potier was the first to develop a genetic selection for reciprocal translocations at chosen sites on the yeast genome (Potier et al. 1982). Jean-Luc Souciet cloned the *URA2* gene and sequenced the majority of it (Souciet et al. 1987).

Besides the main work on pyrimidines, I have maintained some students with an interest on RNA maturation and biosynthesis. With Francine Exinger, we obtained dominant and semidominant mutations that were impaired not in the biosynthesis of the ribosomal RNAs, as we first described, but in their maturation. In a search for mutants blocked in the addition of the poly(A) tail to mRNAs, Fabienne Perrin and Barbara Winsor obtained temperature-sensitive mutants in which the level of poly(A)$^+$ RNAs dropped quickly after a shift to the nonpermissive temperature. Jean-Claude Bloch showed that it was not the addition of the poly(A) tail which was impaired but rather the stability of the poly(A) tail (Bloch et al. 1978).

After my return to Gif in the summer 1987, I left the continuation of the study of pyrimidines to my collaborators who remained in Strasbourg. I decided to focus on the role of the poly(A) tail on mRNA stability.

## ACKNOWLEDGMENT

I extend my appreciation to Patrick Linder and Mike Hall for their help in improving both the English and the manuscript.

## REFERENCES


Aigle M. and F. Lacroute. 1975. Genetical aspects of (URE3) a non-mitochondrial, cytoplasmically inherited mutation in yeast. *Mol. Gen. Genet.* **136:** 327–335.

Bach, M.L. and F. Lacroute. 1972. Direct selective techniques for the isolation of pyrimidine auxotrophs in yeast. *Mol. Gen. Genet.* **115:** 126–130.

Bach, M.L., F. Lacroute, and D. Botstein. 1979. Evidence for transcriptional regulation of orotidine-5′-phosphate decarboxylase in yeast by hybridization of mRNA to the yeast structural gene cloned in *Escherichia coli. Genetics* **76:** 386–390.

Bloch, J.C., F. Perrin, and F. Lacroute. 1978. Yeast temperature sensitive mutants specifically impaired in processing of poly(A)-containing RNAs. *Mol. Gen. Genet.* **165:** 123–127.

Champe, S.P. and S. Benzer. 1962. Reversal of mutant phenotypes by 5-fluorouracil: An approach to nucleotide sequences in messenger-RNA. *Genetics* **48:** 532–546.

Davis, R.H. 1963. *Neurospora* mutant lacking an arginine specific carbamyl phosphokinase. *Science* **142:** 1652–1654.

Jones, E.W. 1972. Fine structure analysis of the *ade3* locus in *Saccharomyces cerevisiae. Genetics* **70:** 233–250.

Jund, R. and F. Lacroute. 1970. Genetic and physiological aspects of resistance to 5-fluoropyrimidines in *Saccharomyces cerevisiae. J. Bacteriol.* **102:** 607–615.

———. 1972. Regulation of orotidylic acid pyrophosphorylase in *Saccharomyces cerevisiae. J. Bacteriol.* **109:** 196–202.

Kammerer, B., A. Guyonvarch, and J.C. Hubert. 1984. Yeast regulatory gene *PPRl. J. Mol. Biol.* **180:** 239–250.

Karst, F. and F. Lacroute. 1977. Ergosterol biosynthesis in *Saccharomyces cerevisiae. Mol. Gen. Genet.* **154:** 269–277.

Lacroute, F. 1964a. Régulation des enzymes de biosynthèse de l'uracile chez la levure. *C.R. Acad. Sci.* **258:** 2884–2886.

———. 1964b. Un cas de double rétrocontrôle: La chaîne de biosynthèse de l'uracile chez la levure. *C.R. Acad. Sci.* **259:** 1357–1359.

———. 1968. Regulation of pyrimidine biosynthesis in *Saccharomyces cerevisiae. J. Bacteriol.* **95:** 824–832.

———. 1973. RNA and protein elongation rates in *Saccharomyces cerevisiae. Mol. Gen. Genet.* **125:** 319–327.

Lacroute, F. and G.S. Stent. 1968. Peptide chain growth of β-galactosidase in *Escherichia coli. J. Mol. Biol.* **35:** 165–173.

Lacroute F., A. Piérard, M. Grenson, and J.M. Wiame. 1965. The biosynthesis of carbamoyl phosphate in *Saccharomyces cerevisiae. J. Gen. Microbiol.* **40:** 127–142.

Liljelund, P., R. Losson, B. Kammerer, and F. Lacroute. 1984. Yeast regulatory gene *PPRl. J. Mol. Biol.* **180:** 251–265.

Loison, G., R. Losson, and F. Lacroute. 1980. Constitutive mutants for orotidine 5 phosphate decarboxylase and dihydroorotic acid dehydrogenase in *Saccharomyces cerevisiae. Curr. Genet.* **2:** 39–44.

Losson, R. and F. Lacroute. 1979. Interference of nonsense mutations with eukaryotic messenger RNA stability. *Cell Biol.* **76:** 5134–5137.

———. 1981. Cloning of a eukaryotic regulatory gene. *Mol. Gen. Genet.* **184:** 394–399.

———. 1983. Plasmids carrying the yeast OMP decarboxylase structural and regulatory genes: Transcription regulation in a foreign environment. *Cell* **32:** 371–377.

Losson, R., R.P.P. Fuchs, and F. Lacroute. 1983. *In vivo* transcription of a eukaryotic regulatory gene. *EMBO J.* **2:** 2179–2184.

Meuris, P., F. Lacroute, and P. Slonimski. 1967. Etude systématique de mutants inhibés par leurs propres métabolites chez la levure *Saccharomyces cerevisiae. Genetics* **56:** 149–161.

Potier, S., B. Winsor, and F. Lacroute. 1982. Genetic selection for reciprocal translocation at chosen chromosomal sites in *Saccharomyces cerevisiae. Mol. Cell. Biol.* **2:** 1025–1032.

Rose, M., P. Grisafi, and D. Botstein. 1984. Structure and function of the yeast URA3 gene: Expression in *Escherichia coli. Gene* **29:** 113–124.

Souciet, J.L., S. Potier, J.C. Hubert, and F. Lacroute. 1987. Nucleotide sequence of the pyrimidine specific carbamoyl phosphate synthetase, a part of the yeast multifunctional protein encoded by the *URA2* gene. *Mol. Gen. Genet.* **207:** 314–319.

Toussaint, A. and M. Faelen. 1973. Connecting two unrelated DNA sequences with a Mu dimer. *Nat. New. Biol.* **242:** 1–4.

Weber, E., M.R. Chevallier, and R. Jund. 1988. Evolutionary relationship and secondary structure predictions in four transport proteins of *Saccharomyces cerevisiae*. *J. Mol. Evol.* **27:** 341–350.

# INSTITUTIONS

G. Fink (CSHL 1981)

F. Sherman (CSHL 1979)

H. Roman (Louvain-la-Neuve, Belgium 1980)

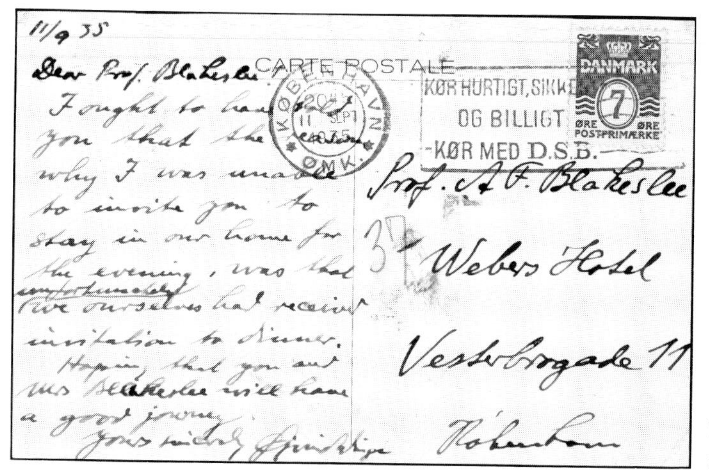
11/9 35
Dear Prof. Blakeslee
I ought to have told you that the reason why I was unable to invite you to stay in our home for the evening, was that ~~unfortunately~~ we ourselves had received invitation to dinner.
Hoping that you and Mrs Blakeslee will have a good journey.
Yours sincerely Øjvind Winge

CARTE POSTALE
KØBENHAVN Ø.MK. 11 SEPT 1935
KØR HURTIGT, SIKKERT OG BILLIGT KØR MED D.S.B.
DANMARK 7 ØRE POSTPRIMÆRKE

Prof. A.F. Blakeslee
Webers Hotel
Vesterbrogade 11
København

Postcard to A. Blakeslee from Ø. Winge (1935)

Front of postcard showing Carlsberg Laboratory

R.E. Esposito and H. Roman (Louvain-la-Neuve, Belgium 1980)

# Humble Beginnings

ROCHELLE EASTON ESPOSITO
*University of Chicago*
*Chicago, Illinois 60637*

## PART 1: A UNIQUE VIEW OF YEAST GENETICS FROM THE LADIES' ROOM AT BROOKLYN COLLEGE

"Watch that door carefully and make sure you get an accurate count of every woman who uses the ladies' room! The future of yeast genetics at Brooklyn depends on it." With this admonition about the seriousness of my task, Sy Fogel placed a colony counter in my hand and walked away, leaving me to contemplate the path that led me to what was until then the high point in my scientific career. It was early 1961, I was an undergraduate, and this was my first day on the job as a research technician in the laboratory of Seymour Fogel and Don Hurst. I was very young and very earnest, feeling very responsible and very excited. My life is really beginning I thought, as my eyes fixed on the ladies' room door and I listened intently to every sound. Periodically, Sy came back to check on me. "How many so far?" he'd ask. "Not very many," I'd answer, a little disappointed, "just three in the last hour." "Fantastic, fabulous," he'd bellow, "you're doing great," and then scurry away leaving me to puzzle over the deeper implications of it all. It became much clearer several weeks later when he announced to me and Mrs. Spates (an elderly lady in their employ as a dishwasher and media maker), "we're taking over the ladies' room, gals!" Sy had used my first set of results as a professional geneticist to convince the administration that the ladies' room in the basement of Ingersall Hall was dispensable and would serve a higher, more noble purpose as a new laboratory for their budding yeast genetics program. It was thus that I contributed to the early growth of our field. (*Historical note:* It was a sign of the times that we never thought to check the traffic to the men's room!)

My acquaintance with the "power of yeast genetics" had actually begun several months earlier, in the fall of 1960. The laboratory section of the basic genetics course was having its first meeting. Don Hurst strode into the room carrying a rack of test tubes containing some cloudy liquid material. "These are yeast cultures," he announced to the class with a hint of amusement dancing across his eyes. "You have four months in which to do experiments and it's up to you to decide what you want to find out about them. The lab sessions are four hours each week starting right now, so we have the whole afternoon to discuss this. Does anyone have any ideas?" With that statement, he placed his pipe into his mouth and coolly appraised the impact of his words as he stared at each of the students one by one. For most of us, this was our first advanced course as undergraduate biology majors and his challenge promptly plunged us into a state of shock. "How can we decide what we want to know, when we don't know anything about yeast? Professors are supposed to tell us what experiments to do and what's important to learn. It's their job!" While our logical minds screamed this thought in silent unison, a few of us managed to stammer in a barely audible voice "what exactly do you mean?"

*The Early Days of Yeast Genetics*

"Exactly what I said," he replied and patiently waited for a response. All told, we sat gazing at each other for 4 very long painful hours, interrupted perhaps every 15 minutes or so by a brave student who not being able to stand the protracted silence any longer offered a suggestion that was usually impossible, if not highly unrealistic, to accomplish. Had we been more experienced, I'm not sure we would have fared much better. The sad commentary on our education was that we had all been well trained in how to answer questions but not how to pose them. Although those of us who squirmed under his scrutiny did not appreciate it at the time, most of us came to realize that Don Hurst was an exceptional teacher. He demanded that students think for themselves and refused to take the easier path of telling us what to do. I was to encounter this same frustrating quality again in my graduate advisor, Herschel Roman, but I don't want to get ahead of myself.

When that first laboratory class finally ended, I made a beeline for the library. There was no way that I was going to endure another 4 hours like the session we had just gone through! All of the genetics books owned by the library seemed to collectively sense my need and flew off the shelves into my eagerly waiting hands. I quickly leafed through one book after another searching for a clue as to what we could do. Then one miraculously appeared—a remarkable book titled *The Chemical Basis of Heredity,* edited by William McElroy and Bentley Glass, consisting of a series of articles from a symposium held at Johns Hopkins University in 1956. At first it seemed like the book would be of no use. There was no entry in the index under yeast, nor was there one under *Saccharomyces* or fungi. (*Historical note:* The absence of these entries was another sign of the times!) Nevertheless, I soon became fascinated as I browsed through the articles. Although my comprehension of the material was limited, it was clear that something important was going on in the conceptualization of the gene. Many of the articles were devoted to analysis of the behavior of independent mutations within a single gene. The pioneering work of George Beadle and Edward Tatum on developing methods to recover large numbers of nutritional mutants in *Neurospora,* the rising popularity of microbial systems for genetic studies, and the breakthrough discovery by James Watson and Francis Crick of the structure of DNA set the stage for a burst of intense activity on the fine structure of the gene. The first article, written by Beadle, painted the picture in broad strokes, and subsequent articles by Seymour Benzer and others provided different perspectives. Something unusual seemed to happen when two mutants in the same gene were crossed: Wild-type recombinants were produced, an unexpected finding I gathered, since prior to the isolation of "allelic" mutations, the gene had been considered to be a single unit. Even more surprising was the fact that the recombination that occurred within genes did not result in reciprocal products as did exchange between genes. The new phenomenon, called gene conversion by some and transmutation by others, indicated that the gene was divisible. Genetics was starting to hook me; this was pretty interesting stuff!

An article by Herschel K. Mitchell, in particular, captured my attention. Although gene conversion had been shown to be associated with reciprocal crossing-over, the frequency could be altered independently by temperature shocks in *Neurospora.* It was therefore proposed that different mechanisms were responsible for these events and that they were probably correlated because each process required chromosome pairing. Mitchell argued that "gene conversion may be particularly important in understanding how genetic material can be transferred from one cell to another at the molecular level." He postulated that "if other influencing factors such as radiation and chemical treatments can be shown to produce independent effects on conversion and crossing-over, a generalized procedure for distinguishing the two processes might result." Here was my sought after clue! A seed began to sprout in my head, and I thought perhaps the class could examine the effect of radiation on gene conversion in the "awesome yeast cultures." Of course I hadn't the foggiest idea of "exactly" how to execute such an experiment, but I was convinced from my reading that it was possible, and with a sense of excitement, I finally left the library clutching what was to become my bible. I spent the next week pouring over *The Chemical Basis of Heredity* and other genetics texts trying to assimilate what I could. Unfortunately, much of what I read was way over my head, out there somewhere with the mysterious radiation I was hoping to study. Undaunted, full of naive optimism and bursting with expectation, I braced for the next encounter with Don Hurst. (*Historical note:* These traits seem to be uniquely intertwined in one's youth!) To my great relief, he responded enthusiastically to the proposal. He said he could provide us with a UV source and a dosimeter to monitor the level of radiation. He taught us how to grow yeast and count and plate them, and helped us to design an experiment using heteroallelic markers to monitor conversion using the *ade2* red/white signaling system. We were off and running and I was entranced. At the end of the course, when finals were over and grades had been entered, I shyly asked him if I might continue for another few weeks to complete the analysis of the convertants that were recovered. "I have a better idea," he said, "how would you like to work as a part-time technician for me and my colleague Sy Fogel? We'll even pay you!" Although I had no idea of what type of research they did, I was thrilled and could not believe my good fortune, for had I the means, I would have happily paid them to work in their laboratory. It was with some surprise that I later learned that their own research focused on yeast. With some embarrassment at my own naiveté, I discovered that they were actually engaged in studying gene conversion, an ironic happenstance that Don Hurst chose not to reveal to me "until you completed the course and we decided if we would invite you to work in the lab."

What I also didn't know then was that Fogel and Hurst were to influence the course of my life profoundly as they would many other young students who came under their spell. Working in their laboratory was a joy! They both took great pleasure in nurturing the intellectual development of young souls and delighted in imparting their collective wisdom on a variety of other matters ("don't get married until you are at least 30!"). Our yearnings and dreams were listened to seriously and we were encouraged to recognize our potential. We were taught to think critically and work precisely and to believe in ourselves. Each Friday evening, undergraduate biology majors gathered at Don Hurst's home to discuss readings on the philosophy of science. At these times, I was often invited to dine with his family and was carefully observed, with some amusement I might add, as I politely accepted and consumed the alcoholic beverage Don insisted I have before dinner, to teach me how to handle myself socially once I left their protection and entered the outside world!

A brief digression into the general intellectual climate in the New York City Colleges is warranted here to place this story into proper context. The City

Colleges for many years charged no tuition and operated essentially as free schools, supported primarily by the state. The only cost was a student activities fee of $8 per year plus expenses for books. The noble goal of the state government was to provide an opportunity for lower-income families living in New York City to educate their children. Implementation of this ideal allowed many of us access to a world that was otherwise unavailable. Because the schools were free and there were a large number of applicants, the admission standards were necessarily stringent. As a consequence, the classrooms were almost exclusively filled with smart but poor students. The air was saturated with enthusiasm, vitality, and purposefulness, spiced with generous doses of humility and gratitude, which diffused through diverse disciplines and captured the imaginations of those exposed to it. It should not be overlooked that these factors were present in addition to the normal atmospheric vibrations caused by the general hyperactive behavior and intensity of New Yorkers. The result was that both faculty and students approached their respective missions, to educate and be educated, with extraordinary commitment and zeal. Sy Fogel and Don Hurst were part of this tradition. Many years later, in pondering the special environment in these Colleges, I realized the obvious, poor + bright = high motivation, and was able to better appreciate the more subtle forces contributing to the unique interactions that occurred there. Most of us were hopeful that if we worked hard, life would be better for us than it was for our parents, many of whose dreams had been tragically shattered by events of The Great Depression and World War II. The faculty responded to the optimism of their willing disciples by encouraging us to continue through advanced studies.

Sy Fogel's specific encouragement to me took the form of "suggesting" that I go to Seattle and get a Ph.D. with Herschel Roman. "What you should do, Sheldon," he said (he found it amusing to address me in this way), "is break up with your boyfriend, leave your family here, and do something really worthwhile with your life! The best person in the country for you to work with is Herschel Roman. We went to graduate school together. He's an outstanding scientist and one of the finest human beings I know. Herschel's truly a prince, a prince among men! A wonderful person and a real mensch. I trust him, he'll train you well and treat you with respect. You'll get a marvelous education in Seattle, only don't expect him to answer any questions for you! He'll just answer you with another question, just like the old man, L.J. Stadler! You'll learn how to be a real scientist." As he talked, I nodded my head, listening and trying to absorb what he was saying to me. The conversation was moving at supersonic speed. "I'll call him right now and tell him you'll be there in the fall." "Nod." With that he picked up the phone and called Herschel. "Hersch, Sy here," was all I heard. The rest of his words faded into the background as it sunk into my subconscious that something momentous appeared to be happening and shock took hold of my system. A few minutes later the audio turned on again in my head and I heard him say, "Well, great! He has a place for you. It's all settled. You made the right decision."

That evening I heard myself saying, "Mom, I'm going to Seattle to get a Ph.D with Herschel Roman." "With who?" she said, and then added "why do you want to go to Alaska? It's so cold there!" "He studies gene conversion," I replied.

An idea had been planted in my head and I was set upon a course that I knew I had to pursue. Ralph Waldo Emerson once said, "there's a right for you that needs no choice." I believe that's the way it was for me. To begin with, I had considered myself exceptionally lucky to have found my way into college, which was quite accidental, and now I was going off to graduate school. Life certainly seemed unpredictable (or was it?). My family were Sephardic Jews who had migrated from Spain to Turkey during the Spanish Inquisition and finally to the United States around 1915. My grandparents, like other immigrants before them, had worked hard to survive in the new country. Although education in general was valued, it was considered a male prerogative. Regrettably, because of the many social and financial burdens of the times, it was also an option that was difficult to exercise. Education for women was quite another matter. Even if it were feasible, it was considered unnecessary. Why would a young woman who would eventually become a wife and mother need or want a college education? Wasn't it a waste of time and money? The expression of such interests evoked fear and concern; it was considered a sign of "strangeness" and a sure indication of failure on the feminine front. This attitude was held by nearly everyone except for my dear maternal grandmother, who secretly whispered to me, in part out of her own frustration, "you be different than the others and go to college, listen to me," adding a couple of quick winks for emphasis.

Nevertheless, despite my own rebellious inclinations, I never really thought seriously about continuing my studies after high school. But since I had to earn an income and since I had loved biology in school, I decided that becoming a medical technician might be within reach. As graduation neared, I set my sights on a 9-month training program offered by the local Community College. Unfortunately (or fortunately), it cost $600. "Sorry" my dad said, "we don't have the money." Distraught, holding back tears, I went to my high school biology teacher (another Seymour, Mr. Seymour Stein), who had always been encouraging to me, for advice. "Why don't you go to Brooklyn College," he said kindly. "It's free. It will take a little longer since it's a four-year program, but you'll have the credentials to become a medical technologist, and if you change your mind you'll have other options as well." "I'm not sure I can do it," I said. Then he turned his palm upward and said, "look, some people go from here to there in a straight line," tracing a path from his thumb to his ring finger, "others may take a more roundabout route," and he zigzagged around his palm, winding up at the ring finger again, "but they can get to the same place if they want to." So I proceeded to college because it was free, despite my family's misgivings about the whole mishegoss, contemplating my palms the whole way. (*Historical note:* In the past, I attributed the path of my life to serendipity, but I now realize that most of the time I was in fact actively making choices, in the Emerson mode, even though it seemed that I was simply doing what those I respected told me to do.) Caryl Roman, Herschel's wife, once said to me "I believe that people do exactly what they want to do." Over the years I have become convinced that she was right. In the end, I believe a stubborn spark within me drove me to Seattle to do exactly what I wanted to do.

"What is this college after college?" my grandfather said as he shook his head at me with anger and confusion in his voice, and a touch of sadness in his eyes, when I told him what I was planning for my future. Clearly, I was straying ever farther from the path and he was losing all hope for me. "It's okay, grandpa, I know what I'm doing!" I answered, a bit exasperated. I realized then that, although he loved me, he would never understand and I would always be a disappointment to him. I believe at that point, a giant generation gap opened and swallowed us up.

I spent my last year in the ladies' room at Brooklyn College, doing a senior research project, whose importance I did not fully appreciate until several years later. Much of the analysis of the properties of gene conversion, until then, was performed with prototrophic (wild-type) recombinants derived from vegetative cells, since large numbers could be easily selected and examined. Sy was troubled that the properties of the conversion process might be missed by this approach. He was often fond of telling me that "in genetics you get exactly what you ask for, but you may not ask for the right thing." Much later, after studying the papers of L.J. Stadler, I realized that this concern probably stemmed from his graduate training with Stadler, who early on was keenly aware of how current concepts of the gene were limited by operational definitions, e.g., by the mutations that were recovered in it. Sy thus felt that it was critical to look at unselected tetrads, and so we began a pilot study to determine the full spectrum of conversion events that could occur besides those leading to prototrophy. This approach eventually became a major endeavor of the laboratory and took many years and the dissection of thousands of tetrads to complete (Fogel and Hurst 1967; Fogel et al. 1971; cf. also Fogel et al. 1979). Sy obtained particular pleasure from dissecting tetrads and prided himself on being able to dissect one nearly every 30 or 40 seconds. He tried to encourage my skills in this area by engaging me in dissection races which he broadcast throughout the laboratory in a running narrative: "I am now picking up ascus number three hundred and twenty one, dropping off spore one, spore two, spore three, and spore four. And now, following closely behind, ascus number three hundred and twenty two, spore one, etc.," he would say all too quickly. (Although amusing, this was, in fact, highly discouraging!)

Sy had great admiration for Herschel and expressed this by posting a sign on the inside door of the media cabinet. Each time any of us required media, we were confronted with the words ROMAN IS NEVER WRONG! Sometimes these words were incorporated into a holy "Gregorian Genetic Chant" that we periodically sung together, usually while dissecting asci or dissecting snails to make glusulase for dissecting asci. The chant started with *In principio erat Mendel. Mendel habat Pisum. Pisum sativum. Pisum alba et Pisum verde. Pisum rotundum*, etc. (followed by several low throaty melodic hummmms). The incantation usually went on to include the contributions of Morgan, Beadle and Tatum, and Watson and Crick, among others. Sy was greatly stimulated in these creative musings by the Latin expertise and tenor voice of Mike Esposito, who was also doing a research project in the laboratory at this time. The annual preparation of glusulase was a particularly festive occasion in the laboratory, in which we often invited in guests from other departments to sing the chant with us. A variety of snails in stock at the Brooklyn Terminal market had been sampled before we finally settled on our favorite, *Tupetella*. On the day of the "crop harvesting" (*Tupetella*'s crops of course), which was dependent on the day they arrived fresh at the market, we made sure that we had plenty of garlic and parsley on hand to season the delicacies we eventually cooked after our work was done. These we supplemented with smoked oysters, which we elegantly served to everyone on petri dishes with toothpicks. (*Historical note:* I'm not sure how animal rights activists today would view this enterprise, but perhaps it would be more palatable if I emphasized that the snails were being sold at the market specifically for that purpose!)

But I'm straying from the point. The sign went up after Don and Sy summarized some of their data on the association between gene conversion and reciprocal recombination. For several years, Herschel had been leaning toward the hypothesis that gene conversion and reciprocal recombination arose from different mechanisms while requiring common preconditions. His view was based on that fact that the events were dissociable. For example, he found that reciprocal recombination among selected gene convertants occurred less frequently in vegetative cells than in asci. Moreover, the correlation between conversion and reciprocal exchange became even weaker with increasing doses when conversion events were stimulated by UV light (Roman 1956; Roman and Jacob 1959). At the time, Herschel favored the idea that conversion might occur by a mechanism akin to bacterial transformation in which a donor segment of DNA from one molecule "simply" replaced a homologous segment in a recipient molecule, perhaps during DNA replication. He imagined that reciprocal exchange involved a more complicated classical breakage and reunion process occurring between chromatids after DNA replication. Don and Sy were skeptical of Herschel's data on spontaneous and UV-induced mitotic recombination. They were particularly concerned that the outside markers used in the experiments, although the most convenient ones available at the time, were too far away from the conversion site to give an accurate picture of the correlation effect, due to the possibility of multiple exchanges. Based on meiotic data, they actually favored the notion that both events resulted from the same mechanism. They therefore repeated the mitotic experiments with different markers. Confirmation of Herschel's earlier UV work led to the sacred parchment on the media cabinet door.

## PART 2: LEARNING YEAST GENETICS THE CORNY WAY

I first met Herschel Roman in the spring of 1962, when he made a brief visit to Brooklyn College. Sy's description of him, together with the daily subliminal impact of the sacred parchment, had led me to form a vision in my mind of a great sage, 10 feet tall, with a long white beard that bespoke unfathomable wisdom and intelligence. On the day of his visit, there was a gentle knock on the door to the laboratory, and when I opened it, I remember consciously looking up expecting to see the giant I imagined and then slowly bringing my gaze down to eye level to see a kind friendly face peering at me curiously from behind his glasses, announcing in a soft voice , "I'm Herschel Roman and you must be Sheldon." I didn't realize then what an important role Herschel was to play in my life, as well as the lives of many others, but more about that later.

My next encounter with the Seattle "yeast people" was in Colorado, at a Genetics Society meeting held in August of that summer. Don and Sy had suggested I attend the meeting with them and afterward drive up to Seattle with the other graduate students. All I really remember of the meeting was meeting the legendary Don Hawthorne, whose contributions were well known and highly revered in Flatbush. My first impressions were perplexing. I had gone to a session in which he was speaking, and although it appeared that he was giving a very serious talk, a group of students toward the back of the room were giggling and periodically burst out laughing. I thought this was highly disrespectful, but much to my surprise, Don seemed to enjoy it, glancing up at the students every now and then with twinkle in his eyes! I was later to learn from the students that he had discretely provoked the whole incident and that he loved to subtly egg them on with his wry sense of humor. This often took the form of presenting his work in the form of a riddle or mysterious detective story that tickled everyone except Herschel, who constantly admonished him to present his work in a clearer fashion. "Don, you simply have to be more serious." This of course encouraged more of the same to everyone's delight, once they caught on to the gist of the game. At one of the first faculty research seminars I attended, Don spoke to the department on the brilliant work he was doing on mating-type switching in *D* gene (now called *HO*) strains (Winge and Roberts 1949; Hawthorne 1963). Herschel introduced him and then chided "now Don, I want you to give this talk so that *even Shelly* will understand it" (I've always assumed that this comment was because I was a first-year student). This put a great burden on me for Don proceeded to direct the entire talk in my direction, and I felt obliged to nod frequently to communicate my comprehension and to encourage him. But at some point in the middle of describing what was happening to the third bud's bud and the first bud's bud's bud, I began to lose it despite all my efforts to concentrate, and alas I could no longer continue my nodding in earnest. It was clear that Don perceived this transition, for he monitored me just a bit longer, then turned away after giving me one last hopeful glance and delivered the rest of the talk to the ceiling.

We (the students and probably everyone else) were never sure of when Don was toying with us, which of course was part of the fun. For example, when we received strains from him to make our first crosses, many of us, by a strange coincidence, encountered great difficulties in these initial efforts, such as the appearance and disappearance of markers among the segregants, spores not growing on complete medium, etc. (*Historical note:* These problems were typically due to the segregation of suppressors that Don was in the process of discovering but had yet to publish, and the presence of a $can^R$ arginine permease mutation in one parent and an *arg* marker in the other leading to failure of arginine-requiring strains to grow on complete medium.) In the end, we either derived the genetic systems from first principles or threw up our hands in exasperation and used other strains! Of course we could not help but wonder if this was a secret rite of passage we all had to pass to prove our genetic calling. In truth, after all these years, I still don't know the answer, for when I've mentioned the possibility to Don, he looks at me a bit bemused while denying any complicity (never very convincingly I might add). The net effect was that we all learned genetics *very very very* well.

The Department of Genetics at the University of Washington was formed in late 1959. When I arrived in Seattle in 1962, the faculty consisted of Howard Douglas, John Gallant, Stan Gartler, Don Hawthorne, Arno Motulsky, Larry Sandler, and David Stadler, with Herschel as Chairman. Three graduate students (Beth Jones, Charley Reichelderfer, and Satya Kakar) and a postdoctoral fellow (Jørgen Friis) in Herschel's laboratory, plus Don Hawthorne, Howard Douglas, and of course Herschel formed the yeast group. A few days after I arrived, there was a going-away party for Satya Kakar, who was Herschel's first Ph.D. student. I remember it vividly because I ruined the party. We all went out to the Century tavern, located exactly 1 mile from the University (alcoholic beverages were not allowed to be sold any closer to the campus), to celebrate. Despite the fact that I was then 21, and carried my birth certificate with me as proof of my age, the tavern would not serve me; apparently three pieces of identification or a Washington State Drinking License (obtained at a small fee to the state) was required. The Century tavern refused not only to serve me, but also, much to my embarrassment, to serve anyone who was with me. I offered to leave, but my new friends would hear nothing of it and so we all left! During that first year, three of us from the entering class (George Mund, Mike Esposito, and myself) joined Herschel's laboratory, and a year later another postdoctoral fellow, Alberta Herman, working on homothallism and mating-type switching in *S. lactis* came on board. The whole department was housed in half a floor of the Botany Building, otherwise known as Johnson Hall. Although none of the other faculty were actually working with yeast, they were highly interested in what was happening in the yeast group and occasionally attended our yeast meetings. Due to our close proximity and small size, we all functioned as a sort of family. Most of the graduate students were unmarried and inseparable. We spent long hours in the laboratory and then usually had dinner together. At the "end of the day" (typically two in the morning), we would exhaustedly traipse to the Hasty Tasty, a greasy spoon restaurant across the street from the Century tavern, only to spend another hour or two together unwinding before catching some sleep and starting another "day." During this period, the coffee room in Johnson Hall, which was a little bigger than a closet, became the primary site of our intellectual interactions. It was here that Larry Sandler would hold forth on a variety of topics and nucleate numerous debates. It was also at this time that my interest in the genetic control of meiosis was peaked specifically by Larry's inquiries into how we thought meiosis might have evolved.

The first year that I was in Seattle, Robin Holliday was also on sabbatical there, working on the fidelity of gene conversion in yeast. He was measuring the reversion frequencies of the auxotrophic spores in 3+:1– asci to see if he could distinguish the convertant spore. Much to my good fortune, Herschel, who was extremely preoccupied with getting the new department under way, suggested I work with Robin for the year he would be in Seattle. Robin and I then decided to initiate an independent project in *Ustilago* (the corn smut fungus) based on some of his previous work and ideas about the relationship between DNA synthesis and recombination. Robin had argued that the effect of UV light in stimulating mitotic recombination might be due to its ability to temporarily block DNA synthesis. He supposed that the inhibition of DNA synthesis triggered an unbalanced growth that caused the induction of chromosome pairing, thereby leading to elevated exchange (Holliday 1961). (*Real historical note:* He was influenced in his thinking by Darlington's earlier notions that a delay in chromosomal duplication stimulated chromosome pairing during meiosis.) Robin suggested that this idea might be further tested by determining whether inhibition of DNA synthesis by other means would also stimulate exchange. Under his guidance, I set out to examine this question by treating synchronized cultures of *Ustilago* with the drug FUdR. We demonstrated that inhibition of DNA synthesis by FUdR was indeed recombinogenic (Esposito and Holliday 1964); Robin later showed that mitomycin C had the same effect (Holliday 1964a). Intrigued by these results, John Gallant, who had been studying thymine-less death in *E. coli* down the hall from us, then determined that thymine starvation induced recombination in this organism as well. These studies eventually led me to my thesis work, which further explored the relationship between DNA replication and recombination in synchronized cultures of yeast.

Working with Robin was among the most stimulating and intellectually engaging experiences I had in graduate school. He was forever placing papers on my desk that he thought might interest me and that he hoped I would read so we could critically discuss them. We had conversation after conversation about genetic recombination and how it might proceed, the implications of which I did not fully appreciate until after he went back to England to write his now famous recombination model (Holliday 1964b). Robin urged me to pursue my studies in Seattle in the British tradition: "spend as much time as you can in the lab, not in formal courses, that's where you'll learn the really meaningful things that you'll remember." Since I was accustomed to doing what I was told, I happily proceeded to spend all my time in the laboratory. When Robin was about to leave Seattle, I discussed with him my desire to continue working with *Ustilago*. He told me that he felt it would be a mistake to do so. He argued that I was in one of the best places in the world to do yeast research, with many excellent people around to interact with and consult for advice. He believed that I would be isolated if I pursued a thesis in another organism that was not being studied by anyone else at Seattle. (*Historical note:* Robin had single-handedly developed *Ustilago maydis* for genetic studies and was among the very few geneticists anywhere studying it at the time.) And so, with great reluctance, I left smut and returned to yeast!

Herschel's entire laboratory was focused on understanding the basis of mutation and recombination (Roman 1957), which were among the central issues in genetics at the time. Beth and Mike were performing fine-structure analysis on several of the *ade* loci that Herschel had earlier identified when he discovered the red/white system (Roman 1956). Beth's studies were aimed at determining the basis for differential spontaneous mutation frequencies at *ade3* and *ade6* (Jones 1972), and Mike's studies were directed at a detailed comparison of X-ray and spontaneous meiotic maps of the *ade8* locus (M.S. Esposito 1967). As a result of my work with Robin, and of Herschel's own studies on the relationship between gene conversion and reciprocal recombination, I chose to focus my thesis research on the issue of whether both types of recombination events derived from the same or independent mechanisms. The approach I took was to determine whether nonlethal doses of agents that stimulated recombination, such as UV light and X-rays, induced gene conversion and reciprocal exchange in a cell-cycle-specific manner and, if the induction of the two events occurred at the same or different stages. The issue that I wanted to address was whether the induction of gene conversion occurred during the period of DNA synthesis (a popular idea at the time) and reciprocal exchange later after replication. I also examined whether there was a temporal progression of gene conversion along a chromosome arm, expected if conversion occurred during the S phase and replication proceeded sequentially along a chromosome arm (R.E. Esposito 1967a,b). I was greatly aided during the early stages of my thesis research by the presence of Don Williamson, who worked at the same institute in England (the John Innes Institute) as Robin and who came to spend a year's sabbatical in Seattle after Robin left. Don had been previously developing procedures for synchronizing budding yeast. His method involved growing yeast to late stationary phase, removing smaller cells from the population, and subjecting the remaining cells to a feeding-starving regimen, which after a few cycles caused them to bud synchronously. I was able to adopt this procedure for my purposes after first constructing an appropriately marked parent strain (A364A), in which a high proportion of cells dropped their buds and became single in stationary phase. While modifying Don's synchronization method, I used A364A to make a diploid (Z65) heteroallelic for three markers along the length of chromosome II, and other strains (derived from strain XS380 from Mortimer and Nakai) to make a diploid (Z69) heteroallelic for five markers along the length of chromosome VII. Homoallelic diploids for these same markers were employed as controls to monitor mutation during the cell cycle. Finally, parallel heterozygous diploids were used to follow the timing of reciprocal recombination resulting in sectored colonies for various auxotrophic markers along the chromosome arms. I discovered that these sectors could be readily detected without replica plating by incorporating the dye magdala red into complete plates. This dye had been reported to accumulate in slow-growing auxotrophic cells and gave a precise dark pink/light pink color signal. This allowed identification of sectored colonies for all of the markers and facilitated further genotypic analysis to confirm the reciprocal nature of the event and locate its position on the chromosome.

We were now ready to perform the experiments. The various diploids were treated with either UV light or X-rays at 10-minute intervals throughout two cycles of synchronous division. Much to my surprise, the events were indeed induced in a periodic fashion but not in the pattern I expected. Both types of exchanges were maximally induced at the same time, just prior to the initiation of DNA synthesis; their induction declined throughout DNA synthesis and reached a minimum during mitosis before beginning to rise again. The time of peak induction did not appear to reflect simple variation in the sensitivity of cells to damaging agents, since normalization at comparable levels of lethality at higher doses given throughout the cell cycle still yielded elevated recombination values just prior to DNA synthesis. Another interesting observation was that the damaging agents appeared to provoke a division delay that potentially might explain the data if the delays provided time for repair; however, I found no consistent pattern for X-rays and UV light in the length of these delays that could be correlated with recombination levels. I repeated these experiments at least seven times for each of the diploids and was never able to detect any differences in the timing of reciprocal and nonreciprocal events during the mitotic cell cycle or between the different markers along the chromosome arms tested. At some point during this stage of my work, Francis Crick visited Seattle and Herschel scheduled a half hour for me to talk with him. I remember him looking at all my carefully obtained graphs and then saying, "only a graduate student would repeat an experiment so many times!"

I was extremely engaged (obsessed might be a more appropriate adjective) in my thesis project and spent many long hours puzzling over the meaning of these results. It seemed to me that although the lesions induced in the DNA just prior to replication likely inhibited replication, eventually, DNA synthesis might proceed past the lesion leaving a gap or mismatch. Perhaps such an unrepaired site stimulated the recombination system and caused the exchange events. The lesions induced in replicated DNA, might then result in fewer recombinants because there was more time to repair the lesions in this DNA by excision repair, prior to the next round of replication. Ben Hall, who had arrived in Seattle midway through my graduate career, was on my thesis committee and was intrigued by these ideas. He was an important influence in encouraging me to develop a molecular model to explain the results and strongly disagreed with Herschel, who felt my hypothesis was too speculative to be included in my thesis. Ben insisted that it remain and won the argument (R.E. Esposito 1967a). However, Herschel persevered and in the end convinced me that it was too risky to include in the paper, which was eventually published on the subject in *Genetics* (R.E. Esposito 1967b).

The atmosphere during my stay in Seattle was quite exciting, with scientists from all over the world coming to visit on a regular basis! In my second year while Don Williamson was in the laboratory, Ethel Moustacchi arrived from France to spend a year's sabbatical as well. Together, she and Don initiated studies to isolate and characterize the various species of yeast DNA on CsCl gradients, starting some of the early molecular work with yeast. Herschel had a large NIH grant at the time that he called "*Saccharomyces* Studies," whose purpose was to develop yeast as an experimental organism and to train new scientists to work in yeast methodologies. As a consequence, many others, in addition to those I mentioned, came to Seattle to study with him and to consult about what was happening in this new system. Herschel's laboratory was a mecca for all those working with yeast, as well as for those not working with yeast but interested in genetics. The growth in the yeast field that took place at that time occurred in an era of unique governmental support of scientific research that many of us now refer to as "The Golden Age." There were sufficient funds available for everyone to do their work, to be creative, and to try new things. (*Historical note:* Many of the breakthroughs of the 1970s specifically derived from this deliberate encouragement of basic genetics a decade before.)

While I was in Seattle as one of Herschel's own graduate students, I had an opportunity first hand to observe and experience the pattern of his thought processes. I came to admire most his ability to reduce complex issues to simple alternatives and his judgment in solving problems of both a scientific and nonscientific nature. Herschel was an exceptionally clear thinker and unusually perceptive about a variety of matters. He had extraordinary integrity and was truly devoted to uncovering the answer to a problem; he was never satisfied with superficial solutions. He delighted in conversing with students and colleagues about all sorts of topics from genetics and other scientific matters to art, literature, and politics. It was plainly evident that his interest and curiosity about the opinions of others were genuine; he loved to debate and probe the depths of one's thinking and thoroughly enjoyed interacting with all those with the courage to spar with him. He often expressed that he wanted the students to consider him one of their own. However, his position as Chairman of the Department, and his stature in the field of genetics, set up an invisible barrier that he constantly had to overcome. For example, he would complain to me "I'm always the last one in the Department to find out all the latest news, everyone else always knows what's going on before me!" But once the students overcame their own fears, gained confidence, and began to talk to him, they never stopped going back for a private chat. There was always a steady parade outside his office door. We all took council with him on scientific issues and personal ones. We even borrowed money from him from a special fund he would keep for us for emergencies. And we all received his famous little squeeze on the elbow, urging us on and encouraging us when we needed it. He accorded each of us the highest respect by paying attention to us and taking what we said seriously, sometimes even more seriously that we did ourselves. In the end, when those of us who had been students in the Department of Genetics had left Seattle, it was clear to most of us that we had received far more than an education in genetics.

There were, of course, times when dealing with Herschel was somewhat frustrating. He was absolutely uncompromising when he felt errors in judgment were being made, especially when he thought that we were not showing proper respect to our common mission. It was an extreme source of annoyance to him if anyone (faculty or students) missed the weekly Departmental Journal Club or Yeast Meetings, and he never hesitated to make his disapproval clear. "What else is more important than attending a journal club or yeast meeting?" Herschel never asked more of others than he demanded of himself. He worked hard and he expected everyone else to do so. He showed up at the laboratory every night, sometime between 7 and 11 o'clock. I remember my extreme embarrassment when I didn't return to the laboratory one evening. The next morning he casually said to me "You know it doesn't cost us anything extra to keep on the lights at night!" We met formally to discuss the progress of my research efforts usually once, at most twice, a year. I would spend a good deal of time, carefully preparing graphs and tables that summarized my data in various ways before each of these meetings. Typically, he got to the heart of the matter within minutes, asking me critical questions that I had not considered, much to my chagrin. (*Historical note:* I later learned that most mentors seem to be endowed with this talent, although Herschel's abilities in this regard were truly extraordinary!) At the height of my dismay, he'd say gently but firmly, "next time you come in, be more prepared."

Which finally brings me to corn! It wasn't until many years later that I realized that my annual meeting with Herschel must have been very similar to the training that he had received as a graduate student working with L.J. Stadler. With corn, he had only one (at most two) planting seasons per year to execute his studies, so it was extremely important to think things through carefully and to design the experiments, including all possible controls, or much valuable time would be lost. The rigorous thinking that corn genetics demanded was an integral part of Herschel's approach to his work with yeast, which I always felt had a certain undefined aesthetic quality or economy of thought to it. The emphasis was clearly on "thinking before doing!" I believe this corny approach formed the underpinnings of much of the elegant work that has occurred in the yeast field over the years. As Sy warned, when I had a question, Herschel would generally answer it with another question, refraining from doing my thinking for me. If he thought that a proposed experiment would not provide the data I wanted, he would say so, but then add, "why don't you try it to satisfy yourself." He respected and regarded his students' thesis work as their own. Indeed, he felt so strongly about this point that he would never consider putting his name on a student's thesis publication(s) unless he had done some of the analysis by his own hand. Although I was disappointed that he was not an author on my thesis paper, I eventually came to appreciate his views, which I think encouraged us to be independent scientists. (*Historical note:* Regretfully, current demands on scientists to publish not only quality, but quality in great quantity make this laudable attitude difficult to pursue.) On the day I left Seattle to begin my postdoctoral studies, I went to his office once more to take counsel with him, asking if he had any advice to give me upon my departure. He thought for a moment and then choosing his words carefully said, "Never publish anything you are not sure of. That's the best advise I can give you."

Mike Esposito and I left Seattle in April of 1967. We spent the next year and a half as postdoctoral fellows working with Harlyn Halvorson at the University of Wisconsin, before taking up residence at the University of Chicago. It was at Wisconsin that we finally initiated our studies on the identification of genes required for meiotic development in yeast. I say finally because we had been planning the analysis for nearly 3 years. As I mentioned earlier, I had become interested in meiosis almost from the beginning of my tenure at Seattle, primarily due to the influence of Larry Sandler. It was hard not to become interested in almost anything that Larry chose to talk about, because he had such a commanding presence: He was brilliant, funny, articulate, and lovable. The students were drawn to him as if to a magnet. If Herschel was the nucleus, then Larry Sandler was the cytoplasm of the department, providing the molecular matrix through which we all interacted. Larry had an enormous impact on me and I think on the development of all of the yeast geneticists at Seattle. When I first met him I was a bit intimidated, for it was difficult to keep up with his verbal virtuosity even though we both spoke the same dialect of Brooklynese. Larry hammered away at my brain, pushing, prodding, and forcing me to explore my own creative potential. In the end, he became the advisor I turned to between my yearly meetings with Herschel.

Early on, Mike and I adopted a habit of having lunch with Larry every day in his office. When Bruce Baker and Adelaide Carpenter joined Larry's group, they too became regulars at our luncheons and solidified the intellectual wedding that was taking place between yeast and *Drosophila* genetics. Meiosis was a frequent topic and the blackboard in Larry's small office became completely filled with our genetic musings about the process of exchange and segregation. During these times, Larry's enthusiasm for both lunch and genetics infected us all. The kosher pickles he brought every day were irresistible. After a while, we all found that it was impossible to think clearly about the problems before us unless we each had a kosher pickle to wave in the air and take a bite of at just the right moment to make our points with appropriate emphasis! Herschel, who generally conducted himself in a more dignified manner, would pass by the office shaking his head disapprovingly: "I don't know, Larry," he'd taunt, "for all your complaints about not being able to get a good bagel in Seattle, you're not as discerning about pickles. Seattle pickles seem to be quite acceptable to you." This challenge called for revenge with more than just words. To get even, we arranged for all 25 graduate students then in residence to be noisily chomping on the largest kosher pickles available in Seattle during Herschel's next graduate student meeting. Much to our unending frustration, however, but true to Herschel's pertinacious nature, he delivered his entire speech and left without acknowledging to any of us the numerous pickles in his presence. (*Historical note:* Caryl Roman recently told me that he did indeed take note of our little escapade, for he described it to her with great amusement when he came home that night!)

Sometime in 1964, Mike and I recognized that yeast was in fact an ideal system in which to study meiosis and thought that we might apply to meiosis the kind of genetic dissection of development that was taking place for the morphogenesis of bacteriophage T4. We subsequently spent numerous hours, over a period of months, discussing ways to conveniently recover mutants to begin dissecting the process. Don Hawthorne's work on mating-type switching in homothallic strains eventually drew our attention to the possibility that *D* gene (*HO*) strains might provide a convenient tool to obtain recessive meiotic mutations in homozygous condition, which was the primary problem in their detection. We even conducted some pilot experiments at the time to prove the feasibility of our ideas before putting the whole project temporarily aside in order to complete our Ph.D. research. When we finally reached Madison, we were able to continue this work and eventually found that indeed we could identify genes controlling meiotic development by this method (Esposito and Esposito 1969; Esposito and Esposito 1972). During this time, Harlyn Halvorson was enormously supportive of our work. We chose to go to his laboratory for postdoctoral studies because we wanted to broaden our genetic background to more biochemical approaches. Harlyn was one of the very few senior investigator's working with yeast at the time who had attempted to address genetic problems from a more molecular point of view. The focus of his interest was in gene regulation in *Saccharomyces* as well as the process of sporulation in *Bacillus subtilis*. Harlyn was enthusiastic and generous in his support. He provided us with a laboratory and a technician and encouraged us to pursue our genetic experiments in a free and unconfining way. We quickly learned more molecular techniques from him and eventually began to analyze acetate utilization and to initiate studies to characterize macromolecular landmarks that occurred during sporulation (Esposito et al. 1969, 1970). Finally, in late 1968, Mike and I left Madison and began our own laboratory at the University of Chicago, contributing to the early growth of the field, which was now poised to enter log phase.

## PART 3: THE LEGACY OF HERSCHEL ROMAN

Herschel Roman was an educator completely devoted to graduate as well as undergraduate training and to developing the new field of yeast genetics. It was of utmost importance to him to provide a stimulating and congenial atmosphere in which we could perform our research. The welfare of students and faculty was his special interest and he wanted us to proceed with clear minds and not be burdened with concerns that would deplete our energies and divert us from our goals. This was paramount to him and he took great pains to ensure that we were not only in an intellectually stimulating environment, but happy in our work. He would say "if you're not happy you can't work well." If he perceived that any of us were having personal or scientific problems, there would be an unobtrusive quiet lunch invitation, "Let's go have some salmon and talk over how things are going." The departmental climate he created in Seattle was highly unusual in being not only scientifically rigorous, but philosophically dedicated to the enjoyment of ideas and one another's intellects. Faculty and students bonded with a sense of equality in taking pleasure in our common avocation. Regardless of how advanced we were in our respective careers, each of us was simply "A Student of Genetics." Herschel nurtured and advanced this attitude, supported by the other faculty in the department. There were no cultural strata to break through, only things to be learned and fascinating problems in genetics to think about and solve. The fabric of the group he created at Seattle was based on a holistic view toward both science and human behavior which involved a simple principle: respect for our work, and respect for one another. These principles were carried over to the emerging field of yeast genetics, in which he was a driving force. A unique character became associated with the field of yeast genetics early on because of him, the hallmark of which is a special camaraderie among us and shared pleasure in studying every aspect of the organism.

It is difficult to describe Herschel's courage after his stroke some years ago. I can only say that he once again became my teacher, inspiring new respect and admiration. During all of the years that I knew him, which now seem far too brief a time in my mind, I came to learn that Sy was right, Herschel Roman was indeed a prince! When Herschel's daughter, Ann, called in July of 1989, to tell me he had passed away, she asked if I might inform a few close colleagues for her. I picked up the phone and called Sy Fogel. We communicated our grief silently to one another for a few moments after I stammered out the sad news, and then Sy quietly said to me "Herschel has cast a very long shadow that most of us are still standing in." As for myself, I am grateful to have had an opportunity to grow in his shade.

## ACKNOWLEDGMENTS

This paper is dedicated to the two Seymours, Don Hurst, and Herschel Roman. I want to especially thank Robin Holliday, Don Hawthorne, Larry Sandler, and Harlyn Halvorson, and the rest of the original faculty of the Department of Genetics at the University of Washington: Gus Doerman, Howard Douglas, John Gallant, Stan Gartler, Ben Hall, Arno Motulsky, and David Stadler for their dedication in educating students like me. And to my own students, thanks for teaching me too!

## REFERENCES

Esposito, M.S. 1967. X-ray and meiotic fine structure mapping of the *adenine-8* locus in *Saccharomyces cerevisiae*. *Genetics* **58:** 507–527.

Esposito, M.S. and R.E. Esposito. 1969. The genetic control of sporulation in *Saccharomyces*. I. The isolation of temperature-sensitive sporulation-deficient mutants. *Genetics* **61:** 79–89.

Esposito, M.S., R.E. Esposito, M. Arnaud, and H.O. Halvorson. 1969. Acetate utilization and macromolecular synthesis during sporulation of yeast. *J. Bacteriol.* **100:** 180–186.

———. 1970. Conditional mutants of meiosis in yeast. *J. Bacteriol.* **104:** 202–210.

Esposito, R.E. 1967a. "Genetic recombination in synchronized cultures of *Saccharomyces cerevisiae*." Ph.D. thesis, University of Washington, Seattle.

———. 1967b. Genetic recombination in synchronized cultures of *Saccharomyces cerevisiae*. *Genetics* **59:** 191–210.

Esposito, R.E. and M.S. Esposito. 1972. The genetic control of sporulation in *Sac-*

*charomyces*. II. Dominance and complementation of mutants of meiosis and spore formation. *Mol. Gen. Genet.* **114:** 241–248.
Esposito, R.E. and R. Holliday. 1964. The effect of 5-fluorodeoxyuridine on genetic replication and mitotic crossing over in synchronized cultures of *Ustilago maydis*. *Genetics* **50:** 1009–1017.
Fogel, S. and D.D. Hurst. 1967. Meiotic gene conversion in yeast tetrads and the theory of recombination. *Genetics* **57:** 455–481.
Fogel, S., D.D. Hurst, and R.K. Mortimer. 1971. Gene conversion in unselected tetrads from multipoint crosses. *Stadler Genet. Symp.* **2:** 89–110.
Fogel, S., R. Mortimer, K. Lusnak, and F. Tavares. 1979. Meiotic gene conversion: A signal of the basic recombination event in yeast. *Cold Spring Harbor Symp. Quant. Biol.* **43:** 1325–1341.
Hawthorne, D.C. 1963. Directed mutation of the mating type allele as an explanation of homothallism in yeast. *Proc. Int. Congr. Genet.* **1:** 34–35. (Abstr.).
Holliday, R. 1961. Induced mitotic crossing over in *Ustilago maydis*. *Genet. Res.* **2:** 231–248.
———. 1964a. The induction of mitotic recombination by mitomycin C in *Ustilago* and *Saccharomyces*. *Genetics* **50:** 323–335.
———. 1964b. A mechanism for gene conversion in fungi. *Genet. Res.* **5:** 282–304.
Jones, E.W. 1972. Fine structure analysis of the *ade3* locus in *Saccharomyces cerevisiae*. *Genetics* **70:** 233–250.
Roman, H. 1956. A system selective for mutations affecting synthesis of adenine in yeast. *C.R. Trav. Lab. Carlsberg Ser. Physiol.* **26:** 299–314.
———. 1957. Studies of gene mutation in *Saccharomyces*. *Cold Spring Harbor Symp. Quant. Biol.* **21:** 175–185.
Roman, H. and F. Jacob. 1959. A comparison of spontaneous and ultraviolet-induced allelic recombination with reference to the recombination of outside markers. *Cold Spring Harbor Symp. Quant. Biol.* **23:** 155–160.
Winge, Ø. and C. Roberts. 1949. A gene for diploidization in yeast. *C.R. Trav. Lab. Carlsberg Ser. Physiol.* **24:** 341–346.

# The Double Entendre

GERALD R. FINK
*Whitehead Institute*
*Massachusetts Institute of Technology*
*Cambridge, Massachusetts 02142*

I never saw Fred Sherman in a sober moment. He was always intoxicated by his own slapstick sense of humor. Of course, my only real chance to observe him closely was each summer for the 17 years that we taught the Cold Spring Harbor Yeast Course together. Perhaps in other surroundings, away from the heat and humidity on Long Island, Fred is as solemn and dignified as any other university professor. But at Cold Spring Harbor for the 3 weeks of the Yeast Course, Fred was a one-man show. During those 17 years, the supporting cast changed—in the first year, we were assisted by Bruce Lukins and then by Chris Lawrence, Tom Petes, and in the later years by Jim Hicks. But none of us knowingly provoked Fred's antics. In fact, willingly or unwillingly I often served as his straight man.

To be accurate, Fred could be serious for short periods. These interludes usually occurred early in the morning when, not fully awake, he was presenting a lecture on genetic nomenclature or the arcane aspects of mitochondrial inheritance. His presentation always began, "Briefly,..." but, inevitably, he spoke for many hours. Delivered in a formal, somber academic style, these talks seemed to project from a surgically dehumored Fred. But, as the morning's coffee purged him of the anesthetic residue of sleep, and he became fully alert, all pretense of seriousness vaporized. There were moments of mock sobriety, but these were setups for impending humor.

The yeast course instructors. Fred Sherman (*left*) and Gerry Fink (*right*) in 1974.

Despite his zany antics, Fred was deadly serious about the Yeast Course itself. Throughout the winter months, he pondered logistical and equipment problems. He was constantly inventing inexpensive gadgets for the course to substitute for the costly ones used in research laboratories. His pride and joy was a device designed to examine the cytochrome content in wild-type and *petite* mutants. To replace the top of the line Carey spectrophotometer used in most laboratories, Fred assembled surrogate machinery consisting of a hand-held spectroscope and a cheap high-intensity lamp. The light was focused through a hole drilled in an aluminum cooking pot to illuminate yeast cells frozen in liquid nitrogen. Although the light intensity was difficult to control, this pots and pans contraption had remarkable resolution. Those students who were not temporarily blinded by the intense beam of the lamp were amazed by the clarity of the cytochrome spectra.

Probably his most useful invention was the Sherman micromanipulator. Early in the history of the course, we borrowed micromanipulators from Brooklyn College. They had only a few to lend and these were on their last legs. When we mistakenly left them at Cold Spring Harbor at the end of one summer, none survived winter use by the permanent residents as doorstops or beaker supports. The situation that faced us the next summer was bleak. Cold Spring Harbor could not afford the $5000 needed for even one Mortimer or de Fonbrune manipulator, and we needed eight of them. Faced with a frustrating but definable problem, Fred showed the ingenuity that had gained him celebrity as a scientist. Using elements from microchip assembly devices and various microscope parts, he invented a cheap micromanipulator that is now used throughout the world.

Each summer, we would arrive several days before the course began and work furiously to clean Davenport, usually with the help of graduate students from Fred's laboratory or mine. We would ferret out microscopes and media and set up workstations with a complete set of supplies and materials. These materials included an elaborate course manual (prepared during the winter months), a detailed description of each day's experiments covering the entire 3 weeks, and a set of strains for each pair of students (there were always eight pairs). The strains were updated each year to include new and useful genotypes from our own and other laboratories.

The laboratory manual, *Methods in Yeast Genetics*, assembled in one volume the standard methods and media for growing yeast as well as the latest techniques used in yeast biology. Our colleagues sent us detailed protocols, often before they were published, to be included in the most current version of the manual. This generosity was a sign of the collective trust and support of the yeast community for the course. These techniques as well as reagents and the yeast strains were given to the students with the idea that they could take them back to their own laboratories to expedite their research. The free distribution of strains and other information in the course contributed to the remarkably cooperative atmosphere that has existed in the yeast field.

The students arrived the day before the official opening, surreptitiously observing Fred's pre-course ritual of aligning microscopes, directing the assistants, and checking supplies. The whole scene had a business-like atmosphere portending a no-nonsense course. This serious façade was probably the ballast that steadied many of the students for the weeks ahead. It must have been daunting for some to realize that they were now thrown "back in school" competing with the likes of Frank Stahl, Julius Marmur, Gottfried Schatz, or Clint Ballou. Even the more seasoned scientists had reason for nervous anticipation. They had given up the comfort of their own laboratories and homes to rough it for 3 weeks, with neither technicians nor students to do their bidding. Moreover, all students regardless of seniority slept in the Page Motel, a Spartan assembly of attached wooden cabins with the appearance of an evangelical campground. Each small room in Page was shared by two students and a cloud of marauding insects driven up the hill by the summer heat. It was rumored that Jim Watson ran the summer courses to feed his pet mosquitos in Page.

The discomfort of the living arrangements and the sense of isolation were intensified by the novelty of the new techniques and the difficulty of executing them in the environment of Davenport Laboratory (now called Delbrück Laboratory). In the first few days, the students struggled to master microdissection (picking up the yeast ascospores and separating them on the agar). They usually worked until 2 a.m. or later the first few nights just learning how to make a microneedle and to manipulate the spores. What they did not know was that the strain Fred brought for this experiment sporulated terribly and never yielded many of the prized four-spored tetrads. So, of course, they had trouble finding them. Although the strain he gave them consistently sporulated poorly, Fred always claimed that "sporulation is better this year than it has ever been, so you should find it easy to dissect."

Even if the frequency of sporulation had been reasonable, dissection would have been difficult because the micromanipulators and microscopes were old and often out of alignment. Moreover, the students' first attempts at making

Davenport Laboratory at Cold Spring Harbor as it appeared in the 1960s and early 1970s.

dissecting needles often resulted in a deformed spike that gouged the agar, entombing the prized spores beneath the surface. Armed with a defective needle and blurry microscope, it could take an hour to find a tetrad and move it up the agar. Davenport Laboratory had only a few window air conditioners to relieve the heat, but the thump of the compressors made the dissecting needles vibrate so wildly that even an expert could not harness the flailing tool.

So unnerving was the vibration of the air conditioners that they were often turned off even on the most stifling summer day. However, this adjustment gave rise to a new terror. The increase in heat and humidity seemed to spawn small aggressive sand fleas that were reborn each morning and mercilessly bit any uncovered appendage. Despite the faulty equipment and voracious insects, many of the students achieved mastery of tetrad dissection. Then they faced a new menace. Each summer, the assistants poured between 10,000 and 15,000 petri dishes, more than half of which became contaminated with a rapidly growing, florid, white mold. Davenport had once been a dusty old boat house and the opportunistic spores seemed to lurk in every crack, ready to sail into a petri dish. Given this atmosphere, it was not surprising that even those students who had weathered the technical difficulties and actually dissected spores would find the next morning that their prize dissection had been overgrown by a vicious mold. Fred's comment "Each student must dissect and analyze at least 10 complete tetrads to 'graduate' from the course. If you don't do so by the time the course is over, you can stay an extra week, or you can come back next year to finish up," seemed to push the weary students to the brink.

I have spent some time discussing the various irritations of a summer course at Cold Spring Harbor because the anxiety evoked by this environment created the perfect backdrop for Fred's humor. The students worked extremely hard and any diversion relieved the pressure and the routine. Each morning at 9 o'clock there was an organizational meeting concerning the progress of the experiments, followed by a 1–2-hour lecture by Fred or me on some area of yeast biology. After lunch, the students would work on an experiment until 2 o'clock when we would trot up to James Laboratory for a talk by a visiting scientist. Fred and I had lined up a lecture series by the most productive yeast biologists. After the lecture was over, the students went back to Davenport to work on experiments until dinner at 6. As the day was filled with lectures and discussion, most of the actual laboratory work was done between 6 p.m. and midnight.

But this prosaic description of a typical day does not convey the spirit of the course. Fred, clad in ragged cutoffs, greeted the students each morning with a deep bow. Then, from a balletic pose that tilted somewhere between first and third position, he would grasp a student's hand and escort him or her into the laboratory in a mock *pas de deux*. Once Fred began, there was no escape from his repartee. Even the most innocent greeting turned into a comedy routine.

*Fred:* How are you doing?
*Student:* Fine. How are you?
*Fred:* Well I think I'm fantastic. But not everyone agrees with me.

The organizational meeting that preceded the lecture at 10 o'clock was replete with surprises. A typical problem solved in the very first meeting of the course was the choice of laboratory partner. The 16 students were asked to pair off as partners for the duration of the course. In Fred's hands, this decision was treated as an extremely serious endeavor. His prologue pointed out the importance of "choice of partner" as a critical educational experience. In view of this, the students were asked to give a brief explanation of their research interests to provide the others with the basis for making a choice. Just prior to the tense moment when students would make their fateful decision, Fred offered an important criterion for the choice: "Try to pick someone who complements your expertise so that you learn something. For example, if you are a biochemist choose a geneticist for a partner. If you are a geneticist, pick a biochemist. If you think you are smart, pick someone who is stupid."

As the course progressed, the morning sessions provided a forum for discussing both the latest published work and the lore, the unpublished anecdotal observations critical for the success of an experimental scientist. In the early days, as now, there were many alluring papers with provocative claims based on shaky data. In the spirit of protecting our wards, Fred and I liberally debunked these papers. So consistent was one author's trail of unsubstantiated claims that Fred balanced his cynicism with the comment, "Just because X has published it doesn't mean it isn't true."

The only part of the day that did not regularly feature a wacky rendition by Fred was the guest lectures at 2 o'clock. But even these were not immune. Fred always wanted to introduce the speaker. His introductions usually began with a biography of the speaker presented in a dry, formal style, but gradually they would wander into saccharine hyperbole (recalling Oscar Wilde's witticism, "Biography poses a new threat to death"). A typical introduction might go something like this:

> This is Lee Hartwell from the University of Washington. He did his degree with Boris Magasanik etc., etc. Hartwell figured out the yeast cell cycle. He is a wonderful geneticist, simply fantastic, a remarkable human being and he is going to give an unforgettable talk. I just can't wait to hear it.

As the hoopla became more outrageous, Fred would appear to become possessed—roll his eyes, heave with rapid short breaths, and stumble off the podium. When Bob Mortimer spoke, Fred would introduce him by reciting a rhyme in a singsong voice:

> *Saccharomyces* on the plate,
> What's the latest thing you ate?
> Was it sugar?
> Was it good?
> Did you do the thing you should?

This doggerel was apparently composed by a student of Mortimer's, Lee Gunther. When asked about its origins, Fred would answer that the "poet" was the most publicized yeast geneticist in the world. This remark was astonishing because no one had ever heard of Gunther. However, Fred was right. Gunther's name (spelled Guenther) appears on every bottle of Difco Yeast

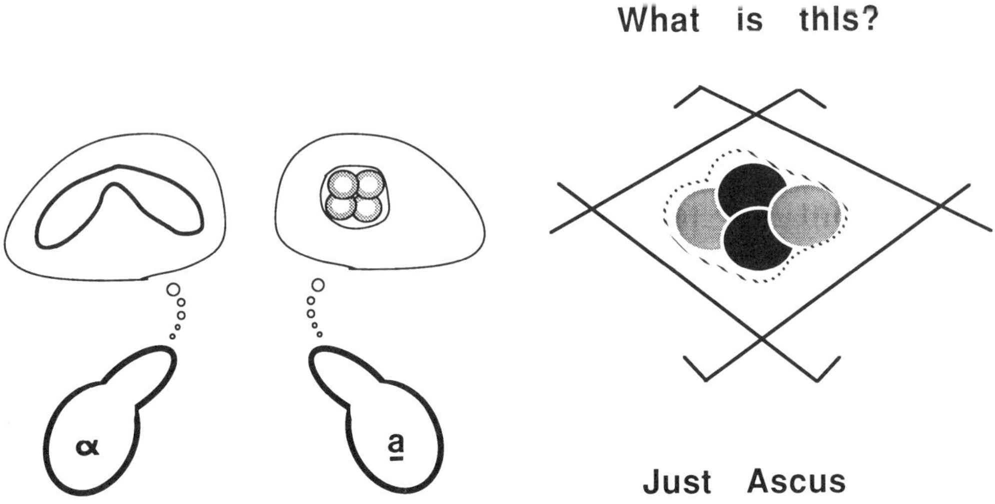

Two of the tee shirt designs from the Cold Spring Harbor Yeast Course.

Nitrogen Base w/o amino acids. Difco originally produced only *Difco Yeast Nitrogen base with three amino acids*. This configuration was extremely inconvenient for use as a minimal medium. Gunther convinced Difco to leave out the three amino acids and has become immortalized on each Difco label, "prepared according to Guenther's modification..."

The lectures were attended not only by the students, but also by Barbara McClintock, Jim Watson, and, in the first few years, Alfred Hershey and Max Delbrück. Delbrück had the habit of getting up in the middle of a lecture and then abruptly bolting out of the room. Max was a tall imposing figure with a shock of white hair, whose exit could not go unnoticed. Rumor had it that Delbrück could not sit through a lecture if the speaker did not get right to the point, or if he considered the point unimportant. The presence of these luminaries, and their apparent interest in this emerging field, seemed to improve the quality of lectures and to encourage even the most secretive speaker to spill out the latest data.

At night, Fred and I would wander about the laboratory discussing experiments with students and making sure that all the necessary reagents were available. Around 10 o'clock, Fred would disappear only to return an hour later washed and dressed for an evening of entertainment. He would usually enter the laboratory with a flourish, make a few balletic turns and announce "Let's dance!" As the students came to understand, this ritual was the prelude to a full night at Chelsey's, a bar in Huntington that provided live music by Little Wilson. At first, only a small coterie of devoted revelers followed the pied piper of yeastdom out of the laboratory, but, as the course proceeded, even the shy and decorous followed him. Chelsey's offered a respite from the commune-like atmosphere at Cold Spring Harbor Laboratory. Those who went seemed to have contact with the outside world and came back imbued with lively tales about the escapades of the previous evening. Each morning, stories of Fred's indefatigable dancing, one time with a biker's girl and another with a bagel, seemed to buoy everyone's spirits and raise expectations for some new and unexpected happening the next night. Although these festive nights lasted well into the morning hours, all hands were on deck for the 9:00 lecture.

In the 17 years that we taught the Cold Spring Harbor Yeast Course (1970–1987, with a sabbatical in 1980 for a cameo performance in Brazil), Fred and I never had a personal argument. There were, however, many disagreements about scientific issues, often to the delight of the students. One of the disagreements had important consequences for the progress of yeast biology.

In the early 1970s, there was considerable opposition to the notion that yeast had any relevance to other systems. In fact, many scientists felt that yeast was an atypical organism, one that had wandered off the evolutionary tree. This notion was fostered by some yeast geneticists who exaggerated the difficulty of tetrad dissection in order to keep the field to themselves. Both Fred and I had determined that the Yeast Course would be dedicated to eradicating this view.

I was therefore surprised and angry when I heard Fred declare, "You cannot obtain deletions of a yeast gene by in vivo mutagenesis." As deletions

The class of 1982. Gerry Fink is jumping and Fred Sherman is crouching to his left. Jim Hicks is behind Fink's left shoulder.

were commonly obtained in bacteria and *Drosophila*, I thought that this was a preposterous statement. Furthermore, it seemed a violation of our pact to get yeast into the mainstream of science. I was piqued by Fred's assertion that yeast was different from the other "good genetic organisms." Fred regaled me with the details of extraordinary experiments involving the use of exotic high-energy particles to produce deletions in *CYC1*, at that time the best-studied gene in yeast. All had failed to produce deletions. I countered with speculations about the possible inviability of *cyc1* deletions and other arguments that seemed plausible at the time. The liveliness of the debate began to involve the students and they took sides. After realizing how heated the discussion was getting, Fred seemed to concede that maybe I was right. He concluded with the pronouncement: "It may be possible to get deletions in yeast, ...but not on this planet."

Determined to prove him wrong, I set out to find deletions at *HIS4* the moment I returned to my laboratory at Cornell. After isolating and characterizing several thousand EMS, UV, and spontaneous mutants, I came up empty-handed; not a single deletion was uncovered. It seemed that Fred was right after all. I was about to abandon these experiments when I heard that John Carbon had cloned the first yeast gene, *LEU2*. The availability of this gene seemed to provide an opportunity to set up a transformation system in yeast; the one hitch was that there were no *leu2* deletion mutants that could serve as recipients for the *LEU2* DNA that Carbon had cloned. Without a stable *leu2* mutant, it would be impossible to tell $Leu^+$ transformants from $Leu^+$ revertants. I was about to abandon the transformation idea when it occurred to me that it might be possible to construct an ersatz deletion by making a double mutant. In practice, a *leu2* double mutant would be as stable as a deletion and make a perfectly acceptable recipient. When a new postdoctoral fellow, Albert Hinnen, arrived from Switzerland, he agreed to try to construct such a double-mutant strain. The only thing going for us in this experiment was the abundance of *leu* mutants that I had collected in my abortive quest for deletions. The idea was to cross two of them, *leu2-3* and *leu2-112*, and to search through the meiotic recombinants to find the rare *leu2-3 leu2-112* double mutant. The key to the search was a recombination test in which each of the meiotic progeny was crossed by the single *leu2-3* and *leu2-112* strains. The vast majority of the population, the single mutant progeny, would recombine with one or another of the parents, whereas the desired double mutant would recombine with neither. Although there was nothing novel in the method, it had the tedious aspect that every meiotic spore had to be crossed by four strains to determine whether it was one of the parents or the desired double mutant. I warned Albert that he would have to screen lots of progeny because the double mutants were likely to be rare, perhaps as infrequent as 1 in 20,000 progeny. If so, this would mean 80,000 tests.

Albert was a thorough scientist and laboriously collected the large number of progeny I had told him were required to produce a winner. When I inquired about his progress, he replied curtly that although he needed only one double mutant, he had isolated a bushel full. It turned out that Albert was not only thorough, but very orderly as well. He had first collected what I had opined were the requisite number of progeny and only subsequently tested their genotype. My underestimate of the frequency of recombination at *LEU2* meant that he had needlessly processed tens of thousands of progeny. Once in possession of this double mutant, however, Albert was able to proceed in his quest for transformation without the confusion of revertants and, also, without the benefit of further statistical projections from me. Indeed, Albert appeared at Cold Spring Harbor one day in August 1977 at the end of the course to announce that he and Jim Hicks, another postdoc, had achieved transformation of yeast. In the midst of my elation, I asked how frequent it was. Albert, with a rather deadpan expression, said he did not know because he had obtained only one transformant. One "transformant"! How could he be sure that this one colony was not a contaminant? Unperturbed by my obvious agitation, Albert replied that DNA obtained from this single transformant (and not from the untransformed strain) hybridized to the bacterial ColE1 sequences of the vector. This simple experiment showed beyond a doubt that he had achieved transformation and convinced me of the power of the new technology.

As I reflect on my years at Cold Spring Harbor, I wonder why I continued to teach the course for so long. Someone recently asked me why I was willing to sacrifice a month each summer away from my family. Actually, my family always accompanied me to Cold Spring Harbor. It was a wonderful vacation for them, and in a sense, they became part of the course. Walking down Bungtown Road from the beach, blond, tanned, and carefree, Rosalie, with my daughters Jennifer and Julia strutting behind, radiated a sense of calm domesticity; they provided a reminder to the students inside Davenport that there was another world on the outside.

Actually, the longevity of the course required considerable political skill on our part. Every year for as long as I can remember, Jim Watson would call Fred and me into his office and declare: "This is the last year of the course. Too few applicants means there is no interest in yeast. We can't afford it." This pronouncement was Jim's method of quality control. In the psychodrama that followed, Fred and I would plead the case for yeast to an ever more skeptical

The younger generation of yeast geneticists conferring at a Cold Spring Harbor course picnic: (*left* to *right*) Jennifer Fink, Julia Fink, Rhea Sherman, Joanna Sambrook.

grand inquisitor. Inevitably, these sessions ended with Watson's comment: "Well, we'll try it for one more year. But, if the number of applicants doesn't go up, next year will be the last year."

It strikes me that what drew me back each year was the allure and excitement of starting something new and the prospect of remarkable discoveries. There was the shared sense that we, both the students and the teachers, were standing at the threshold of a gold lode—we knew that yeast had great potential and we were eager to set about mining it. In the early years, the techniques we were teaching could not be learned at most universities. As a consequence, we teachers had the best of all possible situations: We had students desperate to learn. We in turn had to stay a step ahead of them. Wonderful students asked questions that sent me straight to the library—Randy Schekman wanted to know about yeast membranes, Ira Herskowitz about mating-type switching, David Botstein about sugar catabolism, Gottfried Schatz about mitochondrial biogenesis. The isolated environment of Cold Spring Harbor contributed a mystical quality that seemed simultaneously to expect and portend great discoveries. Although the description of the course advertised by Cold Spring Harbor was virtually the same for 17 years, each year was in reality a new course. Only Fred and I were a constant.

Each summer when the course began, I would wonder whether Fred was still enthusiastic. I never had to ask him. I could tell from the Cheshire grin that squeaked out as he partnered one of the students into the laboratory that he was up for another year.

# The Carlsberg Laboratory: Historical Retrospect and Personal Reminiscence

JØRGEN FRIIS
*Department of Molecular Biology*
*Odense Universitet*
*Odense M, Denmark, DK-5230*

## THE LAB

"Carlsberg Laboratorium" is a name that makes immediate associations with beer, especially because of the slogan used in the worldwide advertisement "probably the best beer in the world." However, on second reflection and among the scientific community, the name indicates a highly respected laboratory from which numerous reports have made important contributions to basic science. One of the reasons for the remarkable success of the Carlsberg Laboratory is the unique working conditions that it offers scientists, being a purely research institution. But this is obviously not the only reason for the Laboratory's reputation. Another and principal reason is that the Board of Trustees of the Carlsberg Foundation has always tried to attract the very best scientists to the Carlsberg Laboratory in Copenhagen, Denmark. That the Board has succeeded in this objective can best be judged from the circle of scientists who have carried out research on the premises of the Laboratory. Without making any evaluation between them, I will mention the names of just two scientists. The choice of the first is based on the supposition that the vast majority of individuals occupied in the scientific disciplines have so been in connection with the notion of pH and Sørensen buffer solutions. The person behind such fundamental concepts is Søren P.L. Sørensen, who acted as the head of the Chemical Department at the Carlsberg Laboratory from 1901 until to 1938. The second scientist is Øjvind Winge, who was appointed the head of the Physiological Department in 1933, a position he held until his retirement in 1956. Among Winge's many significant publications in the field of genetics, his paper from 1935 on the alternation of haplophase and diplophase in different yeasts, *Saccharomycetes*, makes him the founder of yeast genetics (Winge 1935). As Winge's contributions to the field of yeast genetics will be covered elsewhere in this volume, I will mention briefly just two of his many contributions. One was "polymeric" genes, a term used to describe the finding that several genes can affect the same character (Winge and Roberts 1952). The other was the discovery of the gene controlling homothallism, *D*, later renamed *HO* (Winge and Roberts 1949).

What were the conditions that made it possible for the Carlsberg Laboratory to develop into such an eminent research institution? It all started in 1875, when the owner of the Carlsberg Brewery, the brewer J.C. Jacobsen, established the Chemical and Physiological Laboratory at the Carlsberg Brewery. The Laboratory was to conduct independent investigations and to

*The Early Days of Yeast Genetics*

provide a fully scientific basis for the operations of malting, brewing, and fermentation. Out of a very intense interest in securing the continuation of brewing by scientifically founded methods, he established the Carlsberg Foundation in 1876, making it the first of its kind internationally. Jacobsen's desire to secure independence from mercantile interests was so deeply rooted in his philosophy that he entrusted the Foundation to continue and extend the activities at the Carlsberg Laboratory. In addition, it was strongly emphasized that "no result of the activities of the institute which is of theoretical or practical importance may be kept secret." Thus, the best possible conditions for scientific investigations were established.

Having been invited to write a chapter on the Carlsberg Laboratory, it was obvious to me that I had to make selections and decisions on which investigations I found of interest for geneticists and especially for those studying yeasts. Part of this account will be based on personal recollection, and I will not make any attempt to review the history of the Laboratory after the year of 1972 when profound changes were made. The continually increasing expenses connected with operation of the Laboratory and the recurrent wish from both society and industry that science should be guided in a more applicable direction made certain changes necessary. Shortly after the retirement of Heinz Holter, the head of the Physiological Department from 1956 to 1971, the Carlsberg Foundation and the Carlsberg Breweries agreed that the Carlsberg Laboratory should be transferred to the Breweries. In the agreement, it was stipulated that the operation of the Laboratory should continue under the rules from the original statutes of the Foundation. The Carlsberg Research Center was thus established in 1972.

In closing this first section, I would like to draw attention, as a service to those readers who would like to obtain a broader knowledge of the Carlsberg Laboratory, to the book *The Carlsberg Laboratory 1876/1976* (Holter and Møller 1976).

## GERMINATION

Emil C. Hansen was appointed the head of the Physiological Department in 1879, a position he held until his death in 1909. He was fully aware that an absolute prerequisite for carrying out his physiological investigations on yeasts was the availability of pure cultures. He was thus the first to establish a pure culture of yeast (Hansen 1883). To obtain such a culture, an obvious method would be to inoculate sterilized growth medium with a single cell, taking precautions that no foreign organisms would contaminate the culture. Hansen suspended cells in water and then counted them under a microscope using a hemocytometer. By making the appropriate dilutions, he obtained a yeast suspension that contained 0.5 cells per cubic centimeter. Using 1 cc of this suspension, he inoculated many culture flasks and then observed them for growth of colonies. He found that approximately half of the flasks would contain one colony, a few flasks two or more, and the rest would be without colonies. To substantiate his findings, reconstruction experiments were carried out by mixing two different yeast species that could be differentiated on the basis of cell morphology and fermenting ability. By making the same manipulation with this mixture as he did with the single strain, Hansen obtained one-colony flasks in which the cells were either of one kind or the other. This left no doubt about the usefulness of his method. One important consequence of Hansen's work was that it laid the foundation for brewing beer starting with a pure culture of yeast, a procedure that was introduced by the brewer J.C. Jacobsen at the Carlsberg Brewery in 1883. This was a success for the Brewery and the economical potential was great, but due to Jacobsen's philosophy that discoveries made at the Carlsberg Laboratory should not be kept secret, the procedure was implemented throughout the brewing industry.

As a passing remark in the above-mentioned publication, Hansen describes a simple and easy method for transporting the samples of yeast that he collected during his journeys to other research laboratories. The yeast is poured onto filter paper, and the excess liquid is removed with more filter paper. When sufficiently dry, the preparation is wrapped in filter paper. The yeast will then survive storage in the air at room temperature for up to 20 months. Hansen points out the advantage offered by the method is that an envelope containing the sample can be posted with ease and at small cost.

In publications dated 1905 and 1907 (Hansen 1911a,b), Hansen attempted to solve the long-standing question of whether the two brewing strains, top yeast and bottom yeast, are independent or whether transition from one to the other can occur. By establishing pure cultures originating from single cells isolated under the microscope from gelatin plates, he was able to show that transitions can occur between the two. The effort that Hansen put into this project was immense, as he in one experiment had to isolate 9948 top yeast cells in order to find a single bottom yeast culture. Other experiments required several thousand single-cell isolations. The results demonstrated moreover that a transition happened more easily in bottom yeast, whereas the top yeast was much more stable. From this, Hansen concluded that top yeast was older than bottom yeast. Hansen was not able to give an explanation for the transitions, but he suggested the hypothesis that the reason behind the alternations between the two physiological forms should be put in the category of mutation. He thus foresaw the linking together of physiology and genetics.

## MATING AND SPORULATION

Just as the Carlsberg Laboratory was the setting for the birth of *Saccharomycetes* genetics (Winge 1935), it gave shelter to the beginning of *Schizosaccharomyces pombe* genetics. The father was Urs Leupold, who during a research stay in Winge's department, described the mating system in this yeast (Leupold 1950). The techniques that Leupold used were developed in the 1930s, due mainly to the outstanding technical skill possessed by Winge's collaborator O. Laustsen. The micromanipulator and microscope were the prime instruments, but mass-matings were also tried in Leupold's experiments. From observations regarding mating properties and sporulation ability, he concluded that homothallism and heterothallism exist in this fission yeast. The two parental types segregated 2:2 during tetrad analysis, providing evidence that the genes controlling the sexual behavior represented a series of alleles at the same locus. In addition, transitions between alleles were observed as sectors during vegetative growth.

In *S. pombe*, vegetative growth is normally as haploid cells but occasional mitotic divisions can be observed in cells that are presumed to be diploids. After appropriate crossings with such presumed diploids, tetrad analysis gave results characteristic of triploid segregations, proving that the parent was indeed diploid and that polyploidy occurs in *Schizosaccharomyces* (Leupold 1956). This fact was further supported by the isolation of an X-ray-induced "gigas" mutant that was demonstrated to be diploid by genetic analysis (Ditlevsen and Hartelius 1956).

Some investigations on cell wall properties were also carried out during the period when Winge was managing the Physiological Department. The phenomenon that was studied was flocculence as it occurs in *Saccharomyces cerevisiae* (Thorne 1951). It was found that flocculence is under the control of a polymeric set of three genes, the character of which was dominant. The data obtained by following the change in phenotype from flocculence to that of nonflocculence showed that 6% of the spores, which on a theoretical basis possessed a gene for flocculence, had mutated to the recessive form. The high spontaneous mutation rate was not influenced by X-ray irradiation, and transition from nonflocculence to flocculence was also observed to happen spontaneously. The author states that the genes have such a high mutability that they are perhaps among the most unstable genes so far encountered in any organism. After Winge's retirement, the cell wall of yeast became the focus of the research in the same laboratories.

## PLATING

When Winge retired in 1956, Heinz Holter was appointed head of the Department of Physiology, a position he held until 1971. A new period began when research interests were shifted from genetics to that of cell physiology, and the main organism studied was no longer yeast but amoeba, especially the giant amoeba *Chaos chaos*. One of the subjects studied with the amoeba was the uptake of substances through the "drinking" of solutions, pinocytosis. However, Holter felt that he had an obligation to take up research on yeast, a responsibility that he thought was connected with his appointment. It was only natural, given the existing interest on uptake, to focus on the wall and plasma membrane of yeast.

The last part of my contribution will be based mainly on some personal experiences. I mention some studies concerning the enzymes responsible for the fermentation of sucrose and melibiose. These enzymes could be localized to the cell wall in yeast strains, where the fermenting ability was under the control of a single gene (Friis and Ottolenghi 1959a,b). In both cases, the experiments involved the use of protoplasts, and, because of their very limited access, the enzyme activities were in both cases measured with the Cartesian diver respirometer (Holter and Linderstrøm-Lang 1943), a micromethod developed at the Carlsberg Laboratory for cytochemical investigations.

During the same period, rumors began to circulate that transformation with higher cells, including yeasts, had succeeded (Oppenoorth 1960). Work with the amoebae had shown that they were impenetrable to glucose, but the sugar could enter the cell by pinocytosis when the process was induced by addition of protein to the solution. Thus, the setting was ripe to see if it would be possible to induce uptake of DNA into yeast cells by addition of protein. During the planning of such experiments, however, it was soon realized that the genetic character that had been used to follow transformation in yeast were not appropriate for further investigations. The genes that were being studied controlled the fermentation of different sugars, and the methodology required assaying carbon dioxide accumulation in small inverted glass tubes (Durham tubes). Moreover, the assays were not too reliable as it would be necessary in certain cases to incubate the tubes for up to 2 weeks before any bubbles were visible. Additional difficulties that were anticipated were the high number of cells expected to be handled and the unavailability of nutritional markers to verify transformants if any were obtained.

The problem was clearly a total lack of experience with the new methods developed and used in other laboratories working in the field of yeast genetics. The experimental techniques available to us at the time were those that Winge had relied on in his eminent contributions to yeast genetics. Winge and his collaborators had used chiefly genetic markers controlling fermentation (appearance of gas bubbles) or giant colony morphology, which took 30 days to develop. Tetrad analysis was used to follow the segregation of genes. Asci were dissected in a moist chamber under the microscope and the spores were placed in wort droplets, which when growth had been observed were transferred to culture flasks with small pieces of filter paper. The method was cumbersome, and a good days work would amount to the isolation of approximately 70 spores. Due to Winge's scepticism about crosses performed by mass-matings, zygote formation was followed under the microscope after placement of spores/cells next to each other in a wort droplet. All in all, the methods were unfit for transformation research.

In light of the situation, Holter invited the founder of the yeast genetics laboratory in Seattle, Washington, Herschel Roman, to the Carlsberg Laboratory in the summer of 1960. During his stay, many attempts to induce transformation were performed using recipients marked with mutations in several nutritional genes. The mutant recipients were treated during different phases of the cell cycle with isolated high-molecular-weight DNA from a wild-type donor strain. When it was possible to obtain protoplasts, these would be suspended in DNA just as intact cells and spores would be. The results from the experiments were never made public, and it will suffice to say briefly that it took several more years after these early attempts before transformation in yeast was reported (Hinnen et al. 1978).

During Roman's stay, it became clearly evident that the demands on equipment arising from experiments involving quantitative yeast genetics could not be met off-hand. I recall the situation concerning pipettes. The ones that were available were of high quality, meaning that the calibration would be ruined during autoclaving. Thus, a supply of 20 0.2-ml bacteriological pipettes were purchased for the yeast group. This meager supply of pipettes was constantly being washed and sterilized. In another incident concerning the woeful state of our pipettes, I remember a big smile appearing on Roman's face when he took a volumetric pipette on which was carefully engraved 5 ml but on which there were no markings indicating to where the liquid should be sucked up. This period was like an adventure; however, a direct consequence of the visit was that I was eventually able to learn from and publish with such outstanding geneticists as Donald C. Hawthorne, Urs Leupold, and Herschel Roman.

The uptake of different compounds was still of major interest at the Carlsberg Laboratory, and the dye Alcian Blue was used to obtain insight into the composition of the yeast cell wall as it would dye some strains and not others. Chemical fractionation of cells revealed that the binding ability resided in the total mannan fraction of the wall. The binding ability was under the control of a single gene, and tetrad analysis gave data which showed that it was linked to *ura1* (Friis and Ottolenghi 1970). The symbol for deficient binding is now *dbl* and the map position is XIL.

During the last years before Holter's retirement, rumors began to circulate among the employees that some major changes affecting the operation of the Carlsberg Laboratory would be undertaken. This affected the scientific milieu of the Department because unavoidable personal problems arising from an impending closing down interfered with the working zest. In the end, all scientists from the Physiological Department found new challenges elsewhere.

## NEW GROWTH

After the transfer of the Carlsberg Laboratory to the breweries, Diter von Wettstein was appointed the acting head of the Physiological Department. In the ensuing period, he was entrusted to assemble a new staff and to build up an institution with new research programs. In the year 1975, he was appointed head of the department which in 1987 made a "bud" by establishing the Department of Yeast Genetics. Morten Kielland-Brandt was appointed the head of this department, and Winge's dream of being able to improve brewery strains is now within reach, not by the classical techniques of breeding but with the newest "genetic engineering" that is practiced at the Carlsberg Research Center.

## ACKNOWLEDGMENT

The author wishes to express his sincere thanks to the Carlsberg Foundation for the support he has received throughout the years.

## REFERENCES

Ditlevsen, E. and V. Hartelius. 1956. A "gigas" mutant in *Schizosaccharomyces pombe* induced by X-rays. *C.R. Trav. Lab. Carlsberg Ser. Physiol.* **26:** 41–49.

Friis, J. and P. Ottolenghi. 1959a. Localization of invertase in a strain of yeast. *C.R. Trav. Lab. Carlsberg* **31:** 259–271.

———. 1959b. Localization of melibiase in a strain of yeast. *C.R. Trav. Lab. Carlsberg* **31:** 272–281.

———. 1970. The genetically determined binding of Alcian Blue by a minor fraction of yeast cell walls. *C.R. Trav. Lab. Carlsberg* **37:** 327–341.

Hansen, E.C. 1883. Undersøgelser over Alkoholgjærsvampenes Fysiologi og Morfologi. II. Om Askosporedannelsen hos Slægten *Saccharomyces*. *Medd. Carlsberg Lab.* **2:** 29–86.

———. 1911a. Overgær og Undergær. Studier over Variation og Arvelighed. *Medd. Carlsberg Lab.* **9:** 62–72.

———. 1911b. Overgær og Undergær. Studier over Variation og Arvelighed.II. *Medd. Carlsberg Lab.* **9:** 73–86.

Hinnen, A., J.B. Hicks, and G.R. Fink 1978. Transformation of yeast. *Proc. Natl. Acad. Sci.* **75:** 1929–1933.

Holter, H. and K. Linderstrøm-Lang 1943. On the Cartesian diver. *C.R. Trav. Lab. Carlsberg Ser. Chim.* **24:** 333–478.

Holter, H. and K.M. Møller, eds. 1976. *The Carlsberg Laboratory 1876/1976*. The Carlsberg Foundation, Copenhagen.

Leupold, U. 1950. Die Vererbung von Homothallie und Heterothallie bei *Schizosac-*

*charomyces pombe. C.R. Trav. Lab. Carlsberg Ser. Physiol.* **24:** 381–480.

———. 1956. Some data on polyploid inheritance in *Schizosaccharomyces pombe. C.R. Trav. Lab. Carlsberg Ser. Physiol.* **26:** 221–251.

Oppenoorth, W.F.F. 1960. Modification of the hereditary character of yeast by ingestion of cell-free extracts. *Antonie Leeuwenhoek* **26:** 129–168.

Thorne, R.S.W. 1951. The genetics of flocculence in *Saccharomyces cerevisiae. C.R. Trav. Lab. Carlsberg Ser. Physiol.* **25:** 101–140.

Winge, Ø. 1935. On haplophase and diplophase in some *Saccharomycetes. C.R. Trav, Lab. Carlsberg Ser. Physiol.* **21:** 77–109.

Winge, Ø. and C. Roberts. 1949. A gene for diploidization in yeast. *C.R. Trav. Lab. Carlsberg Ser. Physiol.* **24:** 341–346.

———. 1952. The relation between the polymeric genes for maltose, raffinose and sucrose fermentation in yeast. *C.R. Trav. Lab. Carlsberg Ser. Physiol.* **26:** 141–173.

# The International Yeast Community

ROBERT C. "JACK" VON BORSTEL
*Department of Genetics*
*University of Alberta, Edmonton*
*Alberta, Canada T6G 2E9*

The International Yeast Conferences have been held approximately every 2 years since the first group of investigators from different countries gathered in Carbondale, Illinois, in 1961. The Carbondale meeting was held for a very special reason: Mutant strains of *Saccharomyces cerevisiae* were being created in seven principal laboratories dedicated to yeast genetics, but the nomenclature for individual genes was not uniform from laboratory to laboratory. Eleven yeast investigators from Italy, Canada, and the United States attended this tiny meeting, which succeeded in standardizing the nomenclature of mutant genes.

The second conference on yeast genetics was convened in France, at Gif-sur-Yvette, just prior to the International Congress of Genetics at The Hague. Fifty-three yeast investigators attended, and all aspects of yeast genetics were discussed. This informal meeting was truly the genesis of what came to be biennial meetings, for it pointed the way to yeast becoming the most important organism for moving the molecular findings from bacteria into eukaryotic organisms. Even then, yeast was recognized as an effective bridge, with the possibility of becoming the most important one.

The meeting at Gif-sur-Yvette set a pattern that was followed until 1974, when the meetings became so large that more formality of presentation was required. Further increases in the population of yeast geneticists required another change in format in 1978. This change was the advent of plenary sessions, multiple poster sessions, and workshops. There is an emphasis on plenary presentations by leading senior and young investigators who have exciting new results that can be presented in light of a review of the entire field. The conferences have maintained essentially this format since 1978. The year that each of the International Conferences was held, the place, the responsible organizers, and the attendance are shown in Table 1. Photographs of participants at the Osaka '68, Chalk River '70, Pisa '72, and Sussex '74 meetings are shown in Figures 1 through 4.

A gap of 3 years occurred between the second and third meetings, but then meetings were so ad hoc in nature that no one was really aware that there might be another meeting. The Chalk River conference marked the first meeting that was called by a number, the Fifth International Conference of Yeast Genetics, and where plans were made for the next one.

In 1976, most of those in attendance at the conference in Schliersee provided abstracts. Terry Cooper took it upon himself to gather these abstracts,

*Table 1* The International Conferences on Yeast Genetics

| | Year | Location | Organizers | Participants/Abstracts |
|---|---|---|---|---|
| 1 | 1961 | Carbondale, Illinois, USA | S. Fogel, R.C. von Borstel | 11 |
| 2 | 1963 | Gif-sur-Yvette, France | P.P. Slonimski | 53 |
| 3 | 1966 | Seattle, Washington, USA | H.K. Roman | 64 |
| 4 | 1968 | Osaka, Japan | T. Takahashi | 83 |
| 5 | 1970 | Chalk River, Ontario, Canada | A. James, J.G. Kaplan | >100 |
| 6 | 1972 | Pisa, Italy | G.E. Magni, N. Loprieno | >150 |
| 7 | 1974 | Sussex, England | A. Bevan, D. Williamson | >200 |
| 8 | 1976 | Schliersee, Germany | F. Kaudewitz | >300 |
| 9 | 1978 | Rochester, New York, USA | F. Sherman | >400/359 |
| 10 | 1980 | Louvain-la-Neuve, Belgium | A. Goffeau | >500/410 |
| 11 | 1982 | Montpellier, France | P. Pajot, P.P. Slonimski | >600/385 |
| 12 | 1984 | Edinburgh, Scotland | I.A. Dawes, P. Fantes | >800/524 |
| 13 | 1986 | Banff, Alberta, Canada | R.C. von Borstel | ~700/442 |
| 14 | 1988 | Helsinki, Finland | M. Korhola | >800/519 |
| 15 | 1990 | The Hague, Netherlands | R.J. Planta | >900/605 |
| 16 | 1992 | Vienna, Austria | M. Breitenbach, R.J. Schweyen | 997/676 |

*Figure 1* Participants at the Fourth International Conference on Yeast Genetics, Osaka, Japan, 1968.

*First row (left to right):* T. Takahashi (standing), H. Kasahara, _______, _______, F. Sherman, _______, J. Ashida, J.G. Kaplan, _______, H. Tamaki, _______

*Second row:* S. Abe. Ö. Strömnaes, R. Snow, H. Heslot, S. Nakai, N. Yanagishima, R.K. Mortimer, ________, A. Nasim, ________, H. Gutz, S. Mori, G.E. Magni, H. Roman, ________, ________, T. Hirano, T. Yamamoto, S. Kamisaka, S. Nagai, K. Wakabayashi, Y. Oshima.

make copies of them, and bundle them up as sort of an informal "Book of Abstracts," which were then sent to all of the yeast investigators who had attended the conference. Since then, a Book of Abstracts has been published for each conference. For the first time in 1986, the Book of Abstracts was published as a supplement to an existing journal, volume 2 of the journal *Yeast*, published by John Wiley and Sons, Ltd. All Books of Abstracts for the conferences since then have been published as supplements to this journal.

As an organization, the community of yeast geneticists and molecular biologists has never been formalized, except for the formation of a Finance and Policy Committee for the International Conferences. A yeast investigator is designated from each nation to try to obtain travel funds from various agencies within his/her country, so that young investigators can travel to the conferences.

There are no dues. Anyone wishing to take responsibility for an activity may do so. For example, Bob Mortimer has taken on the task, as his responsibility to the yeast community, of publishing genetic maps at irregular intervals. From 1961 until 1976, Jack von Borstel and Bob Mortimer have published lists of genes and gene products of yeast. This has helped to standardize the nomenclature. Terry Cooper keeps addresses of the yeast geneticists and molecular biologists up to date and sends address books to everyone in the book. The address books are now published as supplements to the journal

*Figure 2* (*See facing page for legend.*)

*Yeast.* In addition to the conventional scientific papers, the editors of *Yeast* have encouraged the publication of mapping data and the results of sequencing of genomic DNA.

Fred Sherman and Gerry Fink initiated an annual course on yeast genetics at Cold Spring Harbor Laboratory in 1970. This popular course still continues to initiate investigators into the techniques of yeast genetics and basic molecular biology, and many of the most vigorous and productive yeast investigators began yeast research within this course.

Also in 1970, in the second edition of Jim Watson's book, *Molecular Biology of the Gene,* he mentioned that "this may be the time for many more biologists to work with organisms like yeast." This thought was already in the works, but his statement probably had an effect on investigators and students who were looking for a new frontier. Jim's comment gave him the vision to encourage yeast research on an enlarged scale at Cold Spring Harbor.

In 1975, annual meetings of yeast molecular biologists were arranged at Cold Spring Harbor and were held in alternate years from the International Conferences. By 1986, these meetings had grown too large for Cold Spring Harbor, and they have since been held as satellite meetings to the Annual Meetings of the Genetics Society of America in a biennial rhythm alternating with the biennial International Yeast Conference. The International and American Yeast Conferences have grown nearly exponentially (see Table 1). In 1986, there were two Conferences in North America, the International Conference in Banff, and 2 months earlier, the first American Conference in conjunction with the Genetics Society of America, held in Illinois. Approximately 700 people attended each of these Conferences, reducing somewhat the attendance at both meetings.

We now are undergoing another shuffling to accommodate the desire to hold the biennial Yeast Cell Biology meeting at Cold Spring Harbor in the

*Figure 2* Participants at the Fifth International Conference on Yeast Genetics, Chalk River, Ontario, Canada, 1970.

*First row (left to right):* ________, D. Hurst, N. Khan, S. Fogel, A. Bevan, A. Marko, H. Roman, M. Esposito

*Second row:* W.B. Lewis, P. Moens, C.F. Robinow, B. Carter, G. Stewart, R. Woods, L. Parks, A. James, M. Brendel

*Third row:* J. Johnston, J. Somers, H. Bussey, B. Johnson, R. Mitchell, W. Duntze, W. Whelan, D. Brusick, J. Warner, D. Bryant

*Fourth row:* E. Tustanoff, V. MacKay, G. Colleja, R. Snow, J.F.T. Spencer, ________, ________, Ö. Strömnaes, C.V. Lusena, H. Heick, M. Heick, ________

*Fifth row:* K.T. Cain, D.J. Gottlieb, ________, F. Lacroute, ________, C. McLaughlin, L. Hartwell, J.R. Mattoon, J. Parker, H. Newcombe, S. Nagai, ________, H. Gutz, F.K. Zimmermann

*Sixth row:* R. Piñon, R.E. Esposito, S. Threlkeld, G. Rank, R.C. von Borstel, U. Leupold, H. Heslot, B. Kilbey, R.K. Mortimer, F. Sherman, E. Moustacchi, G.E. Magni, H.C. Birnboim, C. Shimoda, N. Yanagishima

*Seventh row:* D. Duphil, H. Halvorson, M. Masselot, P. Meuris, J. Lemontt, N. Loprieno, R. Holliday, R. Hill, M. Resnick, W. Laskowski, I.L. Ophel, W.F. Baldwin, A. Henaut, ________, ________

*Eighth row:* ________, W. Fangman, J.G. Kaplan, P.T. Magee, A. Nasim, B. Cox, T. Ito, ________, ________

*Figure 3* Participants at the Sixth International Congress on Yeast Genetics, Pisa, Italy, 1972.

*First row (left to right):* S. Fogel, R.C. von Borstel, A. Bevan, G.E. Magni, P. Slonimski, H. Roman, N. Loprieno, M. Resnick, D. Wilkie, J. Mitchell, ________, ________

*Second row:* G. Bronzetti, A. Henaut, L. Herrera, ________, M. Luzzatti, ________, H. Halvorson, C. Lawrence, G. Fink, R.E. Esposito, F. Sherman

*Third row:* G. Michaelis, R. Woods, T. Petes, H. Mori, R.K. Mortimer, ________, J.R. Matoon, M. Esposito, N. Yanagishima, ________, M. Masselot, G. Simchen

*Fourth row:* J. Friis, G. Morpurgo, F.K. Zimmermann, ________, A. James, ________, M.E. Schweingruber, N. Gunge, I. Takano, ________, E. Moustacchi

*Fifth row:* ________, ________, ________, H. Heslot, ________, A. Nasim, L. Grivell, W. Duntze, B. Johnson, D. Williamson

*Figure 4* Participants at the Seventh International Conference on Yeast Genetics, Sussex, England, 1974.

*Participants include (in alphabetical order):* W. Bandlow, G. Banks, G. Bernardi, T. Bilinski, B. Carter, T. Cooper, J. Cosson, B. Cox, J. Davies, I. Dawes, E. Dubois, B. Dujon, D. Duphil, M. Esposito, R.E. Esposito, P. Fantes, G. Faye, G. Fink, L. Frontali, C. Gaillardin, J. Game, A. Goffeau, H. Gutz, D. Hawthorne, A. Herring, H. Heslot, R. Holliday, A. Hopper, E. Jones, F. Kaudewitz, Z. Kotylak, R. Labbe, T.M. Lachowicz, F. Lacroute, A. Linnane, G. Little, M. Luzzati, G.E. Magni, F. Messenguy, M. Minet, M. Mitchison, R.K. Mortimer, J.C. Mounolou, E. Moustacchi, P. Nurse, S. Oliver, T. Petes, M. Plischke, J. Pringle, J. Pugh, M. Resnick, V. Richmond, H. Roman, J. Rytker, R. Schweyen, A. Scragg, A. Sels, L. Senna, F. Sherman, G. Simchen, P. Slonimski, R. Snow, D. Spencer, J.F.T. Spencer, B. Stevens, J. Stratford, D. Thomas, P. Thuriaux, P. Unrau, R.C. von Borstel, R. Wheatcroft, D. Wilkie, D. Williamson, G. Yarranton, I. Zhakarov, F.K. Zimmermann.

same year as the International Yeast Conference so that the North Americans can attend conferences in the United States in alternate years. This will tend to make the International Yeast Conference a Eurasian Conference, but in 1993, the International Yeast Conference and the North American Yeast Conference will meet jointly in the United States.

Three yeast investigators, Jeff Strathern, Beth Jones, and Jim Broach, decided at the 1979 Cold Spring Harbor Yeast meeting to edit a book on yeast molecular biology. Two volumes appeared, and these have been extremely important in consolidating the literature for everyone in the field. Volume 1 of the second edition of three volumes on yeast molecular and cellular biology appeared in 1991; volumes 2 and 3 will be published in 1992 and 1993, respectively.

Stability was given to the yeast genetics and molecular biology community by Herschel Roman, who presided as the nominal chairman of the International Conferences. He retired as the "Pope" in 1984; the mantle was assumed by Piotr Slonimski.

The yeast geneticists and molecular biologists are now the most vital community representing any organism. The quality of scientists and the science they perform is remarkable. Many of the functions of mammalian genes are known only because they resemble genes in yeast. Yeast is now the organism of choice for gene expression. Shuttle vectors incorporating genes from viruses to bacteria to humans are placed in yeast in order to study gene expression. Nearly every cell biologist must turn to yeast to complete studies begun on mammalian cells. Moreover, much biotechnology could not be done if it were not for the information derived from the investigations on yeast genetics and molecular biology.

The tradition is clear. Yeast is responsible for the bread we eat, and the beer, wine, and sake we drink. The procedures for large-scale cultivation and growth of yeast have accrued from the traditions of the prehistoric days when yeast became the first domesticated organism. The vast modern fermentation industry stems from these ancient origins, and the new genetic engineering and recombinant DNA technologies profit from and use the fermentation techniques devised for the brewing of beer.

Finally, it is important to note that yeasts other than *Saccharomyces cerevisiae*, particularly *Schizosaccharomyces pombe*, are represented by a number of plenary speakers at the International Conferences. There is a good deal of comparative genetics and molecular biology done among the yeasts in order to define which phenomena are general and which are specific to a species. As each species of yeast gathers its own principal investigators, new satellite meetings to the International Conferences begin to emerge, each for presentations of problems unique to the species under consideration.

# Name Index

# Subject Index